THEORY OF LAND LOCOMOTION

THEORY OF LAND LOCOMOTION

THE MECHANICS OF VEHICLE MOBILITY

by M. G. Bekker

ANN ARBOR
THE UNIVERSITY OF MICHIGAN PRESS

LIBRARY OF CONGRESS CATALOG CARD NUMBER 56–10101

Paper back ISBN: 978-0-472-75020-7

PRINTED AND BOUND BY CPI GROUP (UK) LTD, CROYDON, CR0 4YY

BY N. V. DRUKKERIJ G. J. THIEME, NIJMEGEN

PREFACE

The purpose of the present outline of the theory of land locomotion is to provide automotive engineers with a comprehensive source of information now available on the physical relationship between a motor vehicle and the environment of its operation, particularly in off-the-road locomotion. It is hoped that the approach made in this book will help to further the growing interest in a better understanding of land mobility problems. It is also hoped that this work may assist in the establishment of what may become a new type of applied mechanics, without which further progress in land locomotion would be greatly handicapped.

This book is based on class notes of a course which the writer was teaching at the Graduate School of the Stevens Institute of Technology during the academic years of 1950–52. The large portion of original work discussed in this volume was performed by the author between 1943–46 as a contribution to the Allied war effort under the auspices of the Canadian Department of National Defense, the United States Army Ordnance Corps, and the Canadian National Research Council. The compilation of pertinent information produced by other workers is based on the References listed. Although almost five years have elapsed since the draft of this book was completed, and considerable progress has been made in the discussed field since that time, it is believed that the presented material will still be of some aid in any generalized study of land locomotion.

The writer is indebted to Mr. Clifford J. Nuttall, formerly of the Experimental Towing Tank, Stevens Institute of Technology, and now with Wilson, Nuttall, Raimond Engrs., Inc., for Chapter XI on scale-model testing, and for his remarks on amphibians. Mr. Milton Morrison of Stevens Institute assisted in the outline of the statistical problems discussed in that chapter.

To Professor Dr. Ing. W. I. E. Kamm, formerly of the Forschungsinstitute für Kraftfahrwesen of the Stuttgart Institute of Technology and now with the Battelle Memorial Institute, the writer owes his gratitude for much constructive criticism and the review of the manuscript.

The author is also indebted to Professor E. T. Vincent of the University of Michigan for his stimulating support which contributed to the publication of this work, to Mr. H. J. Hamblin of the National Institute of Agricultural Engineering in England, and to Professors Joseph T. Thompson of Johns Hopkins University, Daniel C. Drucker of Brown University, Newman A. Hall of New York University, and K. S. M. Davidson of the Stevens Institute of Technology for their valued comments and encouragement.

MIECZYSLAW GREGORY BEKKER

Birmingham, Michigan
December, 1955

CONTENTS

THEORY OF LAND LOCOMOTION

I. INTRODUCTION

The problems encountered during the 50-year period of motor-vehicle development have embraced an ever-growing scope. This might be illustrated not only by the physical increase in the size and weight of vehicles, but also by the changing nature of the problems involved; the latter have varied from the trivial question of whether a highway speed close to 20 mph could be sustained by the human organism to the modern problem of off-the-road locomotion, in which higher speeds are required.

In order to meet the growing changes, the primitive power train and the running gear of the early automobile have developed into a complex mechanism, the functioning of which largely depends on an elaborate system of intricate elements. If a comparison from the field of biology may be borrowed, it may be said that the original primitive aggregate of motor-vehicle "cells" has become today an advanced organism which performs much more complex functions in a more varied physical environment than ever before. This environment must accordingly affect the development of the organism itself to a greater extent than it did in the past.

Studies of how a motor vehicle changes the social, economic, and other fields of life, and how these reciprocally influence the development of an automobile, have been made repeatedly.[1] An inquiry, however, for a more accurate knowledge of the process in which the whole vehicle and its elements affect the physical medium of their operation, and of the way in which that medium reciprocates through the implication of physical principles essential in a rational design or use of a vehicle, was never made on a large scale. The inadequacy of the existing knowledge in this field was recognized relatively late, and only preliminary attempts have been made in a search for working methods which would bring more rational tools into the assessment of trends and ideas.[2, 3, 4]

Automotive engineers and users appear to be waiting for the development of a general theoretical outline of methods, particularly in the field of off-the-road locomotion, so that long-range policies for developing an economic and rational scheme of transportation can be more quantitatively assessed than is now possible.[5]

II. LOCOMOTION IN NATURE

The concept of land transport was not invented by mankind; it was first materialized by nature in enormous amounts of solids and water which have moved on the surface of the earth since prehistoric time. Perhaps our present land transportation may be considered only as a more differentiated and more sublime form of this original system, whose capacity to move astronomical numbers of tons of matter has remained unmatched by the capacity of all the rail-, high-, and waterways of the globe, taken together.

Flow in a Channel

If it is agreed that geological and technological processes may be compared, and that the work done with modern machines is a continuation of the work originally performed by geological factors,[6] then a closer study of problems related to moving rivers may be of interest since it presents a general background of modern problems of land transport.

Consider first the water flowing in a river. It is subjected to the force of gravity which supplies the motive power and to resistance forces occasioned by the river bed. If a prism of flowing water is considered, as shown in Figure 1, then the force of movement resistance R equals the component of prism weight W, parallel to the bottom once the water is assumed to flow in uniform motion:

$$R = W \sin \alpha \,. \tag{1}$$

The discussed weight of the prism may be determined in terms of the notation shown in Figure 1:

$$W = \gamma \, bdl \,,$$

where γ is the specific weight of water. Since $\sin \alpha$ is usually very small, it may be replaced in formula (1) by the tangent of the slope $S = h/l$ of the river bed. The resistance to flow, then, for a unit of length l of the prism will be

$$R = \gamma \, dSb \,. \tag{2}$$

It is known from hydraulics [7] that the speed v of water flowing in the channel shown in Figure 1 may be satisfactorily expressed by means of the semi-empirical formula

$$v = \frac{1.486}{n} d^{2/3} S^{1/2}, \tag{3}$$

where n is a constant depending on the characteristics of the channel walls.

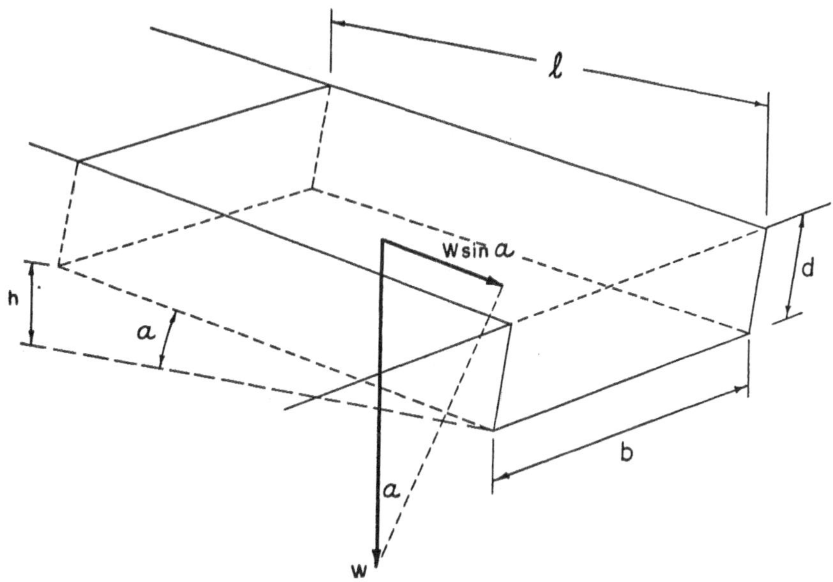

Fig. 1

Substituting the value of S determined from equation (3) into equation (2) gives

$$R = \frac{n^2 \gamma b v^2}{(1.486)^2 d^{1/3}}. \tag{4}$$

Therefore, the power $P = Rv$ consumed to overcome the resistance to flow will be

$$P = \frac{n^2 \gamma b v^3}{(1.486)^2 d^{1/3}}. \tag{5}$$

Since the unit weight of the unit water prism is $W = \gamma\, db$ (lb/ft), the power per unit of weight required to have the water flow at a given speed is, according to equation (5),

$$\frac{P}{W} = \frac{n^2 v^3}{(1.486)^2\, d^{4/3}}, \tag{6}$$

or when denoting

$$K = \frac{n^2}{(1.486)^2\, d^{4/3}}, \tag{7}$$

$$\frac{P}{W} = K v^3 . \tag{8}$$

It may be seen that the rivers present a considerable movement resistance since the power consumed to overcome this resistance increases with the third power of the speed of transport. This may be seen in Figure 2, where lines of power consumption with reference to the speed were computed. The graph is plotted in hp/ton instead of lb ft/sec per lb as shown in equation (7), and refers to the rivers whose dimensions are marked beside each line.

Assume that the flowing prism of water is a section of an imaginary train "rolling" down the river. It is interesting to note that the unit power required to propel such a "train" increases rapidly with a reduction in the dimensions of the cross section of the moving "cars." This example illustrates the original relationship between vehicle form and economy of transport and suggests that if this relationship is implicit in a most primitive transportation system, it must be more explicit in other means of land transport. Thus the importance of morphological studies of man-made machinery is emphasized. Studies in this direction were actually originated by Neesen approximately fifteen years ago [2] with reference to land, air, and water vehicles, and will be discussed later.

Transportation of water mass by flow in a channel may be compared to the transport of water by a weightless train, the locomotive of which is replaced by the mechanism of gravity forces. Such a train is an irreversible one because its locomotive cannot back. Traffic of this kind would have ceased long ago on the earth's surface after exhausting transportable water mass if there were not in existence at the same time another transportation system acting in the opposite direction: the evaporation of water and its uplift by means of hydrostatic forces. This

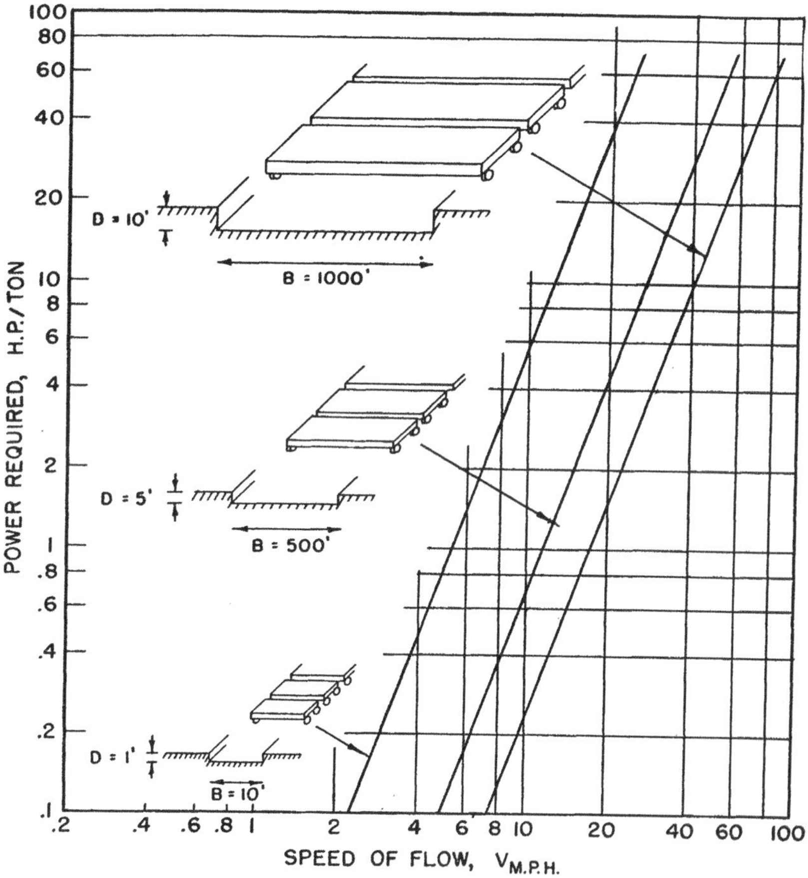

Fig. 2

phase of the whole cycle of the movement of water is not the subject of our considerations since it does not relate directly to land transport. It is interesting to note, however, that the origin of hydrostatic forces also is gravitational, and that the mechanism of the whole cycle is thus practically the same, with the sun furnishing the energy required to upset the status quo and set gravity to work indirectly when moving the masses of water upwards.

No purely gravitational power system can operate on a constant altitude; its main characteristic is the production of vertical motion. This is the main disadvantage of such a transport scheme, because, in it, the movement in a horizontal direction is inseparably the one of the two components of a general motion which exists only when a vertical component of translation is not equal to zero.

Locomotion of Animals

The independence of locomotion upon the ground surface from gravitational forces was not achieved until the first appearance of animals equipped with a lever mechanism which could exercise upon the ground reversible forces in the horizontal direction. Thus, it is the animal world which first solved the real problem of land locomotion, and indirectly the problem of transport, by gaining an almost unlimited freedom of movement in place of the restricted, irreversible fall of a body under the action of the gravitational field. It is interesting to note that this achievement, of course, materialized without the help of a wheel.

As Rashevsky has pointed out, the locomotion of all organisms, with the exception of protozoa, is produced by a system of levers. In the case of snakes or caterpillars, the whole body is a lever; in the case of other animals, their extremities are systems of levers.[8]

There are several types of animal locomotion, depending on the type of mechanism used. A survey of the more important types may be not only interesting from a general point of view, but may shed some light upon the historical meaning and rationality of lever systems applied to the locomotion of man-made machines.

Consider first the locomotion of those species which have an effective system of extremities constituting the lever mechanism of versatile locomotion. Basically, these species embrace all quadrupeds which can jump or run.

Running and Jumping

Following Rashevsky's suggestion, assume that running is essentially a series of consecutive jumps. During each jump, the animal is for a while completely without contact with the ground, and its center of gravity describes a parabola. In a series of consecutive jumps, each starting from a standstill, the total kinetic energy is lost. Thus, a kangaroo, or a grasshopper, has to impart a large force in order to cover the respective losses of energy.

The process of running is not much different. In this case, however, it may be assumed that only the vertical component of kinetic energy is lost, and the animal runs with a more or less uniform speed. Thus it would appear that the running of a horse is more economical than the leaping of a frog.

The following is a more detailed examination of this process. It must be made clear, however, that the assumptions made for the sake of simplicity lead to approximate results only. If the animal's center of gravity moves along a parabola as shown in Figure 3, and if the energy spent in lifting the center of gravity from point 0 to 0′ during the unfolding of the extremities is neglected, then the total work expended for covering the distance in one leap will be equal to the kinetic energy $mv_o^2/2$, where m is the mass of the animal and v_o the initial speed imparted by the muscles.

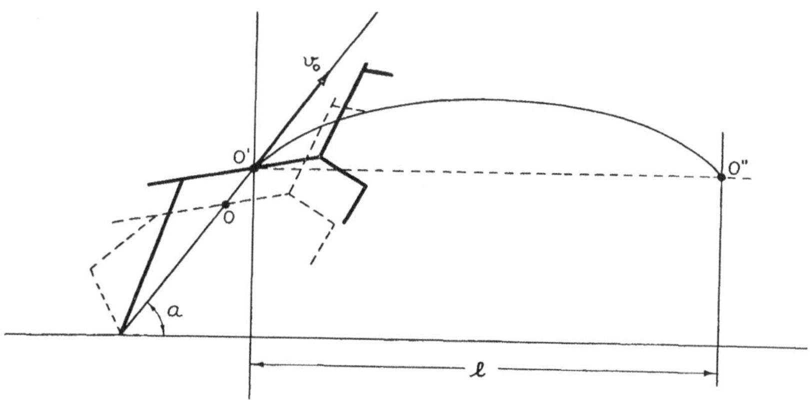

Fig. 3

The animal may now be considered as a projectile subjected to the rules of external ballistics; hence, the time of "flight" from point 0′ to point 0″ equals [9]

$$t = \frac{l}{v_o \cos \alpha}, \tag{9}$$

whereas the distance of leaping l is

$$l = \frac{2v_o^2 \sin \alpha \cos \alpha}{g}. \tag{10}$$

The time of "flight" may be expressed by combining equations (9) and (10):

$$t = \frac{2v_o \sin \alpha}{g}. \tag{11}$$

During the process of leaping, the imparted kinetic energy is spent at the average rate of

$$P = \frac{mv_o^2}{2t}. \tag{12}$$

By substituting into equation (12) the value t of equation (11), the average power consumed may be determined as

$$P = \frac{mgv_o}{4 \sin \alpha}. \tag{13}$$

In other words, power consumed per unit weight for the purpose of overcoming the "resistance" of the leap is

$$\left[\frac{P}{mg}\right]_{\text{leap}} = \frac{v_o}{4 \sin \alpha}. \tag{14}$$

During the process of running, only the vertical component of kinetic energy may be assumed as being lost; hence,

$$P = \frac{mv_o^2 \sin^2 \alpha}{2t}$$

and

$$\left[\frac{P}{mg}\right]_{\text{run}} = \frac{v_o \sin \alpha}{4}. \tag{15}$$

It is interesting to note that the "economy" of leaping increases with the increase of the angle α up to 45°, whereas that of running reduces with the increase of α.

During running, the angle at which the initial speed v_o is sloped should be kept as small as possible, which is in agreement with conclusions reached by Rashevsky.[10] The smaller this angle, the less the initial lifting of the center of gravity O (Figure 3) and the smaller the error which is inherent in formula (15) by the omission of the work spent in that lifting. Photographs of running animals show that the angle α varies from 10°

to 20°. By substituting into formula (15) the value of $\alpha = 10°$ and the speed of approximately 40 ft/sec, which corresponds to the speed of a racing horse, the developed power may be calculated as follows:

$$\left[\frac{P}{mg}\right]_{\text{run}} = \frac{40 \sin 10°}{4} = 1.7 \text{ lb ft/sec lb.}$$

If this value is to be expressed in horsepower per ton of weight, then

$$\left[\frac{P}{mg}\right]_{\text{run}} = \frac{1.7 \times 2000}{550} = 6 \text{ hp/ton}\,.$$

Data referring to the power developed by a running horse contain a few uncertainties. After consulting many available sources, the average figure of from 4 to 5 hp/ton may be agreed upon with reference to a galloping racer. Under these circumstances, the figure determined from equation (15) is quite accurate.

According to formula (14), a leaping animal under the same circumstances ($v_o = 40$ ft/sec) should develop the power of

$$\left[\frac{P}{mg}\right]_{\text{leap}}^{10°} = \frac{40 \times 2000}{4 \sin 10° \times 550} = 200 \text{ hp/ton}\,,$$

which would be excessive. As has been noted, the optimum leaping angle is $\alpha = 45°$. An animal leaping at that angle should develop

$$\left[\frac{P}{mg}\right]_{\text{leap}}^{45°} = \frac{40 \times 2000}{4 \sin 45° \times 550} = 51 \text{ hp/ton}$$

in order to overcome the "external resistance" of locomotion, if no air resistance is accounted for. A leaping rabbit at $v_o = 20$ ft/sec would then develop about 26 hp/ton or, assuming its weight at 10 lb, $26 \times 10/2000 = 0.13$ hp.

The general picture of power required per unit weight (hp/ton) as a function of speed (mph) for both leaping and running animals is shown in Figure 4. This figure gives the data enclosed by values calculated for $\alpha = 10°$ and 45° and may be considered as a fair approximation of the magnitude of the movement resistance encountered by animals propelled by well-developed extremities. It should be noted that the moment of inertia of these extremities and the power for their swinging were not taken into consideration. More details in this respect may be found in References 8 and 11.

Crawling

Besides running or jumping, there is another very common type of animal locomotion: crawling. This type, which consists of consecutive foldings and unfoldings of extremities, is the most primitive method of propulsion. By following Rashevsky, crawling may be presented graphically as in Figure 5a. During the "folding," the "front" end I remains in fixed contact with the ground; during the "unfolding," the rear end II does the same. The locomotion of some caterpillars is of this nature.[8]

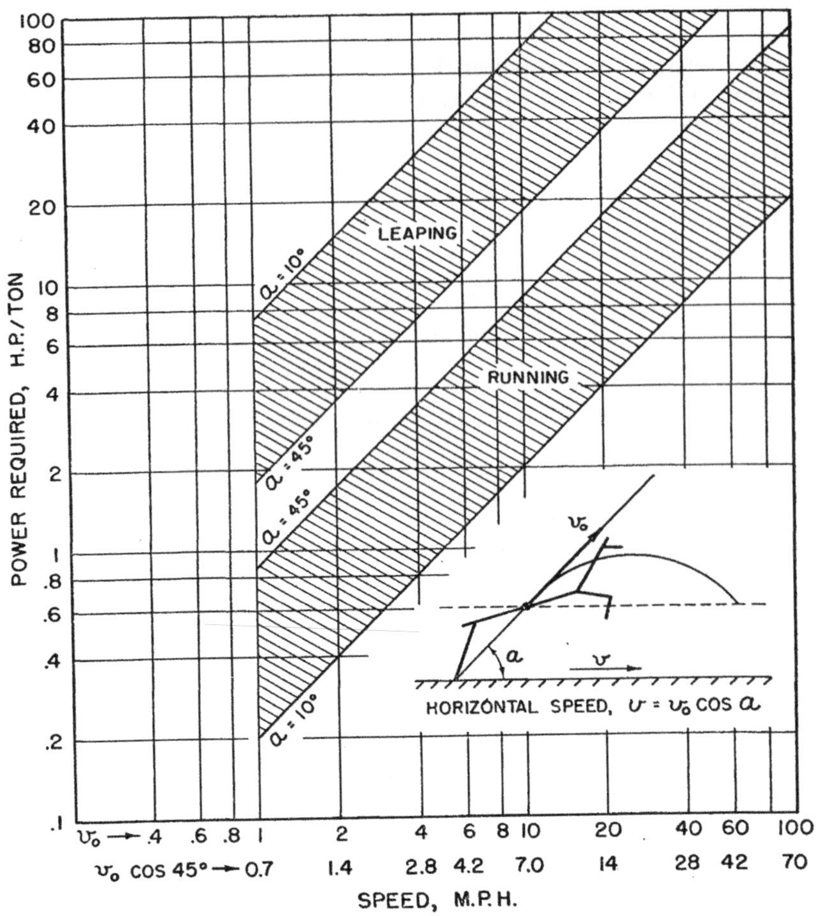

Fig. 4

In the case of crawling, considerable friction may be developed between the body of the animal and the ground; no known data seem to be available regarding the amount of energy required in this type of locomotion. There are probably many different values that depend on the details of this type of movement: the resistance which results from unknown variables such as a partial uplift of the body, coefficient of friction, etc.

In order to determine some figures which would indicate the order of the magnitude of power involved, certain ideal mechanisms are adopted which, although simplified, would bear some similarity to the mechanism of crawling in the animal world. The proposed scheme is shown in Figure 5b.

The main body having mass m is supported by n legs, the total mass of which may be assumed to be equal to km, where k is the coefficient of proportionality. Assume that the legs may be consecutively lifted up and moved forward a distance d, with the center of gravity being raised a height h, and subsequently lowered to the ground, as shown in sketches *a, b, c, d, e* of Figure 5b, so that in a final stage the whole body is moved by distance d. By an analogy with vehicles, it is assumed that the force F opposing the forward movement of a link is proportional to its weight: mgf/n, where f is the coefficient of movement resistance.

The force opposing the lifting of a link is kmg/n. Assume further the crawling to be so slow that the forces of inertia may be neglected and that the speed v_d of moving a particular link forward equals the speed v_h of lifting this link. Let these speeds be denoted by v. Accordingly, the power consumed for moving a link forward is $mgfv/n$. For lifting a link at height h, the power required is $kmgv/n$. The total power required for propulsion is then $P = (mgfv + kmgv)/n$, if the movements of the links are consecutive but do not overlap.

The speed v_a of the "animal" may be determined as follows: The time required for the forward movement of n links is nd/v; that for lifting is nh/v. Since a particular link has to be lowered before the movement of the adjacent one starts, the additional time required for lowering is nh/v, if this process also is performed with speed v. Hence, the time for lifting and lowering is $2\,nh/v$.

The total time needed to close the whole cycle of movement is then

$$t = \frac{nd}{v} + \frac{2nh}{v}. \tag{16}$$

Since, during this time, the main body moves a distance d, the speed of locomotion v_a is

$$v_a = \frac{vd}{nd + 2hn}. \tag{17}$$

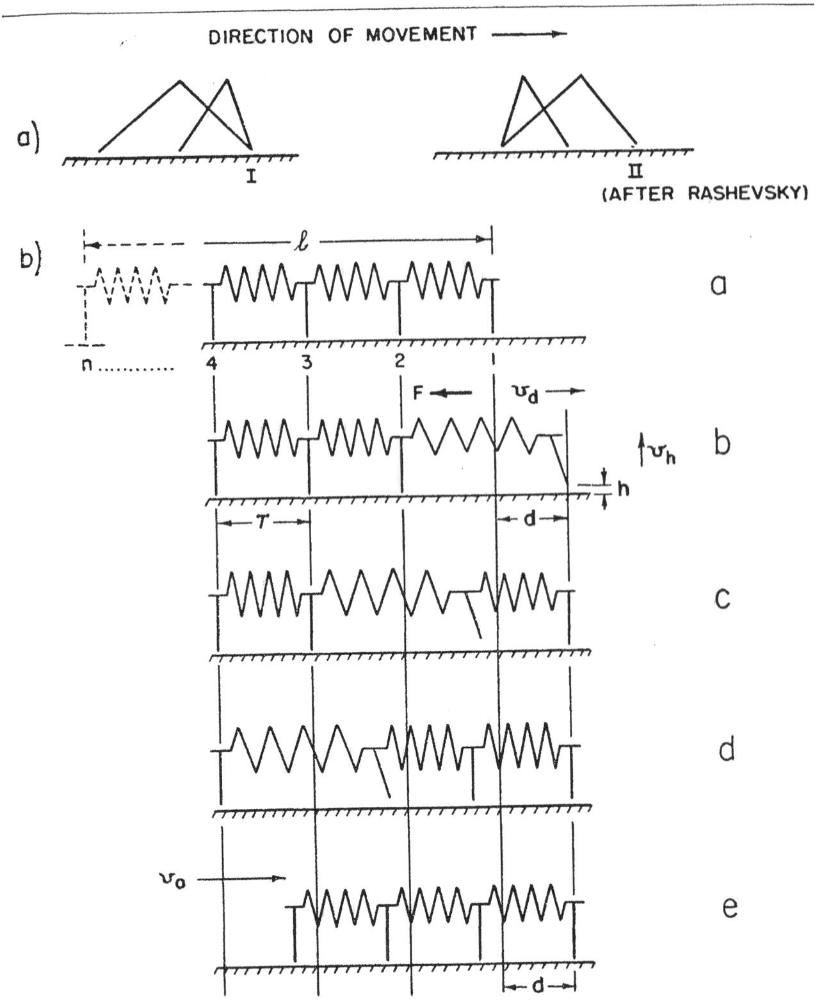

$v_d = v_h = v$

Fig. 5

Now, the previously determined total power required for moving the body, $P = (mgfv + kmgv)/n$, may be determined as a function of the speed of locomotion v_a instead of the speed v related to the propagation of the "wave" of consecutive extensions and contractions of particular links. From equation (17),

$$v = \frac{v_a n\,(d + 2h)}{d} \tag{18}$$

and

$$P = \frac{v(mgf + kmg)}{n} = \frac{mgv_a}{d}\,(d + 2h)\,(f + k)\,. \tag{19}$$

Power per unit weight mg may be obtained directly from equation (19):

$$\left[\frac{P}{mg}\right]_{\text{crawl}} = \frac{v_a}{d}\,(d + 2h)\,(f + k)\,. \tag{20}$$

It is logical to assume that the forward extension d of the "animal's body" remains within a certain ratio η to its link length τ. Since $\tau = l/n$, then $d = \eta l/n$. A plausible assumption would be to take $\eta = 0.3$, and hence $d = 0.3\,l/n$. It is logical to expect further that the "animal" will lift its legs as little as possible in order to reduce the lifting work to a minimum. This is, of course, affected by the roughness of the terrain. The idea of roughness, however, is relative to the size of the animal. It may therefore be useful to express h with reference to l as follows: $h = l/\lambda$. A reasonable λ value would be of the order of 50. The meaning of this assumption is simply that a leg cannot drop or be lifted more than 1/50 of the length of the animal's body.

If the values $d = 0.3\,l/n$ and $h = l/50$ are substituted into equation (20), then

$$\left[\frac{P}{mg}\right]_{\text{crawl}} = v_a\,(0.134\,n + 1)\,(f + k)\,. \tag{21}$$

Assume, for example, that the weight of the extremities is 20% of the weight of the body, i.e., $k = 0.2$, and that the coefficient of movement resistance f equals that of a rolling pneumatic tire upon a concrete highway, i.e., $f = 0.02$.[12] Then formula (21) will be reduced to the following form:

$$\left[\frac{P}{mg}\right]_{\text{crawl}} = 0.22\,v_a\,(0.134\,n + 1)\,. \tag{22}$$

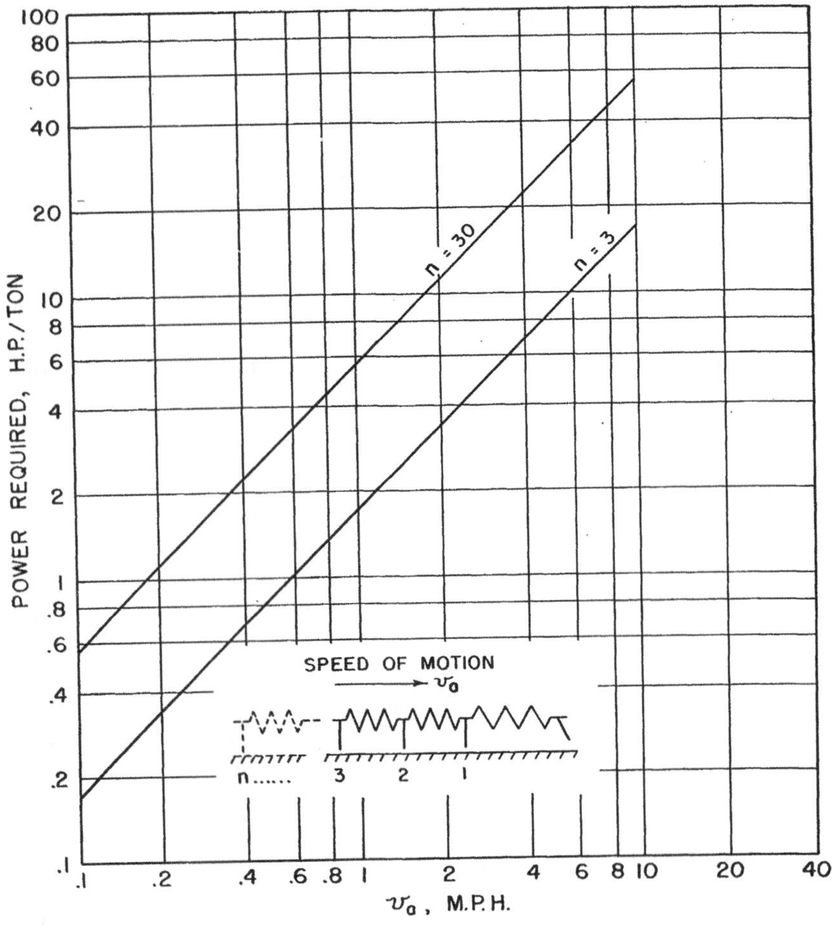

Fig. 6

The power requirements of such an "animal" at various speeds v_a are shown in Figure 6 for the minimum conceivable number of links $n = 3$ and for $n = 30$. It would appear that an increase in the number of links increases the unit power needed for the development of a given speed. This point needs some examination. Since $n = l/\tau$ and $d = 0.3\,\tau$, according to the assumptions previously made, equation (22) may be written in terms of l, d, and v_a as follows:

$$\left[\frac{P}{mg}\right]_{\text{crawl}} = v_a\,(0.04\,\frac{l}{d} + 1)\,(f + k)\,. \tag{23}$$

It will be seen that since $l/d = 3.3\,n$, the case of $n = 3$ corresponds to $l/d = 10$, whereas the case of $n = 30$ involves $l/d = 100$. It may then be concluded that the cause of the power increase is not the increase in the number of links itself, but rather the reduction of the unit stretch d with reference to the total length l. The locomotion is more efficient, the more "elastic" the body of the animal, or the longer the wave of elongation and contraction.

These considerations embark upon a new concept of locomotion by means of a longitudinal oscillatory motion of the body, accompanied by a "wave" of leg movement in the vertical plane. A concept of a similar motion on the surface of the earth is not restricted only to the above-considered mechanism of translation. Among snakes there may be found another type of locomotion which is performed by a similar but transverse oscillatory movement.

Sliding

According to Mosauer,[13] the snake assumes an approximately sinusoidal shape. Forward movement is performed in such a way that the whole body follows a given sinusoidal path. Rashevsky explains the mechanism of such a movement by assuming that the coefficient of friction between the snake's body and the ground is smaller in the direction of the axis of the animal's body than in the direction perpendicular to this axis [8]. It would appear that such an assumption is not essential from a purely mechanical viewpoint, although it is plausible and there are reasons to expect its validity in the animal world.

Consider the zig-zag line in Figure 7 as representing the snake's path. Movement is made possible by force F exercised by the snake in the direction perpendicular to the axis of locomotion. The magnitude of this force is limited by the transverse friction F_t of the body against the ground because force F_v, being a component of F vertical to the body, cannot be larger than F_t. The latter equals $mg \tan \mu_t$, where m is the mass, g is gravity acceleration, and μ_t is the angle of friction in the direction F_t. Hence,

$$F_v = F_t = mg \tan \mu_t\,. \tag{24}$$

The driving force F_d, which moves the body with speed v and which is another component of F, thus equals

$$F_d = F_v \tan \varphi = mg \tan \mu_t \tan \varphi \, . \tag{25}$$

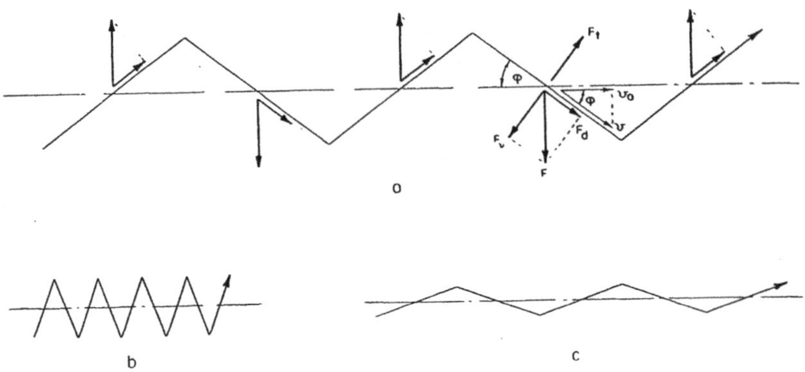

Fig. 7

This force, however, has to overcome the friction $mg \tan \mu_l$ in the longitudinal direction of the snake's body and therefore equals

$$F_d = mg \tan \mu_l \, , \tag{26}$$

where $\tan \mu_l$ is the coefficient of friction in that direction. Hence, combining equations (25) and (26) gives

$$mg \tan \mu_t \tan \varphi = mg \tan \mu_l$$

or

$$\tan \varphi = \tan \mu_l / \tan \mu_t \, . \tag{27}$$

Equation (27) indicates that sliding may occur if $\tan \varphi$ satisfies values of μ_l and μ_t, correlated in equation (27). If $\tan \mu_t < \tan \mu_l$, then $\tan \varphi > 1$ and $\varphi > 45°$. The sinusoid becomes "steep" as shown in Figure 7b. Since the speed of locomotion $v_a = v \cos \varphi$, the resultant movement is very slow. The motion is more "economic" when $\tan \mu_t > \tan \mu_l$. The sinusoid is then more elongated and the losses in speed of translation are smaller. Observations indicate that snakes move at $\varphi \sim 20°$ or less. This would suggest that the ratio of $\tan \mu_l / \tan \mu_t = 0.36$. If the friction coefficient in the longitudinal direction, $\tan \mu_l$, is assumed to reach the value of 0.2, then the transverse friction should amount to $\tan \mu_t = 0.55$.

From a locomotion point of view, the problem of power requirement is one of the most interesting. It has been determined previously that the minimum driving force is $F_d = mg \tan \mu_l$. Since this force acts with speed v, the power considered is

$$P = F_d v = mgv \tan \mu_l \, . \tag{28}$$

It is interesting, however, to determine the power as a function of the speed of locomotion of the animal: $v_a = v \cos \varphi$. By substituting v_a for v in equation (28), the following formula may be obtained:

$$P = \frac{v_a \, mg \tan \mu_l}{\cos \varphi} \, .$$

Since $\cos \varphi = 1/\sqrt{1 + \tan^2 \varphi}$,

$$P = v_a \, mg \tan \mu_l \sqrt{1 + \tan^2 \varphi} \, . \tag{29}$$

It has been previously determined that $\tan \varphi = \tan \mu_l / \tan \mu_t$ [equation (27)]; hence, finally, the unit power of sliding may be obtained directly from equation (29):

$$\left[\frac{P}{mg}\right]_{\text{sliding}} = v_a \frac{\tan \mu_l}{\tan \mu_t} \sqrt{\tan^2 \mu_l + \tan^2 \mu_l} \, . \tag{30}$$

If the values $\tan \mu_l = 0.2$, $\tan \mu_t = 0.55$, and $\varphi = 20°$ are assumed, formula (30) may be written as follows:

$$\frac{P}{mg} = 0.21 \, v_a \, .$$

An increase of motion resistance may be illustrated by reversed values: $\tan \mu_l = 0.55$, $\tan \mu_t = 0.2$, hence $\varphi = 70°$:

$$\frac{P}{mg} = 1.6 \, v_a \, .$$

Curves which show the power unit in hp/ton with reference to speed of locomotion v_a in mph are plotted in Figure 8. This picture illustrates the quantitative relationship between the wave movement and forward motion of a snake. As was mentioned before, the 70° zig-zag line resulted from the friction coefficient $\tan \mu_t < \tan \mu_l$. Such conditions may occur when a "snake" climbs a hill sloped at an angle α. In this case, an addi-

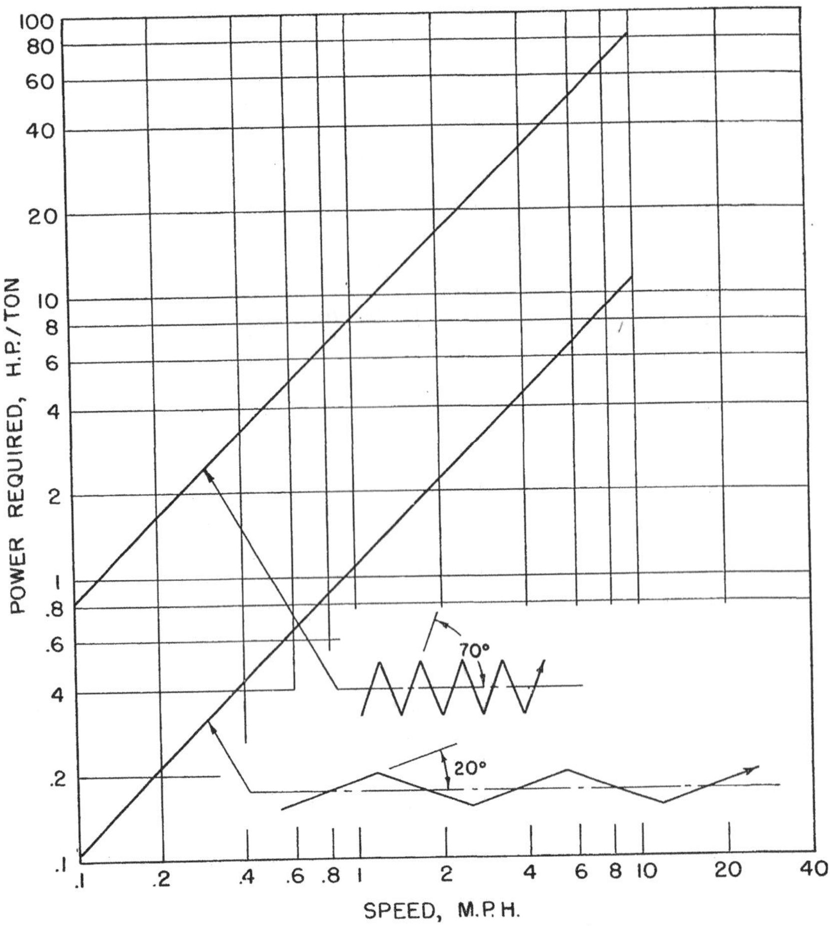

Fig. 8

tional force $mg \sin \alpha$ acts downward along the axis of symmetry of the zig-zag line. This force forms an angle of $(90 - \varphi)$ with the frictional force F_t, and reduces the latter force by $mg \sin \alpha \sin \varphi$. Thus the new reduced driving force is $F_d = (mg \tan \mu_t - mg \sin \alpha \sin \varphi) \tan \varphi$. At the same time, the frictional force $mg \tan \mu_l$ acting in the direction of the sliding body increases by $mg \sin \alpha \cos \varphi$. Finally, then, there is lesser driving force F_d and larger movement resistance. Under these circumstances, the only way of increasing F_d is to increase the angle φ. This has the same effect as if μ_l were increased or μ_t reduced. When a "snake" slides down a hill, conditions are reversed and lower values for φ are permissible. Such a mechanism acts in principle as a torque converter whose function has something in common with the process of driving a vehicle uphill in a zig-zag line instead of straight forward.

The described mechanism of sliding appears to be very close to that of ice- or roller-skating. In this case, skates or rollers form two adjacent portions of a sinusoidal wave which may "elongate" or "contract," depending on the type and speed of movement.

In order to cause a sinusoidal movement, the snake's body must contract the muscles on one side and release them on the other. The propagation of this wave of contraction occurs with velocity $v = v_a/\cos\varphi$. Since $\cos\varphi < 1$, the snake cannot move with reference to the ground with a speed greater than the speed of the contraction wave.

Walking

The mechanism of the walking of a biped is entirely different from the previously described types of locomotion. It is complex and contains many unknown factors. Since our interest, however, is concentrated on simplified schemes which would give only a reasonable order of approximation, the biped type of locomotion is substituted for by some sort of a geometrically simpler pattern. The only pattern of this kind which would be close enough to the mechanics of a walk appears to be the rolling of a polygon, in which the center of gravity is consecutively dropped and lifted up.

In this scheme the center of the polygon would be located in the center of the animal body around which the legs swing, and the number of sides of the polygon has to be adjusted to the length of the step. In conjunction with this assumption, consider Figure 9. It shows a biped in the intermediate position when both legs are symmetrically located with reference to the vertical axis, each at an angle α with that axis. When moving the body forward with speed v, leg E assumes in the final stage of the step a vertical position which is also simultaneously assumed by leg D. In this position, the center of gravity of the body which was at 0 assumes the position 0′, elevated by the distance h with reference to the previous position. Thus, the raising of the center of gravity during one step is

$$h = \overline{BC} = l\,(1 - \cos\alpha)\,. \tag{31}$$

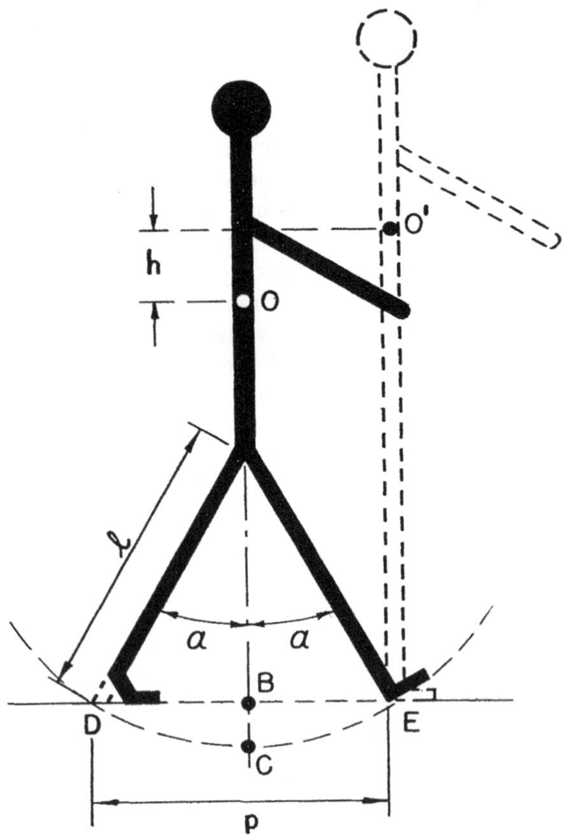

Fig. 9

This movement may be related to length $\overline{DE}$ of the step having speed v. If it is assumed that $\overline{DE} = p$, the time in which one step is made is $t_s = p/v$, and, accordingly, the average speed (per one step) with which the elevation of the center of gravity occurs is $v_h = h/t_s$. By substituting into this equation the previously determined value t_s, the average speed of movement of the center of gravity may be found:

$$v_h = \frac{hv}{p}.$$

Since the force acting with this speed is the mass m times acceleration g, the power required per unit of weight of this type of propulsion is

$$\frac{P}{mg} = v_h = \frac{hv}{p}, \tag{32}$$

or, by substituting into equation (32) the previously determined value of h [formula (31)], the required power is

$$\left[\frac{P}{mg}\right]_{\text{walk}} = \frac{vl}{p}(1 - \cos\alpha). \tag{33}$$

The same relationship may be expressed in terms of the length of the step p. By following Figure 9,

$$p = 2l \sin\alpha. \tag{34}$$

By substituting $\sin\alpha = \sqrt{1 - \cos^2\alpha}$ into equation (34), it is found that

$$\cos\alpha = \sqrt{1 - 0.25\frac{p^2}{l^2}}, \tag{35}$$

and thus, finally, by combining equations (33) and (35), the unit power of walking may be determined as follows:

$$\left[\frac{P}{mg}\right]_{\text{walk}} = \frac{vl}{p}\left(1 - \sqrt{1 - 0.25\frac{p^2}{l^2}}\right).$$

If the average man's walk is assumed as defined by $l/p = 1.65$, then the formula for the unit power required for walking takes the form:

$$\frac{P}{mg} = 0.0775\, v. \tag{36}$$

The curve representing equation (36) in hp/ton is plotted in Figure 10. It may be noted that, for instance, the power of 1.65 hp/ton required at a walking speed of 3 mph is not in basic disagreement with figures quoted elsewhere.[3] This suggests that comparing walking to the rolling of a polygonal prism which has a side length equal to the length of the step is not fundamentally wrong. The smaller the steps, the more sides to the polygon and the closer it approaches a circle. Thus, it may be concluded that in a biped nature almost ideally approached the concept of a wheel as a means of locomotion, although a fully rotating joint was not created by the organic world.

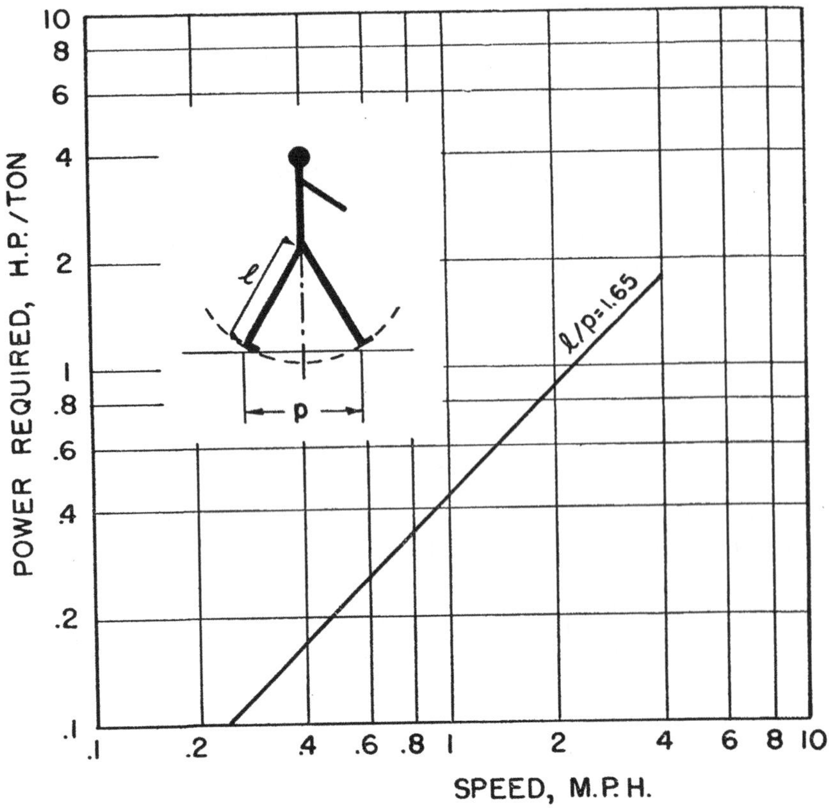

Fig. 10

Movement Resistance and Type of Translation

All the discussed schemes of locomotion have different mechanisms of movement resistance. These mechanisms, in a broad sense, may be related to the action of hydrodynamic, frictional, and gravity forces. The last resist movement either directly, when opposing an upward movement, or indirectly, in the form of a loss in kinetic energy. The resistance of eddies which develop in the case of hydrodynamic flow is complex. It would be a rather crude oversimplification to contemplate any mechanical models in this field, although the formation of rotating cylindroids of fluid moving in the vicinity of channel walls is quite distinguishable. These cylindroids may be seen, for example, in the photographs by Prandtl and Tietjens; [14] it appears as if the transportation of a fluid is also based on the idea of a rotating cylinder even though the latter appears in the intangible form of an eddy.

Locomotion through running and jumping necessitates the swinging action of extremities which may be compared to a multijoint pendulum. Sliding and crawling involve the oscillatory movement of the driving elements in the plane of translation: sliding perpendicular to the axis of movement, and crawling along that axis. Walking, as previously described, may be identified with the rolling of a rimless wheel, with spokes supporting the load.

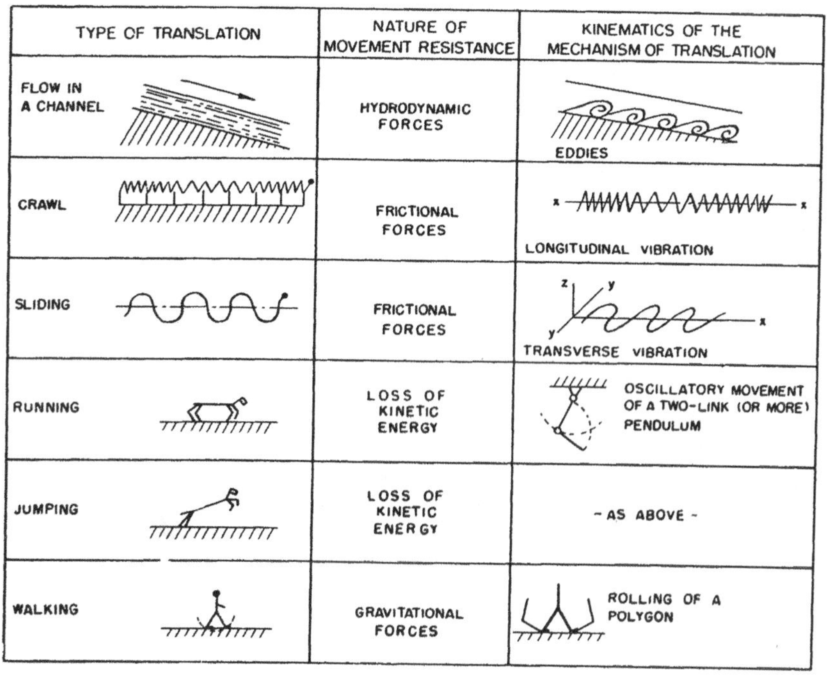

Fig. 11

A graphical presentation of the mechanisms of locomotion in nature is shown in Figure 11. An examination of this picture will undoubtedly suggest the question of whether or not all the theoretically possible ways of movement on the surface of the earth have been included. The answer is certainly negative; for instance, existing vibrational mechanisms of locomotion develop oscillation in the plane of translation, in the direction of the *xx* and *yy* axes only. There seems to be no known vibrational movement in the direction of the *zz* axis, if jumping is not considered as an oscillatory motion. Such a movement in its pure form would not produce longitudinal translation; however, if coupled with horizontal forces, it could move the body in an economic way.

Man-Made and Animal Types of Locomotion

A detailed analysis of the problem might be interesting and far reaching. A study of the locomotion created by nature may elucidate some of the problems of vehicle mobility in the same way as the study of bird flights helped in the understanding of flight in the early beginnings of aeronautics. Particularly, the problems of cross-country operations suggest that an investigation of this subject may be useful, for it is known that the "mobility" of wild and domesticated animals in adverse terrain conditions, in most instances, is superior to the mobility of many modern vehicles. Although respective cases have been often reported, they were never sufficiently explained. For example, some authors referred to the unusually good performance of a horse's or ox's shoe in soft ground, stating that such a performance is achieved through the adjustability of the slope of the shoe-bearing area to the slope of the supporting force of the soil.[15] No closer explanation, however, was given in any known source of information.

The speed of locomotion developed in a difficult terrain frequently reaches higher values in the animal world than in the domain of the machine. Even the "road" speed of some animals, such as the cheetah, may reach 80 mph, which compares most favorably with an automobile, although the mechanics of the locomotion of the animal and of the automobile are entirely different.[16] Some crabs, if assumed to be 1/10-scale models of a "walking machine," may be considered to walk with speeds up to 40 mph. However, no close study elucidating this observation is known.

In a long chain of evolution, nature seems to have developed quite a few types of locomotion on the surface of the earth. It seems to have started with the movement of water under the action of gravity and ended with a final type of locomotion in the form of walking which, in an extreme case, may be compared to the rolling of a wheel.

Only man has developed a wheel. It would offer no resistance to movement if it were perfectly rigid and if it moved on a rigid surface in a vacuum. When moving in a soft medium, as Reynolds pointed out in 1876, a "soft roller" has to slip since: "It appears that there are two independent causes which affect the progress of a roller on a plane: the relative softness of the materials and the diameter of the roller. Of these the curvature of the roller always acts to retard its progress, while the other (relative softness) to retard or accelerate, according as the plane is softer than the cylinder or vice-versa... Thus an iron roller on an India-rubber plane will make less than its geometrical progress; while an India-rubber cylinder on iron plane will make more than, less than, or exactly its geometrical progress according to the relation between its diameter and softness, or what comes to the same thing its weight, which conclusions are borne by experiment." [17] The inherent movement resistance resulting from the above-described slip of a rolling cylinder is augmented by other factors, the physical nature of which is most complex.

Assume, in agreement with common experience, that the rolling resistance of a rigid wheel moving upon a practically rigid surface, steel on steel, is of the order of 5 lb/ton, whereas in the case of a soft surface, it may increase to 300 lb/ton.[12] The respective power requirements for overcoming this resistance are:

$$P_{\text{hard surface}} = \frac{5 \times 5280\, v}{3600 \times 550} = 0.0133\, v \qquad \text{hp/ton}$$

$$P_{\text{soft surface}} = \frac{300 \times 5280\, v}{3600 \times 550} = 0.800\, v \qquad \text{hp/ton}$$

where speed of locomotion v is measured in mph.

Thus, if, as previously, air resistance is neglected, the power requirement for the rolling of a wheel in two extreme types of environment can be presented by the area marked in Figure 12. By plotting the previously determined power requirements for animal locomotion in the same graph, it is evident that the most expensive methods of moving are, of course, crawling and sliding. Flow and wheel rolling on a hard surface require, in certain cases, the same power, although the power of flow in-

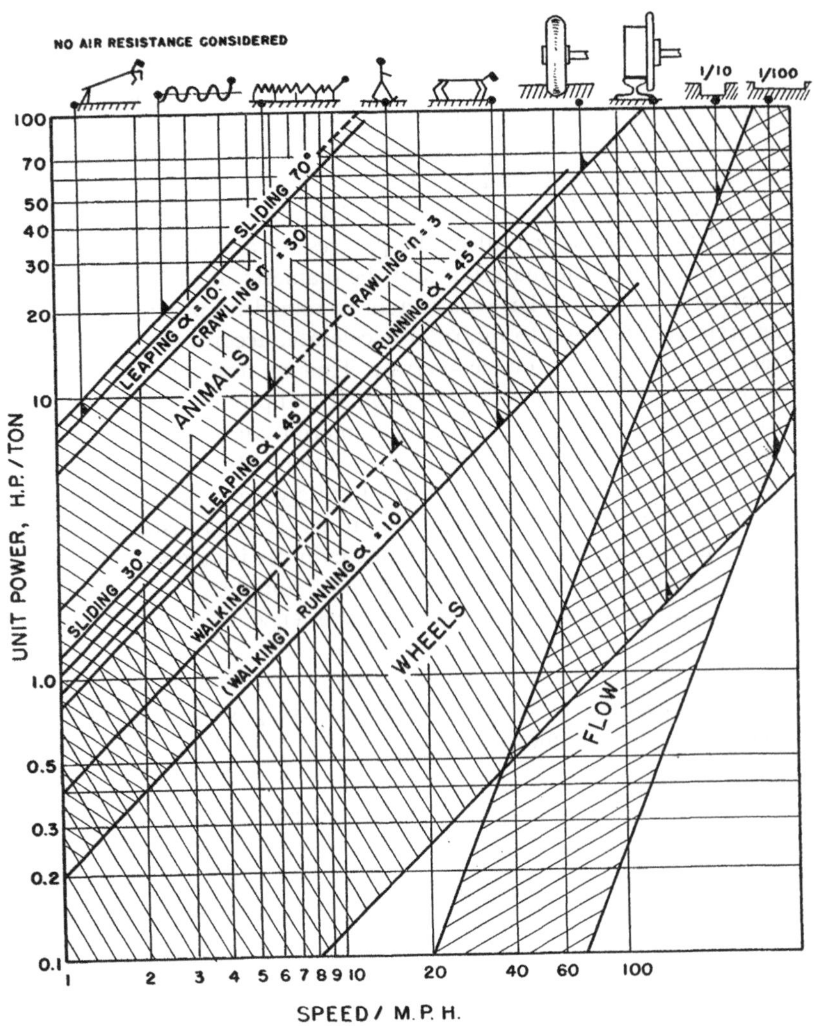

Fig. 12

creases faster with speed than that of rolling, because of its hydrodynamic nature. A wheel driven in a soft terrain may not be as economical as walking or running: it requires more power per unit of weight.

Despite this deficiency, the wheel became the universal means of locomotion for man-made machines. It is the main running gear of not only all the "wheeled" vehicles, but also of the so-called tracked ones. (As a matter of fact, a tracked vehicle may be considered as a purely wheeled vehicle which carries its own road in the form of a track, since nature does not provide roads for locomotion.) Nature, however, wisely developed animals with a walking mechanism which, on a hard surface, may be as economical as a steel wheel on rails, and which, in soft ground, is superior to the pneumatic, within the limits of power available in the animal motor.

The wheel became the universal means of locomotion not because of its over-all efficiency but because it must have been relatively simple and easy to conceive and provide road pavement whenever it was needed. Roads have been built for millennia; real conditions, however, in which heavy loads have to be moved entirely without the help of a hard-surface road, had not existed until recently. Thus, only the recent need of off-the-road transport created the true requirement to devise equipment for moving over soft ground.

Since a wheel applied in this environment is not as efficient a mechanism of locomotion as the other means provided by nature, the whole situation requires at least a survey of original assumptions and facts on which man-made locomotion, i.e., locomotion on wheels, was based. Such a survey, even when not leading to any spectacular development, would clarify many problems and would tell precisely where our modern concepts stand on the broad background provided by nature.

III. LOCOMOTION ON WHEELS

It is not known when the wheel was invented. It was probably conceived independently by many persons, such as was the case for scripture or metallurgy. The motives of invention were probably of a religious nature.[18] Some authors have maintained that the idea of a roller, materialized in a round-shaped stone or in the trunk of a tree moving downhill, did not inspire the invention.[19] Whatever its origin, the wheel was initially a rare object and a luxury. The oldest pictures of wheeled carriages known, dating from about 3200 B.C., depict vehicles built for the transportation of warlords and chiefs.

The Wheel in Ancient Times

Before wheels were invented, the natural transportation system briefly described in the preceding chapter was the only one in existence. In that system, no roads were built and men traversed continents by foot. They undoubtedly experienced in remote times the first difficulties of moving across country during the mass migration of peoples. Steppe, tundra, marshes, desert, mountains, and even plains presented a great variety of obstacles to a pedestrian who, aided by his innate instinct only, moved thousands of miles without any transport equipment.

Many centuries elapsed before the first improvement of this natural transportation system had been achieved. This came not as a result of better knowledge of the relationship between a foot and soil, but through the domestication of the horse. The improvement then was mainly an increase in power available while the mechanism of locomotion remained the same.

Further revolution in transport was brought about not only by the invention of the wheel but also by the invention of the harness, which enabled our ancestors to use the horse as a tractor for towing much greater loads than it could carry.[20] However, carriages which were developed as a consequence of this invention did not have the valuable properties of the horse's shoe or pedestrian's foot. Man must have found immediately that a chariot does not travel as easily in soft terrain as it does over hard ground. Perhaps this first observation of the relationship between a vehicle and soil led to the development of paved roads, which reached a high state of advancement in the ancient Roman Empire.

Road and Wheel in Early Technology

The invention of the highway appears to be the first fruit of primitive soil mechanics as applied to vehicle problems, and the first recognition of the limitations of the wheel. Trends which resulted from this invention continued far into the future. The slow development of a network of roads and tracks with horse-drawn wagons lasted for thousands of years: progress achieved in this period followed the development of handicraft and was insignificant.

New ideas which brought fundamental improvement in means of locomotion had to wait until modern times when the development of transport was made possible by the advent of technology and by the changes in cultural and economic conditions which speeded up the progress of the machine.

Two centuries after the Middle Ages ended, men started to look into the foundations of their individual and social life. The spiritual opposition which had resisted technology had been previously weakened in a long preparatory process that furnished the ground for a gradual technological development.[21] In the eighteenth century, about one hundred outstanding inventions were made. These radically changed the existing way of life.

Application of water power for mass production, which inevitably led to a wider distribution of goods, was an invitation for developing better communication and organization of transport. This is reflected in the invention of wooden railways covered with iron (1716), iron wheels for cars (1775), and also cast-iron rails (1767). The steam carriage (Cugnot, 1769), the steam boat (Joufroy, 1781), and the signal telegraph (Claude Chappe, 1793) are further examples of trends which started in this field as a natural consequence of the increase in production.[22]

The new movement gradually spread over the western world. Apart from economic and political obstacles, the only limitations in speeding the expansion of the Industrial Revolution were geography and climate.[23] Country roads and tracks could not be traversed during the spring thaw or autumn rains, since mud and clay were as unconquerable as human prejudices. Is it peculiar that in this initial period of industrial expansion, and of new social trends, the idea of a vehicle which would have an improved performance when traversing roadless country came to man's mind?

At that time, rails made of wood and later of cast iron were used in mines. They were cumbersome, costly, unreliable, and the possibility of constructing modern railways was not considered for almost a century. When the improvement in existing means of land transport did occur, it had to be based on the experience gained with transportable rail units which were located wherever required by mining conditions.[43]

Invention of the Cross-Country Vehicle

In 1770, a patent in England by Richard Lovell Edgeworth was taken out for a "portable railway" or artificial road, which was to move along with any carriage applied to it. The fact that this invention was inspired by the use of rails in mines is indicated by the following extract from the patent: "The invention consists in making portable railways to wheel carriages so that several pieces of wood are connected to the carriage which it moves in regular succession in such a manner that a sufficient length of railing is constantly at rest for the wheels, to roll upon, and that when the wheels have nearly approached the extremity of this part of the railway, their motions shall lay down a fresh length of rail in front, the weight of which in its descent shall assist in raising such part of the rail as the wheels have already passed over, and thus the pieces of wood which are taken up in the rear are in succession laid in the front, so as to furnish constantly a railway for the wheels to roll on."[24]

It appears from the above description that the pieces of wood were connected in such a manner as to form an endless chain girding the front and rear wheels of a carriage. If such were the case, the Edgeworth patent refers to the idea of a full-track vehicle. It seems probable, however, that this patent might have implied an idea of separate shoes, attached to an individual wheel and moving with the wheel in order to secure a steady support. This solution, which refers to the idea of a shoed wheel, was propounded by many inventors. The link between these two ideas would be the concept of a chain girding a single wheel. This also formed a pattern for the subsequent development of individual wheel tracks which in modern days developed into traction devices. The discussed three trends derived from the Edgeworth patent are shown in Figure 13.

Full-track vehicles attracted the attention of the greatest number of inventors. It would not be easy to enumerate all the patents taken in England alone between 1800–80. A typical invention is the idea by

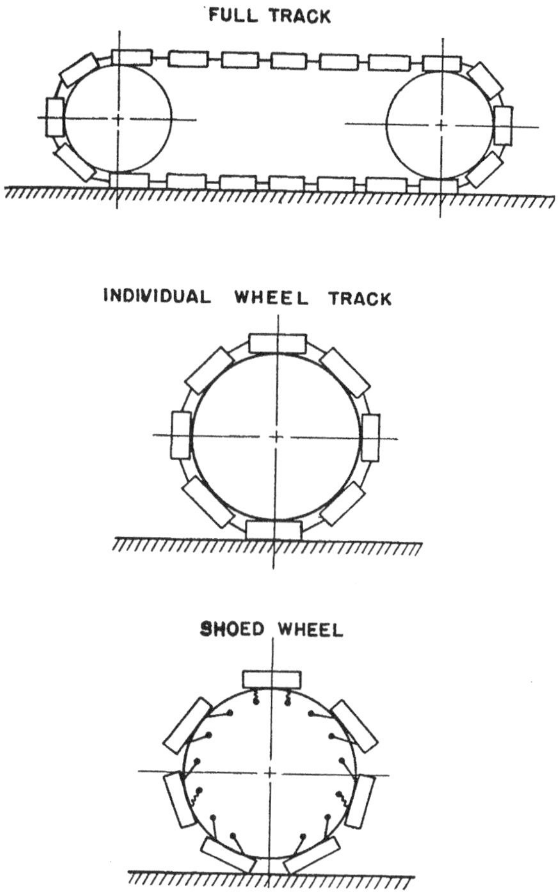

Fig. 13

John Richard Barry, who patented in 1821 "two endless pitched chains stretched out by passing round two chain wheels at each end of the carriage." Unfortunately, none of the inventions of this type were ever employed commercially, although a vehicle designed by Guillaume Fender of Buenos Aires, and patented in 1882 by John Clayton Newburn, had a general layout almost identical to that of modern tracked vehicles (Figure 14a), with the exception of the steering mechanism which was nonexistent. Horses were supposed to propel all the vehicles by drawing the carriage in the required direction.

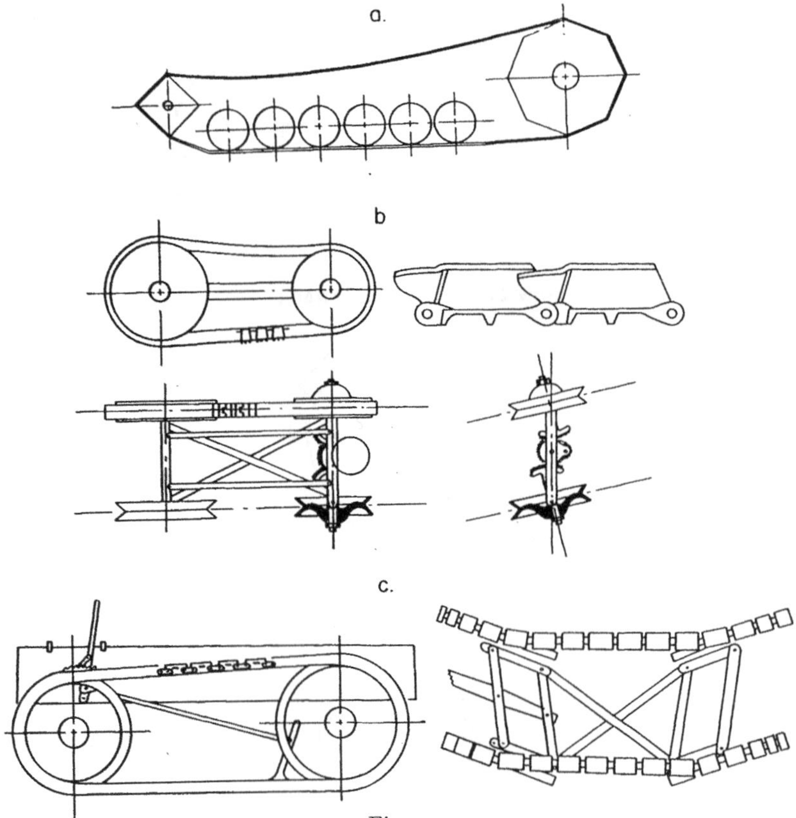

Fig. 14

An interesting idea was patented by G. F. Page of Baltimore in 1884: the vehicle was claimed to have a "chain of links taken over both driving and [as the author terms] pilot or guiding wheels." This claim may be considered as the first attempt at using a horse to steer a vehicle by means of turning the front wheels and thus warping the tracks (Figure 14b). A similar idea was proposed by Justice Johnson in 1896 (Figure 14c), and was revived in modern times in one of the experiments made in Australia and England.

Mechanical propulsion, however, could not have been introduced without solving on a broader basis the principle of steering. Solutions proposed in American patents (Batter, 1888, and Edwards, 1890) refer to a concept in which a track designed for propulsion is located in the rear of a vehicle, whereas the front wheels pivoted around a vertical axis perform the steering. This is the simplest scheme for steering a self-propelled vehicle and was conceived during the era of steam engines (Figure 15).

Individual wheel tracks did not appeal to the imagination of inventors as much as tracks wrapped around the front and rear wheels of a carriage. Two men, H. G. Woodbridge (1882) and W. Applegarth (1886), patented, without much success, shoed chains attached directly to a wheel rim. An ingenious solution had been previously proposed by A. Dunlop (1861) but difficulty was created by the fact that a long pitch chain fixed to the wheel had to be provided with a complicated guiding mechanism. The attempts by Dunlop to design a simpler chain (1874) also were unsuccessful (Figure 16).

It may be surprising that the shoed-wheel devices were the most successful and were commercially used. Such a vehicle developed by Boydell (1864) was, for 20 years, the only cross-country machine. The hinge principle of attaching feet to the wheel rim was very simple, and this was undoubtedly the main reason for its success (Figure 17). According to C.F.T. Young: "Up to the present time (1860), this vehicle can be

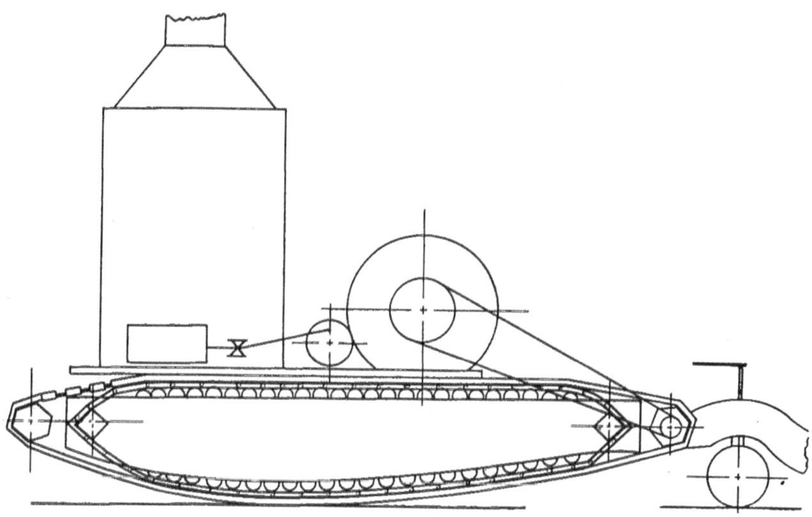

Fig. 15

regularly and profitably worked over the same ground without injuring it, and is the only engine on this principle which has ever been regularly worked."[25] The reference to "injuring the ground" is the first published comment on soil deformation under the action of a vehicle.

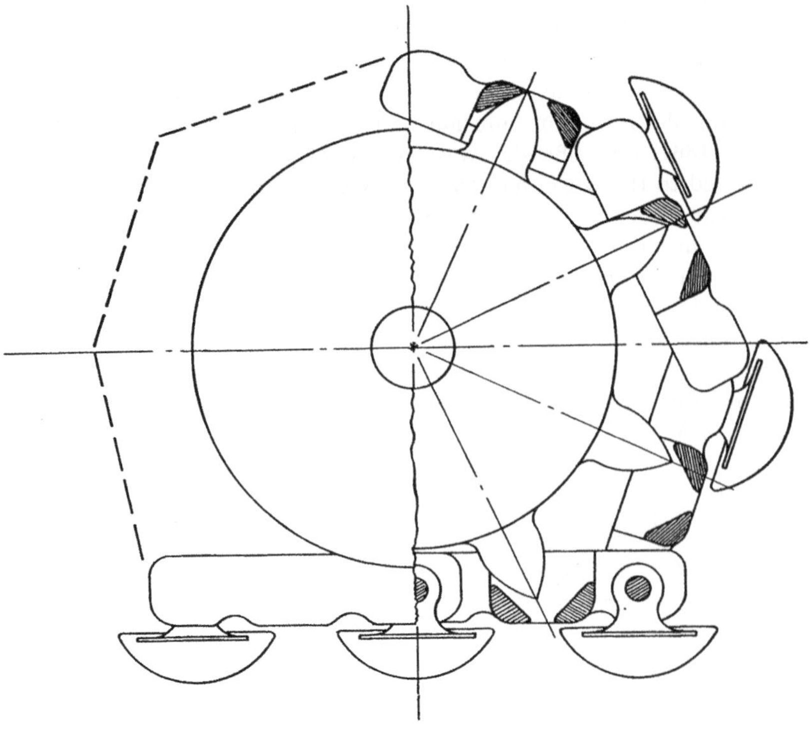

Fig. 16

The failure of all other types of devices and the outstanding experience with Boydell's vehicle inspired B. J. Diplock (1899) to design another "pedrail," the new feature of which was the spring suspension of the portion that carried the wheel. Springs apparently did not come into use on a cross-country vehicle until it actually experienced terrain roughness. Thus, the problem of the surface contour of soil, and its significance in traffic, emerged in a definite form at the middle of the nineteenth century and called attention to the geometrical properties of soil surface. As will be seen later on, the geometry of the ground is as important in off-the-road transport as the physical properties of the soil itself.

At the end of the nineteenth century, the only vehicle which could move across country was that of Boydell. In spite of numerous inventions and attempts, no tracked vehicle in the modern sense had been developed. The cause of this failure is obvious: technology was in its infancy. In most cases, wood and cast iron were the only materials available. Since tracks are very highly stressed elements, an application of such materials would have led to large dimensions of vehicle elements in order to reduce the stresses. This would have increased the weight, and any gain in reduced ground pressure due to the application of tracks would have been cancelled by a simultaneous increase in the load.

The vicious circle was reasonably solved by Boydell, who applied simple wooden shoes, hinged to the wheels. Other solutions had to fail because, at that time, inventors did not know about modern light-weight engines. The only available motive power was that of horses and mules, or steam generated in bulky unreliable boilers.

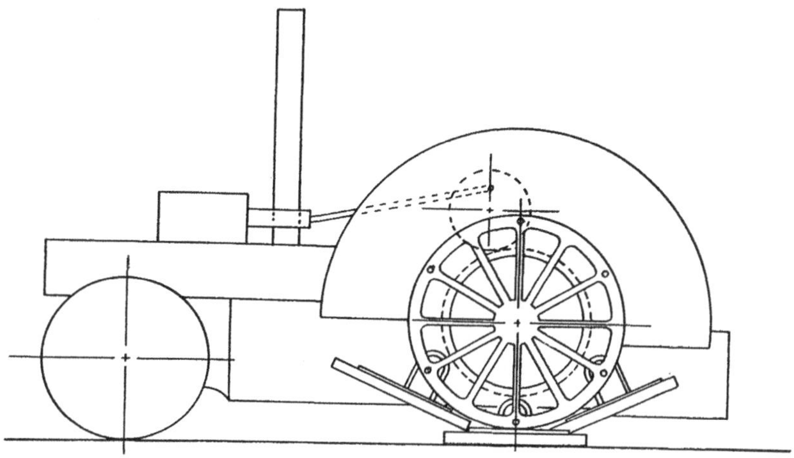

Fig. 17

Beginning of Modern Development

The situation was, however, rapidly changing. The steam carriage (Trevithick, 1801–2) and steam automobile (Hancock, 1827) were followed by the development of railway transport. In 1845, the pneumatic tire was invented (Thomson). A little later, the first internal-combustion engine appeared. After the gas engines by Lenoir (1863) and Otto and Langen (1867) were developed, Daimler (1883) designed a high-speed gasoline motor, and the first automobiles started to conquer the roads.

The improvement of the cross-country vehicle created during the roadless era of Edgeworth, however, was slowed down, and development was concentrated on the automobile and highway. Undoubtedly, it was easier to design a vehicle for cruising on the expanding network of excellent roads than for negotiating a variable terrain which changed properties with rain or wind. The invention of the highway was almost two millennia old, whereas the idea of a cross-country vehicle had been propounded for only about 100 years. The soil mechanics of Coulomb (1776) and Boussinesq (1885) was not intended to improve cross-country vehicles, but was meant to serve civil engineers in devising more monumental earthworks and roads.[26, 27]

In spite of the overwhelming development of the automobile in the early 1900's, the idea of developing a cross-country vehicle was not completely abandoned. Boydell's device was further improved and appeared in the form of Diplock's "pedrail" as some sort of hinged shoes attached to a wheel. The vehicle was built between 1901 and 1908, but its limitations were numerous. It was impossible to increase sizably the dimensions of shoes beyond an area restricted by design. The improved performance obtained with insufficient bearing area must have been very small compared to the simple wheel. As a result, the idea of a full track gained in popularity again.

In 1900, Beadmond proposed to apply tracks to pneumatic tires. The automobile, however, was still too primitive to cope with such a heavy duty, and the idea therefore had to wait for many years, until it was applied in the form of traction chains to wheeled vehicles (Figure 18).

In 1901, the patent taken by the Lombard Co. in the United States claimed a half-tracked steam engine that could be used on packed snow or muddy roads, so that heavy loads of timber could be hauled at all seasons of the year. This vehicle looked like the Batter tractor and was very successful because of the higher level of technology that was available to the manufacturers. The mention in the patent application of snow and soils as media negotiable by a vehicle was the first sign of linking the problem of snow mechanics and soil mechanics as far as trafficability problems were concerned.

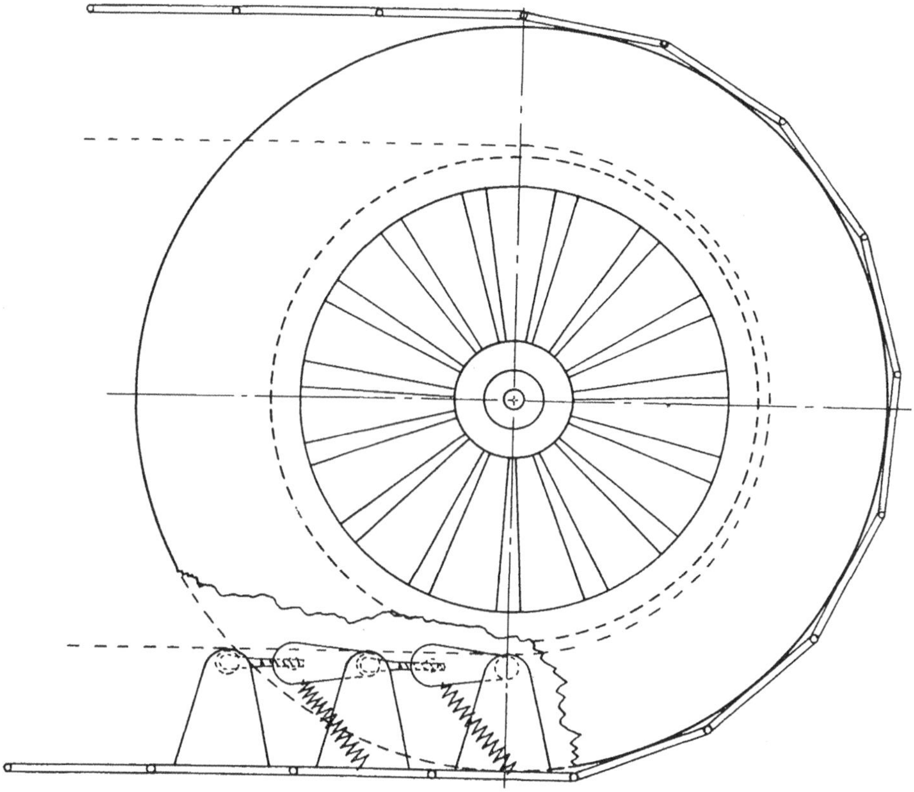

Fig. 18

The turning point in the broadening of the application of cross-country vehicles came with the interest displayed by the British War Office, shortly after various tractors were tested (1908) in a terrain which the ordinary wheeled vehicle could not negotiate. The vehicles tested were those designed by D. Roberts of the firm of Richard Hornsby, and the slow-running, one cylinder gas engine was used for the first time (Figure 19). A new idea of steering was devised, although the general layout of the vehicle was not much different from that proposed by Fender. The steering mechanism showed a definite influence of the automobile development and consisted of an automobile differential with two brakes fixed on the shafts. The few models which were developed between the years 1904–7 proved to be fairly successful.

A further step in the development of tracked vehicles was the application of an automobile engine to the propelling of tracks. This, however, was not made in a revolutionary way: Hornsby Co. purchased a 40-hp Schneider passenger car and, by adding two pairs of wheels, merely adapted tracks of light design. In the light of modern experience, it is

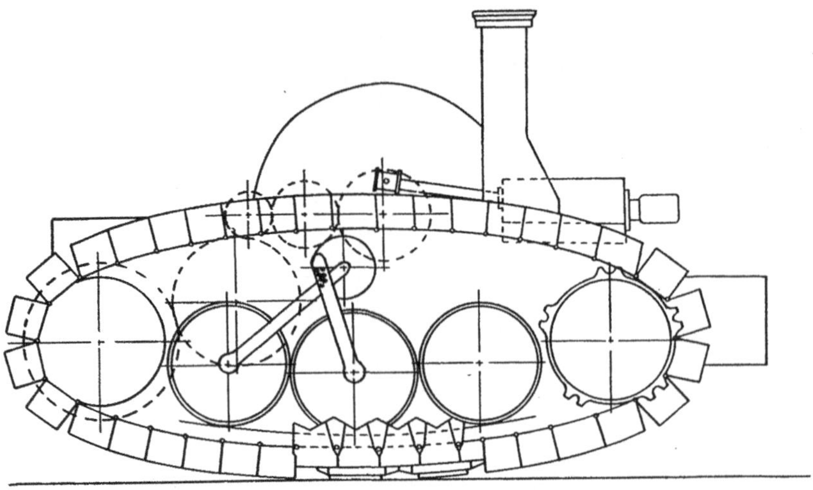

Fig. 19

evident how little might have been expected from a 40-hp tracked vehicle, particularly when built in a primitive way by a hasty conversion of a French passenger car of 1907. Although the performance of Hornsby vehicles was encouraging, it was not satisfactory, and it is natural that the War Office soon lost all interest in further tests.

The only workable machine at that time was a steam tractor built in the United States by the Phoenix Co. and the Holt Manufacturing Co., based on principles laid down by Batter, Edwards, and Lombard. The latter type of tractor was used in Canada around 1908.

In the meantime, Messrs. Hornsby worked on improving the gasoline tractor, which was done by converting a Mercedes passenger car. However, the main achievement of this development is seen in the perfection of the steering mechanism, which, as designed by Roberts and James

(1909), consisted of an automobile differential and dog clutches applied to both shafts (Figure 20). It is evident from the foregoing that the whole work on improving vehicle performance was concentrated on engineering details which were basically developed by the industry for use in automobiles. Problems of the relationship between soil and vehicle were practically forgotten.

After Messrs. Hornsby found that the Mechanical Transport Committee

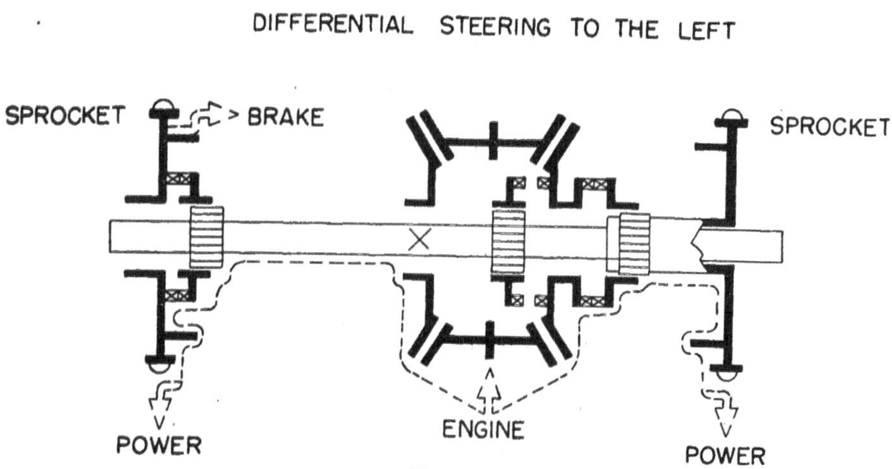

Fig. 20

of the War Office in London did not possess a grant sufficient to proceed further with cross-country vehicles, they decided to abandon the development. In 1912, they sold most of their foreign patents to the Holt Manufacturing Co., Stockton, California, and the Holt Caterpillar Co., New York.

In the meantime, Diplock dropped his idea of the "pedrail" and switched to the design of a track. However, being influenced by his previous inventions, he proposed a complicated track which was fitted with rollers running on an ellipse-like frame. All the known inconveniences of such a system could not be overcome even today, and naturally, it had little success in 1910.

Perhaps the most significant innovation made by Diplock (1908) was his proposal that the vehicle body be divided into two parts, pivoted around a vertical hinge. In this case, steering could be effected by turning both parts of the body around the hinge in a "well-known way," as implied by the inventor, thereby avoiding large loads acting upon the soil and tracks while steering. Thus, the insolvable problem of steering a long but narrow vehicle by means of Hornsby's device was eliminated (Figure 21). However, as far as it is known, this idea has never been seriously contemplated.

Real progress was made by the previously mentioned Holt Co. when it greatly popularized tracked vehicles in the United States. The Holt tractor of 1911 was the predecessor of modern agricultural tractors; its general layout, type of track, and location of major mechanisms have been changed slightly since that time. About 1912, the Schneider Co. of Le Creusot and the Delahaye Co., both in France, built tractors identical with the Holt type.

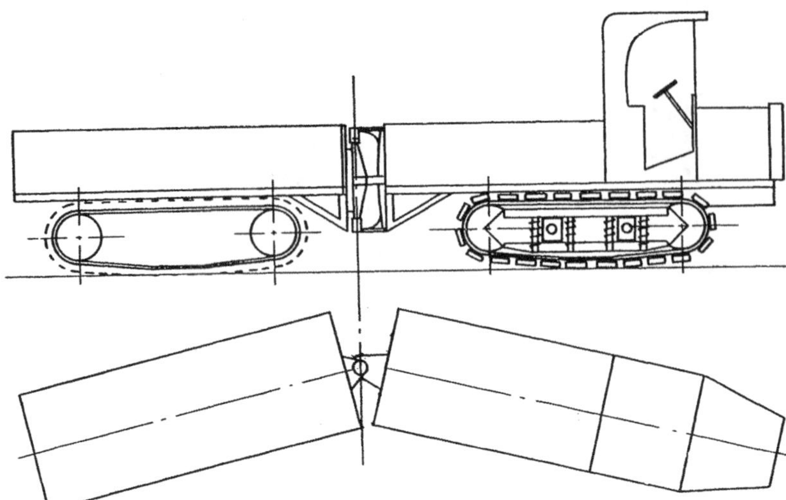

Fig. 21

World War I

At the outbreak of World War I, a tracked vehicle existed in a completed mature form. Its limited use at that time, as compared to today's track-layers, resulted not from a wrong concept which has been gradually developed over a period of almost one and a half centuries, but from the general status of technology and from an insufficient theoretical knowledge of the phenomena involved between soil and vehicle. The principles of soil mechanics, which at that time were first propounded by Terzaghi [28] from a civil-engineering point of view under the pressure of a series of construction catastrophes (for example, accidents during the construction of the Panama and Kiel Canals), were too novel to form a basis for the investigation of soil-vehicle relations.

It should be noted, however, that the idea of a uniform distribution of load over the whole contact surface between track and soil was carefully applied by using a large pitch track supported by several small-diameter bogies. Thus, one of the very essential principles of soil loading was well satisfied. Further refinement waited for new initiative and new requirements. These came after World War I had begun.

On the one hand, there was an American Holt tractor, the only one ready for commercial use, and potentially susceptible to further refinements without changing the main concept (Figure 22a). On the other hand, there happened to be a stalemate on the western front during 1915–18 which stimulated a few far-sighted men in England and France to lay down specifications for a modern tank.[29, 30] In wartime, there is little opportunity for an exhaustive study of new emerging problems, and the adaptation of existing solutions and experience is inevitable, notwithstanding whether they are suitable or not. Thus, Holt became the nucleus of all tank development in Allied and, subsequently, Central Power countries.

The first French tanks, the St. Chammond and the Schneider, were simply redesigned Holt tractors; they resembled what today would be called armored weapon carriers (Figure 23a). Since they exceeded in weight anything previously designed by Holt, a great deal of trouble was encountered, not only from a mechanical point of view, but also as far as soil action was concerned. Apparently, data in regard to the bearing capacity of soil were not available. Costly experience on the battlefields in France later resulted in an increase in the track width by about 35%. This change was not, of course, figured out from the allowable soil pressure, but from the vehicle dimensions, which permitted the designer to correct the error only partially.

Although British tanks were apparently different from the Holt tractor, which, as previously mentioned, was developed from the original idea by Hornsby and Diplock, they were based on the same principles. Since they also entered a much higher weight class, the problem of power transmission and steering was troublesome. Tanks Mark I–VIII (Figure 23b) could not easily adopt the types of automobile gears existing at that

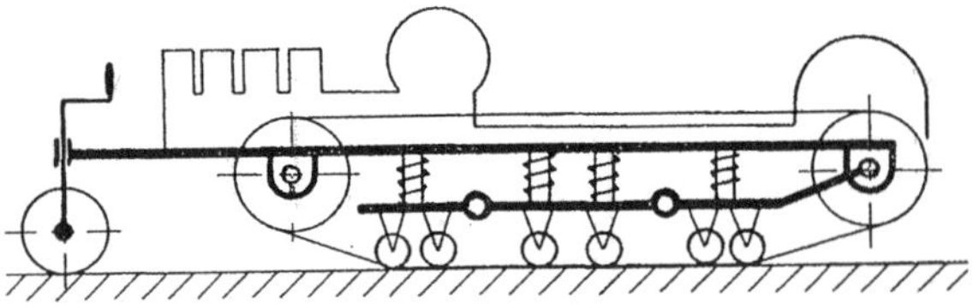

a.

b.

c.

Fig. 22

time. Consequently, their development had to rely on the designing and testing of various transmissions in which mechanical, electrical, and hydraulic mechanisms were conceived. Under these conditions, there was not much time left for solving problems of soil reaction upon the vehicle. The heavy burden of engineering problems overshadowed everything.

The difficulty of these problems may be seen in the fact that toward the end of the war only light-weight tanks of the Renault (Figure 23c) or Whippet class were mainly produced. They were much more successful than heavy tanks, since they directly embodied the purest form of the Holt concept and made full use of available, automobile-like gear boxes, clutches, and brakes. The Germans, who had much less time than the Allies for developing their tanks, merely contemplated converting hundreds of automobiles and building the LKW tanks on their stripped chassis.

It is obvious that World War I opened new fields for studying the relationship between a vehicle and its environment, but the mechanical adjustments of existing mechanisms to new requirements consumed practically all the efforts so that no time was left for basic studies. Consequently, the experience gained in 1915–18 did not contribute much to the broad knowledge of the mechanics of locomotion and of the medium in which it takes place. The experience mainly dealt with engineering questions of design, production, exploitation, and maintenance. It also created a vast field of military problems which were evolved by the new tactical doctrines.[31, 32] Sooner or later all these factors would have undoubtedly led to more fundamental studies if the end of the war had not stopped further development in this field for many years to come.

Period between Two World Wars

New hopes raised by the Treaty of Versailles seemed to promise years of peace. Since the League of Nations was to make the world safe for democracy, the general trend toward disarmament made military-technical development practically nonexistent. Cross-country vehicles which were so widely used during the war were converted to the new civilian applications which were gradually found in the field of agriculture. The development of wheeled and tracked equipment in the United States, England, and France finally became a well-organized basis for progress in which there was ample space for fundamental study.[34] The Germans, who according to the peace treaty could not develop army equipment, also concentrated in the early 1920's on the improvement of agricultural machinery.[34, 35] The Russians, in their efforts to rebuild the country in five-year plans, focused attention on the problems of tractor production while compiling and acquiring all the

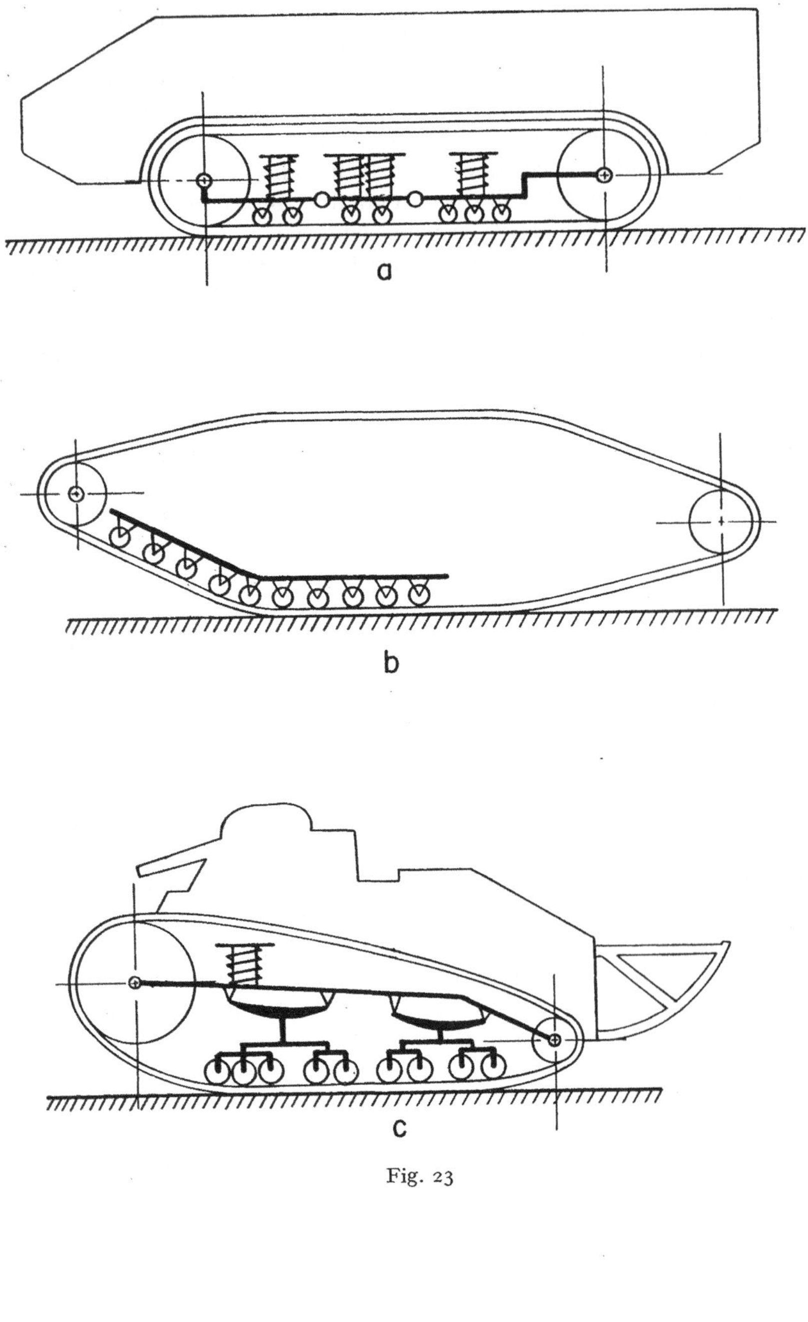

Fig. 23

available information, predominantly from the United States. They published a number of valuable periodicals, monographs, and handbooks.[15, 33, 34]

Systematic tests of agricultural tractors (Nebraska) led to definitions of merit of various types of machinery and defined the idea of certain relationships between soil and vehicle from an agricultural point of view. Many details of locomotion mechanics emerged in their full significance. The effect of the center-of-gravity location of a tractor upon the pressure distribution and tractive effort, and the importance of supporting the greatest number of track links were formulated as early as 1926.[36] Problems of the actions of spuds and grousers as well as questions referring to movement resistance, sinkage, wheel dimensions, etc., were seriously considered around 1935.[37, 38, 39] The idea of the "ground pressure," defined as the result of dividing the total weight of a vehicle by the contact area of the latter with the ground, was widely introduced in the literature. Various figures were quoted which made the design of new equipment less arbitrary. The distribution of ground pressure with reference to the external forces acting upon a vehicle was considered.[33]

By trial and error, the technical standard of agricultural tractors was improved. The introduction of controlled differential steering and Diesel engines was perhaps the most significant mark of progress in the mechanical engineering of these vehicles. This process evolved a standardized type of one-purpose vehicle which embodied the same main concept that characterized the early Holt tractor (Figure 22a, b). Thus, the general line of progress remained unchanged. It was only refined. The new item which was added in the 1930's, in a competitive effort to reduce the exploitation cost, was a wheeled tractor, the steel-rimmed grouser wheels of which were replaced by a giant pneumatic tire (Figure 22c). Progress in this line closely followed the development of automobiles and, like an automobile, had very little to do with soils. Only empirical studies were undertaken. Accordingly, the achievements in the field of agriculture, although very important, have definite limitations. Since they have been concerned with a certain type of work done in cultivated soils only, the results obtained refer to specific conditions encountered. The trial-and-error method resulted in solutions which were valid only in particular cases. It has been almost impossible to make any broader generalizations of these solutions in regard to conditions other than those concerned with agriculture.

The generalizations, whenever made, were very approximate. For instance, the explanation of the grouser action of a spudded wheel was once based on the Rankine theory of passive earth pressure,[37] although even Coulomb's solution would have been inaccurate for practical use. The idea of "ground pressure," extended over types of vehicles and soils other than those contemplated in agriculture, has been misleading as far as proper evaluation of various developments is concerned.

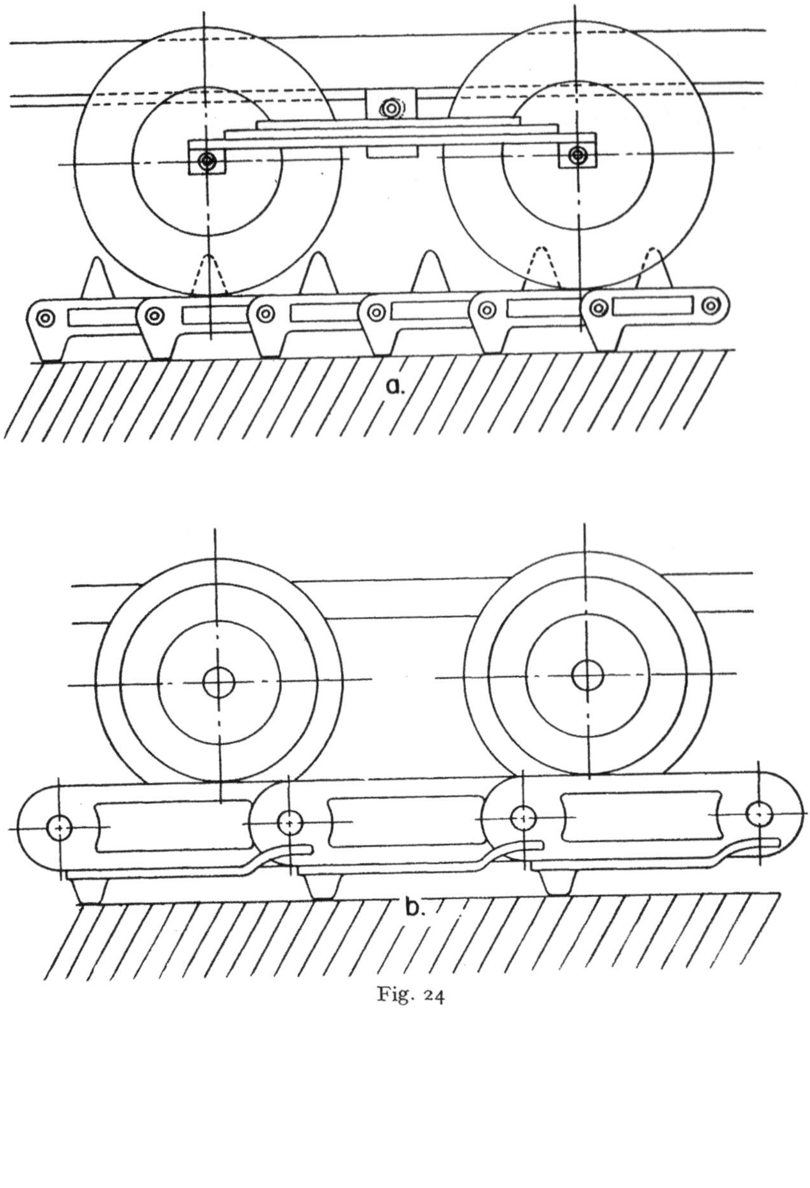

Fig. 24

Although, as was mentioned before, research and design of military vehicles were practically at a standstill in the period following the end of World War I, some innovators proposed various ideas predominantly in regard to the mechanical improvement of a vehicle.[32] The short life of early tanks called for better engineering and materials. This led, for example, to the introduction of rubber-tired bogies by Carden-Lloyd in 1926 (Figure 24a) and of more resilient spring suspension by Vickers at about the same time. The Vickers—Carden-Lloyd principle, with the addition of a light short pitch track, not only increased the life span of a tracked vehicle, but also increased its maximum speed, which was in line with the British tactical concept. These refinements had little relation to the better understanding of soil. However, the internal movement resistance of vehicles at higher speeds was considerably reduced. The introduction by Citroen-Kegresse (1923–28) of the rubber track (Figure 25) in half-track vehicles illustrates further the same trend which brought into the picture the previously described Vickers—Carden-Lloyd design.

The French tactical concept based on the idea of a heavy slow tank for supporting the infantry in an attack did not readily adopt the Vickers—Carden-Lloyd running gear. An example of a vehicle based on this concept was the 70-ton Char 2C and 3C (1925–28) which was much larger than its British contemporaries. Since at that time there were two tactical doctrines, one of which called for a light fast vehicle and the other for a heavy slow vehicle, the main problem was how to reconcile these two extremes. Thus, technological design problems of weight (armor and gun) and speed (engine) again overshadowed the question of vehicle-soil relationship.

Undoubtedly, the most interesting example of a light speedy vehicle is the one designed in the United States by Christie, between 1921–28 (Figure 26a). By combining extremely well-sprung, rubber-tired bogies of large diameter with a long pitch track in which guiding lugs were driven by means of rollers located on the sprocket, Christie claimed fantastic speeds. This, however, was not the most significant feature of his design, if looked upon from a general point of view. The combination of a large-diameter wheel with a long pitch track provided the more uniform load distribution similar to that achieved by agricultural tractors. Because of the ingenious method of track driving as well as the resilient suspension, speeds which are denied to the commercial tractors were also developed. At that time, this fact was not properly assessed. For a long time, Christie's ideas remained unaccepted in spite of the fact that around 1935 the Russians

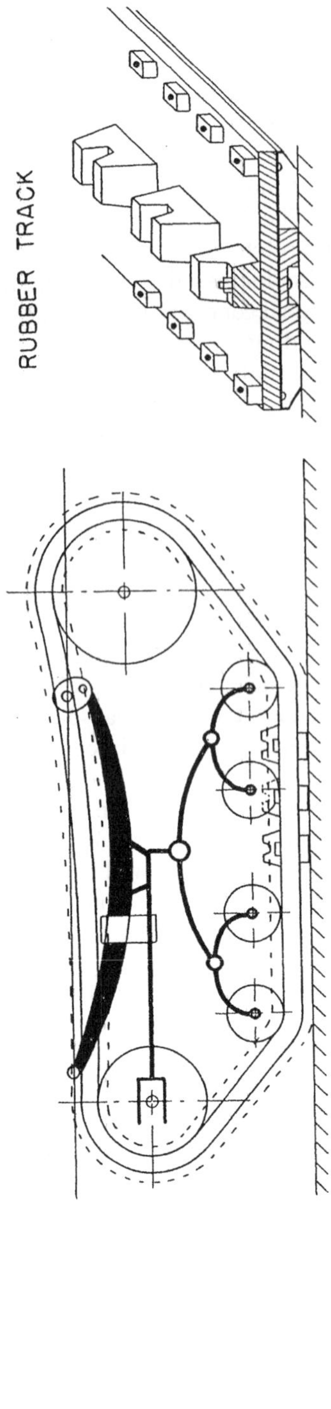

Fig. 25

produced large quantities of modified Christie tanks. These tanks later evolved into the well-known Russian T-34 which outmatched German tanks during the early stages of World War II and which found their use in Korea in 1950. British and German tanks built during the last war borrowed a great deal of Christie's concept. Today its main features have been accepted generally.

Tracked vehicles of the 1930's had poor performance in what was called at that time strategic or operational mobility.[32] The high ground pressure exercised upon paved roads by the knife-like spuds which were supposed to develop better tractive effort (Figure 24b) damaged the roads excessively. Vehicles had to be transported to the field of operation by trucks or rail in order to avoid the quick wear of mechanisms and highways. The life span of tracks, suspensions, and transmissions was probably 1/3 of the present one.

In order to improve this situation, many wheel *cum* track vehicles were developed [St. Chammond, 1920 (Figure 26b); Vickers, 1929 (Figure 26c); Landsverk, 1935; Austro-Daimler, 1936; etc.]. It was realized very soon, however, that such a combination gave neither a good wheeled vehicle nor a good tracked one. Thus, the attempt of extending the trafficability range from a paved road up to soft ground by a mechanical combination of two different running gears failed, although a great number of discussions and papers appeared in approximately 1930 and thereafter, and although several attempts were made to clarify design and engineering questions.[34, 40, 41, 42]

It appears that, at that time, the attention of engineers and designers was absorbed by another experience. The progress in the field of antitank gunnery and the problem of armor protection changed the emphasis from mobility of the vehicle to its fire power and armor. The main trend that resulted from this change was to build up more heavily armored and armed vehicles, and to do this as soon as possible, rather than to study vehicle mobility. In view of the lack of fundamental theory, the old concept of "ground pressure" was adopted again merely leading to wider tracks for heavier vehicles. It was soon found out, however, that making vehicles wider and heavier could not be pursued indefinitely.

An interesting achievement from a general point of view was the introduction of overlapping bogie wheels by the Germans (Figure 26d). This provided a more continuous support for the track and lessened the peak values of the actual ground pressure.

It may be concluded that general problems of vehicle performance

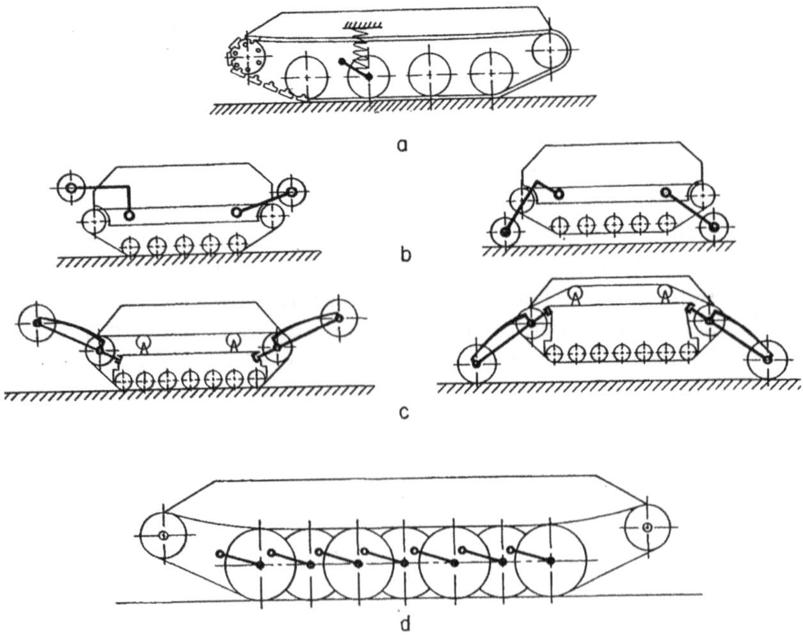

Fig. 26

were only slightly elucidated in spite of the tremendous engineering progress made in the period between the two World Wars. Investigators were busy refining old concepts pioneered in the Holt tractor of the early 1900's. However, they followed too closely the progress in automobile design and merely adopted improved engines and transmissions. The study of the relationship between the vehicle as a whole and the medium in which it moves was forgotten since there was no such problem in automotive research, with the exception of the study of air resistance.

World War II

With the outbreak of World War II, a new development of cross-country vehicles started again in the heat of wartime conditions which were not much different from those that had prevailed during World War I. The Civil War in Spain, in 1936, provided a proving ground in which new and old ideas could have been tested under real field conditions. However, the experience gained concentrated again on limited, though vital, trends in development: a tendency toward devising larger guns and heavier armor.[44]

The race which started around 1939 with 2-pounder and 37-mm guns, matching armor plates from 1 in. to 2 in., led to the use of 17-pounder and over 100-mm guns, matching 4-in. to 5-in. armor, by the end of World War II. This change was by no means easy from an engineering point of view. First of all, shoulder-fired guns had to be mounted in elaborate devices and power drive for turrets had to be developed. Intricate firing and aiming mechanisms, wading provisions, ventilating systems, observation devices, and all sorts of auxiliary equipment required for the operation of such complex tanks were the primary objective of design and development.

The problems of suspension and steering mechanisms also had to be adapted to new conditions.[45] The suspension design, which further remained under the influence of automobile development, predominantly utilized the Christie principle refined by the application of torsion bars, and the overlapping wheels became standard German practice. Steering passed from the stages of refinement of the original automobile differential into the stage of development of planetary-gear mechanisms which were more economical, flexible, and safe at higher speeds and heavy loads. Engines were the logical issue of progress in metallurgy, production methods, and fuel chemistry. The electric drive, which operated the Mark VIII tanks during the first World War and which was dropped because of the excessive weight, was successfully developed during the second World War. It was apparently not adopted only because of other nontechnical reasons.

Again, preoccupation with engineering work on details overshadowed problems of mobility. In spite of all progress, the soft-ground performance of the Renault tank, Model 1917, in many respects surpassed that of modern tanks. In addition, the old formula of "ground pressure," which was successfully workable when applied to the Holt tractor or to its derivative, the Renault tank, proved hopelessly unsatisfactory when applied to Shermans, Crusaders, and Tigers.

Vehicles practically reached the upper limit of weights and dimensions. Large safety factors which covered the uncertainty in calculating "ground pressure" by dividing the weight by the ground contact area could not be applied freely, as they had been in the past. In consequence, the effect of all unknown factors of vehicle performance, hitherto hidden by the large margin of the safety factor, came into the picture. It was discovered that some vehicles with "higher" ground pressure had better performance than those with the "lower" one, and that those with shallow spuds developed a better tractive effort than those with the deep ones.

It finally appeared impossible to increase vehicle mobility by increasing the width of tracks, even for more trivial reasons: the German Tiger had already become so wide that the track had to be removed when the vehicle was transported by railroad. As late as 1940, it was still learned through costly experience and trial and error that the track could not be made longer, because the vehicle could not be steered. A dead end was reached. It was then realized that further improvement should be sought not by merely increasing the track dimensions in accordance with the "ground-pressure" formula but by investigating some other means.

The problems of "flotation" in soft ground and "traction" were tackled first because they seemed to be ready for quick solution. Numerous and costly tests were undertaken. Various types and spacings of grouser plates were explored by "cut-and-try" methods. The results not only were out of proportion with the amount of work spent, but proved to be rather confusing. Similar results were brought about by work on cross-country tires.[5]

It was suspected, for instance, that there was an optimum distance between two grousers or spuds. A few tests proved this fact in some particular cases. However, when the findings were applied to slightly different conditions, the result was opposite. Under these circumstances, the necessity of basic research became more emphasized.

The history of vehicle development seems to be analogous to that of aircraft design. Years ago, people flew without knowing anything, or very little, about the air and the structural strength of the aircraft. In the 1920's, it became evident that further progress depended on a knowledge of air dynamics and other branches of applied mechanics. Perhaps land transport does not encompass such broad problems as flying, and from this standpoint, the analogy is not complete. Nevertheless, an applied mechanics, helped by a statistical knowledge of the conditions related to the use of motor vehicles in various parts of the world and seasons of the year, is today as important in relation to the future development of motor vehicles as aerodynamics and other sciences are in regard to the progress in aviation.

Development of Highway and Off-the-Road Motor Vehicles

It was previously noted that the development of the automobile deeply affected the progress made in the field of off-the-road vehicles.

This situation was beneficial during the early stage of vehicle development. At the beginning of World War II, however, it became clear that the experience gained with an automobile or commercial tractor was insufficient to solve more special problems of cross-country transport. This fact seems to have been forgotten in the years after the war. It was rather tacitly assumed that industry, being successful in the design of regular though specialized commercial equipment, could tackle any special design problems of vehicles which would operate in any conceivable terrain. Recent experience has caused disillusionment in this respect.

A review of automotive technical literature (Figure 27) shows that the predominant theme of study and consideration is the engine as a burner of fuels and user of lubricants.[5] An explanation of this may be sought in the fact that the basic concept of an automobile was frozen in the early 1920's and that automotive engineering has therefore been forced to engage in further refinement of details, in which exploitation and production problems or fashion predominate. In this situation, chemists and physicists of the oil industry, armed with powerful methods of modern science, have had an ample field for studying the complex combustion process. It may be noted marginally that this process occupies a number one position in the automotive world because the cost of fuel and lubricants used by a car during its life span usually surpasses the cost of the vehicle production itself. As a result, it may be assumed logical that the accent in the development of an automobile or an industrial motor vehicle was put on fuels and lubricants. If it is assumed, moreover, that the average speed and mobility of a car will be radically increased not because of the application of an improved principle of locomotion, if the latter may be improved at all, but through the application of better roads, more efficient traffic signalization, etc., then it becomes evident that the future progress depends heavily on further development of super highways and automatic signals and controls which would reduce the number of accidents and increase traffic safety. The experience gained with railroads sets a pattern to follow, particularly because of the present highway situation which creates a paradox: accidents cannot be avoided in any other way than by limiting the speed of a 125-mph "dream" car to 45 mph.

Like a steam locomotive, a highway motor vehicle for years has been more dependent on traffic controls and road development than on the development of the vehicle itself. When using a metaphor, it may be said that an automobile has reached its maximum muscular efficiency,

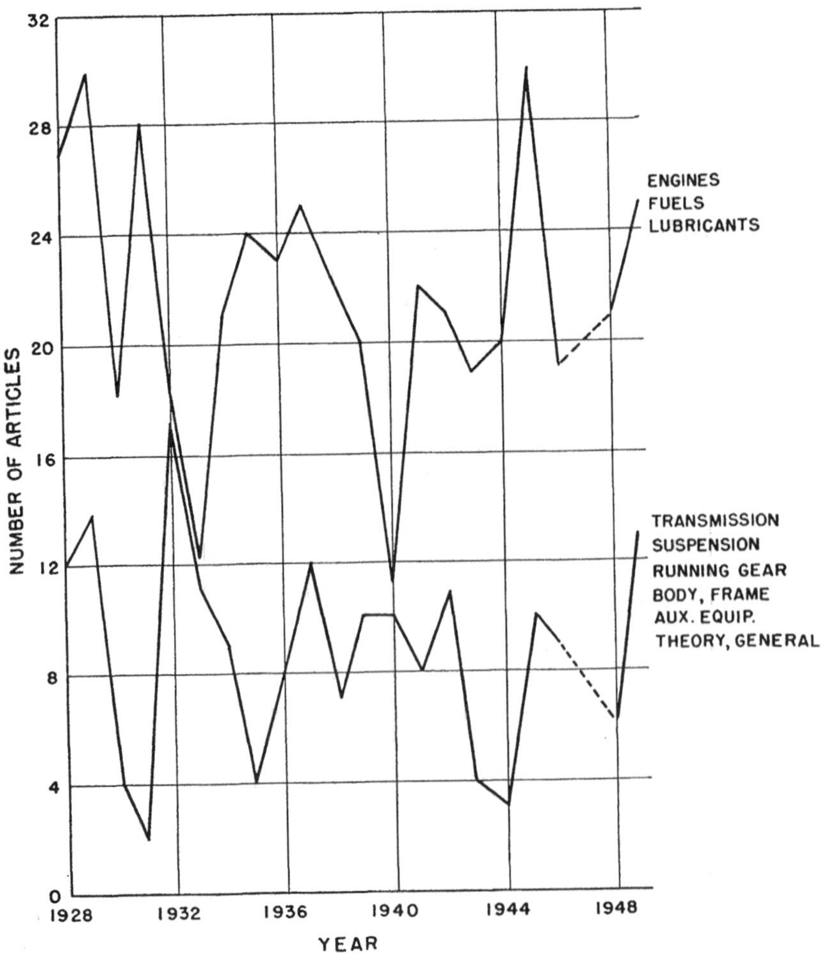

Fig. 27

and that its future improvement lies in the development of the "nervous" system of controls and in the corresponding development of the hard-surface environment in which a car moves.

Thus, there is rather little hope for success in expecting that the development of an automobile will directly foster the development of the off-the-road vehicle which faces an entirely different environment. Nevertheless, the tremendous experience gained by automotive engineers is invaluable in establishing main lines of future work, although the problem faced in the case of cross-country vehicles may be entirely different.

Military and Industrial Requirements

The difference between highway and off-the-road vehicles is particularly distinguishable when military and industrial requirements are confronted. Take, for example, the power (hp) as measured per unit area of ground (sq in.) which is in contact with the vehicle and which is thus

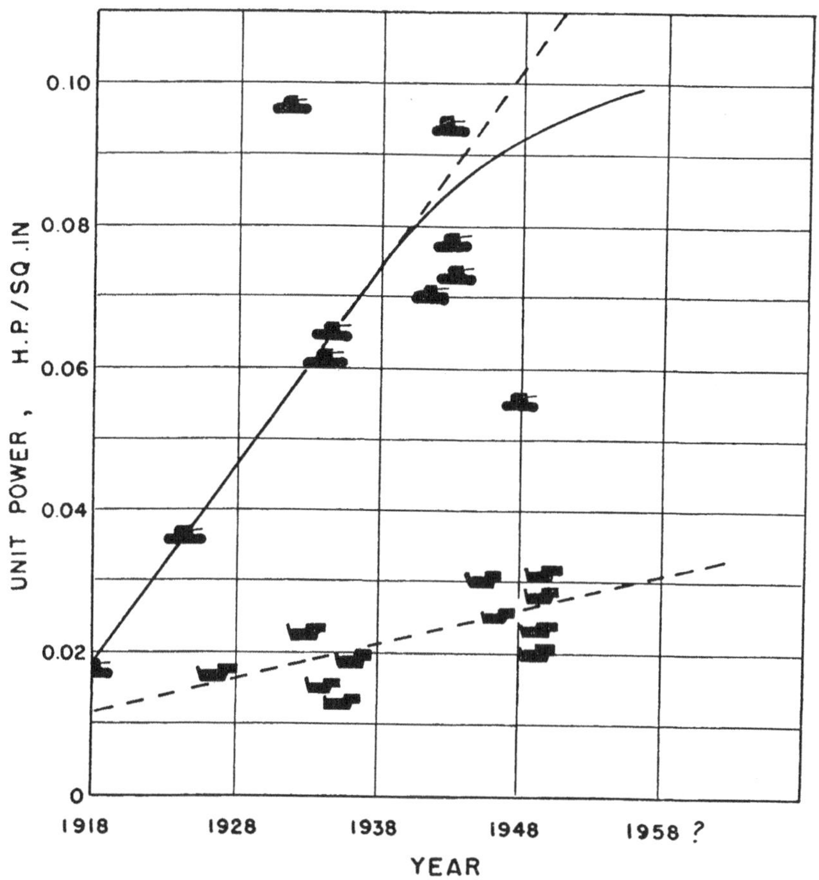

Fig. 28

instrumental in absorbing the propelling and load-carrying forces. Figure 28 indicates that the power absorbed by the soil underneath the tanks rose sharply since 1918, whereas that of agricultural tractors showed a rather insignificant increase. The same trend is reflected in the increase of speed. At the beginning, all tracked vehicles developed the same speed, but since World War I the maximum speed of military vehicles rose faster than that of commercial vehicles (Figure 29).

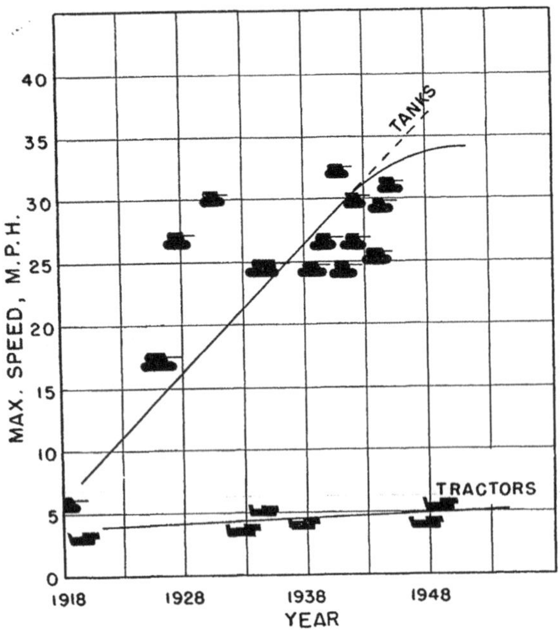

Fig. 29

A plot expressing the relationship between engine power and power absorbed per one square inch of the area obtained as the product of tire width times tire diameter again shows that the difference between the tire loads of an agricultural tractor and those of a military vehicle is remarkably large (Figure 30).

Perhaps the most significant distinction between civilian and military requirements lies in the variety of media in which both types of vehicles are to operate. Arable soils vary slightly from a trafficability point of

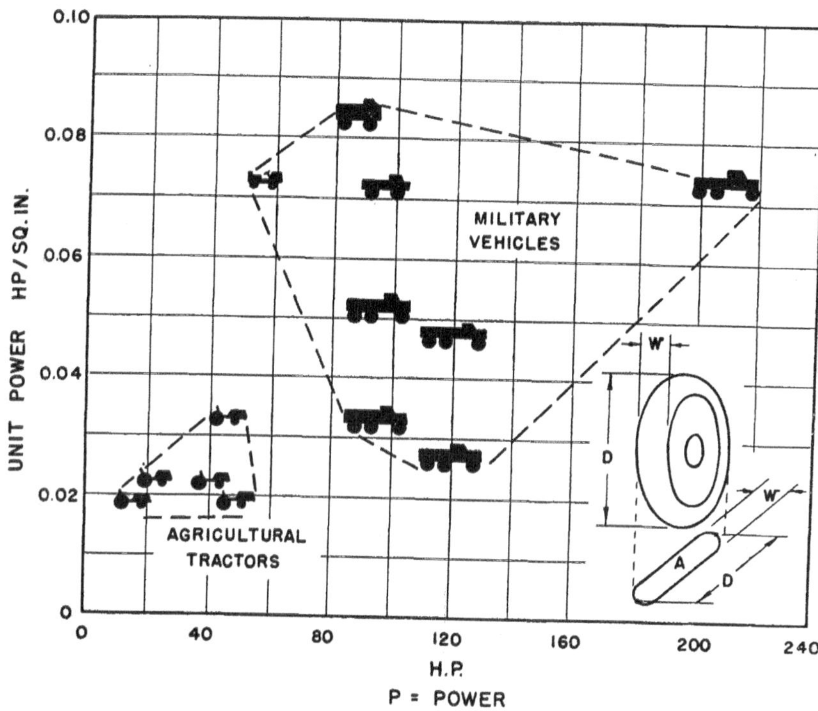

Fig. 30

view; hence, agricultural tractors have to cope with more or less stabilized soil conditions. Various industries operate vehicles in certain standard conditions. If traffic conditions are exceptionally difficult, operation is closed or a temporary road is built. The difficulty is solved without incurring drastic studies of soil-vehicle relationship which would lead to eventual changes of vehicle design.

Army vehicles are not subject to any stabilization of conditions in which they are to be used. If there is any stabilization at all, it results from their limited performance and inability to cope with obstacles encountered rather than from the doctrine of not fighting in mud or snow. As previously mentioned, requirements are growing steadily: a vehicle which, during the first World War, operated in the moderate climate zone only, had to be employed in the tropics during the second World War. Tests are now being conducted in order to develop vehicles which would operate in the Arctic. Only the difference between a jungle and an arctic desert can describe the problems in vehicle design and operation that are faced if these demands are to be satisfied.

Another relatively new military requirement is airportability. Up to now, this requirement was followed by decreasing the weight and dimensions of existing vehicles so that they could fit in an aircraft. Such a course, however, may be justified only as an emergency measure. What is visualized in the long run is the design of airportable equipment which would be in line with the whole philosophy of transporting land vehicles by air. The establishment of this philosophy would be helped tremendously if it were based on eventual new vehicular concepts derived from an imaginative, scientific research, instead of on the presently available ideas of conservative design and use, which were conceived at the time when flying was still a remote possibility. The paradoxical process of flying something that is destined to move on the earth's surface may be justified only by a wholly unorthodox approach which calls not only for the mere reduction of weight and dimensions in order to make the equipment fit in an aircraft, but also for the detailed study of the form, size, and endurance optimums.

This problem is entirely new and is growing fast in importance, not only because of the military significance of off-the-road transport, but also because the extent of using cross-country vehicles in the civilian field is by no means negligible.

In one of his papers on the mechanics of wheels, as used in agriculture,[38] McKibben figured out that 0.4125 ton mile of transportation is required for each acre per 100 lb of machine weight per foot of effective operating width of the equipment. He points out that the use of a tractor is equivalent in average to 8 ton miles per acre, and that Iowa's crop land alone requires a conservatively estimated 50 million ton miles, exclusive of the weight carried on traction and tillage elements.

Although the figures quoted above are based on 1938 statistics and refer to one state only, they clearly indicate that cross-country operations are not a small fraction of the road-bound locomotion even in peacetime economy. In 1955, the value of industrial cross-country equipment production reached the one-billion-dollar mark. If an enormous number of cross-country military vehicles are added to this picture, then the significance of transportation over roadless lands becomes of paramount importance.

At particular instances, the requirements for cross-country locomotion

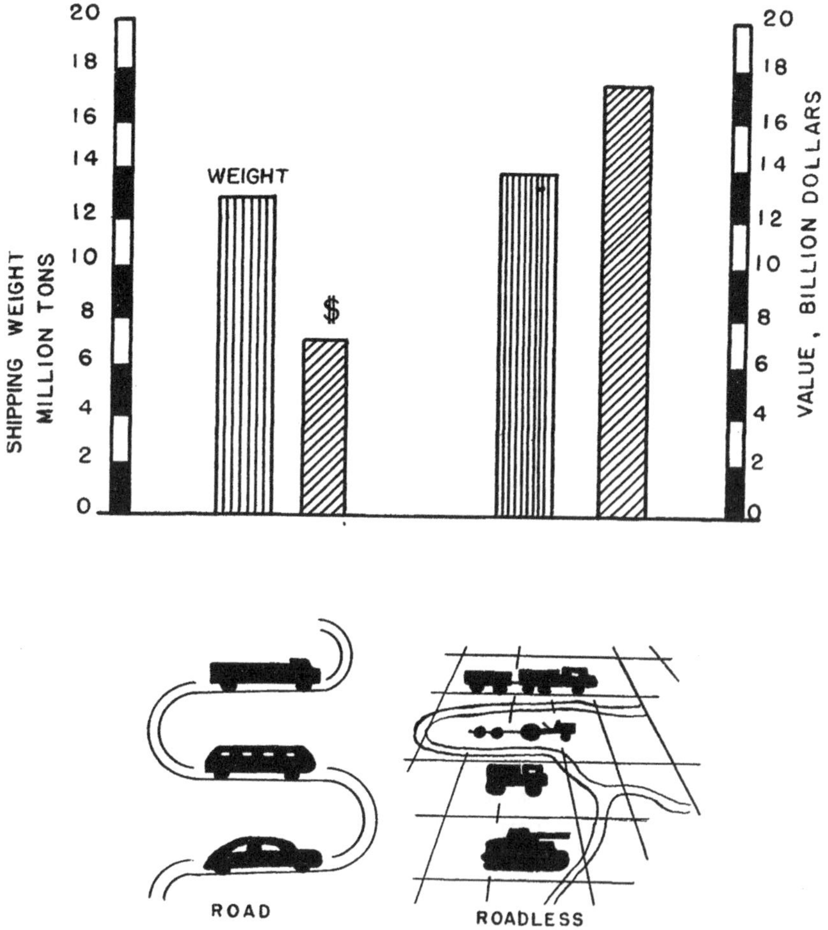

TRANSPORT
(PRODUCTION 1942—1945)

Fig. 31

seem overwhelmingly to exceed those of highway transport. Accurate statistical data are not available. A rough estimate, however, which would at least show the order of magnitude of the figures involved, is shown in Figure 31. On this graph, the shipping weight and value of American highway vehicles, and those designed for off-the-road operation, are compared on the basis of wartime production during the period 1942–45. It will be seen that the production of cross-country vehicles exceeds in capacity and cost that of standard vehicles. It must be remembered that the above occurred during a war which was fought mainly in a part of the globe where the network of highways and roads is highly developed.

Ways of Future Progress

The brief analysis of locomotion on wheels described in this chapter, although incomplete, leads to two conclusions which may be stressed in the following general form: An economic wheel rolling on a highway, or on a steel rail, has reached a certain efficiency, the increase of which depends on the development of the road and traffic controls rather than of the wheel itself. An uneconomic wheel in an off-the-road operation which was never properly studied, in many cases, has not reached the efficiency of the locomotion created by nature.

A fundamental study of this problem is now required in the face of the changing world, in which new requirements exceed those previously known, and in which cross-country locomotion occupies an important place.

It has been shown that the efforts spent on improving locomotion were mostly of a technological and engineering nature. If a radical increase of performance is expected, particularly in off-the-road operations, should these efforts follow the same course in the future? The answer to this question may be found, perhaps, in the following considerations.

As mentioned before, locomotion on wheels may be classed into two groups—one in which a wheel operates on a hard surface like rail or pavement, and the other in which it rolls on virgin soil either directly or through the intermediary of a "portable rail," the track. Figure 32 shows the unit power required for the locomotion of animals (plot 1), cross-country vehicles (plot 2), and road and railroad vehicles (plot 3). Data regarding trains and automobiles were based on figures quoted by Gabrielli and Von Kármán.[3] Also, the lower limit of animal locomotion was not assumed in conformity with the line traced in Figure 12, but in accordance with the data for a horse quoted by the above-mentioned authors. Power-weight ratios and speeds of actual tanks and tracked and wheeled cross-country vehicles were plotted in accordance with available data,[31, 32] and the agglomeration of points thus obtained was enclosed by two lines which approximately delineate plot 2 of cross-country vehicles.

The slope of the lines limiting power against speed in the case of vehicles is steeper than that of similar lines referred to the animal world. This difference is due to the aerodynamic resistance of vehicles which was not included in the case of animals, and which may be neglected for the sake of present considerations. The fact that the Gabrielli and Von Kármán data refer to the actual power installed in vehicles, whereas the power related to zones 1 and 2 was calculated on the basis of the assumed movement resistance as shown in Figure 12, also appears irrelevant if only a general picture of locomotion and general trends are to be discussed.

Figure 32 indicates that the presently built vehicles when operating off the road form an extension of the animal type of locomotion throughout the realm of higher speeds. From the point of view of power requirement, they do not differ substantially from the animal world. Road and railroad vehicles are definitely more efficient and economical. They approach a theoretical border line of power and speed which is typical not only of land vehicles, but also of air and water ships. This line was plotted first by Gabrielli and Von Kármán [3] and is marked in Figure 32 as referred to single vehicles. Trains which have less frontal area per unit of weight, and thus smaller air resistance, fall below the discussed line.

The approximate equation of the border line, which no airplane, ship, or any other vehicle seems to have crossed, and which is closely approached by wheeled vehicles running over hard surface, is as follows:

$$\frac{P}{mg} = 0.001\, v^2 \, .$$

If expressed in dimensionless form, using pounds, feet, and seconds, the equation takes the following form:

$$\frac{P}{mgv} = \frac{v}{7820} \, .$$

As Gabrielli and Von Kármán pointed out, the value 7820 must have the dimension of speed in order to make the equation dimensionally correct. Since it refers to all land, sea, and air vehicles, this value also should be independent of the medium in which a vehicle operates.

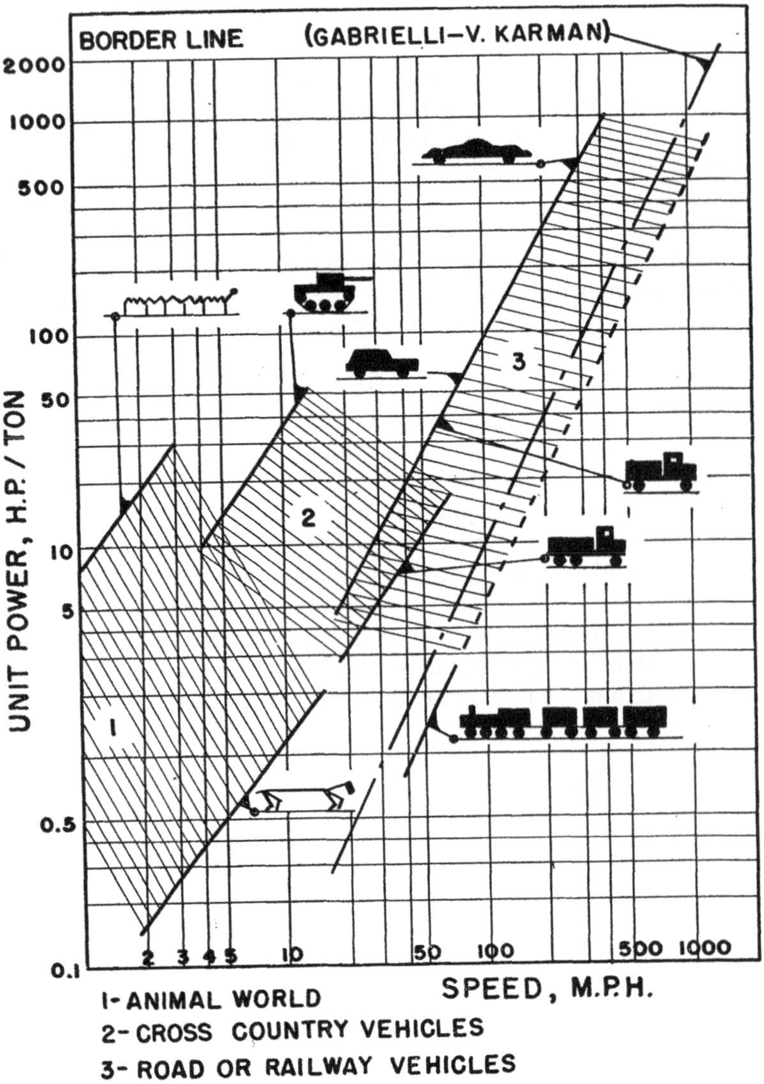
BORDER LINE (GABRIELLI–V. KARMAN)
2000
1000
500
100
50
10
5
0.5
0.1
UNIT POWER, H.P. / TON
1
2
3
2 3 4 5 10 50 100 500 1000
SPEED, M.P.H.
1- ANIMAL WORLD
2- CROSS COUNTRY VEHICLES
3- ROAD OR RAILWAY VEHICLES

Fig. 32

It is logical to assume that the ultimate speed of a vehicle, as referred to power per unit of weight, depends in the final analysis on the strength and weight of material used. Gabrielli and Von Kármán have observed that the ratio of breaking stress σ_u and density ϱ for structural materials such as steel, aluminum, or even plywood is constant within relatively close limits. The average value of this ratio is of the order of $(1000)^2$ $(\text{ft/sec})^2$. The square root of this value would have the dimension of speed and would equal $\sqrt{\sigma_u/\varrho} = 1000$ ft/sec. It may therefore be substituted for the figure of 7820 in the above equation.

$$\frac{P}{mgv} = \frac{v}{\frac{7820}{1000}\sqrt{\frac{\sigma_u}{\varrho}}} = 0.128\frac{v}{\sqrt{\frac{\sigma_u}{\varrho}}}. \tag{37}$$

Accordingly, the location of the border line in Figure 32 might be interpreted in terms of the mechanical properties of materials available for the construction of modern vehicles. It would indicate, for example, that an increase in the ultimate strength of material or a reduction in its density by 4% would cause, for a constant unit power, a speed increase equal to approximately 1%. This interpretation of the limitations in increasing the speed of modern vehicles was given by Gabrielli and Von Kármán with a cautious warning that further study of the problem is necessary in order to arrive at a more definite conclusion as to whether the discussed border line has the assumed meaning. Both authors have indicated in the second publication of their work [125] that instead of the σ_u/ϱ ratio, the square root of the ratio of modulus of elasticity to density may be taken, since it also has the dimension of speed and an almost constant value for many materials, equal to approximately 16,000 ft/sec.

If it is assumed that the problem may be interpreted as conjectured by Gabrielli and Von Kármán, it may be concluded that through the improvement of construction materials the left-hand lines which border zone 2 of cross-country vehicles and also zone 3 of road vehicles cannot be brought substantially closer to the ideal speed-power limit corresponding to all the vehicles of land, sea, and air.

The border lines are basically determined by the movement resistance with which a given medium reacts against the motion of a given vehicle. It thus becomes certain that the study of the mechanism of the resistance and of the rational means in overcoming it is today perhaps the only hope for making transportation more economical, particularly in the off-the-road realm.

If it is realized how much has been accomplished in this respect in the field of aerodynamics and hydromechanics, and how little was done in the field of resistance offered by a cross-country drive, then a partial explanation may be given as to why practically all the ships, aircraft, and railway and racing cars are close to the ideal limit, whereas cross-country vehicles constitute some sort of continuation of the primitive way of locomotion that has existed in nature since the Mesozoic era. Perhaps nothing can be done in this respect. If such is the case, it has to be proved first.

IV. MORPHOLOGY OF MOTOR VEHICLES AND ENVIRONMENT

An object moving in a given medium tends to assume the form which would offer the least resistance to motion. Such a tendency may be seen in the animal world as well as in the world of technology. Accordingly, the evolution of forms of living matter and of man-made vehicles could be considered as a measure of progress in locomotion; there seems to be no better illustration of progress achieved in the field of modern transport than a comparison between the forms of a wooden cart of a prehistoric era and a streamlined racing car.

Form Index

The process of shaping land-vehicle forms has its own history in which rational trends and accidental occurrences have played a complex role. A review of the morphological characteristics of vehicles and their relation to the environment in which they move may be helpful in the evaluation of various aspects of development and certain standards established during the last 25 years.

Consider a vehicle as a rectangular prism. In order to describe its form graphically, two indices are proposed: one expresses the ratio of height to length (h/l), and the other, the ratio of width to length (w/l). The dimension of length is that oriented in the direction of motion. In the discussion, only the maximum over-all dimensions of the vehicle as enclosed by the prism will be considered. The height-length ratio will be called the height aspect and will be denoted by h_a; the width-length ratio will be called the width aspect, denoted by w_a.

Figure 33 shows the relationship between h_a and w_a for various sea, air, and land vehicles. Data on railway trains are not plotted because they fall far below ships, reaching w/l and h/l values of the order of a few hundreds.

A line intersecting the graph at 45° and passing through the beginning of the coordinates determines ratios of $h = w$. Above this line, only vehicles are grouped the width of which predominates over height; below, those with height larger than width.

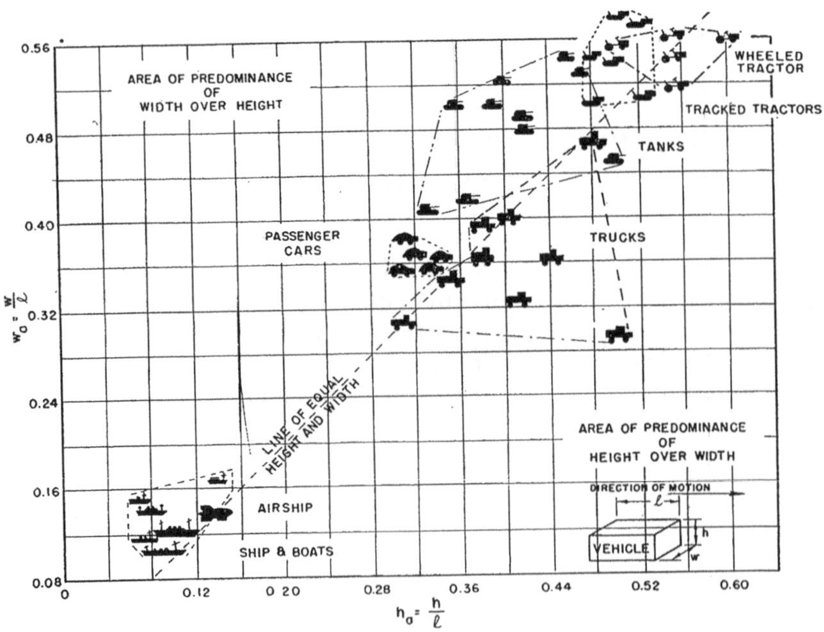

Fig. 33

It may be seen that the majority of vehicles tend to keep close to the $h = w$ line. Tanks and tracked tractors, however, have a definite tendency toward a wider shape. Tracked vehicles, in general, are found on the top of the graph. Trucks and cars show opposite trends; their form approaches closer to the form of ships than that of tractors and tanks. This is understandable since the speeds developed by an automobile are high enough to make air resistance important. Tanks and tractors move so slowly that any "hydrodynamic" shape would appear to be superfluous. However, this statement is not quite correct, because it overlooks the important task of off-the-road vehicles.

Obstacles encountered in a cross-country drive result not only from softness of the ground, which stalls a vehicle, but also from many other ground features which may slow down and stop locomotion, such as the surface contour of the ground and all the hindrances which may be encountered in the form of trees, stones, narrow passes, mountain trails, etc. Such obstacles obviously can be overcome much more easily by a vehicle having smaller frontal area as compared to its length than by those having high w/l and h/l ratios. Thus, the "streamlining" of cross-country vehicles, as expressed by low values of form indices, is, in many cases, as essential as the "streamlining" of fast vehicles moving against a hydrodynamic resistance. Actually, what is the difference between the particles of a fluid or gas and those of a solid, if they all tend to smooth up the shape of a moving body so that it offers the least resistance?

It may seem that agricultural tractors are not the subject of such refinement because they plough and harvest on plains at slow speeds. It should be borne in mind, however, that they must often perform special operations that might be conducted, for instance, in jungles, mountains, forests, or narrow passes; their stubby form cannot be considered practical for these purposes. This situation would appear particularly disadvantageous in regard to heavy tractors, since their width and length aspects are the largest. To summarize, it may be said that the form indices of tracked vehicles, which plot above the indices of all other vehicles, are not favorable.

A similar observation is made in Figure 34, where the frontal area per unit of vehicle length (wh/l) is plotted against the volume of a vehicle (whl). It is seen again that agricultural-type tractors have a tendency toward becoming more stubby, the larger they are. Military vehicles and automobiles show a more stabilized tendency in keeping the same ratio of frontal area with reference to the length at all sizes. This may be explained by a more carefully observed tendency toward reducing the frontal area of a tank in order to reduce the probability of a hit by enemy fire, or of an automobile in order to reduce the resistance of the air. The amount of scattering of the points plotted in Figure 34 indicates a certain lack of uniformity in the application of principles which lead to the idea of a streamlined body. It suggests that a study of the basic forms of motor vehicles would be very useful.

Specific Weight

The idea of the "specific weight" of a vehicle appears to be interesting from a shipping point of view. In the case when a land vehicle is to be transported by air, the bulkiness of the cargo is particularly important. The approximate specific weight of a prism corresponding to a vehicle of maximum length l, height h, and width w is plotted in Figure 35 as a function of vehicle gross weight. It will be seen that tractors are the heaviest with reference to volume of shipping space; in this case, the values of W/whl reach approximately 35 to 40 lb/cu ft. Next are tanks,

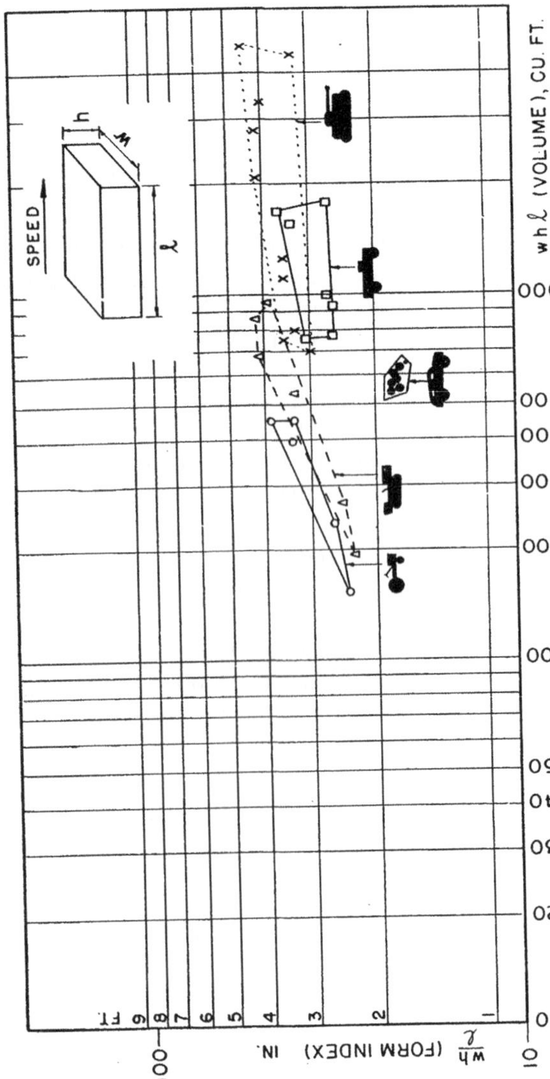
SPEED
h
w
ℓ
wh ℓ (VOLUME), CU. FT.
10
20
30
40
50
100
200
300
400
500
1000
wh/ℓ (FORM INDEX) IN.
10
100
1
2
3
4
5
6
7
8
9
FT.

Fig. 34

followed by trucks. Passenger cars are the most bulky, with specific "shipping" weight of the order of approximately 5 lb/cu ft.

In sea transport, economy in exploiting the available load and space depends on the density of the cargo and vice versa.[46] Ships are built with different forms, depending on whether they will carry oil, hay, grain, or ore. Similar considerations apply to air transport. There is enormous difficulty, however, in economically flying in the same type of aircraft a vehicular cargo whose density may vary from 5 to 50 lb/cu ft. This calls attention to the need for studying the form, size, and weight of motor vehicles from an airportability point of view.

It may be interesting to note that the upper limit of the "shipping" specific weight of motor vehicles (tracked tractors) is close to that of ships (35 to 45 lb/cu ft), whereas the lower limit, as determined by passenger cars, approaches the value of airplane density: 8 to 12 lb/cu ft.[4]

The real specific weights of vehicles, however, are higher because the true volume of vehicles is lower than that enclosed by considered shipping prisms. In a first approximation, it may be assumed that the actual specific weight of vehicles is approximately 35% higher than the figures plotted in Figure 35. It will thus be seen that the specific weight of water

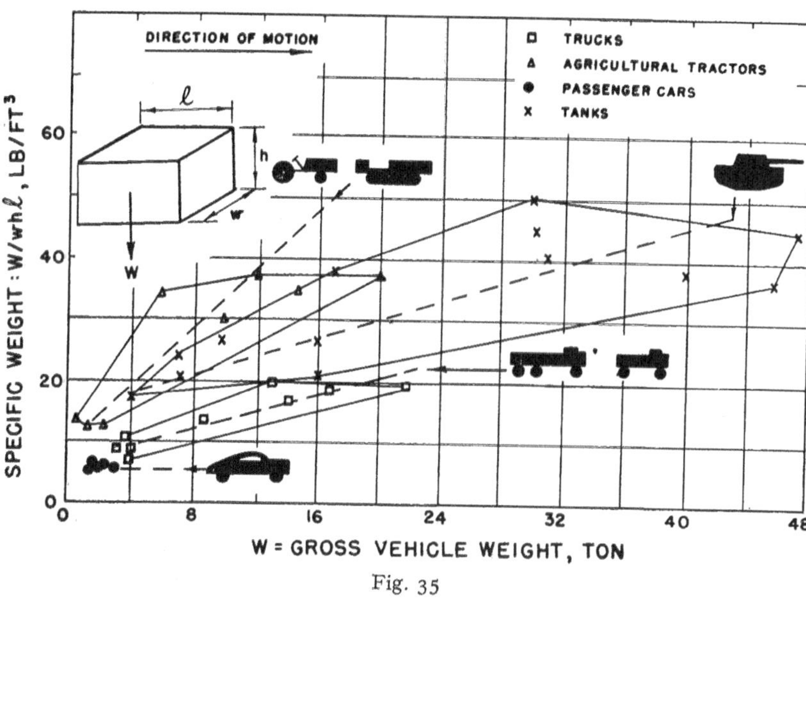

Fig. 35

(64 lb/cu ft) is reached by trucks and light-weight tanks in the 5-ton weight class. This explains why it is relatively easy to make these vehicles amphibious, whereas medium and heavy tanks either have to be submergible or require enormously large floats.

Size and Form

A survey of motor-vehicle forms may appear interesting for other general reasons. From Figures 33, 34, and 35, it may be deduced that not only the form but also the size of standard vehicles appears to be rather rigidly enclosed within certain limits. These limits are shown by plotting particular dimensions with reference to vehicle weight: Figure 36 is a plot of the length and width of automobiles; Figure 37 shows the same dimensions with reference to tanks and tracked tractors, respectively, plotted against the gross weight of given vehicles.

It is seen that highway vehicles, in general, have a constant width which does not depend on vehicle weight, and which may be expressed approximately by the equation:

$$w = \sim 80\,.$$

The width of tracked vehicles, however, increases with an increase in weight, in accordance with the following approximate formulae:

$$w_{\mathrm{tank}} = 90 + 1.0\,W$$

and

$$w_{\mathrm{tractor}} = 49 + 3.75\,W\,.$$

The length of the discussed vehicles may be expressed roughly by the following equations:

$$l_{\mathrm{tank}} = 170 + 1.9\,W$$

and

$$l_{\mathrm{tractor}} = 85 + 6.5\,W\,,$$

where dimensions l and w are measured in inches and gross weight W in tons. When combining the above equations, it will be found that in a first approximation:

$$\frac{l}{w_{\mathrm{tank}}} = 1.9$$

and

$$\frac{l}{w_{\mathrm{tractor}}} = 1.7\,.$$

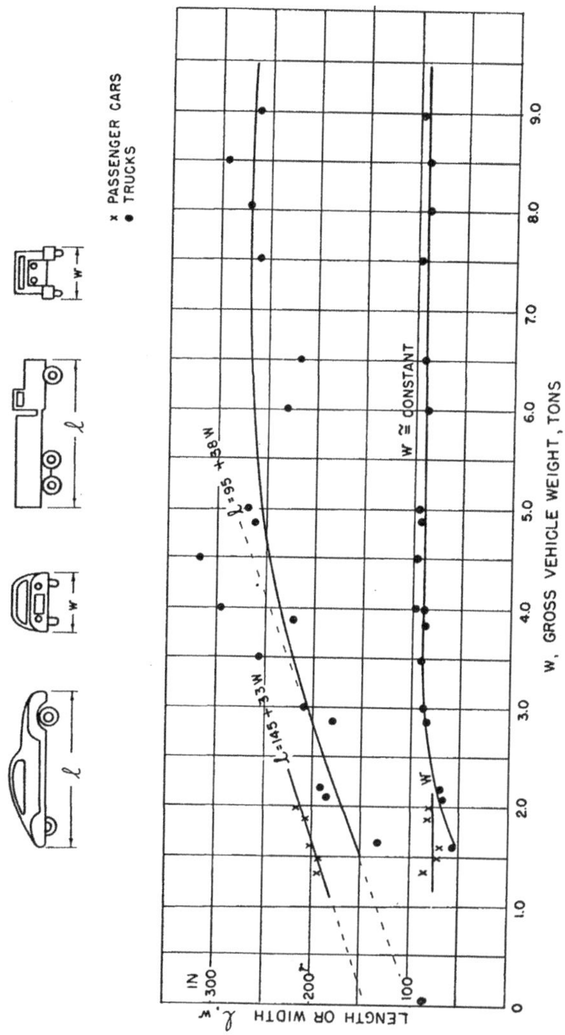
x PASSENGER CARS
• TRUCKS
ℓ=145+33W
ℓ=95+38W
W ≅ CONSTANT
LENGTH OR WIDTH ℓ, w
IN
300
200
100
0
1.0
2.0
3.0
4.0
5.0
6.0
7.0
8.0
9.0
W, GROSS VEHICLE WEIGHT, TONS

Fig. 36

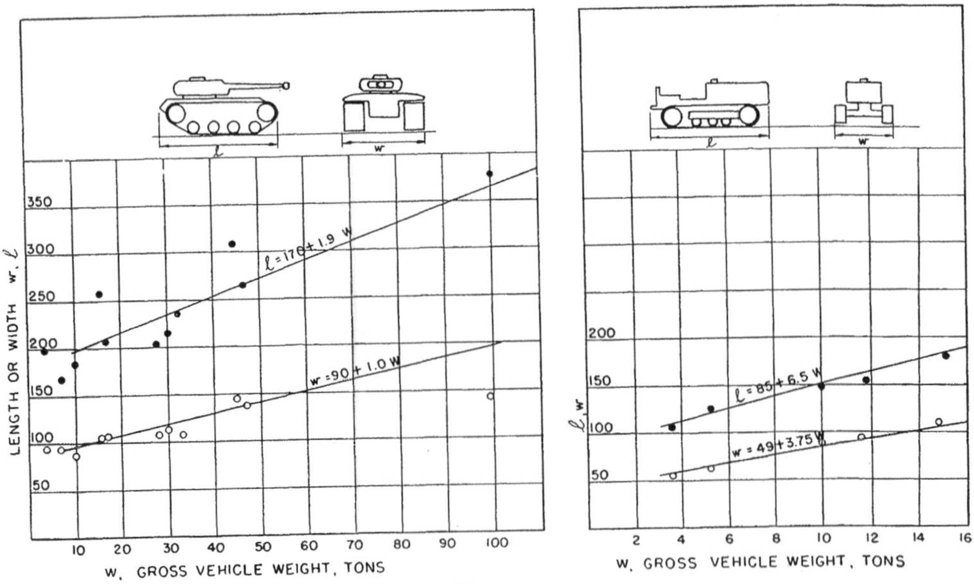

Fig. 37

In other words, it appears that the ratio of length to width of tracked vehicles remains constant at all vehicle weights.

Origin of Forms: Tracked Vehicles

The reason why the width of highway vehicles does not change beyond a certain limit, whereas tracked vehicles follow the rule of a constant width-length ratio, is an interesting problem and throws some light on the way in which the form of the above-mentioned types of motor vehicles has been shaped.

As was mentioned in the preceding chapter, since the time of Hornsby, tracked vehicles steer by the imparting of a slewing moment through the application of uneven forces to each track. Assume that, in a first approximation, a vehicle is moved by driving forces F_1 and F_2 exercised by respective tracks (Figure 38). If $F_2 > F_1$, the vehicle will probably turn around point 0, and the moment of turning resistance M will be overcome by the driving forces F_2 and F_1. From the equation of moments (see equation 395), if motion resistance is neglected:

$$wF_2 = M\ .$$

The moment of turning resistance M is directly proportional to track length l, vehicle weight W, and lateral friction μ_l. It also is proportional to a certain function ζ of the load distribution p upon the track:

$$M = \mu_l W l \zeta .$$

On the other hand, the driving force F_2 is limited by the track load $W/2$ and the coefficient of longitudinal friction μ_t between the track and soil, frequently called "adhesion," which can be developed between the soil and the track without spinning the latter:

$$F_2 = \frac{W}{2} \mu_t .$$

By combining the last three equations, the following formula is obtained:

$$\frac{l}{w} = \frac{\mu_t}{\mu_l} \frac{1}{2\zeta} . \tag{38}$$

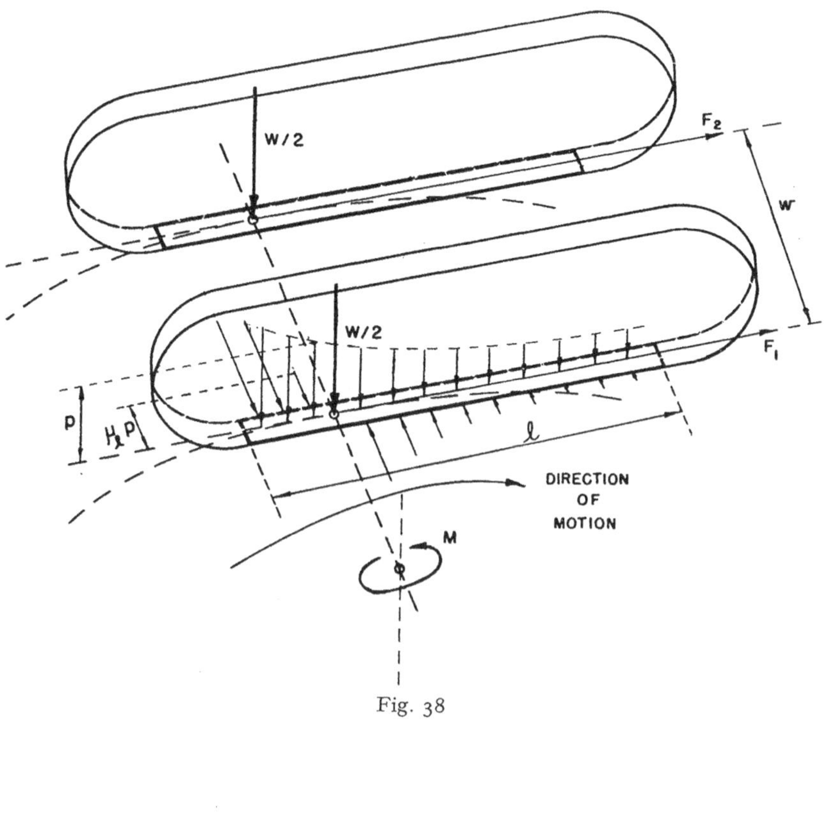

Fig. 38

It indicates that the width-length ratio of the vehicle depends on the frictional coefficients μ_l and μ_t and on the function ζ of load distribution along the track. It will be seen that the coefficients μ_l and μ_t, as well as ζ, are practically invariable in limiting conditions for all tracked vehicles, and thus the steerability of these vehicles is determined by a constant value as defined in formula (38). The numerical value for function (38) was previously quoted. This is the reason why the length-width ratio of tracked vehicles must remain constant if the vehicle is supposed to steer. Values of l/w larger than 1.7 or 1.9 would lead to a noncontrollable vehicle.

In conclusion, it may be said that the form of tracked vehicles was originated by the concept of the steering mechanism that was adopted with the introduction of the automobile differential by Hornsby in the early 1900's. It does not depend on the environment, and this occurrence is in contradiction to the rational development of the form of any moving body.

Origin of Forms: Wheeled Vehicles

The situation found for tracked vehicles does not apply to wheeled vehicles. The length of cars and trucks is not a direct function of vehicle width, as shown in Figures 36 and 37. The practically constant width w expressed by equation on page 72 may be traced with a degree of probability to ancient times, when a horse-drawn carriage was the only means of locomotion. Roads, streets, bridges, and the whole concept of modern locomotion seem to be built up around the dimensions borrowed from the world of animal tractors.

Figure 39 shows approximate-scale sketches of various means of land transport, starting with a four-horse Roman chariot. It is interesting to note that all sorts of animal-drawn carts, whether they are derived from Egypt or China, are practically of the same width and may be compared with their Roman counterpart. A visitor to the Smithsonian Institution will not be surprised to see that a primitive Mexican cart is in the same width class as an early automobile. Also, a modern automobile is not much wider than its predecessor of the early 1900's. A wheeled tractor falls into the same category, and so do practically all buses, trucks, railway cars, and locomotives. Evidently, the form of a vehicle has been closely related to that of a horse or an ox, and the relationship has shaped the gauges of the channels of land transportation for millennia. Thus, not only an automobile, but also a railroad train, and even a tank, indirectly bear the consequences of the domestication of animals by men of the Neolithic age.

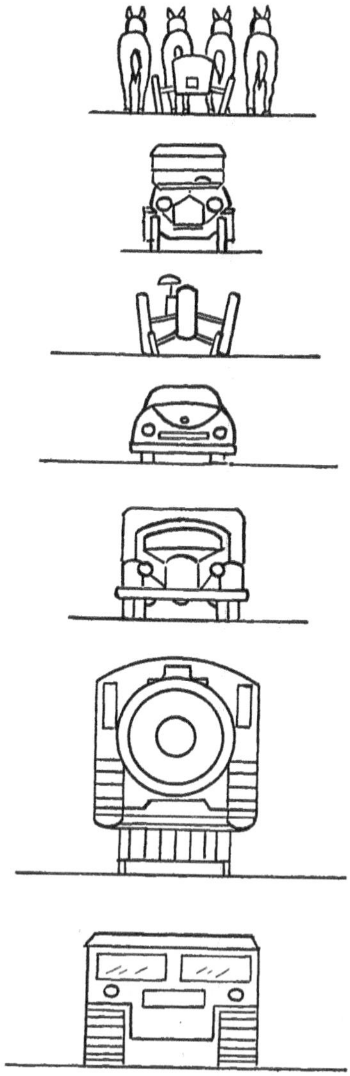

TRENDS IN VEHICLE WIDTH
Fig. 39

Form and Environment

In principle, as was mentioned before, the adaptation of form to environment becomes more distinct, the more advanced the organism. The advancement itself may be considered nothing else but a product of the interaction between the organism and environment. It may be deduced then that the progress in any development is marked primarily by the better adaptation of forms to changing conditions. On the contrary, the lack of such adaptations would mean a fairly primitive stage of development, or a development which represents a dead end of evolution.

It was shown that the length of highway vehicles is in a certain sense free from the natural and artificial gauges of environment. Figure 36 shows that vehicle length grows freely until vehicle weight reaches approximately 5 tons. Thereafter, however, it again becomes stabilized, since apparently road clearance, as measured by accepted radii of curvature, and first of all, by maneuverability on the streets, starts to play a decisive role at that vehicle size. Attempts toward the standardization of vehicle dimensions and form for military and economic reasons[47] indicate that the morphological problem of vehicles encroaches not only upon the questions of safety of travel and traffic control, but also on production and cost. The effect of environment reflects upon the morphology of vehicles in a complex way indeed.

Some simple relationships between the form of cross-country vehicles and the medium of their operation may be determined in the case of obstacle performance. When negotiating a narrow tortuous pass (Figure 40a), a vehicle having length l may stall. If, however, it has the form of a tractor-trailer unit, each having length $l/2$, it would move. Thus, the fundamental problem of vehicle form cannot be answered rationally without determining statistically the geometrical characteristics of the terrain. The single vehicle versus train of vehicles is particularly related to this point.

Similar considerations apply when a vehicle crosses a bush or woods. For instance, a vehicle having width w_1 would have to knock down three tree trunks simultaneously, which may stall the locomotion, whereas a narrower vehicle of width w_2 would encounter only two trees, and could move freely (Figure 40b). In this case, some application of botanical science enters into the picture of morphology if a "wood-cracker" type of vehicle is contemplated.

The facts just mentioned are obvious. Their quotation might be con-

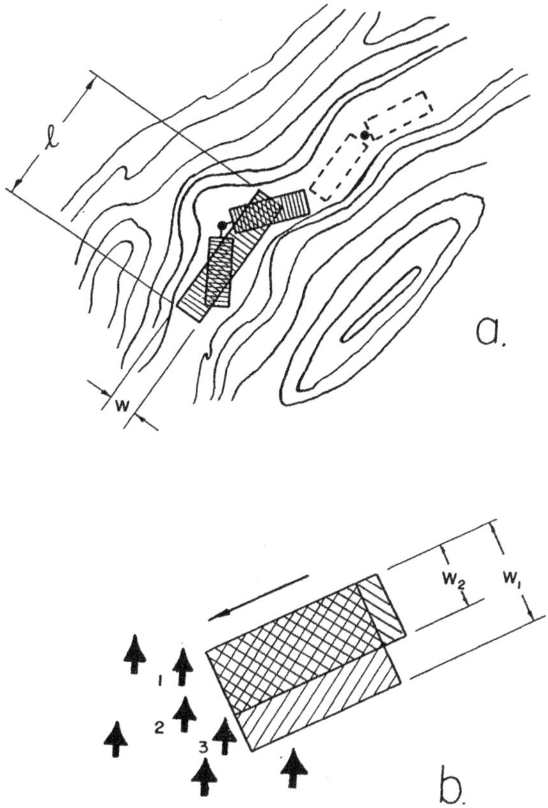

Fig. 40

sidered as superfluous if there were not many examples of ignoring their existence. Unfortunately, a quantitative study of the geometrical proportions of the environment has never been sufficiently advanced to furnish adequate data of statistical nature in respect to a rational vehicle form.

Similar conclusions refer to the obstacles formed by an uneven surface of the ground. The obvious effect of such unevenness is the bouncing, pitching, etc., of a vehicle. Ground roughness not only limits the speed, but also, under certain circumstances, may immobilize a vehicle. This will happen, for instance, if the length of the "ground wave" λ is larger than that (l) of a vehicle, and if the shape of the vehicle form is approximately that of a rectangle, having a height close to the amplitude of the

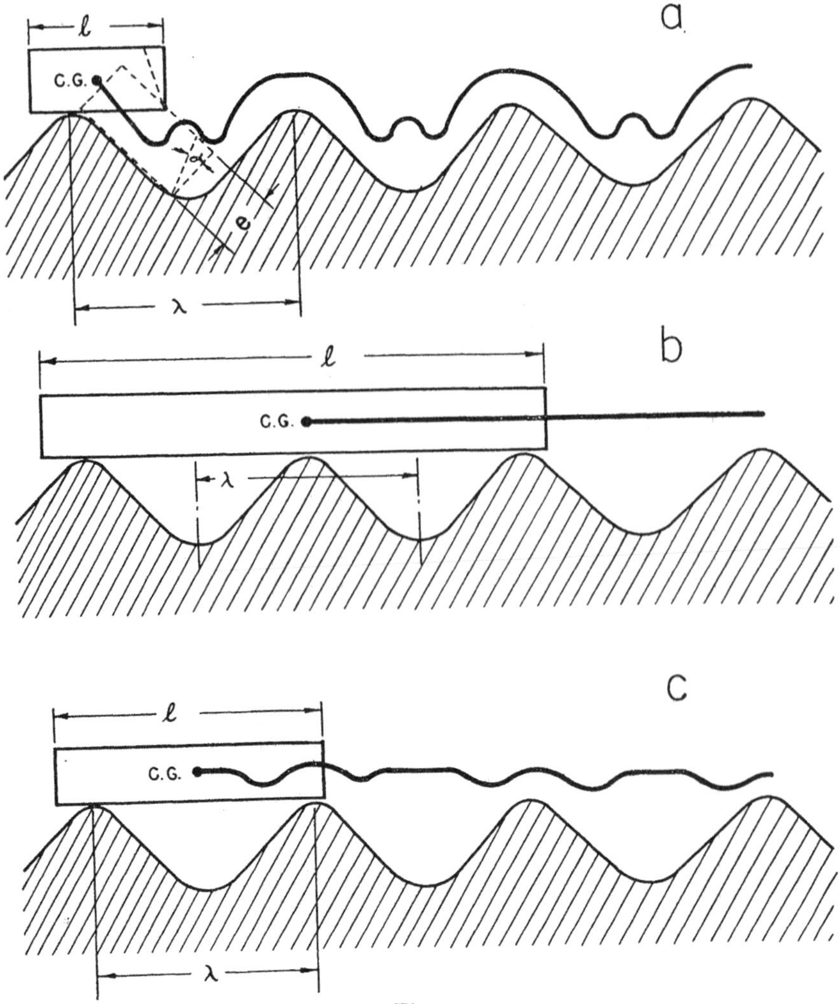

Fig. 41

"wave" (Figure 41a). If, however, the body is tapered by an angle α, then its side will not dig into the ground and the vehicle will move. This kind of correlation between a vehicle form and ground unevenness does not appear to be fully stressed, although the first lessons of this kind were learned when the St. Chammond tank failed on a battlefield of

France during the first World War because a protruding portion of the armor in the front of the tracks prevented the crossing of trenches. A general impression of the smoothness of a vehicle ride, as indicated by the vertical motion of its center of gravity and by the ratio of vehicle length l to ground wave λ, is shown in Figure 41b, c.

The forms of the running gear and vehicle "underbelly" are particularly important. For instance, a vehicle provided with four wheels will be immobilized, no matter how long it is (Figure 42a, c), if the gap enclosed by the wheels and the bottom of the body is insufficient to clear the peak of ground unevenness. A multiwheel vehicle of sufficient length would not only be effective, but would also provide a smooth ride (Figure 42b). Even a six-wheeler, composed of the same elements, still would be mobile (Figure 42d). Better vehicle mobility may be secured if an adequate track is provided. This is a fundamental feature of the tracked-vehicle form which serves to increase the vertical-obstacle performance as measured by the "go and no-go" criterion. One of the main roles of the track, as will be shown later, is the "bridging" of ground unevenness.

The need for "streamlining" the vehicle form is perhaps most emphasized in the case of mud-crossing equipment. An ideal form for a wheeled vehicle would be that shown in Figure 43a. In practice, however, modern cross-country carriers and transporters have different "underbelly" shapes which, as a result (Figure 43b), are responsible for plowing and pushing the dirt to a much greater extent than the form shown in Figure 43a. The bulldozing effect of a standard wheeled vehicle is enormous: it is the main cause for the bogging down of these vehicles.

In the case of tracked vehicles, as in the case of wheeled vehicles, there also is a danger that the vehicle will inevitably be immobilized if the mud or snow depth is greater than the "ground clearance." Then the bottom supports the weight and the tracks turn loose without any load, and hence without traction (Figure 43c). The only solution in this case would be that proposed by Diplock approximately 38 years ago, that is, to widen the track so that it would cover the whole underside of the vehicle, as shown in Figure 43d. Such an arrangement would permit more sinkage so that the running gear could reach the hard pan and keep the vehicle going.

When a vehicle passes a bottomless mud, it is obvious that the vehicle form should be capable of providing enough hydrostatic forces to keep it afloat. The running gear should derive its propelling forces more from the viscous properties of a fluid or half-fluid medium than from the mechanical

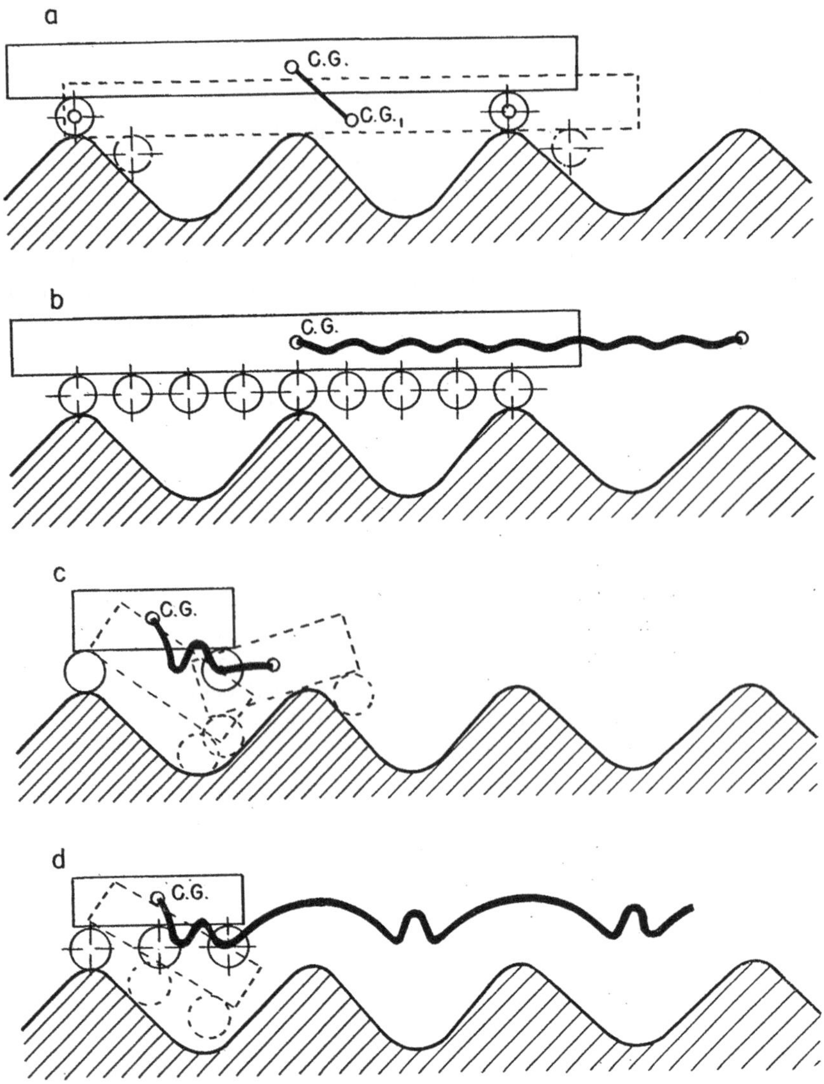

Fig. 42

strength of the ground. Amphibian vehicles for operation in water have been successfully built. Floating vehicles for operation on muskeg or other types of bottomless loose mass have not been developed yet because the basic form of available vehicles is not suitable for that purpose.

A comparison of an early motor vehicle, which was styled after its contemporary horse-drawn coach, and a modern quasi-streamlined car indicates that the latter is better adapted to the development of higher speeds of locomotion, made possible through the elimination of ground unevenness and softness. A plot (Figure 44) of the change of form as a function of maximum speed shows that the tendency is toward streamlining the vehicle body in order to reduce air resistance, which incidentally is the main "highway" resistance. Historically, however, the modern shape of a car body is derived from a tendency to protect the passengers against the air impact rather than from aerodynamic considerations.

The degree of aerodynamic streamlining in modern cars is not absolutely perfect, and different shapes, particularly of the rear portion of the body, may be observed. Some modern trucks are not streamlined

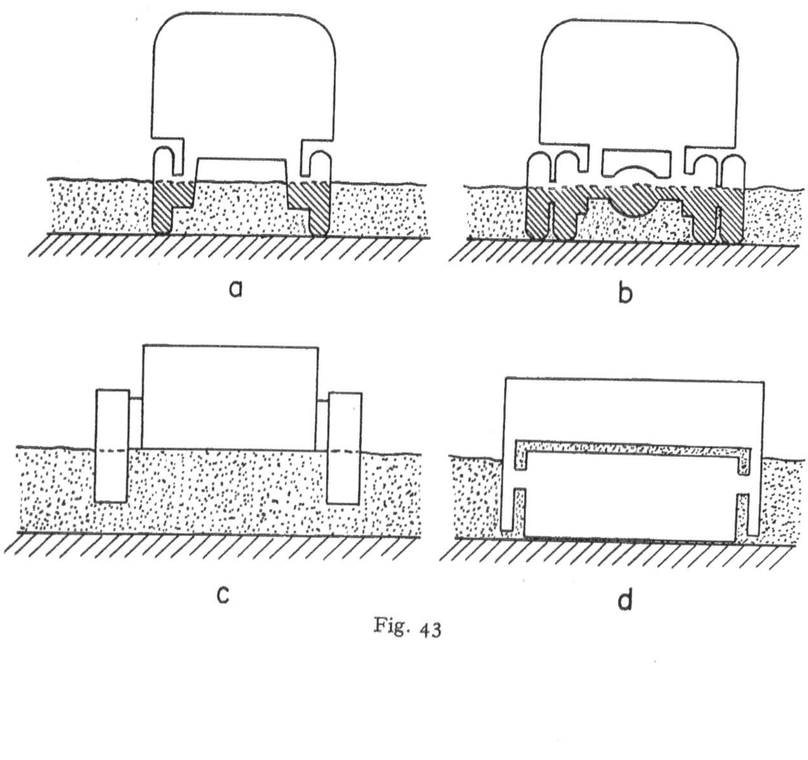

Fig. 43

at all. Fashion and other factors are apparently as important in shaping the body as the laws of aerodynamics. It might be well remembered, however, that fashion and taste change, whereas the physical laws of least resistance of a moving body remain the same.

The effect of air resistance is most accentuated, of course, in shaping the bodies of racing cars (Figure 44, upper row of cars), although the necessity of rational economy dictates the same trend in the design of passenger cars.[48] Experiments with a regular automobile, performed by Kamm shortly before the last war, indicated that a car at high speeds requires stabilizing fins which prevent it from being swept off the road by winds and air currents.[49] Thus, the automobile form is ostensively approaching that of an airship. This development appears logical since the main environment in which fast automobiles operate is the air, and the only difference between the car and an airship is that the former drives right on the boundary formed by the pavement laid on the earth surface, whereas the latter navigates high above. Similar conditions apply to trucks and buses.

When applying the previously defined measure of progress, it may be said that the development of the automobile has surpassed that of

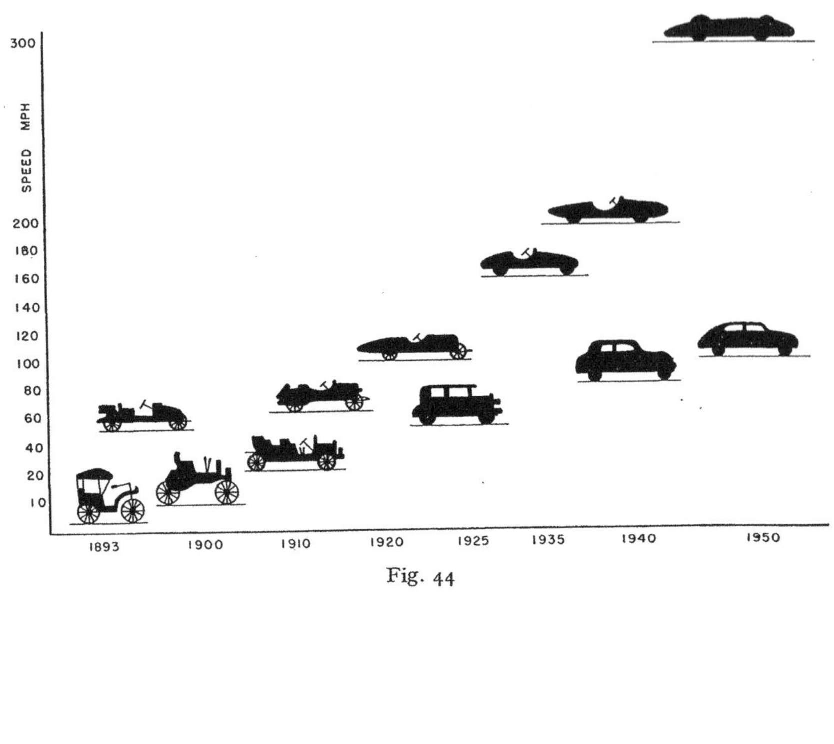

Fig. 44

cross-country vehicles because of the more advanced adaptation of its form to the medium in which it operates. (Its almost perfect shape, however, is waiting for some refinement which would increase the economy of high-speed locomotion.) Since the adaptation of the automobile form to the medium of operation is almost complete, it may be concluded that the process involved is close to completion. This would confirm the conclusions made in Chapter II to the effect that further development of the automobile depends more on the development of highways than on the automobile itself.

As was shown before, the form of cross-country vehicles is not satisfactory. It was shown that the form in question has been affected either by certain concepts of steering adopted in the case of tracked vehicles, or, in a general case, by the factors which shaped man-made means of locomotion when a horse- or ox-drawn carriage was the available equipment. The lack of statistical data in the evaluation of geometrical and certain physical properties of ground surface makes it difficult to formulate quantitatively the requirements regarding the rational form of a vehicle. Once these data, however, are made available the progress in off-the-road locomotion will largely depend on the rational development of vehicle form.

Evaluation of Relationships between Form and Environment

In previous considerations, the adaptation of vehicle form to the medium in which it moves was discussed when taking into account simple features of that medium, for instance, vertical or horizontal obstacles, traffic gauges, ground consistency, and air resistance. This list does not embrace all the factors which may be included in the environment of vehicle operation. As a matter of fact, the air of a battlefield infested with flying anti-tank projectiles constitutes a peculiar type of medium, which requires special attention in the streamlining of a body. Accordingly, the definition of the environment itself is most important if a proper evaluation of form is to be attained.

An assessment of the degree with which a cross-country vehicle form is capable of coping with the environment may be made on the basis of the "go and no-go" criterion. By defining the ground "softness" and obstacle dimensions which immobilize a vehicle, a numerical index of the critical form may be obtained, provided that all other factors affecting performance, such as power train, economy, etc., provide an optimum and are not variables.

Such a test, however, is not complete because it delineates bordering conditions only. In modern times, it is important to know not only the conditions that will stall a vehicle, but also how a vehicle will move. Since speed appears to be the most universal exponent of the quality and quantity of motion, it may enter into many criteria in which the adaptation between form and environment is evaluated.

For the sake of simplicity, a terrain is considered as the sole medium in which a vehicle operates. This may be done if the speeds are below approximately 30 mph, so that air resistance may be neglected. It is also assumed that only natural terrain obstacles are encountered.

Broadly speaking, the average vehicle speed measured during a given operation is affected by the terrain through:

(a) the occurrence of movement resistance and slip which slows down the speed, and

(b) the existence of terrain surface roughness and unevenness which causes an unavoidable slowdown in speed, due to the effect of pitching on the fatigue of the crew or materials, or due to the necessity of bypassing non-negotiable obstacles.

It is known from common experience that the factors listed in point (a) result from what may be called terrain consistency. The physical meaning of this word is closely related to the strength of the ground and will be discussed later on. For the time being, the degree of terrain consistency is recorded qualitatively by using popular terms based on common experience, such as soft, medium, and hard ground. The degree of the terrain surface roughness also will be defined in qualitative terms of smooth, average, and rough ground surface, depending on the extent to which it affects the ride of a vehicle.

Any combination of the discussed terrain features, e.g., softness and hardness, or smoothness and unevenness, may present any conceivable type of ground surface having any degree of strength and roughness.

Assume that both the degree of ground strength and the degree of ground unevenness permit the vehicle to develop a certain speed v that is specific for given terrain conditions (Figure 45). Let v be plotted on the zz-axis. If the degree of ground unevenness is plotted on the xx-axis, and if the degree of ground consistency is plotted on the yy-axis so that point 0 corresponds to both the ideally smooth and hard surface, then any point A_z located on the zz-axis will denote the maximum speed that may be developed by a given vehicle on a hard even surface. Points A_y will determine zero speed, which is due to the stalling of the vehicle by the softness of the ground (slip sinkage). Points A_x will be related to similar conditions caused by an uneven ground surface (a ditch, vertical obstacles, etc.). Vehicle form No. 1 will develop its maximum speed on a hard even surface equal to A_z' and will be stalled by the ground having strength A_y' and bumps A_x'. In any terrain conditions corresponding to the unevenness of the surface U_w and ground strength W_w, speed v will be determined by the distance $P'Q'$, with point Q' located on the surface $A_z'\ A_x'\ A_y'$.

Thus the knowledge of the speed of a vehicle in any type of ground as depicted by Figure 45 will give the quantitative evaluation of vehicle form with reference to various terrain conditions. Any comparison with other vehicles would give a similar picture which might be directly assessed in a strict manner.

Consider, for example, vehicle form No. 2. Its maximum speed on a hard even road will be lower than that of vehicle No. 1 and will be denoted by A_z''. If vehicle No. 2 has, for instance, a "portable rail" which

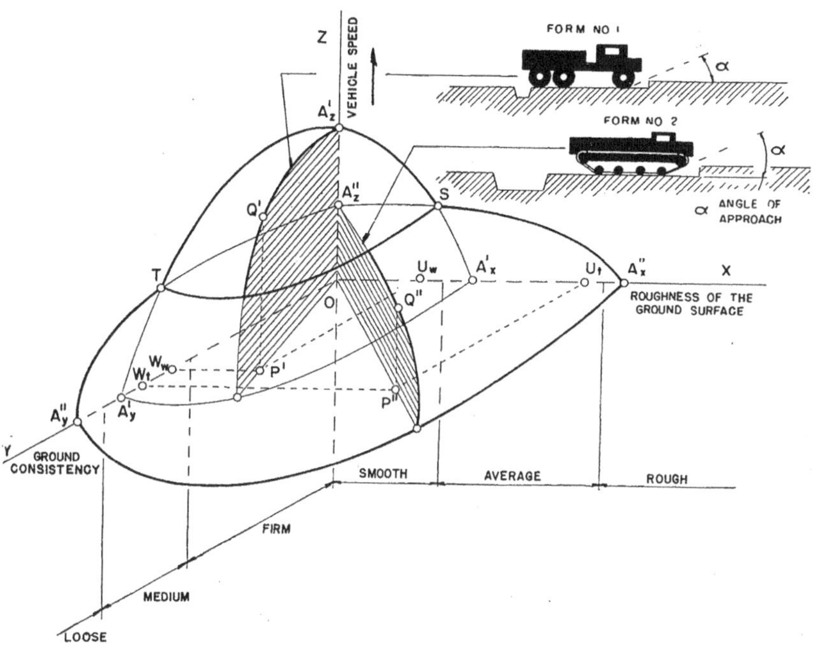

Fig. 45

bridges the ups and downs of the ground-surface wave, and a larger angle of approach of the track which enables it to overcome higher obstacles, then it will be able to cross ditches and other obstacles that would be unpassable by vehicle No. 1. Accordingly, it will be found that the stalling of vehicle No. 2 by ground unevenness will be determined by point A_x'' which is located farther off the center of coordinates 0 than A_x'. Since the track reduces unit loads of single wheels upon the vehicle rolls, point A_y'', corresponding to zero speed caused by a lack of the necessary ground strength, also will be located farther away from point 0 than A_y'. Accordingly, ground of strength W_t and roughness U_t will be covered with speed $P''Q''$, which is inaccessible to vehicle No. 1.

The curved surfaces A_z' A_x' A_y' and A_z'' A_x'' A_y'' may or may not intersect. For the case when they do intersect, the points located along the line TS refer to those terrain conditions in which both vehicles will move with equal speed. Above this line, vehicle No. 1 will be superior to No. 2; below, vice versa. Line TS determines those conditions of the environment in which both forms behave alike.

Thus, a rigorous and direct comparison of different vehicle forms used in defined terrain conditions may be conceived. The problem, however, is complex. The scale for determining a uniform measure of ground consistency, or strength, and surface roughness has not yet been satisfactorily devised. A theoretical investigation of the relationships among speed, ground strength, and unevenness has barely commenced. As mentioned before, statistical data relating to the disposition of various types of soils and their surface contours are nonexistent.

On the basis of the foregoing considerations, it appears that if the problem of form adaptation is reduced to the evaluation of vehicle speed in given conditions, then its quantitative solution may be visualized. It should be stressed again, however, that the speed which is taken as a measure of the correlation of form and environment would not be an instantaneous speed developed at a single test, but the average speed developed over a larger period in a typical zone of operation. In such tests, even the temporary stalling of a vehicle by snow, mud, trench, or stones would have a broader meaning than found by the presently used "go and no-go" criterion, because a vehicle incapable of overcoming these obstacles directly would have to bypass them, thus covering a longer distance as compared to the vehicle capable of going through the obstacles. This would in turn lengthen the time of travel and reduce the operational speed, indicating a rather imperfect vehicle form.

If the impact of the environment upon a vehicle form can be reflected in the value of the operational speed, then the attempt to improve vehicle morphology could be reduced to the study of means which would increase that speed in various terrain conditions, as visualized by the combination of various degrees of surface roughness with various degrees of mechanical strength of soil. This would inevitably require a parallel study in nature of all the terrain combinations which can be traversed by existing vehicles, as well as of those terrain conditions which possibly may be made passable in the future by radically new types of locomotion.

Stabilization of Forms and Vehicle Concept

The stabilization of motor-vehicle forms is to be related not only to the accepted design concepts but also to the stabilization of the type and performance of its basic element. Forms of engines and transmission components have remained practically unchanged for years and unaffected by other developments, such as, for instance, the multiengine drive which was widely introduced in plane and ship propulsion as well as in modern railways.

With the stabilization of components, the concept of the modern vehicle became standardized and may be described as a certain number of combinations of basic mechanisms, namely, running gear, power train, and steering mechanism, which may be obtained by changing the relative location of these elements and/or by applying various numbers of driving components. Typical combinations which are most frequently used on and off the road are shown in Figure 46.

Vehicles are grouped in three categories: wheeled, tracked, and skied. The abundance of wheeled-vehicle forms indicates the flexibility of this type of equipment. Only a few types have their counterparts in track and ski sections. It will be seen that the configuration of the major components in all categories consists of the same pattern, the elements of which are put together in the same way, although differing in location and in number of actuated wheels. Perhaps the pusher type of motor sleds (6C of Figure 46) is one of the few forms departing from the standard form, while the half-tracked snowmobile with a ski in the front (9C) is only a variation of a 6×4 wheeled vehicle (9A), the driving wheels of which are wrapped with a track (9B) while the steering wheel is substituted for by a ski. The lack of more radical forms is characteristic in this column. A form that may be considered radical is illustrated in 14A, which shows a two-engine vehicle in which the front and rear axles

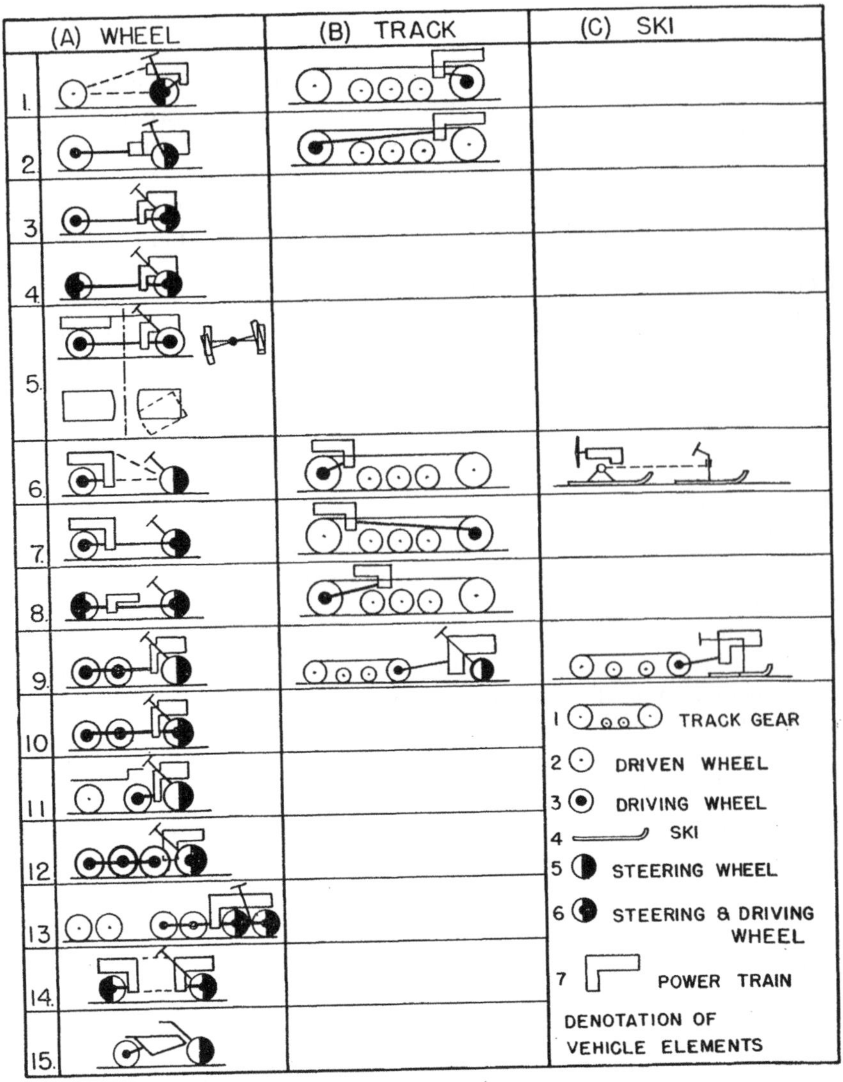
(A) WHEEL
(B) TRACK
(C) SKI
1.
2.
3
4.
5.
6.
7.
8.
9.
10
11
12
13
14.
15.
1 TRACK GEAR
2 DRIVEN WHEEL
3 DRIVING WHEEL
4 SKI
5 STEERING WHEEL
6 STEERING & DRIVING WHEEL
7 POWER TRAIN
DENOTATION OF VEHICLE ELEMENTS

Fig. 46

are propelled independently by two independent power trains, including gear boxes, differentials, etc. A successful vehicle of this type was the German-made "Tempo";[50] however, its ephemeral life did not permit investigations of its possible significance to be made.

Another illustration of a radical departure from a conventional scheme is the previously described articulated vehicle type proposed by Diplock. A more modern version of this vehicle (5A) was designed by Pavesi.[32] His tractor was articulated in the axis vertical to the ground as well as along the longitudinal axis of the vehicle.

By comparing the mechanisms of locomotion shown in Figure 46 with those developed by the animal world and charted in Figure 11, it may be seen that all animal schemes are practically missing. Perhaps this is inevitable. It would be justified to note, however, that there is no convincing evidence as to whether some types of animal locomotion would not be useful in devising transportation machines. Is a walking machine, for example, useful at all? The fact that such machines failed 100 years ago and gave way to a wheel [19] does not mean that modern technology would not cope with the problem in a better way. On the contrary, there is a strong need for such a type of locomotion, for instance, in sugar-cane plantations.[51] A walking vehicle, built for this purpose in the United States, is not only a radical departure from the time-seasoned tradition, but is also an example of the search for new concepts in the narrow realm of present vehicle forms. An isolated attempt, however, aiming at the design of a vehicle destined to operate in a specific type of environment, will not give a general answer to these questions.

Morphological Studies

Morphological studies which would lead to more fundamental answers are tedious [4] and must be based on the consideration that all design assumptions are the same. Such conditions usually are not fulfilled, and vehicles under comparison may behave in a different way in the same environment because of structural or purely engineering variants. Any such irregularity must be carefully considered in experiments designed for the study of form and, if possible, a comparable result must be extrapolated. A morphological investigation of vehicles that is performed in the design stage is more reliable and much more free from the influence of nonessential factors than an analysis of vehicles available in the market.

An excellent and unique published example of such an analysis was made a little more than a decade ago by Neesen. The study was prompted by the lack of general comparable characteristics of various means of land, sea, and air transport, which would enable a rational planning to be made of a broad policy in the development of respective networks of transport.[2]

By starting with uniform assumptions and design procedure, which made a complete comparison of various vehicle features possible, Neesen showed the true relationship between the form of a vehicle and speed. Some of his graphs, which relate to cars, buses, and trucks, are reproduced in Figures 47 through 52.

Figure 47 shows the change in weight, length, and power with reference to the maximum road speed of a small-size, 4-passenger car. Length

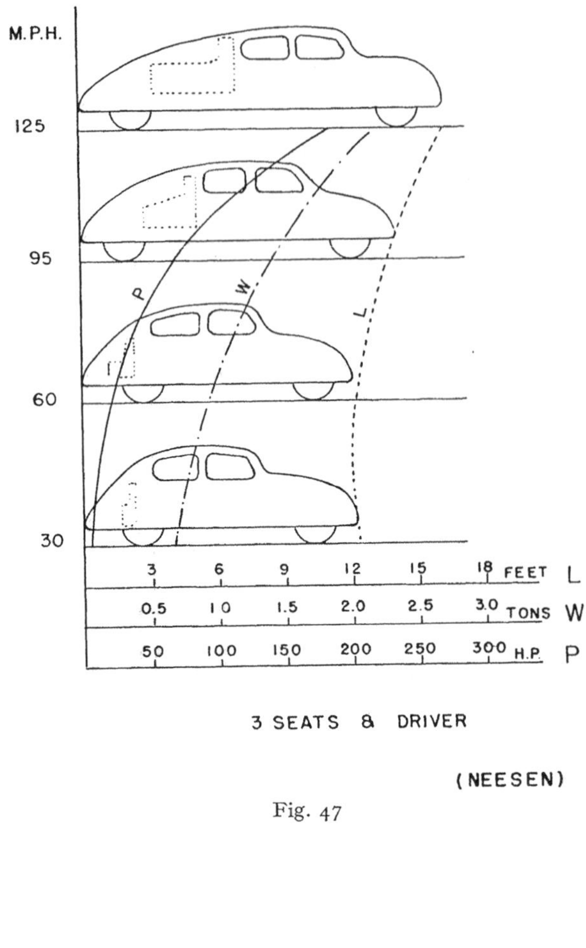

Fig. 47

increases only very slightly with an increase in power and weight as required by higher speeds. Height and width remain practically unchanged. The same occurs in a more accentuated way, however, for a 7-passenger car with a front engine (Figure 48). The dimensions and related power-weight ratio of a 40-passenger bus show a very moderate tendency to

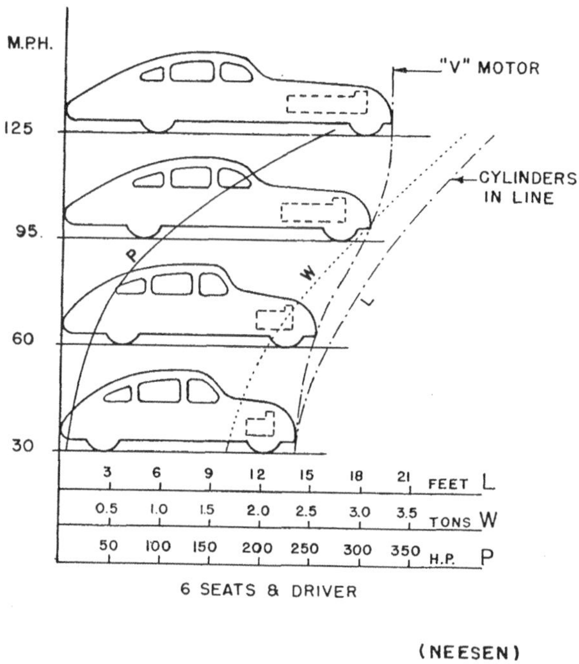

Fig. 48

increase with an increase in speed (Figure 49). Thus, bus transportation appears to have a more economical and rational development, in which speed increase is planned, than transportation by individual cars. The heaviest buses, however, show a sharper tendency to change their morphology, dimensions, and power with increasing speed (Figure 50). The tendency of size, weight, and power to change with an increase in speed for a 2 1/2-ton truck and for a 20-ton road train is shown in Figures 51 and 52, respectively.

It would be most interesting to trace similar diagrams for other types of vehicles, particularly those of the off-the-road class, in order to see the limits in present design concepts. The execution of such a task appears essential in making a broad policy for long-range vehicle de-

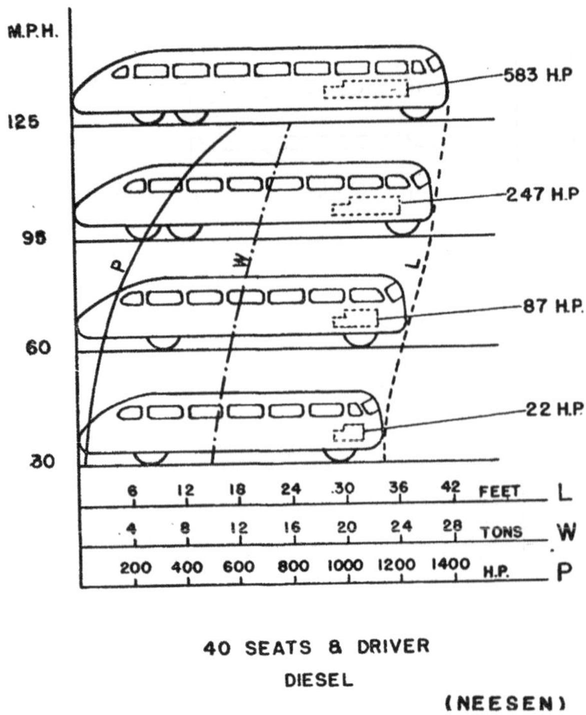

Fig. 49

velopment. The difficulty lies in the assumption of proper environment. This assumption was simple in the cases considered by Neesen: the environment was highway and air. In the case of cross-country vehicles, however, a medium typical for a given operation has to be contemplated. In any case, the above-reviewed method may be useful. A review of similar studies in aeronautics also may be of interest.[52, 53]

An example of the discussed assessment of basic vehicle features is illustrated in Figure 53, which shows two approximately scaled sketches—one represents the form and power of a truck which is adapted to the environment in accordance with rational requirements, and the other shows the same factors evolved from a utilitarian development carried out while neglecting the principle of least resistance.

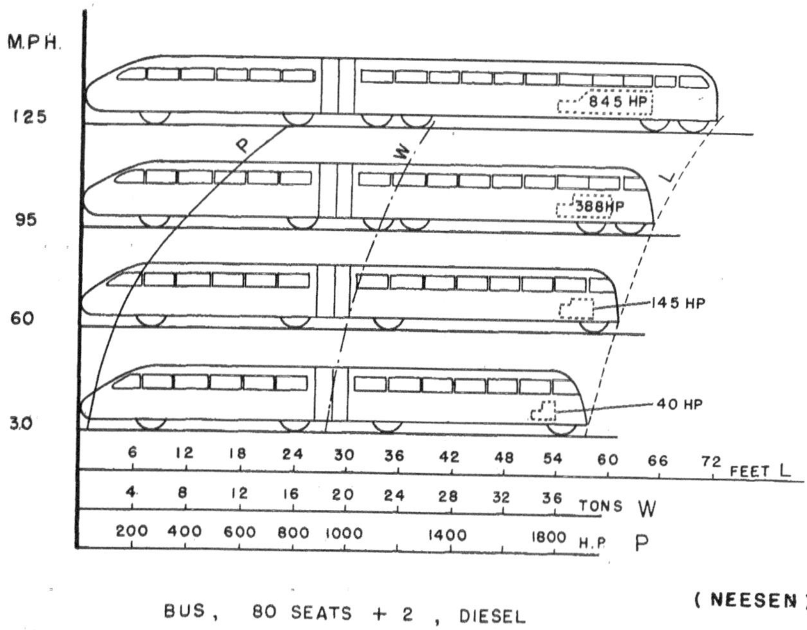

Fig. 50

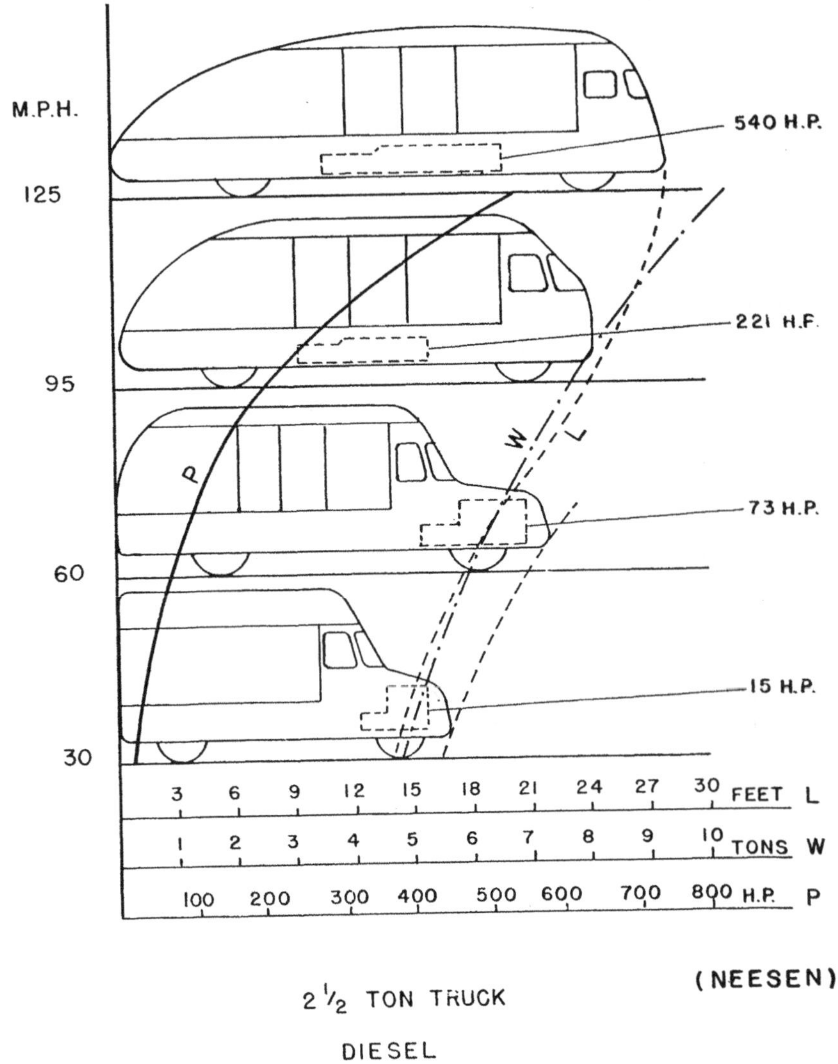
M.P.H.
540 H.P.
125
221 H.P.
95
W
L
P
73 H.P.
60
15 H.P.
30
3 6 9 12 15 18 21 24 27 30 FEET L
1 2 3 4 5 6 7 8 9 10 TONS W
100 200 300 400 500 600 700 800 H.P. P
(NEESEN)
2 1/2 TON TRUCK
DIESEL

Fig. 51

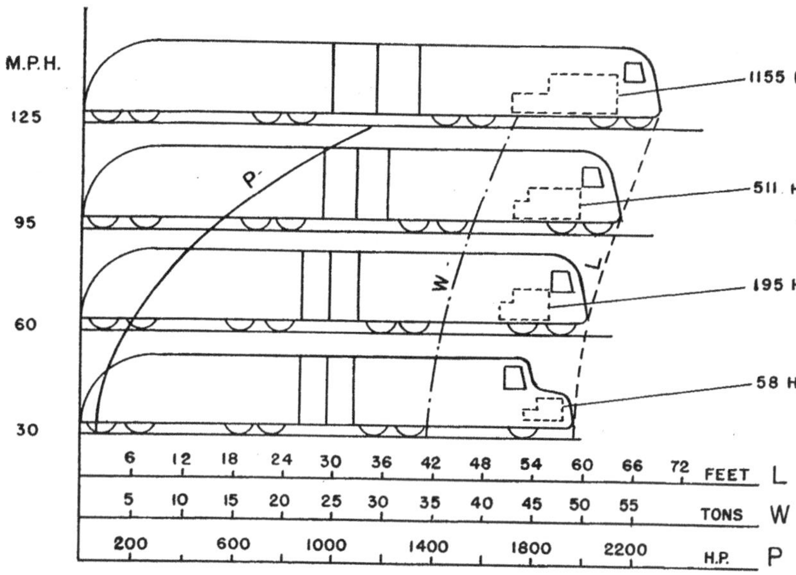

20 TON DIESEL (NEESEN)

Fig. 52

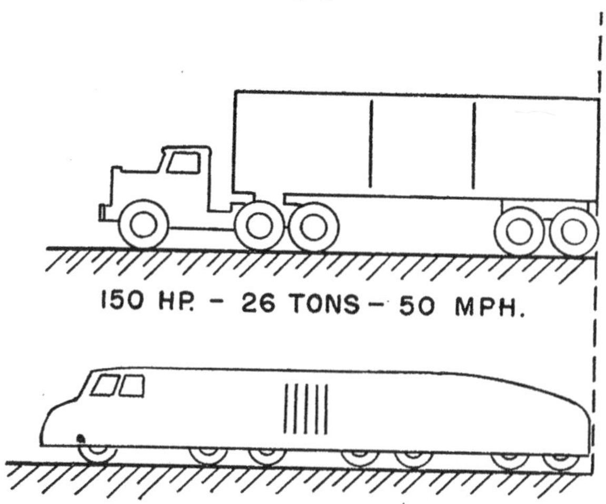

Fig. 53

Critical Size

The weight and dimensions of a vehicle are finite. Respective limitations depend not only on the strength of materials used, but also on design concepts which predetermine the vehicle form and a standard arrangement of structural elements, as well as on the type of environment.

A study of this kind would be complex if it were to include all the factors involved.[4] If, however, only the one factor that appears to be most critical is considered, then the answer to the maximum-size—form question may be solved in a way similar to that in mathematical biology when the ultimate size of various species of animals is considered.[8, 122]

If, for instance, a graph is made in which vehicle weight is related to the ground pressure it exercises upon the soil, it will be seen that when the vehicle size is increased within its accepted regular form, the pressure will reach a certain point which approaches the ultimate values of the bearing power of soil, for example, those adopted in building codes. Thus, a deduction might be made as to what limit in weight, as determined by the present vehicle form, may be reached. Since a given weight must be enclosed by a specific form, the ultimate size also may be defined.

Theoretically, the minimum conceivable pressure exercised by a vehicle may be assumed by dividing the vehicle weight by the area equal to the vertical projection of a rectangle having the same length and width as the maximum length and width of the vehicle. Such a pressure would express the unit load of the ground which would be obtained if the whole "vehicle block" were supported by the ground. It represents an ideal case which might be materialized if, for instance, the tracks were as long and as wide as the maximum length l and width w of the vehicle.

In reality, this scheme has not been materialized. The ground contact area of wheeled vehicles is several times smaller than the projected area of $(lw)_{\max}$, whereas that of tracked vehicles is approximately 1/3 of the $(lw)_{\max}$ area.

Figure 54 shows the values of weight W and W/wl plotted for various types of vehicles. The points are scattered in such a way that a definite upward trend may be seen, which indicates the existence of the tendency toward an increase in "ground pressure" with an increase in vehicle weight. The mean line indicates that the vehicle weight as presently designed tends to grow faster than the area supporting the weight. If this trend holds, and there seems to be no hope that it will change unless, for example, ultra-light structural materials are discovered, then by extrapolation of the mean line, a 300,000-lb vehicle will exercise a minimum pressure of about 5 psi. In reality, as was mentioned before, this pressure would be 3 times larger because only 1/3 of the projected area of tracked vehicles, which are on the top of the discussed line, supports the weight. Accordingly, a 150-ton vehicle would reach a pressure of 14 psi, as indicated by the intersection with the upper line showing an actual track pressure.

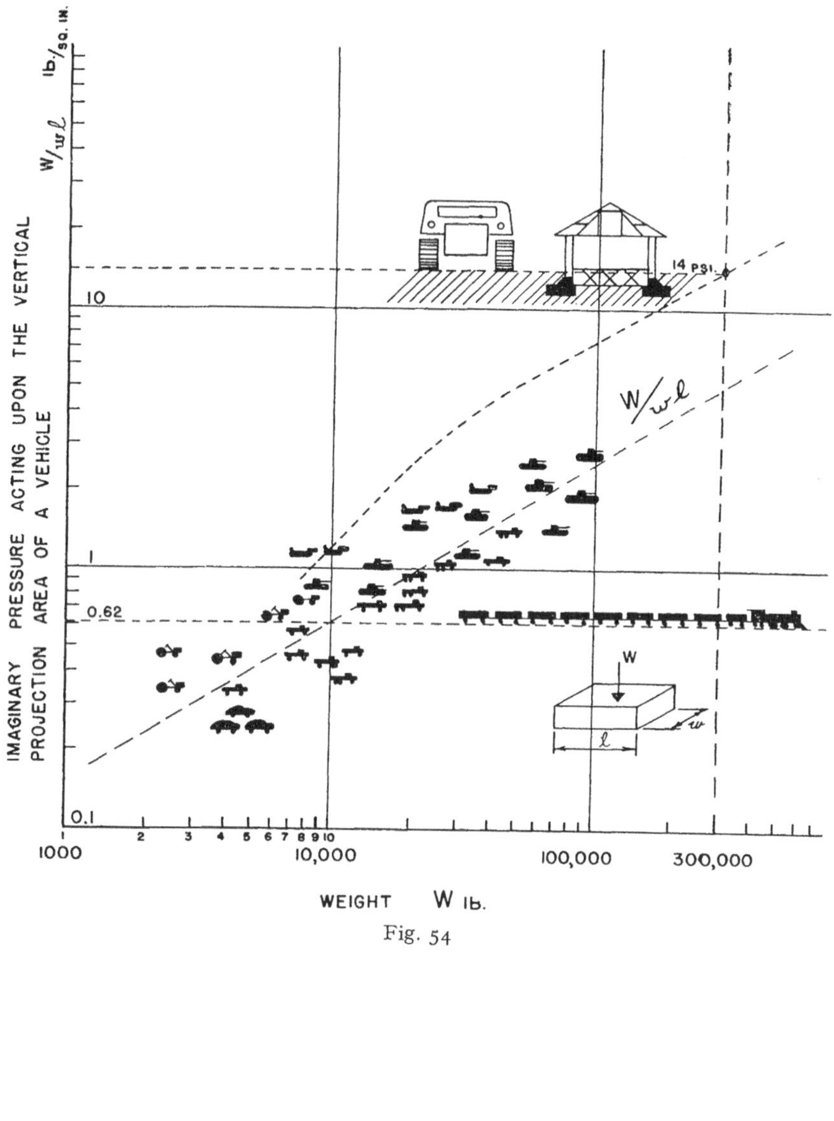

Fig. 54

The New York Building Code regulates the bearing power of soil, specifying that, for instance, for a soft clay, the maximum allowable pressure is 14 psi. Hence, the vehicle would reach the value of pressure which, although safe for a stationary building, certainly would be unsafe for a fast-moving vehicle.

If, therefore, the criterion of the safe load for clay is assumed as a gauge of the capacity of the environment, then vehicles designed in accordance with present trends could not be heavier than 150 tons because they would exercise higher pressures than buildings of a corresponding weight, and thus would be impractical. Accordingly, it may be conjectured that the upper limit in the size of a cross-country vehicle seems to have been almost reached since 100-tons tanks were built during the last war.

If tracked vehicles are made so that their total projection of *wl* area is supported, then, of course, the critical limit may be increased very considerably. It therefore would appear reasonable to expect that the size of the "belly" of tracked vehicles will be reduced rapidly until it disappears, if vehicles heavier than those existing are to be built.

The line shown in Figure 54 which describes the relationships among weight, form, and size of motor vehicles may be expressed by the following equation:

$$wl = 481\ W^{3/8}\ ,$$

where W is measured in pounds and wl is expressed in square inches. In square feet,

$$wl = 3.35\ W^{3/8}\ .$$

In order to know how large the 150-ton vehicle will be if its dimensions are limited by the New York City Building Code, based on a bearing power of clay, $W = 300{,}000$ lb is substituted in the previously deduced equation:

$$wl = 335\ \sqrt[8]{2.7} = 380 \text{ sq ft.}$$

Since the average w/l value of tracked vehicles may be assumed to be 0.55 [mean value of equations (43) and (44)], then

$$0.55\ l^2 = 380$$

$$l = \sqrt{380/0.55} = 26 \text{ ft.}$$

In other words, a tank of that weight would be shorter than a bus. It would be much wider, however, with $w = 26 \times 0.55 = 14$ ft. These dimensions are close to those of the 100-ton German "Maus" which is considered to be the largest vehicle of its kind ever tried.

Train Concept

Figure 54 shows another interesting aspect of vehicle development. It is obvious that the "ground pressure" of a railway train is constant and equal to the "ground pressure" of an individual train member, provided that railway cars are assumed to be identical in form, size, and weight, which basically is true. Thus, for instance, at $wl = 96{,}000$ sq in. a 30-ton train element will exercise a "ground pressure" of about 0.63 psi. Obviously, if a train is comprised of 10 elements, or in other words, if a long snakelike articulated vehicle weighs 300 tons, then the "ground pressure" will remain the same.

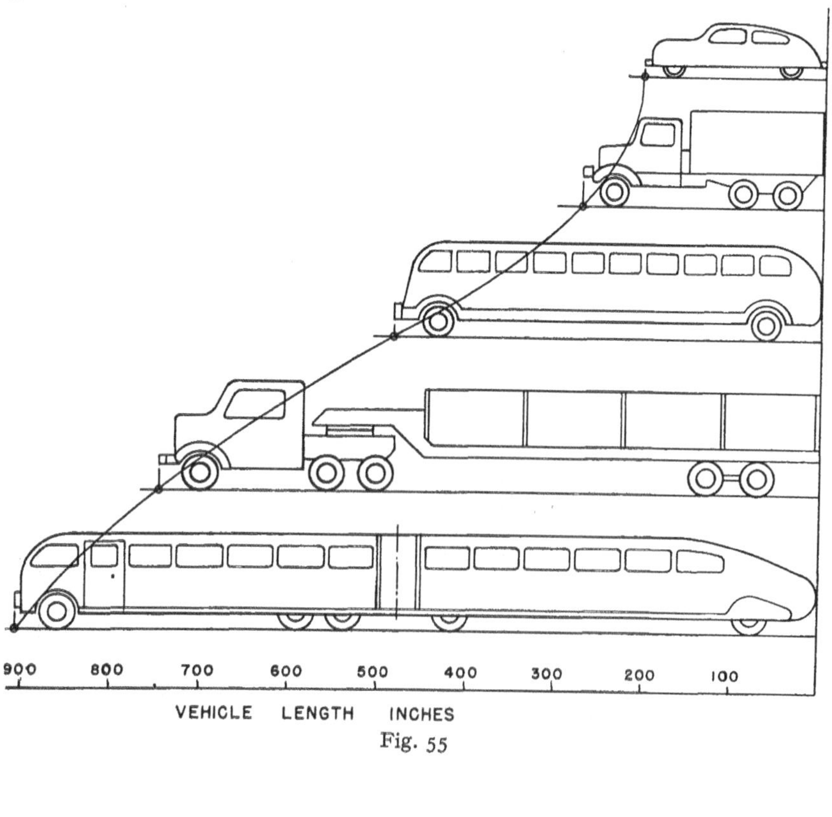

Fig. 55

This seems to be the greatest asset in railway-train development. If railroad equipment had been subjected to the same limitations of form and shape as those existing in the development of road and cross-country vehicles, then the railroads certainly would not have developed to the present extent. It may be interesting to know why this asset has not been considered seriously in the development of land vehicles because it appears that further increase in size and weight of land vehicles would require the introduction of "train concept" in order to forestall the inevitable consequences of ground-pressure increase, as implicit in present design trends.

Initial attempts in this direction were made for other reasons. Figure 55 shows that at a length of 500 in., road vehicles are articulated, and thus better adapted to the traffic gauges as discussed before. Similar though more differentiated articulation exists in the animal world, for example among snakes, apparently for the same reason. The morphology of animals and that of man-made vehicles apparently have some common points in their development.

The present vehicle forms, however, do not seem to evolve any more. The prevailing concepts of design as well as the stability of forms of vehicle components lead to a rather limited number of schemes according to which new designs are conceived. Serious difficulties recently reported in off-the-road operations indicate that a better adaptation of vehicle form to the environment is needed. A study of morphological characteristics of vehicles and the medium in which they operate indicates that this field opens a large number of problems, the solution of which may perhaps radically change the present static conditions.

V. SOME PROBLEMS OF SOIL AND SNOW MECHANICS

Any rational investigation of the relationship between a vehicle and the ground concerned with its locomotion must inevitably lead to a study of the mechanics of the ground and of the vehicle itself. In this chapter, a brief review of soil and snow mechanics will be made with the purpose of stressing only those problems which may bear directly upon vehicle performance according to the present limited knowledge of the subject. In further chapters, an approach to the mechanics of a vehicle will be attempted so that a more rigorous picture of soil-vehicle relationship may be obtained.

Stability and Elasticity Problems

Following Terzaghi's outline of soil mechanics, it may be assumed that an analysis of vehicle performance may be divided, in this study, into two groups, i.e., a study of the stability and elasticity problems. The latter extends over the study of plasticity.

"The stability problems deal with the conditions for the equilibrium of ideal soils immediately preceding ultimate failure by plastic flow. The most important question in this category is the computation of the minimum pressure exerted by a mass of soil on a lateral support, [and] the computation of the ultimate resistance of the soil against external forces, such as the vertical pressure exerted on the soil by a loaded footing [bearing capacity problem].... In order to solve these problems it is sufficient to know the stress conditions for the failure of the soil. No consideration need be given to the corresponding state of strains unless there are certain limitations.... Even if such limitations exist, it is sufficient to consider them in a general way without attempting a quantitative analysis of the corresponding strain effects.

"Elasticity problems deal with the deformation of the soil due to its own weight, or due to external forces such as the weight of buildings. All settlement problems belong to this category. In order to solve these problems we must know the relationship between stress and strain for the soil, but stress conditions for failure do not enter into the analysis." [54]

Although this statement obviously refers to the questions related to civil engineering, it contains the fundamental directive for the discussed study when it is interpreted in terms of vehicle problems.

Thus, the stability problem, which is to be solved in regard to vehicle performance, should deal first with the equilibrium conditions of ideal soils before their ultimate failure by plastic flow. The solution obtained will determine the safe vehicle loads or bearing capacity of soil under vehicle action. It will be applied in those cases in which the vehicle moves on the soil surface at the very moment when an infinitesimal overloading would cause the failure of soil surface and the sinkage of the vehicle. In order to explore this problem, only stress conditions will have to be investigated.

A real vehicle, however, only seldom moves without breaking the soil surface. Approximately 75% of the weak grounds encountered will never sustain the loads of modern traffic without failing through plastic flow and without undergoing extensive deformations. Under these conditions, the plasticity problem becomes a natural extension of the elasticity problem, which embraces all the practical questions of vehicle sinkage, slip, movement resistance, etc.

It will be shown in the course of this chapter that there is little theoretical information available in regard to the solution of the elasticity and plasticity problems of vehicles. Much more has been done in regard to the solution of the stability problem, which accordingly will be the main theme of this work. The emphasis, however, should be put particularly on the future study of elasto-plastic problems since vehicle performance enters more critical stages when the vehicle crosses a very soft ground, with considerable sinkage, slip, and movement resistance, than when it operates on the top of soil, with correspondingly less deformation of the ground.

Time Factor

An interpretation of the soil-vehicle relationship in terms of stability and plasticity problems, as outlined above, appears to be somewhat different, at least from one point of view, if compared to the interpretation of the phenomena encountered in civil engineering. This point of difference, which makes vehicle problems look simpler, relates to the time factor.

Soil deformation under the action of buildings and similar structures may undergo a process half a century long, whereas similar phenomena caused by moving vehicles do not last longer than fragments of seconds, or seconds. Such a short duration of the phenomena simplifies the problem, because an assumption of the invariability of the mechanical properties of the soil involved is quite tolerable. Similar assumptions are not acceptable in the evaluation of the stress-strain relations between soil and a building or a dam.

The consolidation of the ground under long-lasting loads and fundamental changes in its structure, which result from such a slow physical process, do not enter in a critical way into our problems, at least at the present stage of our knowledge. Thus, all the difficulties which stem from the unstable character of soil properties seem to be greatly alleviated if the relationship between the vehicle and the ground is considered.

The time factor, however, has no absolute value and its relativity has to be often recognized in order to account for any changes which may occur in the structure of particular types of soil during the period when a vehicle passes over.

Stress Function in the Two-Dimensional Problem: Cartesian Coordinates

In conformity with the previous discussions, the problem which bears upon the investigation of the soil-vehicle relationship reduces itself to the study of stress and strain patterns under vehicle action.

As is known, the most general, though not quite complete, solution of this problem may be obtained by means of methods established in the theory of elasticity when dealing with an ideal homogeneous isotropic medium which is perfectly elastic.[55, 56, 57, 58]

The assumption of such a medium in our considerations may appear unwarranted since soils and snow are ostensively nonhomogeneous, unisotropic, and inelastic. It will be realized, however, that the complications involved in the nonhomogeneity of the structure of soil mass under consideration do not change the general character of the conclusions reached by means of the theory of elasticity more than similar complications do when a solid is investigated. Moreover, most soils and snow, before breaking through plastic flow, do behave like an elastic medium and even visual observations "...indicate the existence of relation between stress and strain similar to those which are known to exist for solid bodies."[28] Also, since the elastic state precedes the failure, the investigation of this state will give information regarding critical loads, deformation, etc.

Take a semi-infinite ideal soil mass limited only by the surface perpendicular to the plane of the drawing, and an infinitesimal soil prism having a rectangular cross section $dx\,dz$ subjected to stresses as shown in Figure 56a. If ϱ is the density of the mass and g the acceleration gravity, then the gravity force acting per unit of prism volume is ϱg, as plotted in the drawing.

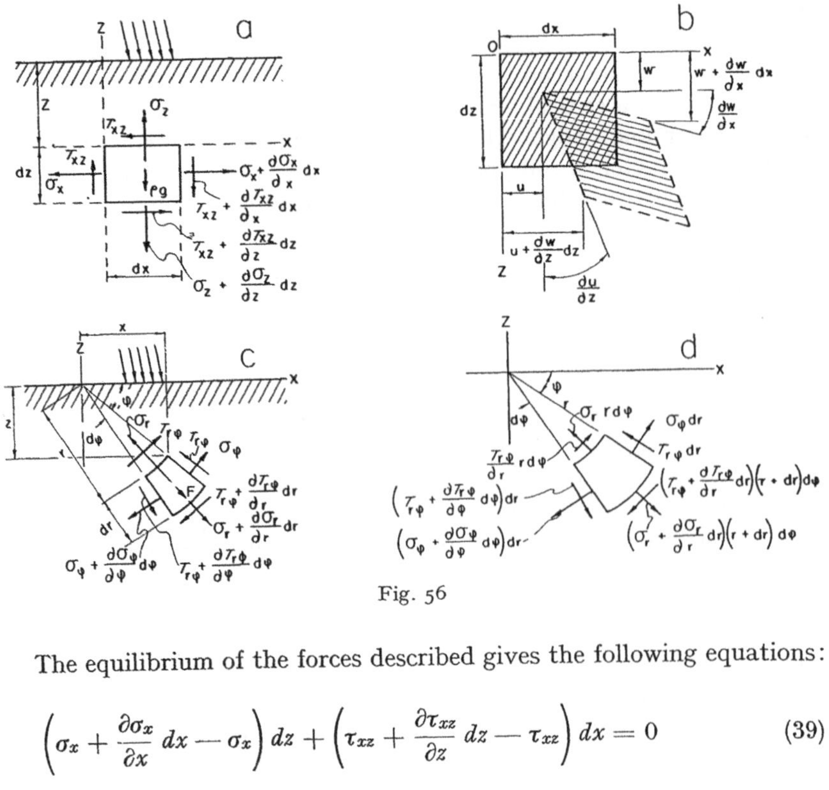

Fig. 56

The equilibrium of the forces described gives the following equations:

$$\left(\sigma_x + \frac{\partial \sigma_x}{\partial x}\,dx - \sigma_x\right) dz + \left(\tau_{xz} + \frac{\partial \tau_{xz}}{\partial z}\,dz - \tau_{xz}\right) dx = 0 \tag{39}$$

$$\left(\sigma_z + \frac{\partial \sigma_z}{\partial z}\,dz - \sigma_z\right) dx + \left(\tau_{xz} + \frac{\partial \tau_{xz}}{\partial x}\,dx - \tau_{xz}\right) dz = -\varrho g\,dx\,dz\,. \tag{40}$$

Equations (39) and (40) may be simplified in the following form:

$$\frac{\partial \sigma_x}{\partial x} + \frac{\partial \tau_{xz}}{\partial z} = 0 \tag{41}$$

$$\frac{\partial \sigma_z}{\partial z} + \frac{\partial \tau_{xz}}{\partial x} = -\varrho g \tag{42}$$

and express the condition of equilibrium. This condition must be preserved throughout the whole mass under consideration, and the stresses at the soil surface should be in equilibrium with the forces acting on it. The condition of equilibrium of surface loads and stresses, called the boundary condition, must then be satisfied in conjunction with equations (41) and (42).

The latter equations, however, together with the equations that determine the boundary condition, do not solve the problem which, as it is known, is statically undetermined. A new condition is to be introduced in order to determine the stress pattern in soil under given external loads if gravity is the only body force, as shown in Figure 56a.

Consider a deformation of the soil prism, the equilibrium of which is determined by equations (41) and (42). This deformation, in a general case, will consist of linear elongations u and w (Figure 56b) and displacements $(\partial u/\partial z)dz$ and $(\partial w/\partial x)dx$ due to the angular distortion of the prism.

The angular distortion $e_{xz} = (\partial u/\partial_z) + (\partial w/\partial x)$ is the shearing strain between the plane xy and zy. If $\partial w/\partial z$ is denoted by e_z and $\partial u/\partial x$ by e_x, then the unit elongations and the unit shear strains are as follows:

$$e_x = \frac{\partial u}{\partial x}\,;\; e_z = \frac{\partial w}{\partial z}\,;\; e_{xz} = \frac{\partial u}{\partial z} + \frac{\partial w}{\partial x}\,.$$

Differentiation of these equations gives:

$$\frac{\partial^2 e_x}{\partial z^2} = \frac{\partial^3 u}{\partial x\, \partial z^2}$$

$$\frac{\partial^2 e_z}{\partial x^2} = \frac{\partial^3 w}{\partial z\, \partial x^2}$$

$$\frac{\partial^2 e_{xz}}{\partial x\, \partial z} = \frac{\partial^3 u}{\partial x^2\, \partial z} + \frac{\partial^3 w}{\partial z^2\, \partial x}$$

or

$$\frac{\partial^2 e_x}{\partial z^2} + \frac{\partial^2 e_z}{\partial x^2} = \frac{\partial^2 e_{xz}}{\partial x\, \partial z}\,. \tag{43}$$

Hooke's law has established that individual elastic deformations are related to the properties of a given material by the following equations:

$$e_x = \frac{\sigma_x}{E}\,; \quad e_x = -\frac{\surd \sigma_z}{E}\,,$$

where E is the modulus of elasticity and $\surd$ the Poisson ratio which determines the contraction of the prism in the xx direction during the compression by stresses σ_z. Experiments show that the unit deformation which occurs when two stresses σ_x and σ_z act simultaneously is obtained by adding individual strains:

$$e_x = \frac{\sigma_x}{E} - \frac{\surd \sigma_z}{E} = \frac{1}{E}(\sigma_x - \surd \sigma_z) \tag{44}$$

$$e_z = \frac{\sigma_z}{E} - \frac{\surd \sigma_x}{E} = \frac{1}{E}(\sigma_z - \surd \sigma_x)\,. \tag{45}$$

According to definition, the unit shear distortion $e_{xz} = \tau_{xz}/G$, where $G = E/2(1 + \surd)$, and is the so-called modulus of rigidity [56]

$$e_{xz} = \frac{2(1 + \surd)\tau_{xz}}{E}\,. \tag{46}$$

If equations (44) and (45) are differentiated twice with reference to z and x, respectively, and the same operation is performed on equation (46) with reference to z and x, then

$$\frac{\partial^2 e_x}{\partial z^2} = \frac{1}{E}\left(\frac{\partial^2 \sigma_x}{\partial z^2} - \frac{\surd\, \partial^2 \sigma_z}{\partial z^2}\right) \tag{47}$$

$$\frac{\partial^2 e_z}{\partial x^2} = \frac{1}{E}\left(\frac{\partial^2 \sigma_z}{\partial x^2} - \frac{\surd\, \partial^2 \sigma_x}{\partial x^2}\right) \tag{48}$$

$$\frac{\partial^2 e_{xz}}{\partial x\, \partial z} = \frac{2\,(1 + \surd)}{E}\,\frac{\partial^2 \tau_{xz}}{\partial x\, \partial z}\,. \tag{49}$$

When the values of the differentials, as determined by equations (47), (48), and (49), are substituted into equation (43), the following will be obtained:

$$\frac{\partial^2 \sigma_x}{\partial z^2} - \frac{\surd\, \partial^2 \sigma_z}{\partial z^2} + \frac{\partial^2 \sigma_z}{\partial x^2} - \frac{\surd\, \partial^2 \sigma_x}{\partial x^2} = 2(1 + \surd)\,\frac{\partial^2 \tau_{xz}}{\partial x\, \partial z}\,. \tag{50}$$

On the other hand, differentiating equations (41) and (42) with reference to x and z, respectively, and adding, gives:

$$\frac{\partial^2 \sigma_x}{\partial x^2} + \frac{\partial^2 \tau_{xz}}{\partial z\, \partial x} = 0$$

$$\frac{\partial^2 \sigma_z}{\partial z^2} + \frac{\partial^2 \tau_{xz}}{\partial x\, \partial z} = 0$$

$$\frac{\partial^2 \sigma_x}{\partial x^2} + \frac{\partial^2 \sigma_z}{\partial z^2} = -2\, \frac{\partial^2 \tau_{xz}}{\partial x\, \partial z}\,. \qquad (51)$$

Accordingly, equation (51) substituted into equation (50) will reduce to the following form:

$$\frac{\partial^2 \sigma_x}{\partial z^2} + \frac{\partial^2 \sigma_z}{\partial x^2} + \frac{\partial^2 \sigma_x}{\partial x^2} + \frac{\partial^2 \sigma_z}{\partial z^2} = 0$$

or

$$\left(\frac{\partial^2}{\partial x^2} + \frac{\partial^2}{\partial z^2}\right)(\sigma_x + \sigma_z) = 0\,. \qquad (52)$$

The conditions of equilibrium (41) (42), the equation of compatibility (52), and the boundary condition present a complete solution of the two-dimensional problem in which weight ϱg is the only body force.

The solution of the problem is performed by the introduction of a function $\Phi(xy)$ which is called the stress function or Airy function, named after G. B. Airy, who first proposed it. The procedure is applied as follows: If any function $\Phi(xy)$ is so assumed that it satisfies the following equations:

$$\left.\begin{aligned} \frac{\partial^2 \Phi}{\partial z^2} &= \sigma_x \\ \frac{\partial^2 \Phi}{\partial x^2} &= \sigma_z \\ -\frac{\partial^2 \Phi}{\partial x\, \partial z} - \varrho g x &= \tau_{xz} \end{aligned}\right\} \qquad (53)$$

then by obtaining the differentials $\partial\sigma_x/\partial x$, $\partial\sigma_z/\partial z$, $\partial\tau_{xz}/\partial x$, and $\partial\tau_{xz}/\partial z$' and by substituting these differentials into the equations of equilibrium

(41) (42), it may be seen that the latter also are satisfied. Since the equation of compatibility (52) must also be satisfied when using the function Φ, the general equation may be found by differentiating equation (53): $\partial^2\sigma_x/\partial x^2 = \partial^4\Phi/\partial^2 z\,\partial^2 x$, $\partial^2\sigma_z/\partial z^2 = \partial^4\Phi/\partial x^2\,\partial z^2$, etc., and substituting these differentials into equation (52). This operation gives:

$$\frac{\partial^4\Phi}{\partial x^4} + 2\frac{\partial^4\Phi}{\partial x^2\,\partial z^2} + \frac{\partial^4\Phi}{\partial z^4} = 0\,. \tag{54}$$

In conclusion, then, the solution of the problem is reduced to the solution of equation (54), provided that the boundary conditions are satisfied. The condition of equilibrium of forces is satisfied, since equations (53) satisfy equations (41) and (42) as previously mentioned. Thus, the stresses σ_x, σ_z, and τ_{xz} may be found if the stress function, Φ, is determined.

The determination of Φ in most cases is not easy. In the theory of elasticity, some solutions of the stress function, as applied to the determination of stresses of simple structures, have been worked out by several investigators.[56] In other cases, ingenious semi-empirical methods of membrane, soap film, and sand-heap analogues have been devised in order to determine more complex stress distributions.[59] Finally, the methods of photoelasticity have produced a powerful tool which, through visual observation of stress distribution, leads to the computation of their values.[60, 61]

Stress Function in Polar Coordinates; Plane Problem

The study of the elastic behavior of soil mass which precedes the failure by plastic flow may be more conveniently made if stress equations (41), (42), and (54) are expressed in polar coordinates instead of in Cartesian coordinates.

The construction of the equations of equilibrium in polar coordinates, which would correspond to equations (41) and (42), may be made in accordance with the stress distribution shown in Figure 56c. If there is no body force, the equilibrium of forces acting on a soil prism, as shown in Figure 56d, will lead directly to the following equations: [56]

$$\frac{\partial\sigma_r}{\partial r} + \frac{1}{r}\frac{\partial\tau_{r\varphi}}{\partial\varphi} + \frac{\sigma_r - \sigma_\varphi}{r} = 0 \tag{55}$$

$$\frac{1}{r}\frac{\partial\sigma_\varphi}{\partial\varphi} + \frac{\partial\tau_{r\varphi}}{\partial r} + \frac{2\tau_{r\varphi}}{r} = 0\,. \tag{56}$$

These correspond to the equations of equilibrium (41) and (42). The condition of compatibility, as expressed by equation (54), may be obtained if the Cartesian coordinates of equation (54) are superseded by the polar ones, in accordance with the following relations, which may be directly read from Figure 56c:

$$r^2 = x^2 + z^2 \; ; \quad \tan \varphi = z/x \, .$$

Accordingly,

$$\partial r/\partial x = x/r = \cos \varphi \; ; \quad \partial r/\partial z = z/r = \sin \varphi \; ;$$

$$\partial \varphi/\partial x = -z/r^2 = -(\sin \varphi) r \; ; \quad \partial \varphi/\partial z = x/r^2 = (\cos \varphi)/r \, .$$

Since, in this case, it is assumed that the stress function, Φ, is a function of r and φ, then

$$\left.\begin{aligned} \frac{\partial \Phi}{\partial x} &= \frac{\partial \Phi}{\partial r}\frac{\partial r}{\partial x} + \frac{\partial \Phi}{\partial \varphi}\frac{\partial \varphi}{\partial x} \\ \frac{\partial \Phi}{\partial z} &= \frac{\partial \Phi}{\partial r}\frac{\partial r}{\partial z} + \frac{\partial \Phi}{\partial \varphi}\frac{\partial \varphi}{\partial z} \end{aligned}\right\} \qquad (57)$$

and the previously determined differentials $\partial r/\partial x$, $\partial \varphi/\partial x$, $\partial r/\partial z$, and $\partial \varphi/\partial z$ may be substituted into equation (57). This will determine $\partial \Phi/\partial x$ and $\partial \Phi/\partial z$. In a subsequent differentiation, the values $\partial^2 \Phi/\partial x^2$ and $\partial^2 \Phi/\partial z^2$ will be obtained. By adding these values, it will be found that

$$\frac{\partial^2 \Phi}{\partial x^2} + \frac{\partial^2 \Phi}{\partial z^2} = \frac{\partial^2 \Phi}{\partial r^2} + \frac{1}{r}\frac{\partial \Phi}{\partial r} + \frac{1}{r^2}\frac{\partial^2 \Phi}{\partial \varphi^2}$$

or

$$\frac{\partial^2}{\partial x^2} + \frac{\partial^2}{\partial z^2} = \frac{\partial}{\partial r^2} + \frac{1}{r}\frac{\partial}{\partial r} + \frac{1}{r^2}\frac{\partial}{\partial \varphi} \, . \qquad (58)$$

On the other hand, equation (54) may be written in the following form:

$$\frac{\partial^4 \Phi}{\partial x^2} + 2\frac{\partial^4 \Phi}{\partial x^2\, \partial z^2} + \frac{\partial^4 \Phi}{\partial z^2} = \left(\frac{\partial^2}{\partial x^2} + \frac{\partial^2}{\partial z^2}\right)\left(\frac{\partial^2 \Phi}{\partial x^2} + \frac{\partial^2 \Phi}{\partial z^2}\right) . \qquad (59)$$

Thus, by combining equations (58) and (59), the final equation for a two-dimensional problem will be obtained:

$$\left(\frac{\partial^2}{\partial r^2} + \frac{1}{r}\frac{\partial}{\partial r} + \frac{1}{r^2}\frac{\partial^2}{\partial \varphi^2}\right)\left(\frac{\partial^2 \Phi}{\partial r^2} + \frac{1}{r}\frac{\partial \Phi}{\partial r} + \frac{1}{r^2}\frac{\partial^2 \Phi}{\partial \varphi^2}\right) = 0 \, . \qquad (60)$$

Equation (60) satisfies the condition of equilibrium (55) and (56), which corresponds to the same condition expressed by equations (41) and (42).

It may be proved by substituting the stresses σ_r, σ_φ, and $\tau_{r\varphi}$ into equations (55), (56), and (60) that these equations are satisfied if these stresses are the following functions of Φ:

$$\sigma_r = \frac{1}{r}\frac{\partial \Phi}{\partial r} + \frac{1}{r^2}\frac{\partial^2 \Phi}{\partial \varphi^2} \tag{61}$$

$$\sigma_\varphi = \frac{\partial^2 \Phi}{\partial r^2} \tag{62}$$

$$\tau_{r\varphi} = \frac{1}{r^2}\frac{\partial \Phi}{\partial \varphi} - \frac{1}{r}\frac{\partial^2 \Phi}{\partial r\,\partial \varphi}\,. \tag{63}$$

Thus, when the stress function Φ, which satisfies equation (60) and the boundary conditions, is known, then the stresses σ_r, σ_φ, and $\tau_{r\varphi}$ may be considered as being determined.

Vertical-Line Load and Stress Distribution

Consider a linear load, V lb/in., acting upon the boundary of the soil mass (Figure 57a). It will be found by inspection that the stress function Φ which satisfies equation (60) is

$$\Phi = -\frac{V}{\pi} r\varphi \sin\varphi\,.$$

Accordingly,

$$\left.\begin{aligned}
\frac{\partial \Phi}{\partial r} &= -\frac{V}{\pi}\varphi\sin\varphi \\
\frac{\partial^2 \Phi}{\partial r^2} &= 0 \\
\frac{\partial \Phi}{\partial \varphi} &= -\frac{V}{\pi} r\sin\varphi - \frac{V}{\pi} r\varphi\cos\varphi \\
\frac{\partial^2 \Phi}{\partial \varphi^2} &= -\frac{2V}{\pi} r\cos\varphi + \frac{V}{\pi} r\varphi\sin\varphi \\
\frac{\partial^2 \Phi}{\partial r\,\partial \varphi} &= -\frac{V}{\pi}\sin\varphi - \frac{V}{\pi}\varphi\cos\varphi\,.
\end{aligned}\right\} \tag{64}$$

By substituting respective values of equations (64) into equations (61), (62), and (63), it will be found that

$$\left.\begin{aligned} \sigma_r &= -\frac{2V}{\pi r}\cos\varphi \\ \sigma_\varphi &= 0 \\ \tau_{r\varphi} &= 0\,. \end{aligned}\right\} \tag{65}$$

This solution is in agreement with the boundary condition since σ_r vanishes at the surface $\varphi = \pi/2$ where no forces act beyond the linear load, V.

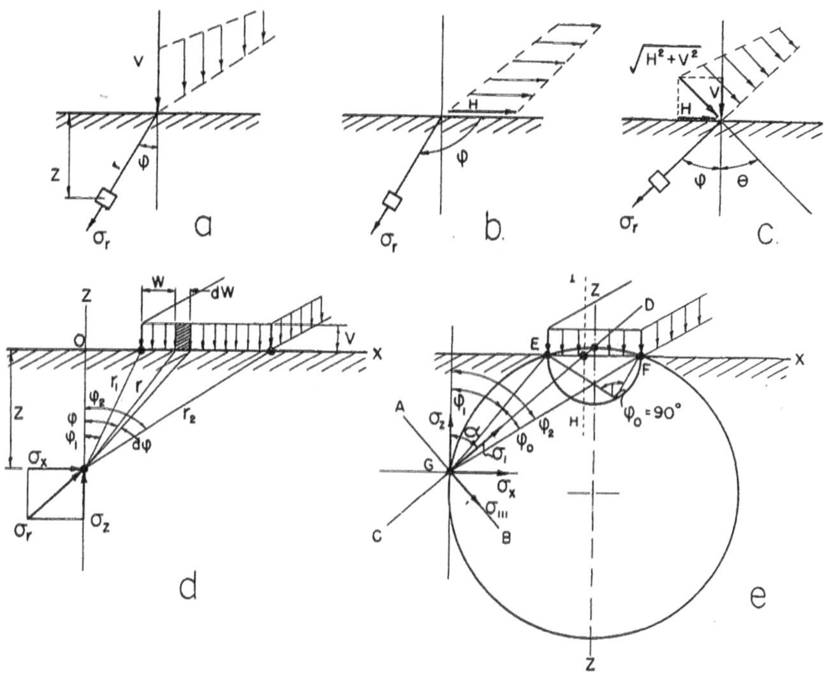

Fig 57

Stresses under a Line Load Having Any Horizontal and Vertical Components

When vehicles act upon the soil, they do not exercise vertical loads only. Horizontal loads H have to be exercised not only in order to pull an agricultural implement or trailer, but also in order to keep a vehicle moving. Thus, the case of a horizontal force H (Figure 57b) is of interest.

It will be seen that this case is identical with the one just cited, provided that an angle φ is measured, as previously, between the direction of the force and the given radius r. In the same way, any force $R = \sqrt{V^2 + H^2}$ sloped to the vertical at an angle θ (Figure 57c) will produce stresses which may be determined by means of equation (65), in which the angle enclosed by radius r and the direction of the force is $\theta + \varphi$:

$$\sigma_r = -\frac{2\sqrt{V^2 + H^2}}{\pi r} \cos(\theta + \varphi) . \tag{66}$$

By projecting the radial stresses σ_r expressed by equations (65) and (66) upon respective areas, the stresses σ_x, σ_z, and τ_{xz} of Cartesian coordinates may be obtained for line loads:

$$\left.\begin{aligned} \sigma_x &= -\sigma_r \sin^2 \varphi \\ \sigma_z &= -\sigma_r \cos^2 \varphi \\ \tau_{xz} &= -\sigma_r \sin \varphi \cos \varphi . \end{aligned}\right\} \tag{67}$$

Stress under the Action of a Strip Load

Another case of interest from a vehicle point of view is the stress distribution under the action of a strip load (Figure 57d). Such strip loads may be visualized, for instance, as loads exercised by the caterpillars of tracked vehicles.

The case of a strip load has been solved by means of the stress function Φ having the following form:[62]

$$\Phi = \frac{V}{2\pi} (r_1^2 \varphi_1 - r_2^2 \varphi_2) .$$

The solution of this problem also may be obtained directly by integrating the stresses previously calculated for the line load.

Consider Figure 57d. According to formulas (65) and (67),

$$\left.\begin{aligned}\sigma_x &= -\sigma_r \sin^2\varphi = \frac{2V}{\pi r}\cos\varphi\sin^2\varphi\\ \sigma_z &= -\sigma_r \cos^2\varphi = \frac{2V}{\pi r}\cos^3\varphi\\ \tau_{xz} &= -\sigma_r \sin\varphi\cos\varphi = \frac{2V}{\pi r}\sin\varphi\cos^2\varphi\end{aligned}\right\} \tag{68}$$

or, if depth z is introduced into equations (68), assuming that $z = r\cos\varphi$,

$$\left.\begin{aligned}\sigma_x &= \frac{2V}{\pi z}\sin^2\varphi\cos^2\varphi\\ \sigma_z &= \frac{2V}{\pi z}\cos^4\varphi\\ \tau_{xz} &= \frac{2V}{\pi z}\cos^3\varphi\sin\varphi\,.\end{aligned}\right\} \tag{69}$$

The load W acting upon the left-hand portion of the strip determined by the angles $\varphi - \varphi_1$ equals

$$W = V(r\sin\varphi - r_1\sin\varphi_1)\,,$$

or since $r = z/\cos\varphi$ and $r_1 = z/\cos\varphi_1$,

$$W = zV(\tan\varphi - \tan\varphi_1)\,. \tag{70}$$

By differentiating equation (70), it will be found that

$$dW = zV\frac{d\varphi}{\cos^2\varphi}\,. \tag{71}$$

An infinitesimal load dW [equation (71)] acting upon the strip enclosed by the angle $d\varphi$ may be considered as a line load dV and may be substituted into equations (69). Accordingly,

$$\left.\begin{aligned}d\sigma_x &= \frac{2\,dW\cos^4\varphi}{\pi z} = \frac{2\,V\sin^2\varphi}{\pi}\,d\varphi\\ d\sigma_z &= \frac{2\,dW\cos^2\varphi\sin^2\varphi}{\pi z} = \frac{2\,V\cos^2\varphi}{\pi}\,d\varphi\\ \tau_{xz} &= \frac{2\,dW\cos^3\varphi\sin\varphi}{\pi z} = \frac{2\,V\cos\varphi\sin\varphi}{\pi}\,d\varphi\,.\end{aligned}\right\} \tag{72}$$

The integration of equations (72) will give

$$\sigma_x = \frac{2V}{\pi}\int_{\varphi_1}^{\varphi_2} \sin^2\varphi \, d\varphi = \frac{V}{\pi}(\varphi_2 - \varphi_1 + \sin\varphi_2\cos\varphi_2 - \sin\varphi_1\cos\varphi_2) \quad (73)$$

$$\sigma_z = \frac{2V}{\pi}\int_{\varphi_1}^{\varphi_2} \cos^2\varphi \, d\varphi = \frac{V}{\pi}(\varphi_2 - \varphi_1 - \sin\varphi_2\cos\varphi_2 + \sin\varphi_1\cos\varphi_2) \quad (74)$$

$$\tau_{xz} = \frac{2V}{\pi}\int_{\varphi_1}^{\varphi_2} \cos\varphi\sin\varphi \, d\varphi = \frac{V}{\pi}(\sin^2\varphi_2 - \sin^2\varphi_1) \, . \quad (75)$$

Principal Stresses and the Strip Load

In the investigation of the state of soil stresses, it is of interest to know the values of σ_r, σ_φ, $\tau_{r\varphi}$, or σ_x, σ_z, and τ_{xz} as well as the directions and values of the principal stresses prevailing at a given point.

Principal stresses are, by definition, normal stresses that do not produce shear in the planes of their application.[63] Accordingly, if one of the planes CD of action of the principal stress is sloped to the vertical at an angle α (Figure 57e), then the resultant stresses produced in the perpendicular plane AB by the projection of the stresses which are due to the action of portions of the strip load located respectively on the right- and left-hand side of the plane HI must vanish. This condition is satisfied if

$$\alpha = \frac{\varphi_2 + \varphi_1}{2} \, .$$

It will be seen then that the angle α (Figure 57e) is the bisector of the angle $\varphi_o = (\varphi_2 - \varphi_1)/2$ and that the direction of the principal stresses $\sigma_{,}$ and $\sigma_{,,,}$ is determined by the planes CD and AB, respectively. The value of the principal stresses may be determined by projecting σ_x, σ_z, and τ_{xz} in the direction of the planes in which those stresses act. Following Figure 58a, it will be seen that

$$\sigma_x = \sigma_{,} \sin^2\alpha + \sigma_{,,,} \cos^2\alpha \quad (76)$$

$$\sigma_z = \sigma_{,} \cos^2\alpha + \sigma_{,,,} \sin^2\alpha \quad (77)$$

$$\tau_{xz} = \sigma_{,} \sin\alpha\cos\alpha - \sigma_{,,,} \cos\alpha\sin\alpha = \tfrac{1}{2}(\sigma_{,} - \sigma_{,,,}) \sin 2\alpha \, . \quad (78)$$

Adding equations (76) and (77) will produce

$$\sigma_{,} + \sigma_{,,,} = \sigma_x + \sigma_z \, . \quad (79)$$

From equations (73) and (74),

$$\sigma_x + \sigma_z = \frac{V}{\pi}(\varphi_2 - \varphi_1) = \frac{V}{\pi}\varphi_o \,. \tag{80}$$

Combining equations (79) and (80) gives

$$\sigma_1 = \frac{V}{\pi}\varphi_o - \sigma_{,,,} \,. \tag{81}$$

When equation (78) is combined with equation (81), assuming that $2\alpha = \varphi_2 + \varphi_1$,

$$\tau_{xz} = \tfrac{1}{2}\left(\frac{V}{\pi}\varphi_o - 2\sigma_{,,,}\right)\sin(\varphi_2 + \varphi_1) \,,$$

or since τ_{xz} is known from formula (75),

$$\frac{V}{\pi}(\sin^2\varphi_2 - \sin^2\varphi_1) = \left(\frac{V}{\pi}\varphi_o - \sigma_{,,,}\right)\sin(\varphi_2 + \varphi_1) \,.$$

Hence,

$$\sigma_{,,,} = \frac{V}{\pi}\left(\varphi_o - \frac{\sin^2\varphi_2 - \sin^2\varphi_1}{\sin(\varphi_2 + \varphi_1)}\right).$$

It may be shown that

$$\sin^2\varphi_2 - \sin^2\varphi_1 = \sin(\varphi_2 + \varphi_1)\sin(\varphi_2 - \varphi_1) \,.$$

Accordingly, the final formula for $\sigma_{,,,}$, assuming that $\varphi_2 - \varphi_1 = \varphi_o$, will take the form of

$$\sigma_{,,,} = \frac{V}{\pi}(\varphi_o - \sin\varphi_o) \,. \tag{82}$$

In a similar way, it will be found that

$$\sigma_{,} = \frac{V}{\pi}(\varphi_o + \sin\varphi_o) \,. \tag{83}$$

Since the value of the principal stresses $\sigma_{,}$ and $\sigma_{,,,}$ depends entirely on the angle φ_o, a circle which passes through points G, E, F (Figure 57e) will be the locus of the constant values of these stresses.

A relationship between the principal stresses and σ_x, σ_z, and τ_{xz}, as expressed by equations (76), (77), and (78), may be presented by means of Mohr's circle,[54, 56] which may be considered as a graphic solution of these equations (Figure 58b).

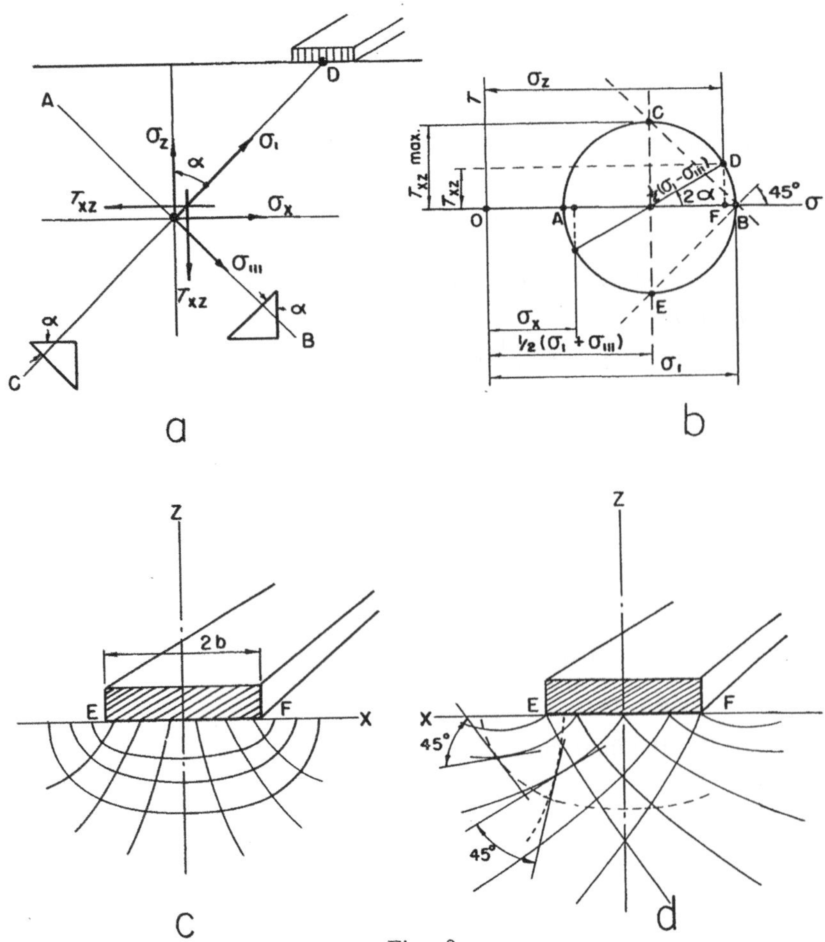

Fig. 58

If, for instance, $\sigma_{,}$ and $\sigma_{,,,}$ are known for a given point of soil mass, then σ_x, σ_z, and τ_{xz} may be found by plotting points A and B, spaced at distances $\overline{OA} = \sigma_{,,,}$ and $\overline{OB} = \sigma_{,}$ from the beginning of coordinates 0. A circle having a diameter $\overline{AB} = \sigma_{,} - \sigma_{,,,}$ will determine the stresses σ_x and σ_z, which correspond to the radius sloped at an angle 2α to the σ axis. In this case, $\overline{FD} = \tau_{xz} = \frac{1}{2}(\sigma_{,} - \sigma_{,,,}) \sin 2\alpha$, which is in agreement with equation (78). Other useful relations between stresses and angles may be readily written on the basis of the denotations shown in Figure 58b:

$$\sigma_x = \tfrac{1}{2}(\sigma_{,} + \sigma_{,,,}) - \tfrac{1}{2}(\sigma_{,} - \sigma_{,,,}) \cos 2\alpha \tag{84}$$

$$\sigma_z = \tfrac{1}{2}(\sigma_{,} + \sigma_{,,,}) + \tfrac{1}{2}(\sigma_{,} - \sigma_{,,,}) \cos 2\alpha \tag{85}$$

$$\sin 2\alpha = \frac{2\,\tau_{xz}}{\sigma_{,} - \sigma_{,,,}} . \tag{86}$$

Equations (84) and (85) lead directly to equation (79). Figure 58b also shows that the value of the maximum shear is

$$(\tau_{xz})_{\max} = \tfrac{1}{2}(\sigma_{,} - \sigma_{,,,}) . \tag{87}$$

From equations (82) and (83), $\sigma_{,} - \sigma_{,,,} = (2\,V \sin \varphi_o)/\pi$. Hence,

$$(\tau_{xz})_{\max} = \frac{V}{\pi} \sin \varphi_o . \tag{88}$$

For $\varphi_o = \pi/2$, τ_{xz} reaches the highest values equal to $V/3.14$. The locus of these values is the circumference of a circle having a strip-load base EF (Figure 57e) as its diameter.

Trajectories of Principal Stresses

The fact that the direction of the principal stresses $\sigma_{,}$ and $\sigma_{,,,}$ is a bisector of the angle EGF (Figure 57e) coincides with the similar property of an ellipse formed by radii $\overline{GE}$ and $\overline{GF}$. If, therefore, a family of confocal ellipses is traced, assuming that the sum of radii $\overline{GE} + \overline{GF} = 2m$, where m is the parameter, then the ellipses in question may be considered as envelopes of the lines AB oriented in the directions of the principal stresses $\sigma_{,,,}$. An orthogonal set of curves which in a similar way determine the direction of the principal stresses $\sigma_{,}$ will be accordingly a family of hyperbolas.

Families of lines that determine the directions of stresses are called the trajectories of principal stresses.[64] Their equation in the case of a strip of width 2b (Figure 58c) is as follows:

$$\frac{x^2}{m^2} \pm \frac{z^2}{m^2 - b^2} = 1 .$$

The foci, according to definition, are located in points E and F.

Experimental methods of finding the trajectories of principal stresses are useful in the determination of stress distribution in more complex

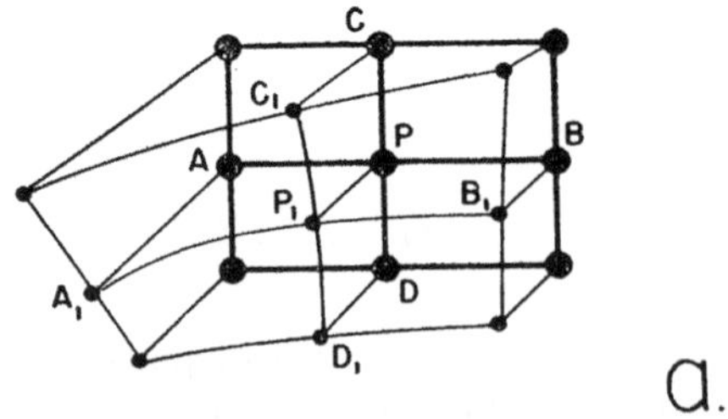

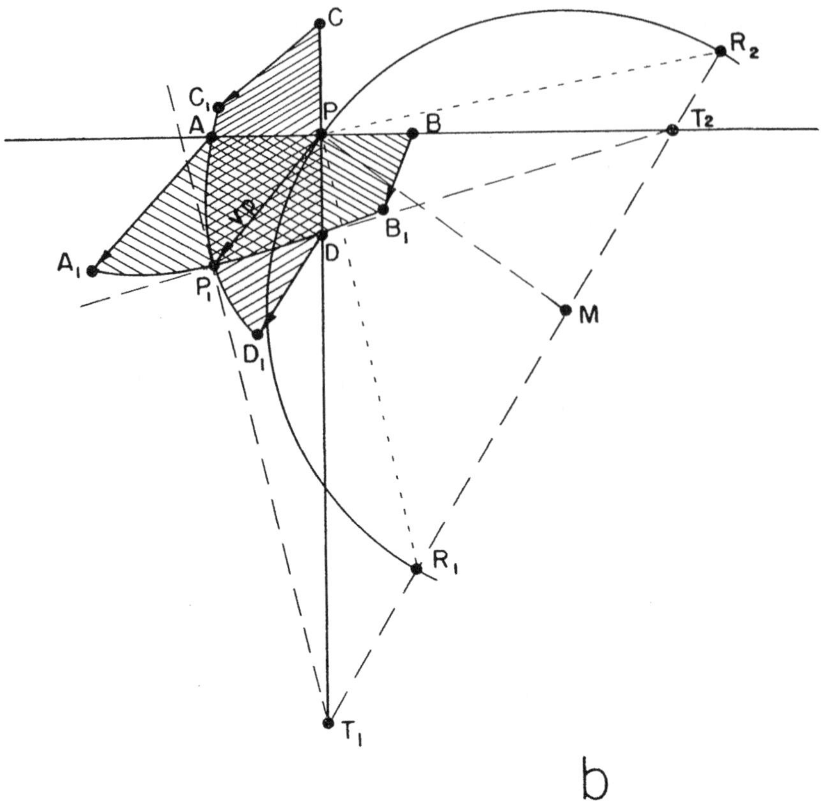

Fig. 59

cases. In addition to the method of photoelasticity [60, 61] and others,[65] a method by Haefeli [66] seems particularly useful in determining the trajectories in soil or snow.

Assume that the soil mass under investigation is marked with a rectangular grid, part of which is shown in Figure 59a. The intersections of a few lines are denoted by points A, B, C, D, and P.

Because of external forces acting upon this mass, a deformation takes place and causes the grid lines to become curved and move in a new position, denoted by points A_1, B_1, C_1, D_1, and P_1.

In order to determine the principal stresses at point P, trace a tangent to line A_1B_1 at point P_1 (Figure 59b). This tangent will give point T_2 at the intersection with line AB. Similarly, the tangent at P_1 to line C_1D_1 will determine point T_1 at the intersection with line CD. After

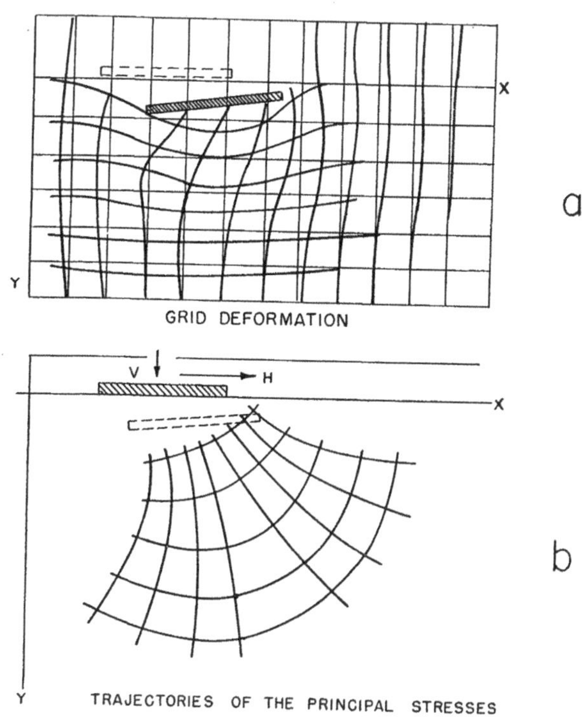

Fig. 60

points T_1 and T_2 are connected by line T_1T_2, a locus of momentary centers of circles, which determine the directions of the principal stresses, will be given.

In order to find these directions at point P, a line perpendicular to the direction PP_1 of flow v_p must be traced. This line, at the point of intersection with T_1T_2, gives the center M of a circle determining the directions of the principal stresses sought. This circle, having radius $\overline{MP}$, defines the discussed directions by means of perpendicular chords R_1P and R_2P.

A graphic determination of the directions of the principal stresses, using the described methods, may be applied to plastic or fragmental materials. To this end, some sort of a rectangular grid has to be painted on the mass, and its deformation is to be determined upon the application of loads.

The trajectories of principal stresses determined by means of Haefeli's method for a plate acting upon the soil under horizontal H and vertical V loads are shown in Figure 60b. The deformation of the grid which leads to the determination of the trajectories is shown in Figure 60a.

Shear Pattern

It was previously shown [equation (87)] that the maximum shear value is

$$(\tau_{xz})_{\max} = \frac{\sigma_{,} - \sigma_{,,,}}{2}.$$

From Mohr's circle (Figure 58b) or equation (78), it is shown that the maximum shear is reached when $\sin 2\alpha = 1$, i.e., when $\alpha = 45°$. Thus, the planes of the maximum shear are inclined at an angle of 45° to the axes of the principal stresses.

Among the many theories on the strength of materials,[67] the theory of maximum shear tends to be considered satisfactorily confirmed by observation of solids. According to this theory, the failure of a given material occurs in the planes in which τ reaches its maximum value. Since these planes are, as previously mentioned, sloped to the axes of principal stresses at an angle of 45°, the planes of slip may be determined once the directions of $\sigma_{,}$ and $\sigma_{,,,}$ are known.

In the case of a strip load, the directions of the slip planes will be determined by the bisectors of the right angles formed by the intersection of the previously discussed confocal ellipses and hyperbolas [equation (88)]. By tracing lines parallel to the directions thus obtained, a series of lines will be derived (Figure 58d). The family of these lines, called the shear pattern, is composed of two sets of curves intersecting each other at right angles, since two directions sloped at 45° to the axes of the principal stresses may be traced at a given point.

It is interesting to note that the directions of the slip planes are determined by lines which may be traced on Mohr's circle (Figure 58b) through points *BC* and *BE*. The particular lines are sloped to the direction of the principal stresses σ at an angle of 45°. This may be seen directly on the drawing. The discussed angles are called the angles of rupture.[54]

Pressure Distribution under Strip Loads

The normal pressure acting on a plane surface located at any depth can be determined from equation (74) or (85). When these pressures are computed, isobars, i.e., lines of equal pressure, may be traced in dimensionless form. Such isobars indicate that at a depth 2b equal to the

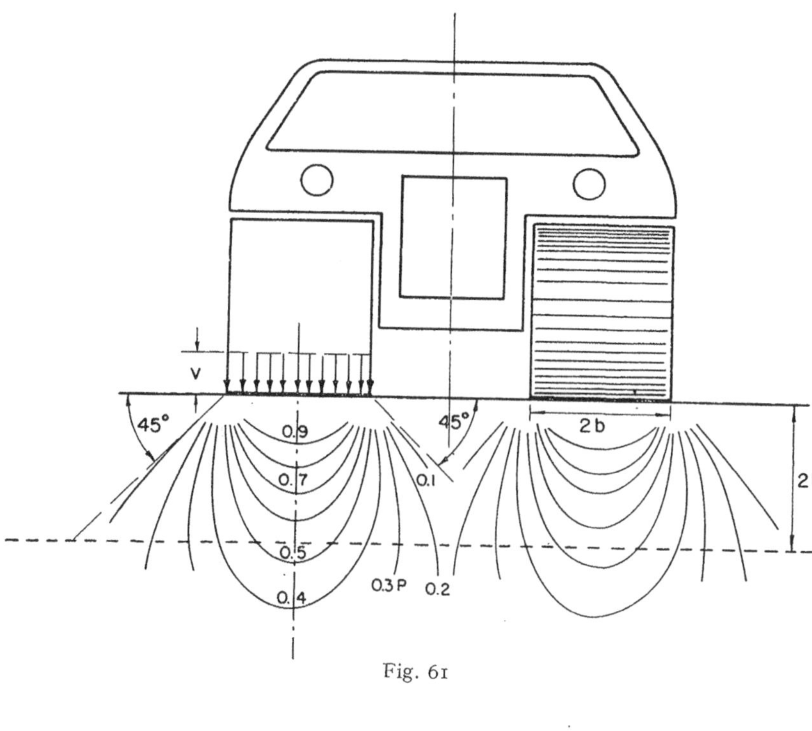

Fig. 61

width of the strip, the load pressure V acting upon the strip load (Figure 61) is reduced by approximately 50% and practically vanishes at a depth 4b equal to twice the width of the strip.

A simplified method of determining pressure distribution and a comparison with the distribution computed by means of a rigorous method using the stress function Φ are given in Reference 68. The planes which limit the stress distribution for all practical purposes may be assumed as being sloped at angles of 45° (Figure 61).

Loads Concentrated in a Point

All the previously considered cases refer to a two-dimensional problem. The general methods discussed in conjunction with the stress distribution apply, however, to three-dimensional problems, in which case the stress-strain relationship in the *yy* direction also should be considered together with the relevant properties of the soil mass along the *xx* and *zz* axes. [55, 56, 58]

Accordingly, Boussinesq determined the stress distribution for a concentrated load, using the stress function. [26, 69] His solution, as referred to such a load, gives the following expression for the normal stress σ_z acting upon various points on a horizontal surface located at depth z (Figure 62):

$$\sigma_z = \frac{3W}{2\pi R^2}\left(\frac{Z}{R}\right)^3 \tag{89}$$

or

$$\sigma_z = \frac{3}{2\pi}\,\frac{1}{\left[1+\left(\frac{r}{z}\right)^2\right]^{5/2}}\,\frac{W}{z^2}, \tag{90}$$

where $r = \sqrt{x^2 + y^2}$ and $R = \sqrt{z^2 + r^2}$. Values of the coefficient enclosed by the brackets in formula (90) are plotted in Figure 62b for various r/z ratios.

If it is assumed that a wheel represents a point load, then the pressure distribution may be computed according to formula (89) or (90), provided that the investigated points are located not too close to the contact area of the wheel and the ground, so that Saint-Venant's principle is satisfied.[56] The soil mass located in the immediate vicinity of this area does not behave elastically. It is called the disturbed zone and its shape depends on the spread of the plastic deformations.[54, 59]

Boussinesq's equations (89) and (90) take the following form when written in polar coordinates (Figure 62a):

$$\sigma_z = \frac{3W}{2\pi} \frac{\cos^3 \varphi}{R^2} . \tag{91}$$

Accordingly, the radial stress σ_r will be

$$\sigma_r = \frac{3W}{2\pi} \frac{\cos \varphi}{R^2} . \tag{92}$$

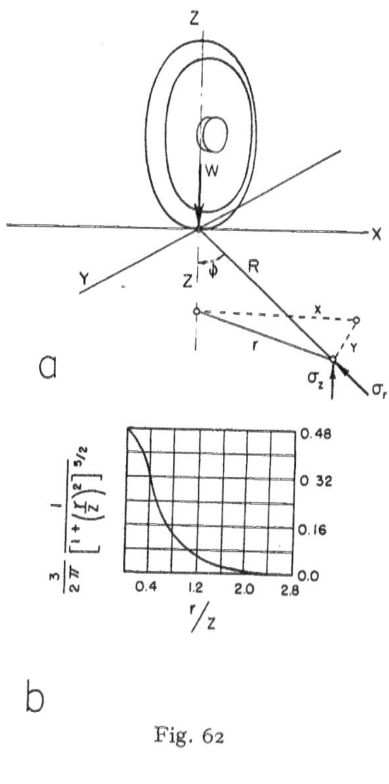

Fig. 62

Pressure under Loads Acting upon Circular, Elliptic, and Rectangular Areas

Stresses acting upon soil under flexible loads applied to circular and rectangular areas may be computed by integration. The term "flexible"

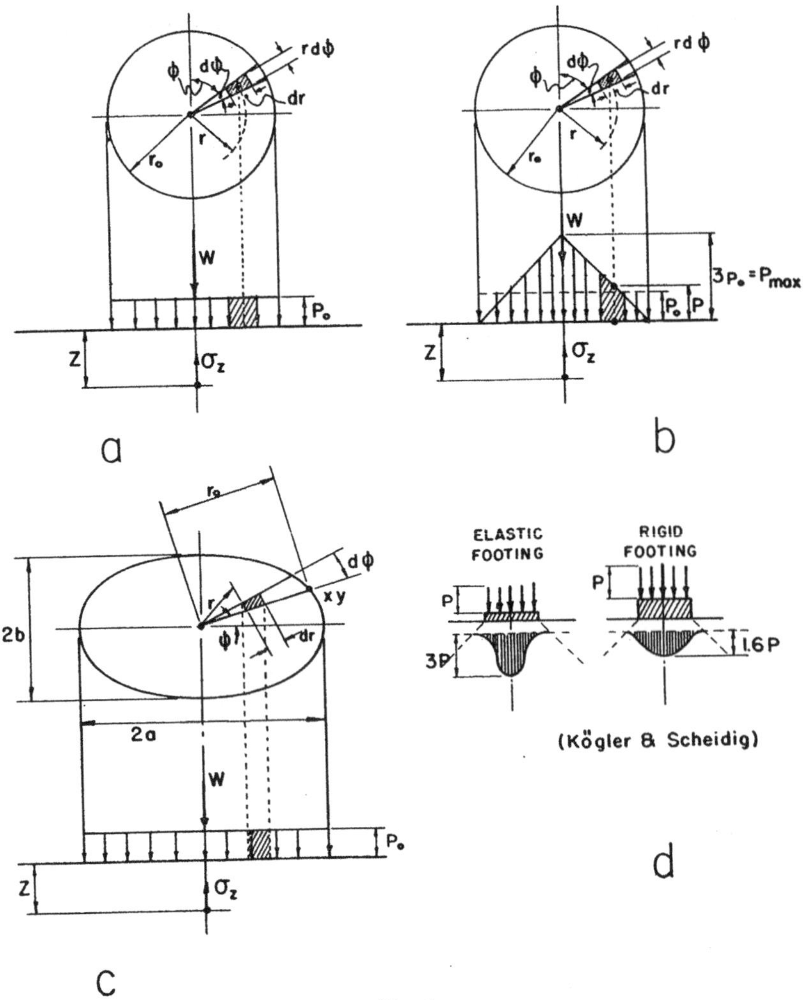

Fig. 63

means that each element of the area deforms freely without rigid connection with other elements.[54]

Take a disc loaded with a uniform pressure p_o and having a radius r_o (Figure 63a). The vertical stress which prevails at depth z below the center of a circular area may be determined as follows.

The elementary load dW acting upon the area $r d\varphi\, dr$ is $dW = p_o r\, dr\, d\varphi$. Hence, in accordance with formula (90), which may be presented in the form of a differential,

$$d\sigma_z = \frac{3}{2\pi} \frac{dW}{\left[1 + \left(\frac{r}{z}\right)^2\right]^{5/2} z^2},$$

the vertical stress σ_z may be expressed by the following equation:

$$\sigma_z = \frac{3}{2\pi} p_o \int_0^{r_o} \int_0^{2\pi} \frac{r\, dr\, d\varphi}{\left[1 + \left(\frac{r}{z}\right)^2\right]^{5/2} z^2} = 3p_o \int_0^{r_o} \frac{r\, dr}{\left[1 + \left(\frac{r}{z}\right)^2\right]^{5/2} z^2}.$$

By substituting $(r/z)^2 = u^2$, it will be found that

$$\sigma_z = 3p_o \int_0^{r_o} \frac{u\, du}{[1 + u^2]^{5/2}} = 3p_o \left| \frac{1}{3[1 + u^2]^{3/2}} \right|_0^{r_o}$$

or

$$\sigma_z = p_o \left[1 - \frac{z^3}{(z^2 + r_o^2)^{3/2}}\right]. \tag{93}$$

Another case of interest may be that of a nonuniform conic load distribution over a circular area (Figure 63b). In this case, the force applied to an element $r\, dr\, d\varphi$ of the area is $p\, r\, dr\, d\varphi$, where p is the pressure acting at a given point. From the relationship between a uniform and conical load distribution, it follows that if the same total weight W is carried in both cases, then $p_{max} = 3\, p_o$ (Figure 63b). Accordingly,

$$\frac{p}{3p_o} = \frac{r_o - r}{r_o}$$

and

$$dW = p\, r\, dr\, d\varphi = \frac{3p_o}{r_o} (r_o - r)\, r\, dr\, d\varphi\,.$$

After substituting dW from the above equation into equation (90),

$$\sigma_z = \frac{9p_o}{2\pi} \int_0^{r_o} \int_0^{2\pi} \frac{(r_o - r)\, r\, d\varphi\, dr}{r_o \left[1 + \left(\frac{r}{z}\right)^2\right]^{5/2} z^2}$$

and

$$\sigma_z = 9p_o \left\{ \int_0^{r_o} \frac{r\,dr}{\left[1+\left(\dfrac{r}{z}\right)^2\right]^{5/2} z^2} - \int_0^{r_o} \frac{r^2\,dr}{r_o\left[1+\left(\dfrac{r}{z}\right)^2\right]^{5/2} z^2} \right\}.$$

The first integral was previously determined when solving equation (93). The second one may be solved by substituting $1 + (r/z)^2 = u^2r^2$. Accordingly,

$$\sigma_z = 3p_o \left\{ 1 - \frac{1}{\left[1+\left(\dfrac{r_o}{z}\right)^2\right]^{3/2}} - \frac{r_o^2}{z^2\left[1+\left(\dfrac{r_o}{z}\right)^2\right]^{3/2}} \right\},$$

or after simplifying,

$$\sigma_z = 3p_o \left[1 - \frac{z}{\sqrt{z^2 + r_o^2}} \right]. \tag{94}$$

Experience indicates that the contact area of a pneumatic tire may be assumed as elliptic in shape. For this reason, the stress distribution under an elliptic area is of special interest.

Take the case of a uniform load distribution (Figure 63c). The load dW acting upon an element of the surface $r\,d\varphi\,dr$ is $dW = p_o\,r\,dr\,d\varphi$, where p_o, as previously, is the unit uniform load. When dW is substituted into formula (90), it will be found that

$$\sigma_z = p_o \int_0^{r_o} \int_0^{2\pi} \frac{z^3\,r\,dr\,d\varphi}{(z^2+r^2)^{5/2}}.$$

The value r_o may be determined from the equation of the ellipse: $x^2/a^2 + z^2/b^2 = 1$, assuming that $r_o = \sqrt{x^2+y^2}$. Accordingly,

$$r_o = \sqrt{\frac{b^2}{a^2}(a^2 - x^2) + x^2}.$$

Since $x = r_o \cos\varphi$,

$$r_o = \frac{b}{\sqrt{1 + [(b^2/a^2) - 1]\cos^2\varphi}},$$

and, upon the first integration, the double integral σ_z becomes

$$\sigma_z = p_o - \frac{p_o}{2\pi} \int_0^{2\pi} \left[\frac{z^2\sin^2\varphi + z^2(b^2/a^2)\cos^2\varphi}{z^2\sin^2\varphi + z^2(b^2/a^2)\cos^2\varphi + b^2} \right]^{3/2} d\varphi. \tag{95}$$

The integration of equation (95) may be performed graphically. It will be seen that the computation of stresses under the load areas is a very cumbersome process and requires a great deal of work.

The computation of stresses acting under rectangular loads may be made under similar assumptions and also requires a very complex procedure which cannot be enclosed in a general formula. Various methods of rigorous or simplified computation and tables, proposed by Love, Newmark, and others, are of invaluable assistance in this respect.[54, 69, 71, 72]

Concentration Factor

The stress computation based on the theory of elasticity, as applied to loose granular masses, gives approximate results only. This has been known since the time of Föppl (1897), who attributed the difference between the theory and experiment to the inelastic behavior of soil.[68]

In conjunction with this fact, various investigators proposed other solutions to comply better with real conditions. Griffith [73] and Fröhlich [74] worked out semiempirical stress equations in which the variable parameter n could be adapted to the changing soil conditions. In this case, the vertical and radial stress formulas for a concentrated load take the following form:

$$\sigma_z = \frac{nW}{2\pi} \frac{\cos^n \varphi}{R^2} \tag{96}$$

$$\sigma_r = \frac{nW}{2\pi} \frac{\cos^{n-2} \varphi}{R^2}. \tag{97}$$

Equations (96) and (97) are identical with the corresponding Boussinesq equations (91) and (92) if n is assumed to be equal to 3.

The value n is called the concentration factor and is larger than 3 in the case of clays. It equals 6 or 7 for coarse sands.

The integration of $d\sigma_z$ under a circular area may be performed in the same way as in the case when Boussinesq's formula, expressed by equation (90), was assumed. In such a case, Griffith-Fröhlich's equation (96) becomes the base of the integration, and the value σ_z for uniform and conic load distribution, previously expressed by equations (93) and (94), respectively, takes the following form, if it is assumed that $n = 6$:[68]

Uniform:

$$\sigma_z = p_o \left\{ 1 - \frac{1}{\left[1 + \left(\frac{r_o}{z}\right)^2\right]^3} \right\} \tag{98}$$

Conic:

$$\sigma_z = 3p_o\left[1 - \frac{z^4}{4(z^2 + r_o^2)} - \frac{3z^3}{8(z^2 + r_o^2)} - \frac{3z}{8r_o}\tan^{-1}\left(\frac{r_o}{z}\right)\right]. \qquad (99)$$

The computation of pressures under elastic foundations based on formulas including the concentration factor is even more involved than that based on Boussinesq's formula. A method proposed by Krynine [75] simplifies the task considerably.

Pressure Distribution in Snow

If it is assumed that snow behaves as an elastic matter, then all the previously discussed formulas may be applied. However, such an assumption is not valid in many cases. Snow often displays the character of a plastic body and must be subject to a different treatment.[76]

Nevertheless, for the determination of the vertical pressure, some Russian authors propose a formula which in its structure resembles formula (90) and, hence, assumes at least to a certain extent some degree of elasticity. According to Kondratiev and his associates,[77] a pressure along the centerline of a circular load

$$\sigma_z = \frac{W}{\left(\frac{z}{r}\right)^2 + k}, \qquad (100)$$

where r is the radius of the contact area and k is a coefficient, the value of which is close to a unit. All the units expressed by formula (100) are in kilograms and centimeters.

Bucher considers the case of snow elasticity as being related to the time factor which determines whether the behavior of snow mass is plastic, elastic, or plasto-elastic.[78] The duration of a stress is directly responsible for the behavior of snow in either way. Dobrowolski [79] called attention to these facts a long time ago. The related problems will be discussed later.

In general, it may be said that the difficulties in determining snow reaction stem from the fact that snow is not a solid, but a mixture of three phases which depend on the thermodynamic equilibrium of solid, liquid, and gaseous states.[79, 80, 81] A mass which changes its properties when under the action of pressure certainly presents considerable discrepancy with an ideal elastic body. Some investigations indicate that formulas (96), (97), (98), and (99) may offer, at least, a first approximation of the order of magnitude of the pressures involved. The value of the concentration factor n, however, requires more study in order to obtain further information on the validity of the discussed formulas.

Pressure Distribution under Rigid Structures

Experiments performed with stress distribution under rigid plates do not conform with equations based on the action of elastic footings. Figure 63d shows that the rigidity of the plate reduces the peak pressure beneath the plate very considerably.

Since there may be no definition of the degree of rigidity of a plate, a rigorous treatment of the problem is difficult. The problem is further complicated by the formation of the "disturbed zone" which changes the load conditions in which the load does not act finally upon the surface as theoretically assumed, but at a certain depth of the soil mass.[54]

Pressure Distribution in a Real Soil

The aeolotropy and nonhomogeneity of soils, which are all-important factors in stress computation for civil engineering purposes, do not appear to be so critical when problems of vehicle mobility are considered. This fortunate coincidence may be explained by the fact that vehicular structures are rather small and, as Figure 61 indicates, the influence of their weight does not reach deeper soil strata. Since the change in soil structure rarely occurs close to the ground surface, and since this change cannot be significant in most cases, the stress propagation may be assumed as taking place in ideal conditions.

The problem of soil stratification and nonhomogeneity may be important, however, in laboratory tests, where an elastic soil layer rests, for instance, on the rigid bottom of a soil bin. In such a case, the pressure distribution previously discussed for a semi-infinite continuum does not apply. As Biot showed, the real pressure at the bottom, for frictionless materials, increases by from 50 to 70% in comparison with the pressure computed by means of Boussinesq's equations (67). The effect of a rigid base and the main conclusions of Biot's work have been discussed by Terzaghi.[54]

The distribution of the normal pressure acting on a rigid base and exercised by an elastic layer which is loaded at the surface depends on the frictional and adhesive forces which develop between the elastic and rigid strata. Also, the Poisson ratio $\sqrt{}$ of the elastic layer enters into the formulas of σ.

If the properties of a nonhomogeneous soil are such that its modulus of elasticity E_v measured in the vertical direction is smaller than the modulus E_h determined for the horizontal direction, then, assuming that $E_v/E_h = k^2$, the pressure σ_z and σ_x of a line load V may be determined by the formula

$$\sigma_x = -\frac{2V}{\pi}\frac{kx^2 z}{r^2 r_1^2} \tag{101}$$

$$\sigma_z = -\frac{2V}{\pi}\frac{kz^3}{r^2 r_1^2}. \tag{102}$$

In this equation, r is the radius of a given point of soil mass located at depth z (compare Figure 57a) and r_1 is the radius of the same point if it were located at depth $z_1 = z/k$; x is the distance between the load and the point, measured in the horizontal direction. For $k = 1$ and $r = r_1$, equations (101) and (102) become Boussinesq's equations (67).

A practical case of this nature may be illustrated by soil mass composed of alternating layers of sandy silt and clay. The moduli of elasticity E_s and E_c may be determined for each of these materials from tests of undisturbed samples.[82] If it is assumed that the thicknesses of the layers are equal, then E_v and E_h for the whole mass will be

$$\left.\begin{aligned} E_h &= \frac{E_s + E_c}{2} \\ E_v &= \frac{2E_s E_c}{E_s + E_c}. \end{aligned}\right\} \tag{103}$$

The maximum stress computed for such a stratified medium in the case of a line load, in accordance with equations (101) and (102), is considerably smaller than that computed from formulas (69) for an isotropic mass (68).

Condition of Plasticity

The stress pattern of an ideally plastic mass which is in a state of flow cannot be determined by means of the equations previously used for the determination of stresses in an elastic state.

Since the plastic state is characterized by stress conditions in which a permanent deformation may occur without fracture, then it becomes apparent that a criterion must be established which would supersede the no longer valid Hooke's law.

The constant value of the specific resistance of deformation which characterizes the deformation of plastic materials is expressed in Huber's theory of constant energy,[83] which also was independently formulated by von Mises and refined by Hencky.[84]

The determination of the state of stresses which define the condition of a plastic flow is based on the assumption of the existence of a specific level of the energy of distortion which delineates the end of elastic and the beginning of plastic deformation. Since this energy does not change during the process of plastic deformation, the condition of plasticity determined by means of this theory is called the condition of a constant energy of distortion and may be derived as follows.

Take a prism of material with coordinates as shown in Figure 56a, with the *yy* axis extending in the direction perpendicular to the plane of the drawing. The state of the stresses-strain relationship of such a prism is characterized by stresses σ_x, σ_y, σ_z, τ_{xy}, τ_{yz}, τ_{xz}, and strains e_x, e_y, e_z and e_{xy}, e_{yz}, e_{zx}. Following Hooke's law, the above relationship may be expressed by the following equations:

$$\left.\begin{aligned}
e_x &= \frac{1}{E}\left[\sigma_x - \nu(\sigma_y + \sigma_z)\right] \\
e_y &= \frac{1}{E}\left[\sigma_y - \nu(\sigma_z + \sigma_x)\right] \\
e_z &= \frac{1}{E}\left[\sigma_z - \nu(\sigma_x + \sigma_y)\right] \\
e_{xy} &= \frac{\tau_{xy}}{G} = \frac{2(1+\nu)}{E}\tau_{xy} \\
e_{yz} &= \frac{\tau_{yz}}{G} = \frac{2(1+\nu)}{E}\tau_{yz} \\
e_{xz} &= \frac{\tau_{xz}}{G} = \frac{2(1+\nu)}{E}\tau_{xz}\,.
\end{aligned}\right\} \tag{104}$$

Since the energy of an elementary deformation equals 1/2 of the product stress $\times$ strain, the total energy of deformation A per unit volume of the undistorted mass may be expressed by the equation

$$A = \tfrac{1}{2}(\sigma_x e_x + \sigma_y e_y + \sigma_z e_z + \tau_{xy} e_{xy} + \tau_{yz} e_{yz} + \tau_{xz} e_{xz})\,. \tag{105}$$

After substituting the values obtained from equations (104) into equation (105), and simplifying, the following equation will be obtained:

$$A = \frac{1 - 2\surd}{6E} (\sigma_x + \sigma_y + \sigma_z)^2 +$$

$$+ \frac{1 + \surd}{6E} [(\sigma_x - \sigma_y)^2 + (\sigma_y - \sigma_z)^2 + (\sigma_z - \sigma_x)^2 +$$

$$+ 6 (\tau^2_{xy} + \tau^2_{yz} + \tau^2_{zx})] \,. \qquad (106)$$

According to the definition of energy of deformation, the first part of equation (106), $\frac{1-2\surd}{6E} (\sigma_x + \sigma_y + \sigma_z)^2$, expresses the energy stored in the volume deformation, whereas the second member of equation (106) defines the energy stored in the distortion of the form.

A pure deformation of the volume cannot be expected to be critical to uniformly stressed material, even for very large stresses. It is therefore logical to assume that the plastic flow depends entirely on form deformation and thus on the energy of form distortion.[83] Once this distortion starts, the energy remains constant. For a pure compression or tension (i.e., when $\sigma_z = \sigma_y = 0$ and when $\tau_{xy} = \tau_{yz} = \tau_{xz} = 0$), the limit-stress value $\sigma_x = \sigma_{pl}$, where σ_{pl} is the yielding stress and

$$\frac{1 + \surd}{6E} [(\sigma_x - \sigma_y)^2 + (\sigma_y - \sigma_z)^2 + (\sigma_z - \sigma_x)^2 + 6(\tau^2_{xy} + \tau^2_{yz} + \tau^2_{xz})]$$

$$= \frac{2(1 + \surd)}{6E} \sigma^2_{pl} = \text{constant}$$

or

$$\sigma^2_{pl} = \tfrac{1}{2} [(\sigma_x - \sigma_y)^2 + (\sigma_y - \sigma_z)^2 + (\sigma_z - \sigma_x)^2 + 6(\tau^2_{xy} + \tau^2_{yz} + \tau^2_{xz})] \,. \qquad (107)$$

Equation (107) expresses the condition of plasticity as based on the Huber-Mises-Hencky theory. In terms of principal stresses ($\tau = 0$), equation (107) may be written as

$$\sigma^2_{pl} = \tfrac{1}{2} [(\sigma_{\prime} - \sigma_{\prime\prime})^2 + (\sigma_{\prime\prime} - \sigma_{\prime\prime\prime})^2 + (\sigma_{\prime\prime\prime} - \sigma_{\prime})^2] \,. \qquad (108)$$

Condition of Plasticity in a Plane Problem

The two-dimensional problem in regard to the condition of plasticity must be considered under two kinds of assumptions: either (1) the strain e_y equals zero, or (2) the stress σ_y equals zero.[70]

In the first case, equation (104) must equal zero. Since, at the moment of plastic flow, the Poisson ratio $\surd = 0.5$, e_y may be written as follows:

$$e_y = \Lambda \, [\sigma_y - \tfrac{1}{2}(\sigma_x + \sigma_z)] = 0 \,, \tag{109}$$

where Λ is an undetermined value which replaces the inversion of the modulus of elasticity E at the moment of plastic flow. From equation (109), for such a plane problem, it results that

$$\sigma_y = \frac{\sigma_x + \sigma_z}{2} \,. \tag{110}$$

Since, for the plane problem, $\tau_{yz} = \tau_{yx} = 0$, equation (110) substituted into equation (107) and simplified will give

$$(\sigma_x - \sigma_z)^2 + 4\tau^2_{xz} = \frac{4\sigma^2_{pl}}{3} \,.$$

If $\sigma_{pl}/\sqrt{3}$ is denoted by k_o, then the condition of plasticity for the plane problem defined by $e_y = 0$ may be written as follows:

$$(\sigma_x - \sigma_z)^2 + 4\tau^2_{xz} = 4k^2_o \,. \tag{111}$$

The second type of plane plasticity condition, determined by $\sigma_y = 0$, may be defined directly from equation (107) when substituting zero values for σ_y, τ_{yz} and τ_{yx}:

$$\sigma^2_x + \sigma^2_z - \sigma_z \, \sigma_x + 3\tau^2_{xz} = \sigma^2_{pl} \,, \tag{112}$$

where σ_{pl}, as previously mentioned, is the yielding stress in pure tension, or compression.

Stress and Shear Pattern of a Plastic Flow

The fundamental equations previously discussed with reference to a two-dimensional problem give a general solution which yields useful solutions in many cases.[59]

The equations of equilibrium (41) and (42),

$$\frac{\partial \sigma_x}{\partial x} + \frac{\partial \tau_{xz}}{\partial z} = 0$$

$$\frac{\partial \sigma_z}{\partial z} + \frac{\partial \tau_{xz}}{\partial x} = 0 \,,$$

and the condition of plasticity, equation (111),

$$(\sigma_x - \sigma_z)^2 + 4\tau^2_{xz} = 4k^2_o \, ,$$

may be transformed in such a way that only τ_{xz} will be left as the variable. To this end, the first equation of equilibrium has to be differentiated with reference to z and the second with reference to x:

$$\frac{\partial^2 \sigma_x}{\partial x \, \partial z} = - \frac{\partial^2 \tau_{xz}}{\partial z^2}$$

$$\frac{\partial^2 \sigma_z}{\partial z \, \partial x} = - \frac{\partial^2 \tau_{xz}}{\partial x^2} \, .$$

The difference between these two partial differentials gives

$$\frac{\partial^2 (\sigma_x - \sigma_z)}{\partial x \, \partial z} = \frac{\partial^2 \tau_{xz}}{\partial x^2} - \frac{\partial^2 \tau_{xz}}{\partial z^2} \, .$$

From the equations of plasticity (111), it follows that

$$\sigma_x - \sigma_z = \pm 2\sqrt{k^2_o - \tau^2_{xz}} \, .$$

When the last two equations are combined, it will be seen that

$$\pm 2 \frac{\partial \sqrt{k^2_o - \tau^2_{xz}}}{\partial x \, \partial z} = \frac{\partial^2 \tau_{xz}}{\partial x^2} - \frac{\partial^2 \tau_{xz}}{\partial z^2} \, . \qquad (113)$$

The solution of equation (113), in conjunction with equations (41) and (42), leads to the determination of stresses.

The shear pattern which is implicit in the equations of equilibrium (41) and (42) and in the condition of plasticity (111) may be determined as follows:

According to equations (84), (85), and (86),

$$\sigma_x = \tfrac{1}{2}(\sigma_{,} + \sigma_{,,,}) - \tfrac{1}{2}(\sigma_{,} - \sigma_{,,,}) \cos 2\alpha$$

$$\sigma_z = \tfrac{1}{2}(\sigma_{,} + \sigma_{,,,}) + \tfrac{1}{2}(\sigma_{,} - \sigma_{,,,}) \cos 2\alpha$$

$$\tau_{xz} = \tfrac{1}{2}(\sigma_{,} - \sigma_{,,,}) \sin 2\alpha \, .$$

The angle α is the angle which the major principal stress forms with σ_z (Figures 58 and 64a). From previous considerations [equation (87)], it is found that the maximum shear value $(\tau_{xz})_{\max} = \frac{1}{2}(\sigma_{,} - \sigma_{,,,})$, which

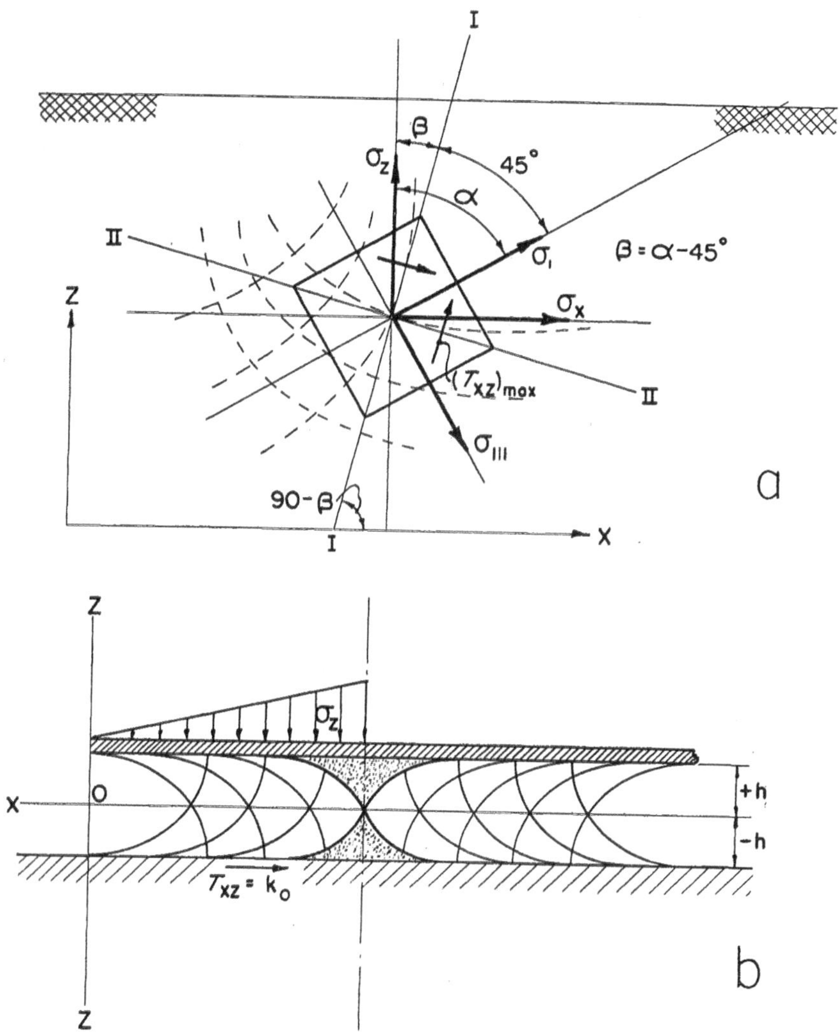

Fig. 64

takes place for $2\alpha = 90°$, or for $\alpha = 45°$. Accordingly, the planes of maximum shear are sloped at angles of 45° to the principal stresses and a tangent to such a plane forms an angle β with the *zz* axis. Figure 64a shows that $\beta = \alpha - 45$. From Mohr's circle (Figure 58b), it follows that in a general case (σ_x ($\lesseqgtr$) σ_z):

$$(\tau_{xz})^2_{max} = \tfrac{1}{4}(\sigma_x - \sigma_z)^2 + \tau^2_{xz} .$$

Since, as previously mentioned, $(\tau_{xz})_{max} = \frac{1}{2}(\sigma_{,} - \sigma_{,,,})$, then

$$\tfrac{1}{2}(\sigma_{,} - \sigma_{,,,}) = \tfrac{1}{2}\sqrt{(\sigma_x - \sigma_z)^2 + 4\tau^2_{xz}} .$$

But from the condition of plasticity, equation (111), the expression

$$\sqrt{(\sigma_x - \sigma_z)^2 + 4\tau_{xz}} = 2k_o ,$$

and therefore

$$\tfrac{1}{2}(\sigma_{,} - \sigma_{,,,}) = k_o . \tag{114}$$

By substituting the values of k_o and $2\alpha = 2\beta + 90$ into equations (84), (85), and (86), the following equations will be obtained:

$$\left.\begin{aligned} \sigma_x &= \tfrac{1}{2}(\sigma_{,} + \sigma_{,,,}) - k_o \sin 2\beta \\ \sigma_z &= \tfrac{1}{2}(\sigma_{,} + \sigma_{,,,}) + k_o \sin 2\beta \\ \tau_{xz} &= - k_o \cos 2\beta . \end{aligned}\right\} \tag{115}$$

The slip lines of the considered mass which form the shear pattern of the flow are tangent to the directions of maximum shear, I-I and II-II, and form two families of orthogonal curves. Their equations are

$$\left.\begin{aligned} \frac{dx}{dz} &= \tan \beta \\ \frac{dx}{dz} &= \tan (90 + \beta) = - \cot \beta . \end{aligned}\right\} \tag{116}$$

The value of the angle β may be determined from the following equation, which may be directly derived from Mohr's circle:

$$\tan 2\alpha = \frac{2\tau_{xz}}{\sigma_z - \sigma_x} .$$

But

$$\tan 2\alpha = \tan (2\beta + 90) = - \frac{1}{\tan 2\beta} ,$$

hence

$$- \tan 2\beta = \frac{\sigma_z - \sigma_x}{2\tau_{xz}} . \tag{117}$$

Among a few cases quoted by Nadai,[59] the following example illustrates the discussed procedure and offers some analogy in the treatment of the flow of a plastic mud mass or snow enclosed, for example, between a hard ground and the rigid flat contact area of a vehicle or of a compacting device. Equation (113) presents an indefinite number of solutions, among which is the one based on the assumption that

$$\tau_{xz} = f_o(z) ,$$

which is of particular interest. In this case, $\partial^2\tau_{xz}/\partial z^2 = 0$ and $\tau_{xz} = cz$, where c is a constant. It will be seen then that this solution satisfies equation (113). According to the condition of plasticity [equation (114)], $(\tau_{xz})_{max}$ cannot be larger than k_o. It may be assumed that there are two parallel lines $z = \pm h$ along which $(\tau_{xz})_{max}$ reaches the k_o value. By taking $c = -k_o/h$ and $\tau_{xz} = -k_o z/h$, it will be seen that $(\tau_{xz})_{max}$ varies between $+k_o$ and $-k_o$ when h varies between $-h$ and $+h$. Accordingly, the two parallel lines having the equations $z = \pm h$ delineate the zone in which equation (113) is practically valid for a plastic flow that takes place under the assumption of a shearing stress $(\tau_{xz}) = f_o(z)$. Beyond these lines, there is no solution of equation (113) that has any physical meaning.

The physical meaning of the discussed lines may be seen in two parallel rigid surfaces which, for example, squeeze the plastic mass. Such surfaces may be assumed as a rigid plate on the top and a solid subgrade on the bottom (Figure 64b).

In the case when the mass in question is compressed, stresses may be calculated from the equations of equilibrium (41) and (42), assuming that $\tau_{xz} = -k_o z/h$:

$$\frac{\partial\sigma_x}{\partial x} = -\frac{\partial\tau_{xz}}{\partial z} = -\frac{\partial}{\partial z}\left(-\frac{k_o z}{h}\right) = \frac{k_o}{h}$$

$$\frac{\partial\sigma_z}{\partial z} = -\frac{\partial\tau_{xz}}{\partial x} = -\frac{\partial}{\partial x}\left(-\frac{k_o z}{h}\right) = 0 .$$

Integration of the above equations gives

$$\left.\begin{aligned} \sigma_x &= \frac{k_o x}{h} + f_1(z) \\ \sigma_z &= f_2(x) . \end{aligned}\right\} \qquad (118)$$

The arbitrary functions $f_1(z)$ and $f_2(x)$ may be determined in such a way that they satisfy the equation of plasticity (111); then, assuming, as previously, $\tau_{xz} = -k_o z/h$,

$$\sigma_x - \sigma_z = +2\sqrt{k_o^2 - \tau^2_{xz}} = \frac{k_o x}{h} + f_1(z) - f_2(x) = +2k_o\sqrt{1 - \left(\frac{z}{h}\right)^2},$$

where the (+) sign corresponds to the case of compression. When this equation is rearranged,

$$f_1(z) - f_2(x) = 2k_o\sqrt{1 - \left(\frac{z}{h}\right)^2} - \frac{k_o}{h}x .$$

Hence,

$$f_1(z) = 2k_o\sqrt{1 - \left(\frac{z}{h}\right)^2} + \text{constant}$$

$$f_2(x) = \frac{k_o}{h}x + \text{constant} .$$

Combining the above equations with equation (118) gives

$$\left.\begin{aligned} \sigma_x &= \frac{k_o}{h}x + 2k_o\sqrt{1 - \left(\frac{z}{h}\right)^2} + \text{constant} \\ \sigma_z &= \frac{k_o}{h}x + \text{constant} \\ \text{and} \quad \tau_{xz} &= -\frac{k_o}{h}z . \end{aligned}\right\} \qquad (119)$$

The shear pattern which determines the flow of the mass under the action of stresses, equation (119), may be determined by the previously discussed way. From equations (115),

$$\tau_{xz} = -k_o \cos 2\beta .$$

By substituting the values obtained from equations (119) into the above equation, it will be found that

$$\frac{z}{h} = \cos 2\beta .$$

Differentiation of the latter gives

$$\frac{dz}{dx} = -2h \sin 2\beta \frac{\partial\beta}{\partial x} . \qquad (120)$$

Combining equations (120) and (116) yields, for the first family of curves,

$$\frac{1}{\tan \beta} = -2h \sin 2\beta \frac{\partial\beta}{\partial x} .$$

Since $\sin 2\beta = 2 \sin \beta \cos \beta$,

$$\partial x = -4h \sin^2 \beta \, \partial\beta ,$$

and finally,

$$\left.\begin{aligned} x &= -h (2\beta - \sin 2\beta) + \text{constant} \\ z &= h \cos 2\beta . \end{aligned}\right\} \qquad (121)$$

The second family of slip lines may be obtained by combining equation (120) with the second formula of equations (116):

$$\frac{1}{\cot \beta} = 2h \sin 2\beta \frac{\partial\beta}{\partial x} ,$$

or, after a transformation similar to that previously done,

$$\partial x = 4h \cos^2 \beta \, \partial\beta$$

and

$$\left.\begin{aligned} x &= h (2\beta + \sin 2\beta) + \text{constant} \\ z &= h \cos^2 \beta . \end{aligned}\right\} \qquad (122)$$

Equations (121) and (122) represent two families of cycloids based on a circle, the diameter of which is $2h$. The shear pattern of the flow near the edges of the plate does not correspond, however, to that expressed by these equations. In the middle part of the mass, shown by a dotted area, the shearing stresses, which balance the friction between the mass and plates, also do not reach the maximum, and the condition of plasticity will not be satisfied.

On the basis of the stress pattern discussed in this section, Jürgenson proposed a squeezing test for determining the shearing stress of plastic masses. The test consists of compressing the material under investigation and measuring the loads at the moment the mass flows.[85]

Shearing Strength of Soil

From a transportation point of view, the soil, or ground, upon which a vehicle moves may refer to any type of material encountered on the upper layer of the earth's crust. It may be stone, ice, or organic matter, which in this case would be classed as a solid having specific mechanical properties. The investigation of the strength of such materials would be similar to that of solids. (The general methods of the theory of elasticity dealing with this subject were discussed previously.) Plastic masses such as saturated clays or certain types of snow also may be encountered. In many cases, these masses may be compared to an ideally plastic medium, a few problems of which were referred to in previous chapters. In the case of fluid muds or rather muddy waters, where hydrostatic forces and viscosity are preponderant, hydromechanics would be the base of study, as found in naval architecture [46] and the construction of amphibian vehicles.[40] Thus, in general, the problems involved in the study of vehicle mobility would embrace not only those strictly belonging to soil mechanics, but also those of other applied mechanics. In this approach, loose granular masses will call for the most consideration, since they cover most of the trafficable earth surface and are the main concern in modern transport requirements.

Granular masses may exhibit cohesive and frictional properties. Cohesion is the bond which cements soil elements irrespective of the pressure exercised by one particle upon the other. Particles of frictional masses can, in principle, then be held together only when a pressure is exercised between them. Thus, the shearing strength of plastic snow, or clay, for example, does not depend theoretically upon the load, whereas that of a dry sand increases with the increase of the load. The above conclusion was confirmed experimentally and is expressed by Coulomb's law, which may be illustrated by the diagram shown in Figure 65a. Two cohesive masses A_c and B_c moving along the plane of shear *m-m* develop a shearing force S_c which does not depend upon the vertical load V. In the case of frictional masses A_f and B_f, the shearing force S_f is a function of V.

Besides purely plastic and frictional soils, there is a great variety of those which combine both properties. In such a case, the shearing stress τ depends on the coefficient of cohesion c and angle of friction ϕ, as shown in Figure 65b. Here, Coulomb's equation reads:

$$\tau = c + \sigma \tan \phi \,. \tag{123}$$

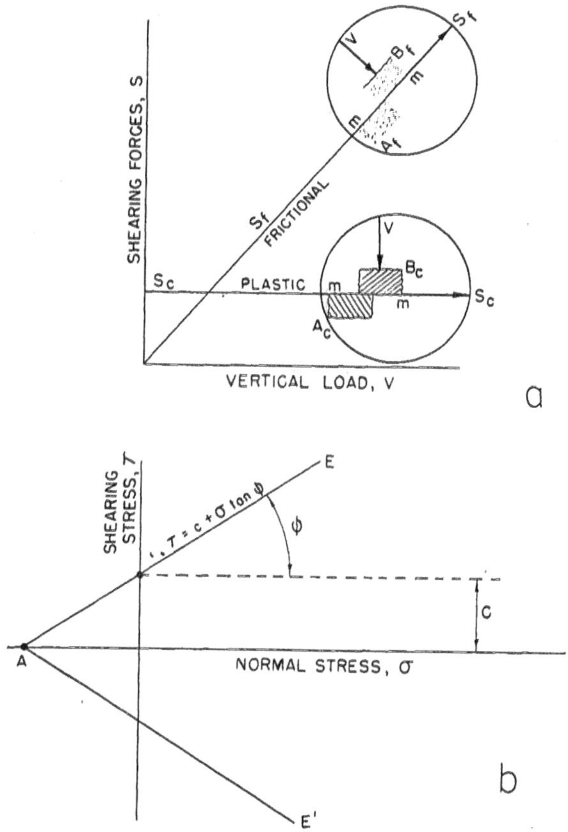

Fig. 65

For dry sands ($c = 0$), this equation will take the form

$$\tau = \sigma \tan \phi \, , \tag{124}$$

and for plastic clays or snow ($\phi = 0$),

$$\tau = c \, . \tag{125}$$

Saturated soils subjected to impact loads applied by running vehicles may carry a considerable part of the stress through the hydrostatic pressure of water enclosed by the pores. In such a case, the values of c and ϕ both depend on the rate of load as well as on the permeability of the soil and on the size of the mass under stress.[54, 86] As experience shows, it may be assumed that the angle of shearing resistance for clays may take any value from 0° up to 20°. For loose dry sand, any value up to 35° or more is possible, depending on the nature of the specimen and the rate of shear.

If it is agreed that the failure of a cross-country vehicle due to the shear of soil by a track, or wheel, takes place at an average speed of from 1 to 3 mph, then the rate of shear for testing purposes may be adjusted accordingly and thus eliminate, to a certain extent, the uncertainty related to the shear rate. In this case, ϕ and c would have specific values for given conditions. As the experience gained with vehicles has so far indicated, the values of ϕ and c determined by means of a quick shear in various soils may be considered to be quite reliable in reference to vehicle performance. Since a shear speed of from 1 to 3 mph, which corresponds to 1.5 to 4.5 ft/sec, presents a high rate of load application, the load may be almost considered as an impact. A detailed analysis of shear tests, in which the meaning and implications of a "quick-shear" test are discussed, may be found in Reference 87.

The so-called triaxial tests, in which a specimen under shear is subjected to the stresses acting along all three axes of the coordinates x, y, and z, are often used for a general study of the problem of failure of materials. It is thought, however, that such a test, although performed in more ideally controlled conditions than a plane shear, may be superseded by an ordinary shear-box test as far as the determination of c and φ for vehicle purposes is concerned. More details regarding the shearing strength of soils and a description of the standard soil-testing procedure are given in an excellent symposium by the American Society for Testing Materials in References 86 and 89.

When determining the friction ϕ and cohesion c of soils, it should always be remembered that changes in the water content produce changes in equation (123). The value $\sigma \tan \phi$ consists of two parts affected by two different causes. The first part is the mechanical friction of rubbing surfaces under the normal stress σ, and the second part is an increase or decrease of cohesion due to the change of water content when the pressure is increased from 0 to σ. This leads to the necessity of planning shear tests which would make the experiment reproducible in full scale. One of the factors involved is the rate of loading, as mentioned before, and the other is the permeability and size of the specimen.

When measuring ϕ and c in the field with the purpose of assessing the shearing strength of soil, the lack of homogeneity of the ground should also be considered. The use of very small samples may be misleading if the soil is in stratified and nonhomogeneous conditions.

To sum up, it may be concluded that c and ϕ present empirical coefficients of a straight-line equation (123) rather than soil constants of any strict physical meaning. As long as these coefficients aid in a better understanding of the relationship between soil and a vehicle, they serve their purpose.

Plastic Equilibrium in a Semi-Infinite Soil Mass

The condition of plastic flow expressed by equation (111) does not give very satisfactory solutions for loose granular masses in which frictional properties follow Coulomb's law: $\tau = \sigma \tan \phi$. In such cases, Mohr's theory of plastic flow has been generally accepted.

Among the problems that may be solved by applying this theory is the problem of equilibrium of soil mass in its natural conditions, i.e., in a semi-infinite continuum limited by the ground surface.

Take Coulomb's equation of shear failure in a general form, as postulated by Mohr. This equation determines the condition of failure by plastic flow at $\tau_{max} = c + \sigma \tan \phi$. As mentioned before, it may be represented by two straight lines $\overline{AE}$ and $\overline{AE}'$ (Figures 65b and 66). Assume that the directions of the principal stresses $\sigma_{\prime}$ and $\sigma_{\prime\prime\prime}$ acting on an arbitrary point P of soil mass are known (Figure 66) and that the value of one of these stresses, for example $\sigma_{\prime}$, also is determined. If Mohr's circle is traced for the state of stresses at point P, in which slip is incipient due to the shearing stress reaching a value expressed by Coulomb's equation, then no point of the corresponding circle may be located beyond the lines $\pm \tau = c + \sigma \tan \phi$ because no shear stress can be larger than that enclosed by the lines $\overline{AE}$ and $\overline{AE}'$. The only circle which will satisfy the condition of equilibrium implicit in Mohr's theory is one which is tangent to the line $\tau = c + \sigma \tan \phi$. Accordingly, when $\sigma_{\prime}$ is known, a circle tangent to $\overline{AE}$ and $\overline{AE}'$ may be traced and $\sigma_{\prime\prime\prime}$ may be determined on the basis of the previously discussed properties of Mohr's circle (Figure 66). In this case, τ_{max} will be reached for $2\alpha = 90 + \phi$ or for $\alpha = 45 + \phi/2$. Since α is the angle enclosed between the principal stress, $\sigma_{\prime}$, and the normal stress, σ, which produces shear in the given plane, it follows that the slip line caused by τ_{max} is sloped to the major principal stress at an angle of $45 - \phi/2$. A similar angle related to the minor principal stress, $\sigma_{\prime\prime\prime}$, equals $45 + \phi/2$.

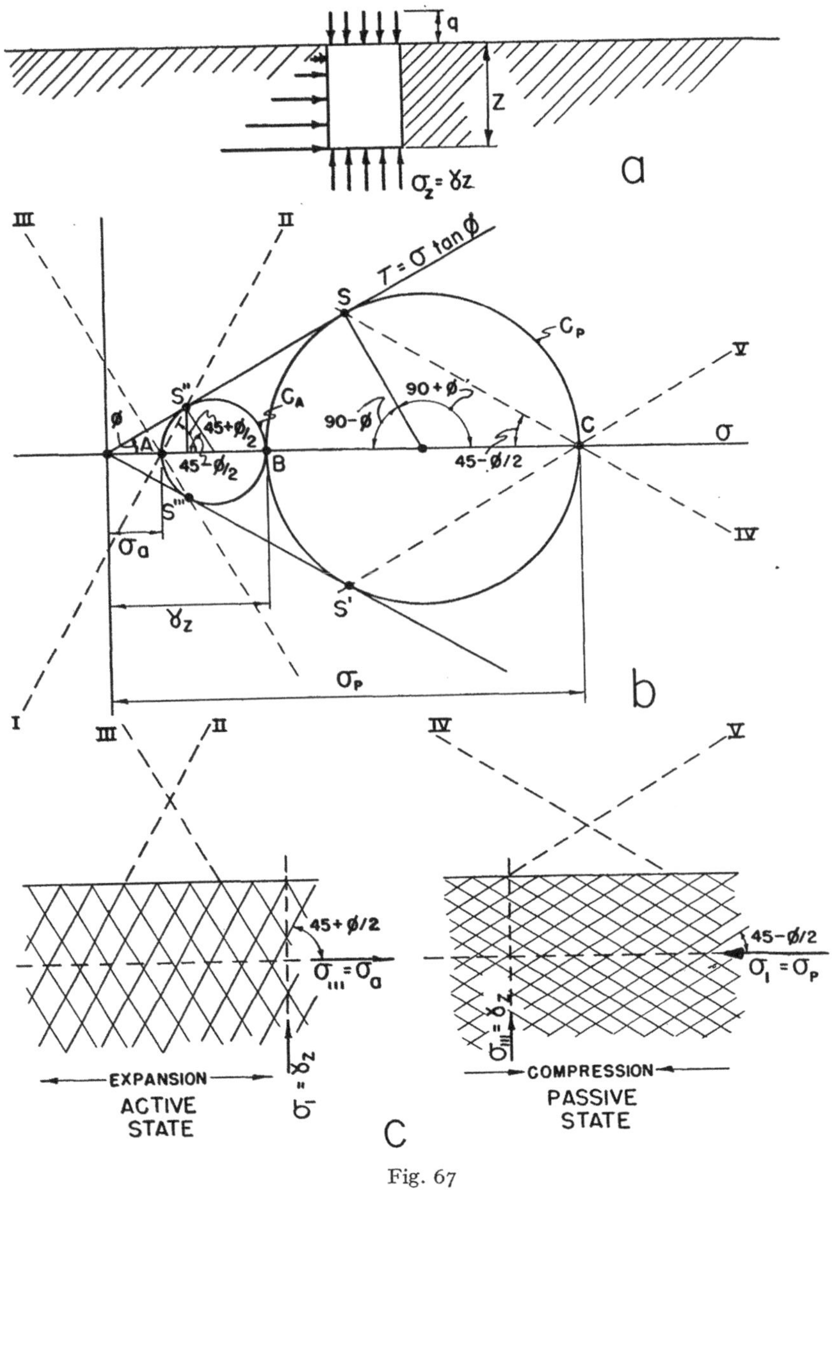
q
Z
$\sigma_z = \gamma z$
a
III
II
$\tau = \sigma \tan \phi$
S
C_P
V
S''
C_A
90 + ø
ø
A
45+ø/2
90−ø
C
σ
B
45−ø/2
45−ø/2
S'''
IV
σ_a
S'
γz
σ_P
b
I
III
II
IV
V
45+ø/2
$\sigma_{III} = \sigma_a$
$\sigma_I = \gamma z$
EXPANSION
ACTIVE STATE
45−ø/2
$\sigma_I = \sigma_P$
$\sigma_{III} = \gamma z$
COMPRESSION
PASSIVE STATE
c

Fig. 67

shearing strength of soil, the lack of homogeneity of the ground should also be considered. The use of very small samples may be misleading if the soil is in stratified and nonhomogeneous conditions.

To sum up, it may be concluded that c and ϕ present empirical coefficients of a straight-line equation (123) rather than soil constants of any strict physical meaning. As long as these coefficients aid in a better understanding of the relationship between soil and a vehicle, they serve their purpose.

Plastic Equilibrium in a Semi-Infinite Soil Mass

The condition of plastic flow expressed by equation (111) does not give very satisfactory solutions for loose granular masses in which frictional properties follow Coulomb's law: $\tau = \sigma \tan \phi$. In such cases, Mohr's theory of plastic flow has been generally accepted.

Among the problems that may be solved by applying this theory is the problem of equilibrium of soil mass in its natural conditions, i.e., in a semi-infinite continuum limited by the ground surface.

Take Coulomb's equation of shear failure in a general form, as postulated by Mohr. This equation determines the condition of failure by plastic flow at $\tau_{max} = c + \sigma \tan \phi$. As mentioned before, it may be represented by two straight lines $\overline{AE}$ and $\overline{AE'}$ (Figures 65b and 66). Assume that the directions of the principal stresses $\sigma_{\prime}$ and $\sigma_{\prime\prime\prime}$ acting on an arbitrary point P of soil mass are known (Figure 66) and that the value of one of these stresses, for example $\sigma_{\prime}$, also is determined. If Mohr's circle is traced for the state of stresses at point P, in which slip is incipient due to the shearing stress reaching a value expressed by Coulomb's equation, then no point of the corresponding circle may be located beyond the lines $\pm \tau = c + \sigma \tan \phi$ because no shear stress can be larger than that enclosed by the lines $\overline{AE}$ and $\overline{AE'}$. The only circle which will satisfy the condition of equilibrium implicit in Mohr's theory is one which is tangent to the line $\tau = c + \sigma \tan \phi$. Accordingly, when $\sigma_{\prime}$ is known, a circle tangent to $\overline{AE}$ and $\overline{AE'}$ may be traced and $\sigma_{\prime\prime\prime}$ may be determined on the basis of the previously discussed properties of Mohr's circle (Figure 66). In this case, τ_{max} will be reached for $2\alpha = 90 + \phi$ or for $\alpha = 45 + \phi/2$. Since α is the angle enclosed between the principal stress, $\sigma_{\prime}$, and the normal stress, σ, which produces shear in the given plane, it follows that the slip line caused by τ_{max} is sloped to the major principal stress at an angle of $45 - \phi/2$. A similar angle related to the minor principal stress, $\sigma_{\prime\prime\prime}$, equals $45 + \phi/2$.

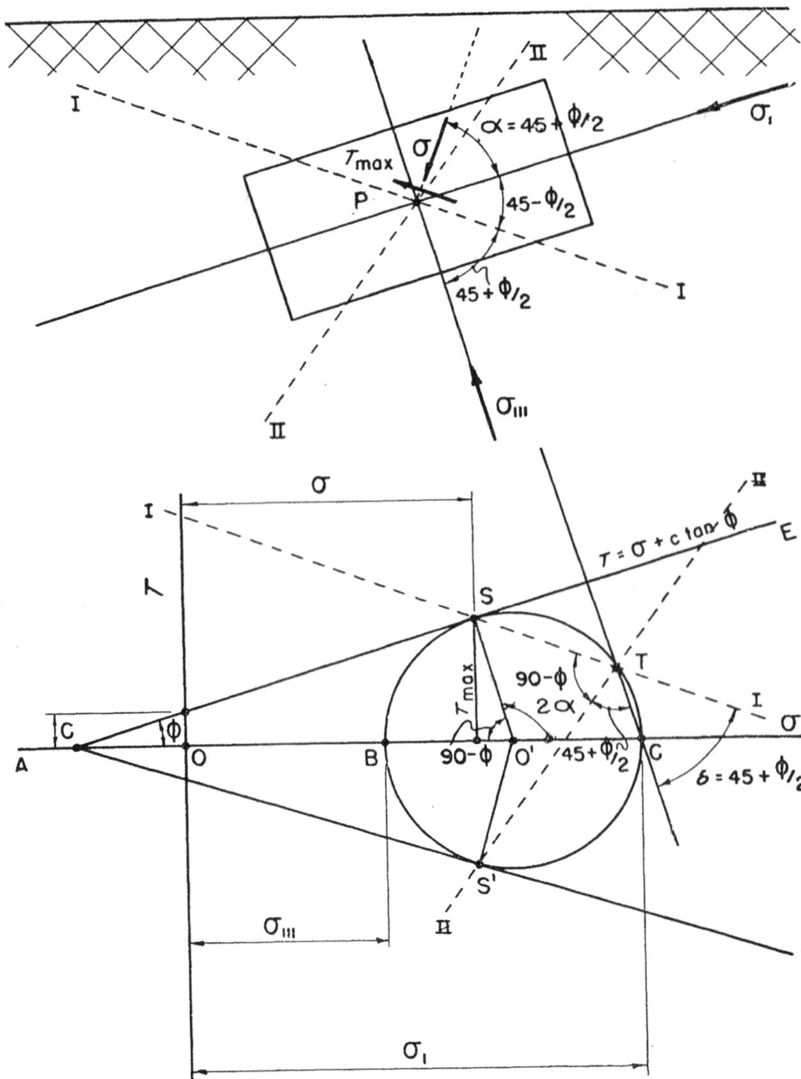

Fig. 66

The directions of the slip planes may be found graphically if a line parallel to the direction of the minor principal stress, σ_{III}, is traced through point C. This line will give point T at the intersection with

Mohr's circle. Point T is called the pole of the diagram. Line $\overline{TS}$ will determine the direction of the first slip plane I-I. The second direction, II-II, will be determined by line $\overline{TS'}$. This method may be checked by the values of the angles involved and denoted in Figure 66. It will be seen that $\delta = 45 + \phi/2$, which means that the direction TS is the direction of one of the slip planes since $\overline{CT}$ is the direction of the minor principal stress, $\sigma_{,,,}$. Similarly, the line TS' determines the direction of the other slip line since it is sloped to the bisector of the angle STS' (direction of $\sigma_{,}$) at an angle $45 - \phi/2$.

From the triangle ASO',

$$\frac{\dfrac{\sigma_{,} - \sigma_{,,,}}{2}}{\dfrac{c}{\tan\phi} + \dfrac{\sigma_{,} + \sigma_{,,,}}{2}} = \sin\phi$$

or

$$\sigma_{,} = 2c\frac{\cos\phi}{1 - \sin\phi} + \sigma_{,,,}\frac{1 + \sin\phi}{1 - \sin\phi}. \qquad (126)$$

Since $\cos\phi/(1 - \sin\phi) = \tan(45 + \phi/2)$, and $(1 + \sin\phi)/(1 - \sin\phi) = \tan^2(45 + \phi/2)$, equation (126) may be written in the following form:

$$\sigma_{,} = 2c\tan(45 + \phi/2) + \sigma_{,,,}\tan^2(45 + \phi/2). \qquad (127)$$

The value $\tan^2(45 + \phi/2)$ will be denoted by N_ϕ and called the flow value. In the case of cohesionless mass, equation (126) gives the following relation of principal stresses ($c = 0$):

$$\frac{\sigma_{,}}{\sigma_{,,,}} = \frac{1 + \sin\phi}{1 - \sin\phi} = \tan^2(45 + \phi/2) = N_\phi. \qquad (128)$$

Equation (126) or (127) expresses the condition of plasticity based on Mohr's theory of plastic flow applied to granular masses.

Consider a plastic equilibrium of a semi-infinite cohesionless mass which satisfies equation (126) or (127). When an imaginary prism located in such a mass is subjected to lateral forces (Figure 67a), it resists those forces by the intermediary of its own weight. If the lateral forces are too large, for example those due to compression, failure will occur; in such a case, it is called the passive failure because the latter is passively

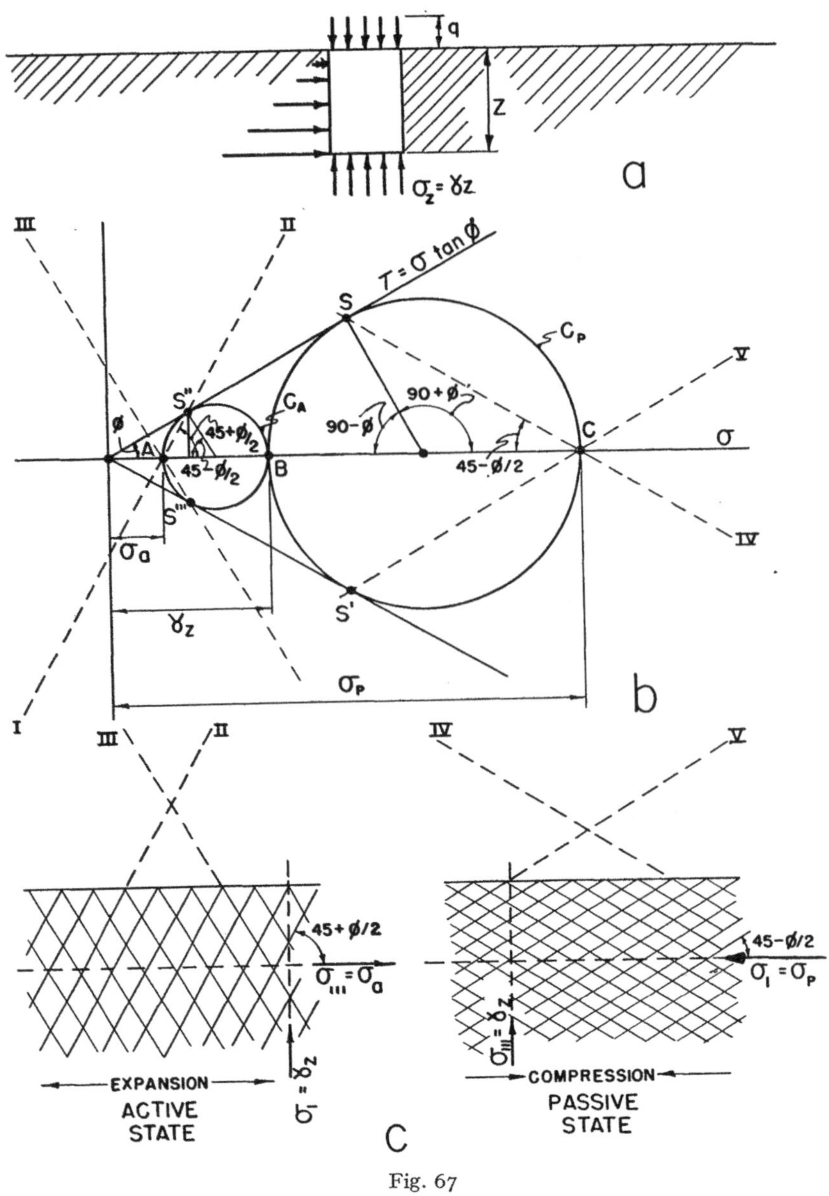
q
Z
$\sigma_z = \gamma z$
a
III
II
$\tau = \sigma \tan \phi$
S
C_P
V
S''
C_A
90+ø
90−ø
45+ø/2
ø
A
B
C
σ
45−ø/2
IV
S'''
σ_a
S'
γz
σ_P
b
I
EXPANSION
ACTIVE STATE
$\sigma_{III} = \sigma_a$
$\sigma_I = \gamma z$
COMPRESSION
PASSIVE STATE
$\sigma_I = \sigma_P$
$\sigma_{III} = \gamma z$
c

Fig. 67

resisted by the soil weight. If the lateral forces that hold the mass of the prism in equilibrium are removed, then the weight of the prism will assist the plastic flow; this case is called the active failure which relates to soil expansion.[54]

Both types of soil failure may be analyzed more closely by means of Mohr's circle. A prism of soil having depth z and width equal to a unit is in the state of plastic equilibrium. Since there can be no shearing stresses along the walls of this prism, the vertical pressure $\sigma = \gamma z$ acting on the bottom and the lateral pressures are the principal stresses. For a cohesionless mass, the condition of failure is given by equation (124), $\tau = \sigma \tan \phi$, and is represented by two lines passing through the beginning of τ and σ coordinates (Figure 67b). Mohr's circle must pass through point B determined by one of the principal stresses $\sigma = \gamma z$. Two possible solutions may be contemplated according to whether the assumption γz is the major or minor principal stress. If γz is the major principal stress, then the lateral forces are due to the expansion or active failure of the soil mass. If γz is the minor principal stress, then the lateral stresses become automatically the major stresses and are to be related to the compression or passive earth pressure.

In the first case, circle C_A will be obtained; in the second, circle C_P. Now, the lines of slip may be determined in the same way as before. In the case of circle C_A, a line corresponding to the line $\overline{CT}$ (shown in Figure 66) and parallel to the minor principal stress, which, in the present case, is horizontal, will intersect the circle at point A, which thus becomes the pole of the diagram. Accordingly, the lines $\overline{AS''}$ and $\overline{AS'''}$ will determine the lines of slip. In the case of circle C_P, the minor principal stress, γz, is vertical; hence, the pole is located in point C and the lines $\overline{CS}$ and $\overline{CS'}$ determine the direction of the slip planes.

Thus the shear pattern for the expansion of soil mass is determined by the smaller circle, in which σ_a is called the active earth pressure, identical with the minor principal stress. In compression, the stress pattern is defined by the larger circle, in which the major principal stress σ_p is called the passive earth pressure. In this case, γz is the minor principal stress.

From the foregoing, it follows that the stress pattern in both cases is formed by a rectangular grid of vertical and horizontal lines, whereas the shear pattern is composed of parallel lines sloped to the horizon at $45 + \phi/2$ in the case of the active state and at $45 - \phi/2$ in the case of the passive state (Figure 67c).

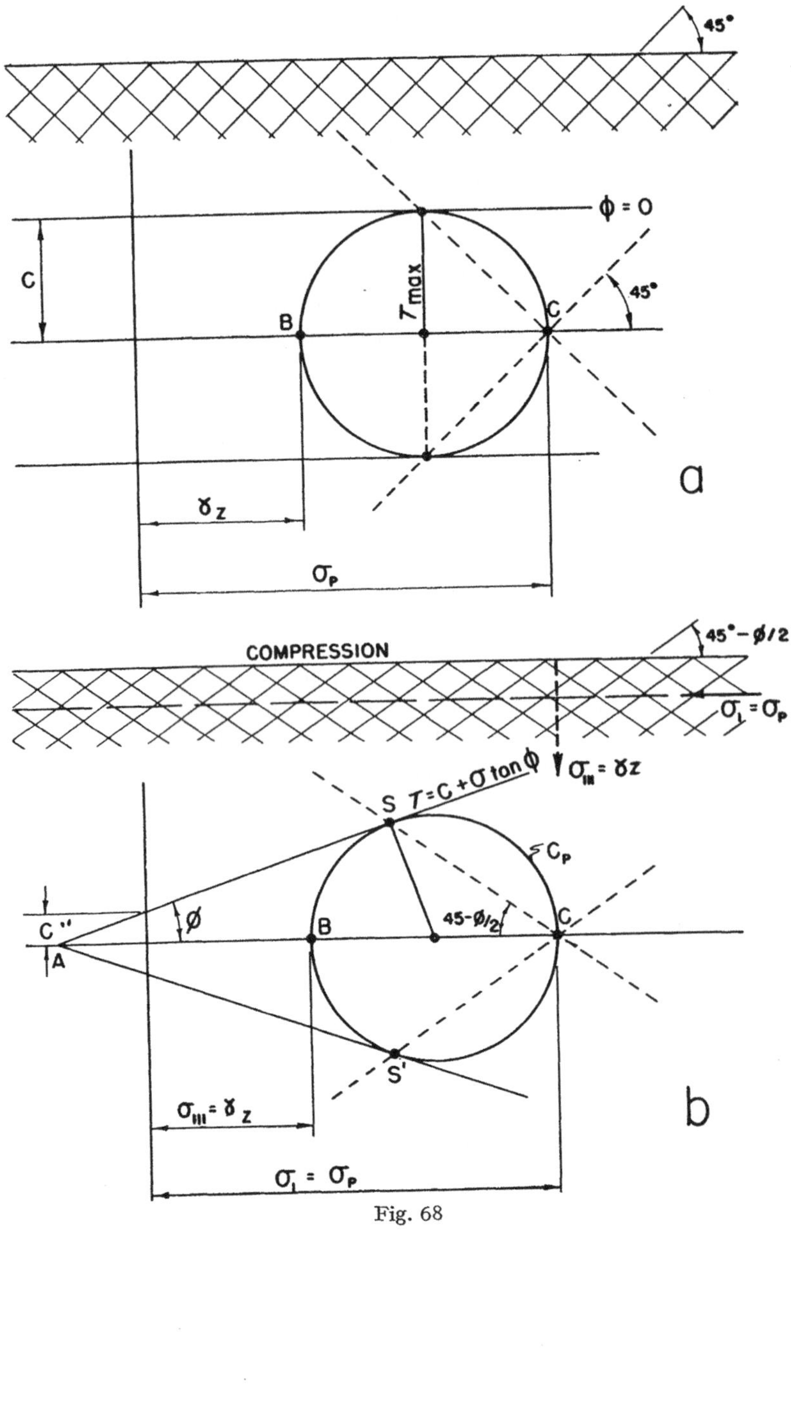
45°
ϕ = 0
C
τmax
B
C
45°
a
γz
σp
45° − ϕ/2
COMPRESSION
σI = σp
τ = C + σ tan ϕ
σIII = γz
S
Cp
ϕ
45-ϕ/2
C''
A
B
C
S'
b
σIII = γz
σI = σp

Fig. 68

For an ideally plastic mass ($\phi = 0$), Mohr's circle will be tangent to two parallel lines (Figure 68a) and $\tau = c$. The shear pattern will be composed of lines sloped to the horizon at 45°, as it was previously determined for an ideally elastic mass (see Figure 58b).

From Mohr's diagram shown in Figure 67b, it results that the active earth pressure σ_a acting upon a vertical section of purely frictional soil mass at depth z is

$$\sigma_a = \gamma z \tan^2 (45 - \phi/2)$$

or, since $\tan^2 (45 - \phi/2) = 1/\tan^2(45 + \phi/2) = 1/N_\phi$,

$$\sigma_a = \frac{\gamma z}{N_\phi} \tag{129}$$

and

$$\frac{\sigma_a}{\gamma z} = \frac{1}{N_\phi} = \text{constant} .$$

This equation indicates that the ratio of horizontal and vertical pressure of a semi-infinite soil mass is independent of depth.

The passive earth pressure prevailing at the moment of failure due to the compression may be determined from Mohr's circle in a similar way (Figure 67b):

$$\sigma_p = \gamma z \tan^2 (45 + \phi/2) = \gamma z N_\phi . \tag{130}$$

It may then be concluded that σ_a and σ_p increase like a hydrostatic pressure (Figure 67a).

The described states of the stresses preceding the plastic flow caused by compression, or the expansion of soil in the direction parallel to the ground surface, are called passive or active Rankine states, respectively.

If the soil surface is loaded with an ideally flexible load, or so-called surcharge q (Figure 67a), then the pressure σ_z at depth z is

$$\sigma_z = q + \gamma z$$

or

$$\sigma_z = \gamma \left(\frac{q}{\gamma} + z\right).$$

Accordingly, since $\sigma_a = \sigma_z / N_\phi$ and $\sigma_p = \sigma_z N_\phi$,

$$\sigma_a = \frac{\gamma}{N_\phi}\left(\frac{q}{\gamma} + z\right) \tag{131}$$

$$\sigma_p = \gamma N_\phi \left(\frac{q}{\gamma} + z\right). \tag{132}$$

The above equations refer to a cohesionless mass. If both cohesion and friction are assumed, then the lines tangent to Mohr's circles will correspond to the equation $\tau = c + \sigma \tan \phi$, which is represented in Figure 68b.

In the case of compression as considered in Figure 68, σ_p is the major principal stress, whereas γz is the minor one. Thus, the previously determined relationship [equation (127)] may be used by substituting σ_p for $\sigma_{\prime}$ and γz for $\sigma_{\prime\prime}$, and

$$\sigma_p = 2c \tan(45 + \phi/2) + \gamma z \tan^2(45 + \phi/2),$$

or, according to the previous denotations,

$$\sigma_p = 2c\sqrt{N_\phi} + \gamma z N_\phi . \tag{133}$$

In view of the fact that soil compression is the major problem, as far as vehicles are concerned, the active ground pressure will not be discussed further. More details in this respect will be found, for example, in the work by Terzaghi.[54]

When summarizing, it may be said that, in accordance with equations (132) and (133), the passive earth pressure σ_p acting at the very moment of soil failure which takes place at surcharge q, friction ϕ, cohesion c, and specific gravity γ may be expressed as follows:

$$\sigma_p = q N_\phi + 2c\sqrt{N_\phi} + \gamma z N_\phi . \tag{134}$$

Earth Pressure on Retaining Walls

Assume that a blade is moved against the ground in such a way that the earth is gradually compressed so that the stress state preceding the plastic flow is reached evenly throughout the relevant soil mass. The experiment shows that the Rankine state of stresses ABC, located close to the soil surface, is reached at the moment of soil failure, defined by an upward movement of the soil mass (Figure 69). The ADB zone is characterized by curved and radial slip surfaces and is called the radial-shear zone.[54] The curvature of the DB surfaces is due to the friction between the blade and the earth. The resultant force P_p' of the passive earth pressure is sloped at the angle of friction μ and, because of the hydrostatic distribution of σ_p, is located at a distance of $2h/3$, where h is the height of the blade.

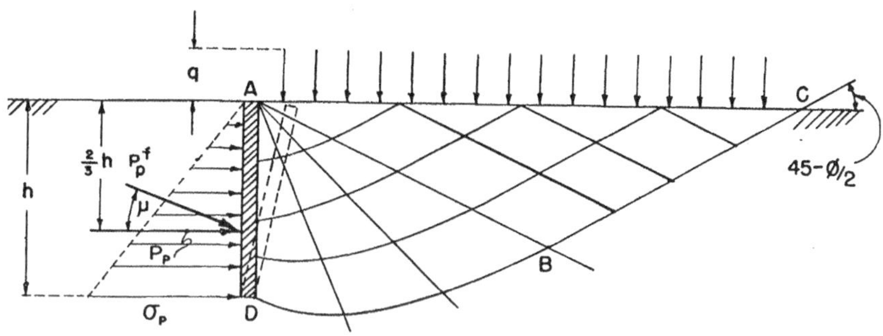

Fig. 69

The value of the force P_p assumed for $\mu = 0$ may be calculated by integrating the pressure σ_p determined by equation (134):

$$P_p = \int_0^h \sigma_p \, dz = \int_0^h (q\, N_\varnothing + 2c\, \sqrt{N_\varnothing} + \gamma z\, N_\varnothing)\, dz$$

and

$$P_p = q\, h\, N_\varnothing + 2ch\, \sqrt{N_\varnothing} + \tfrac{1}{2}\, \gamma h^2\, N_\varnothing\,,$$

where $N_\varnothing$ is a constant previously defined. However, the value of the normal force which results from the passive earth pressure and the frictional forces due to the friction μ existing between the soil and the wall would be different. In the first approximation, the normal passive earth pressure acting upon the vertical wall may be expressed by

$$P_n = q\, K_q + c\, K_c + \gamma z\, K_\gamma\,, \tag{135}$$

where K_c, K_q, and K_γ are constants which do not depend on z and may be computed.[54]

If the retaining wall is sloped to the horizon at an angle α (Figure

70a and b), then the passive earth load P_n normal to the retaining wall will be $P_p/\cos(90 - \alpha)$ and, according to equation (135),

$$P_n = \frac{1}{\sin\alpha}\int_0^h (c\,K_c + q\,K_q + \gamma z\,K_\gamma)\,dz$$

or

$$P_n = \frac{h}{\sin\alpha}(c\,K_c + q\,K_q) + \tfrac{1}{2}\gamma h^2 \frac{K_\gamma}{\sin\alpha}. \qquad (136)$$

It will be seen that the total normal force P_n is composed of two parts. The first,

$$P_n' = \frac{h}{\sin\alpha}(c\,K_c + q\,K_q),$$

is independent of depth. Since it is uniformly distributed, its point of

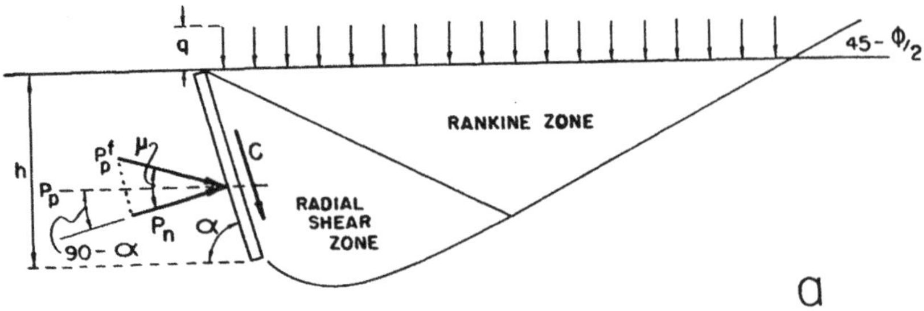

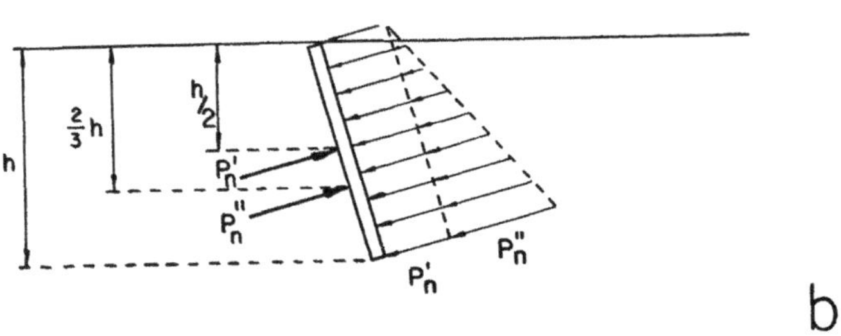

Fig. 70

application will be located in the middle of the retaining wall (Figure 70b). The second part of P_n,

$$P_n'' = \tfrac{1}{2}\gamma h^2 \frac{K_\gamma}{\sin\alpha},$$

depends on depth and increases like hydrostatic pressure. Therefore, its point of application is located beneath the ground surface at a distance equal to $2h/3$. According to equation (136), the earth pressure $P_p{}^f$ equals

$$P_p{}^f = \frac{P_n}{\cos\mu} = \frac{h}{\sin\alpha\cos\mu}(c\,K_c + q\,K_q) + \tfrac{1}{2}\gamma h^2 \frac{K_\gamma}{\sin\alpha\cos\mu}. \quad (137)$$

The problem of determining $P_p{}^f$ can be solved by computing the constants K_c, K_q, and K_γ. Various methods of computing these constants are known in soil mechanics. For instance, the method of the logarithmic

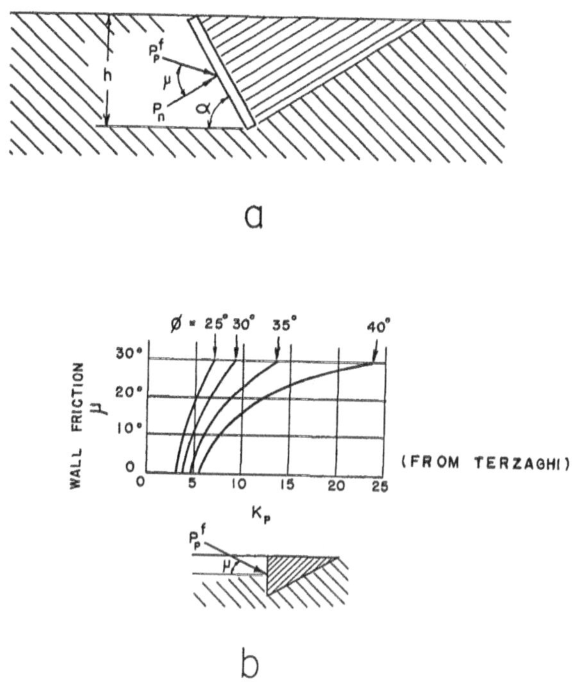

Fig. 71

spiral [54] appears to be very useful and gives accurate results. The simplified method by Coulomb gives only approximate results because it assumes conditions of soil failure in which the surface of sliding consists of a plane passing through the edge of the retaining wall (Figure 71a). In such a case, for an ideal cohesionless mass,

$$P_p{}^f = \tfrac{1}{2}\gamma h^2 \frac{K_p}{\sin\alpha\cos\mu}, \tag{138}$$

where K_p is a constant whose value is computed for various angles of friction μ and ϕ (see Figure 71b).

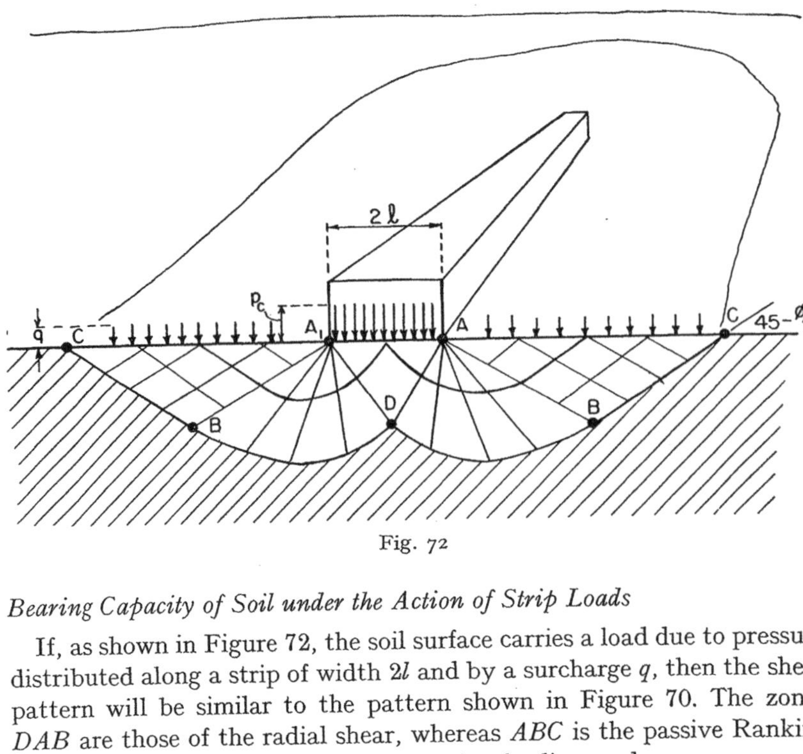

Fig. 72

Bearing Capacity of Soil under the Action of Strip Loads

If, as shown in Figure 72, the soil surface carries a load due to pressure distributed along a strip of width $2l$ and by a surcharge q, then the shear pattern will be similar to the pattern shown in Figure 70. The zones DAB are those of the radial shear, whereas ABC is the passive Rankine zone, the properties of which were previously discussed.

At the moment of failure, the whole structure will sink in such a way that the wedge-shaped soil body ADA will move downward without changing its form, as if it were rigidly attached to the footing.

The form of this body may be directly studied by a photographic method [90] and depends on the angle of friction between the base of the footing and the soil. The value of this angle may be theoretically enclosed between any value of ϕ and $45 + \phi/2$.[54]

Assume that the base angle is ϕ. Assume further that both walls DA of the soil body may be identified with the previously considered retaining walls, the action of which was investigated with the purpose of determining the passive earth pressure acting upon the compressed soil (Figures 70 and 71).

In this case, the walls will be sloped at angles $\alpha = 180 - \phi$ (Figure 73a). The coefficient of friction μ between the soil and the blade will be replaced by ϕ, since an internal sliding of soil on soil takes place during the failure. The force of adhesion C acting along the side of the wall equals

$$C = \frac{lc}{\cos\phi}.$$

The passive earth pressure $P_p{}^f$ deflected by the angle ϕ will be vertical, since the base angle of the triangle $A'AD$ is equal to ϕ. Accordingly, since $\alpha = 180 - \phi$, $\sin(180 - \phi) = \sin\phi$, $h = l\tan\phi$, and $\mu = \phi$, formula (137) may be written as follows:

$$P_p{}^f = \frac{l}{\cos^2\phi}(c\,K_c + q\,K_q) + \tfrac{1}{2}K_\gamma\,\gamma\,l^2\frac{\tan\phi}{\cos^2\phi}. \qquad (139)$$

If the weight per unit of length of the wedge-shaped soil body $A'DA$ (Figure 73) is

$$W_1 = \gamma\,l^2\tan\phi\,, \qquad (140)$$

then the equilibrium of all the forces acting upon that wedge may be expressed by the following equation:

$$V = 2P_p{}^f - W_1 + 2C\sin\phi\,,$$

where $C = lc/\cos\phi$ as previously computed. By substituting into this equation the values of equations (139) and (140), the following will be obtained:

$$V = \frac{2l}{\cos^2\phi}(cK_c + q\,K_q) + 2lc\tan\phi + K_\gamma\gamma l^2\frac{\tan\phi}{\cos^2\phi} - \gamma l^2\tan\phi$$

or

$$V = 2lc\left(\frac{K_c}{\cos^2\phi} + \tan\phi\right) + 2lq\frac{K_q}{\cos^2\phi} + \gamma l^2\tan\phi\left(\frac{K_\gamma}{\cos^2\phi} - 1\right).$$

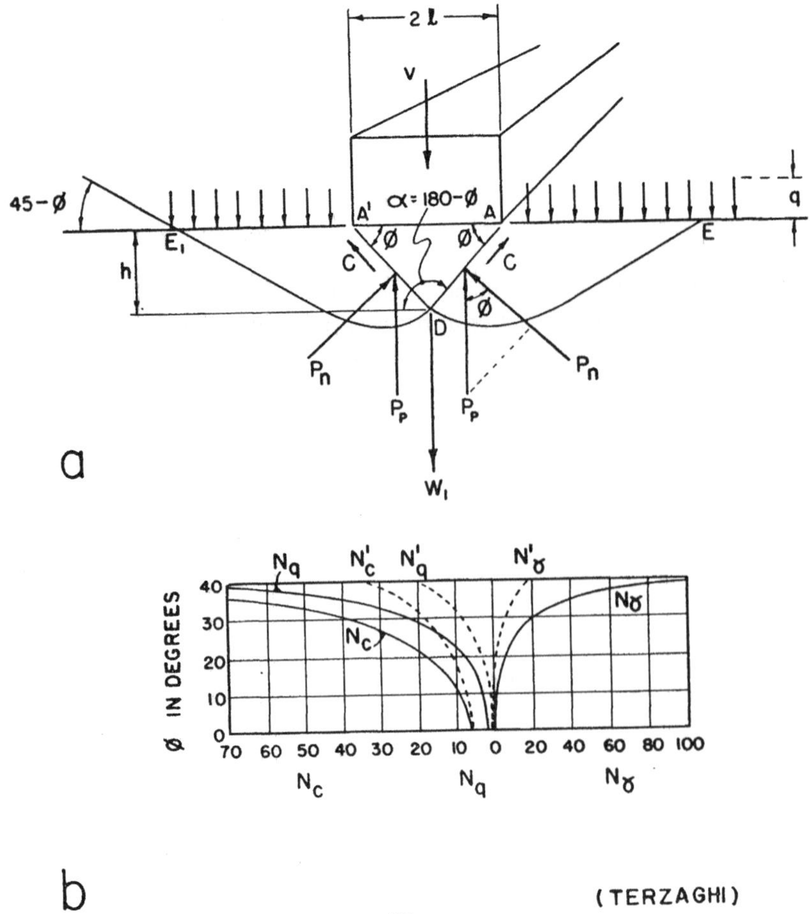

Fig. 73

If it is denoted that

$$\frac{K_c}{\cos^2 \phi} + \tan \phi = N_c$$

$$\frac{K_q}{\cos^2 \phi} = N_q$$

$$\tfrac{1}{2} \tan \phi \left(\frac{K_\gamma}{\cos^2 \phi} - 1 \right) = N_\gamma ,$$

then

$$V = 2lc\, N_c + 2lq\, N_q + 2\gamma l^2\, N_\gamma\,. \tag{141}$$

The values of the coefficients N_c, N_q, and N_γ were computed and are quoted from Reference 54 in Figure 73b. These values determine the bearing capacity of soil for the so-called general shear failure which is characterized by the upward movement of the entire mass $E'A'D$ and EAD (Figure 73a). In some loose soils, such a movement does not occur because the compression of soil caused by the sinking load is not sufficient to reach the outer edges of the soil mass under failure denoted by E and E'. In such a case, the so-called local shear failure takes place, and according to Terzaghi: [54]

$$V = \tfrac{4}{3}lc\, N_c' + 2ql\, N_q' + 2\gamma l^2\, N_\gamma'\,, \tag{142}$$

where N_c', N_q', and N_γ' have smaller values than N_c, N_q, and N_γ, as shown in Figure 73b.

The described method of determining the bearing capacity of soil is not unique and there are a few other methods which serve the same purpose. A review and a comparison of the various formulas evolved have been made by Maag.[91] It will be seen that all the formulas give only approximate values; the discrepancy between the various results obtained illustrates the nature of the problem and the limitations involved. Newer methods developed by modern theories of plasticity appear more promising.[70]

Snow Problem in General

Snow is the type of "soil" with which many vehicles have to cope in modern days. Yet the knowledge of the physical processes which take place in this material is rather limited and the studies of the problem only recently have resulted in establishing the foundations of what is called snow mechanics. One of the first exhaustive studies which embraced the physical aspect of ice and its role in shaping geographical and climatic factors was made by Dobrowolski.[79] Another study which gives general descriptive information about snow was made by Seligman.[92] Tammann founded the physics of ice,[93, 94] and Nakaya pioneered the studies on ice crystals,[95] whereas the Russian school, apparently headed by Weinberg,[96] embraced many practical snow problems extending from hydrology and oversnow transport up to snow removal on highways and airstrips. Snow compaction, in particular, seems to have been one of the main problems of snow mechanics.[97–100]

Special studies on snow electricity and precipitation are to be credited to Langmuir and Schaffer.[101] Bader and Niggli [80] and DeQuervain [102] have contributed to studies on the physics of snow and its metamorphosis. The foundations of snow mechanics linked with soil mechanics were laid down by Haefeli [76] and later followed by Bucher.[78] Aside from Dobrowolski's book,[79] an exhaustive bibliography and source of information regarding ice was assembled in the work by Dorsey.[103]

The present interest in snow mechanics will be limited to those problems which may have some possible connection with the oversnow movement of vehicles. The selection of these problems may prove at this time somewhat arbitrary, since there is very little knowledge of the relationship between the changing properties of snow and the movement of vehicles. Accordingly, the foregoing remarks have to be considered as being introductory.

Shearing Strength of Snow

When dealing with soil, it has been assumed that its strength may be determined as a function of certain "constants" which are called friction, ϕ, cohesion, c, and specific gravity, γ.

In the preceding chapters, it was demonstrated that c, ϕ, and γ are quite sufficient, with the help of theoretical mechanics, to establish an approximate and general theory of soil power in supporting certain types of loads. Since, from the point of view of mechanics, snow may be considered as another type of cohesive and/or frictional mass, there should be little doubt that similar reasoning also should be useful. Although this assumption might be questionable because of the ever-changing properties of snow, it appears to be quite correct as far as the present studies indicate, and may serve the purpose of providing more knowledge of the phenomena with regard to the performance of snow vehicles.

The "constants" ϕ and c which, as was seen in the case of soil, are not real constants, but merely certain coefficients satisfying the equation

$$\tau = c + \sigma \tan \phi ,$$

have to be treated in a similar way in the case of snow. Certainly, in this case, the variability of c and ϕ is more complex because of the snow metamorphosis [80] which causes the state of the stress-strain relationship to depend on temperature, and because it is more sensitive to the rate of load application. The latter undoubtedly also involves more far-going changes in the mechanical properties of snow than the movement of water in the pores of soil under pressure.

Experiments made on the shear strength of various materials indicate that the theoretical function $\tau = c + \sigma \tan \phi$ is practically satisfied without exception, and that the line of rupture obtained experimentally for materials like marble, concrete, and soil [104] is a straight line corresponding to that function (Figure 74a).

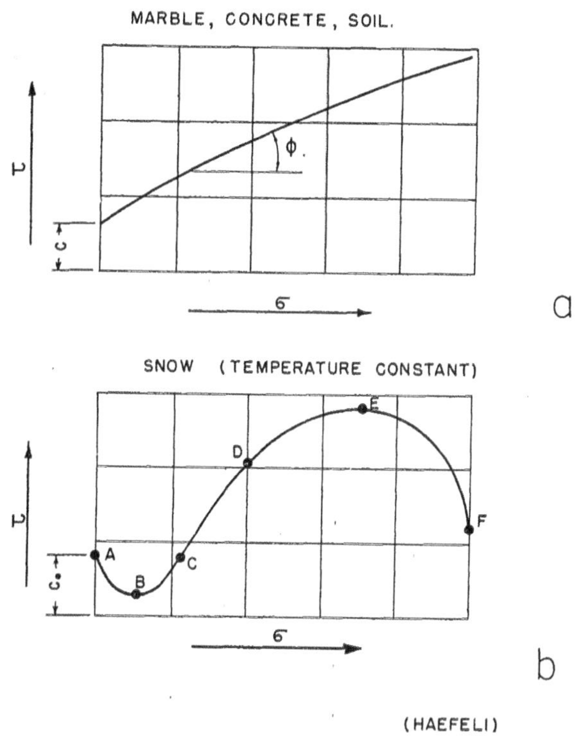

Fig. 74

However, similar lines determined experimentally for snow have a different character (Figure 74b) and are far from satisfying the discussed equation. The work by Haefeli has led to the following conclusions.[76]

The shearing strength of snow is a function of at least three independent factors: pressure, time, and temperature. If temperature is kept constant, the relationship between τ and σ will be as that shown in Figure 74b. A fresh snow sample subjected to pressure σ undergoes a disruption of its structure which is followed by a decrease in τ (line A–B). When the pressure is further increased, an increase in τ will be observed. This phenomenon may be explained by the effect of compaction and changes in snow structure, expressed by the solidification line (B–C–D). Further compression leads to the changing of snow into ice, the strength of which (point E) finally yields into a plastic flow up to point F, from where a complete liquefaction may end the process. The lower the temperature, the higher is the value of E.

The line of rupture of snow $ABCDEF$ cannot be represented, therefore, by the equation $\tau = c + \sigma \tan \phi$, though, at particular instances, it may be interpreted as a certain function of the variable "constants" c and ϕ.

Experiments by the author indicate that the process described by the line ABC in Figure 74b has little practical importance from the point of view of vehicle operation, because it refers to very low pressures which can never be attained in vehicle design. However, within the range of pressures encountered, the line CDE is representative for the types of snow so far tested in conjunction with vehicle performance. The results of the experiments which lead to this conclusion are shown in Figure 75. Various types of snow, designated by numbers 1, 2, 3, and 4, produce al-

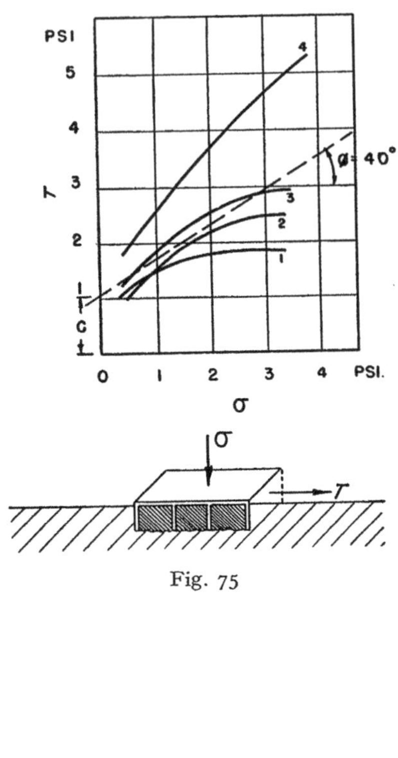

Fig. 75

most a straight-line relationship between τ and σ which relates to pressures enclosed between 1 and 4 psi. This relationship corresponds to that expressed by the line *CDE* of the general curve shown in Figure 74b. Thus, it appears possible to determine snow shear in any specific case, for example, the shear strength of snow 3 as an approximate function $\tau = 1 + \sigma \tan 40°$, in which cohesion $c = 1$ lb/sq in. and friction $\phi = 40°$. Under these assumptions, the previously discussed practical methods of determining the strength of soil should be valid when referred to snow, and the application of general methods similar to those used in soil mechanics for the purpose of determining the state of stresses in a snow cover appears to be justified. The main difficulty is that the changing character of the snow properties requires the adoption of certain average values of coefficients which would cover the field of oversnow vehicle operation.

The Principle of Relaxation as Applied to Snow and Soils

The versatility of snow forms, which may change from a solid icy crust to a soft plastic mass, indicates that the stress-strain relationship of this material may have to be evaluated in terms of both elastic and plastic deformations, particularly if the rate of load application is considered.

The problem is not new. As far back as the 1860's, Thomson and Maxwell were concerned with the elasto-plastic behavior of solids.[105] A model of a body which shows both elastic and plastic (viscous) properties was considered by Houvink [106] and is shown in Figure 76a. If rod *a* is pushed by an impact load *P*, then the fluid enclosed between cylinder *b* and piston *c* will not be able to escape through orifices *d* and only the elastic deflection of spring *f* will characterize the deformation *e*. If load *P* is increased slowly enough, then spring *f* will be practically at rest, and the escaping fluid will lower the original position of rod *a*, thus illustrating a plastic (viscous) deflection. Any other way of applying load *P* will involve both plastic and elastic deformations.

Thomson's theory assumes that the total stress is composed of the sum of plastic and elastic stresses and that, accordingly, there is the following relationship between the shearing strain e_x and stress τ_{xz} of a prism of mass undergoing deformation (Figure 77):

$$\tau_{xz} = G\, e_x + \eta \frac{de_{xz}}{dt}, \tag{143}$$

where G = modulus of rigidity and η = viscosity. Maxwell's concept is

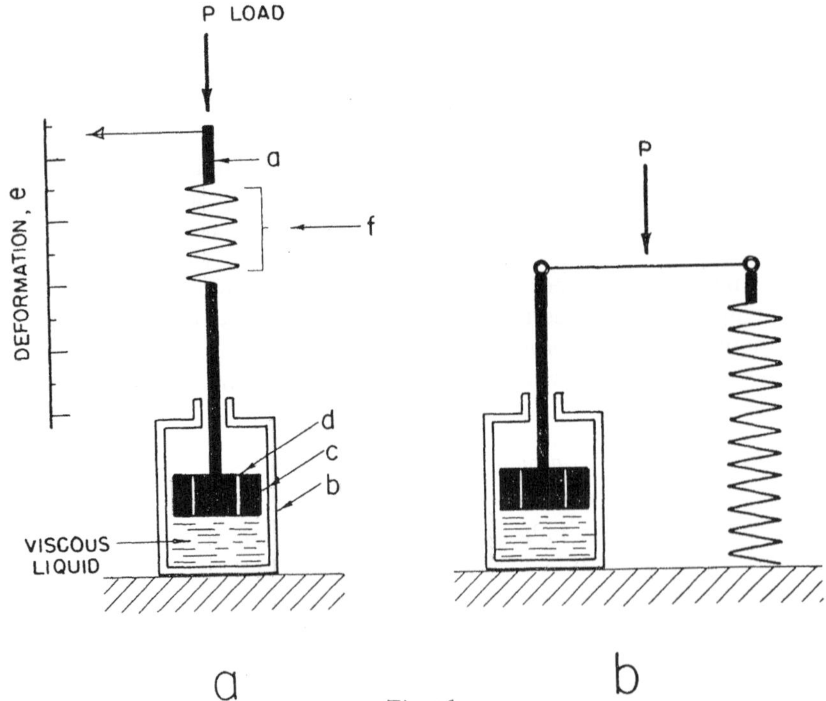

Fig. 76

based on the assumption that elastic and plastic (viscous) strains add, and, according to this, the total stress may be defined as follows:

$$e_x = \frac{\tau_{xz}}{G} + \frac{1}{\eta} \int_0^t \tau_{xz}\, dt$$

or, to present this equation in a form similar to that of equation (143),

$$\tau_{xz} = G\, e_x - \frac{G}{\eta} \int_0^t \tau_{xz}\, dt\,. \tag{144}$$

Equation (144) is represented by the model sketched in Figure 76a, whereas equation (143) is represented by the model in Figure 76b. There is, of course, an open question as to which model, if either, illustrates the elasto-plastic behavior of snow in the best way. Dobrowolski [107] and Dorsey [103] analyzed the implication of both theories. A general discussion of the models representing a given body was made by Freudenthal.[105] Bucher [78] apparently belongs to the followers of Maxwell's concept and of the corresponding principle of relaxation. The latter may be described briefly as follows: if it is assumed that the rate of deformation e_x of a snow prism is constant (Figure 77), then $de_x/dt = 0$. Accordingly, equation (144) may be written as follows:

$$\frac{d\tau_{xz}}{dt}\frac{1}{G}+\frac{\tau_{xz}}{\eta}=0\,.$$

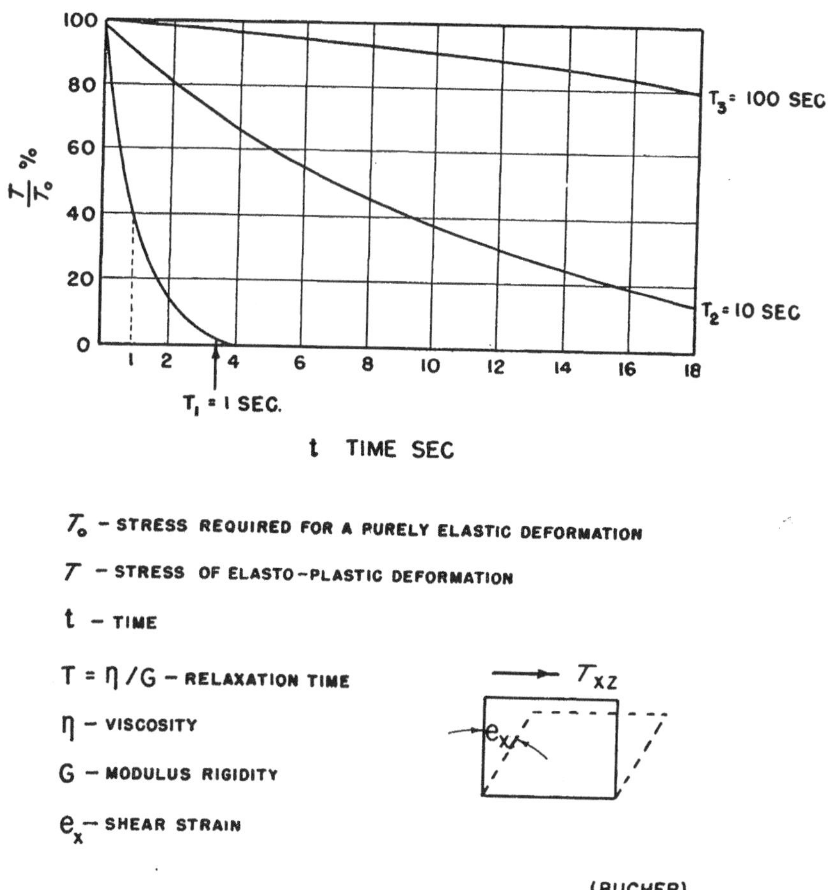

Fig. 77

After integrating,

$$\tau_{xz} = \tau_o \, e^{-\frac{G}{\eta}t}, \tag{145}$$

where τ_o is the shearing stress at time $t = 0$. If the ratio η/G is denoted by T, equation (145) may be written as follows:

$$\tau_{xz} = \tau_o \, e^{-\frac{t}{T}}. \tag{146}$$

This equation expresses the general principle of the relaxation theory: when an aggregate of particles is subjected to the shearing strain e_x by a shearing stress τ_{xz}, it gradually breaks, and relaxes the stress; the broken elements form another aggregate which is of different structure than the original one, and the stress is relieved if not continually renewed. If the rate of relaxation is directly proportional to the strain, then, if left alone, the stress τ_{xz} will decrease with time exponentially as shown by equation (146). The time required to reduce it to $1/e = 0.3679$ of its original value τ_o is called the relaxation time T. The latter, it may be repeated, is the time that is needed to bring the elasto-plastic stress τ_{xz} caused by deformation e_x to the value of $\tau_o \times 0.3679$. A large relaxation time means the preponderance of elastic deformations; a small one, that of plastic deformations.

The relaxation time defined by the formula $T = \eta/G$ requires experimental determination, as was done a long time ago with reference to ice.[107] Quite recently, Bucher, when considering the work by DeQuervain, Kuhn, and Weinberg, arrived at the conclusion that for temperatures between 0° and –40° C, T may be assumed to be between 10 and 10^6 sec. The values of G may be assumed as approximately 10^5 kg/m², whereas η varies between 10^6 and 10^{11} kg sec/m².[78]

The relaxation curves for various relaxation times of a few types of snow are traced in Figure 77. This figure shows that snow T_3 is apparently well hardened, since during 18 sec, its original stress is reduced from 100% to about only 80%. Soft snow T_1, whose relaxation time is 1 sec, relieves stresses up to zero in 4 sec. For the period of 1 sec, however, it retains 40% of the original stress as shown by the dotted line. It may be generally observed that at relatively high temperatures and low speeds of load application, snow will behave like an ideally plastic mass, or a viscous fluid. At dynamic loads, however, elastic effects also may appear, particularly in low temperatures.

It appears justified to expect that forces which are applied to the ground with a speed close to an impact will probably be counteracted with an elastic load reaching, in extreme conditions, perhaps 50% of the total snow stress, whereas the rest of the reaction will be of plastic nature. This would refer to hard snow at low temperatures. Soft, wet snow will inevitably flow under any load because of the overwhelming preponderance of plastic deformations. A more detailed study of the elasto-plastic behavior of solids and theoretical considerations on this subject may be found in References 106 and 105. The discussed inelastic behavior of snow also refers to any other type of soil.

Changes in the Mechanical Properties of Snow

Studies which have been conducted with the purpose of determining trends in the change of the mechanical properties of snow as a function of the various factors involved have been summarized by Bucher as follows: any compression, increase in grain size, or decrease in tempera-

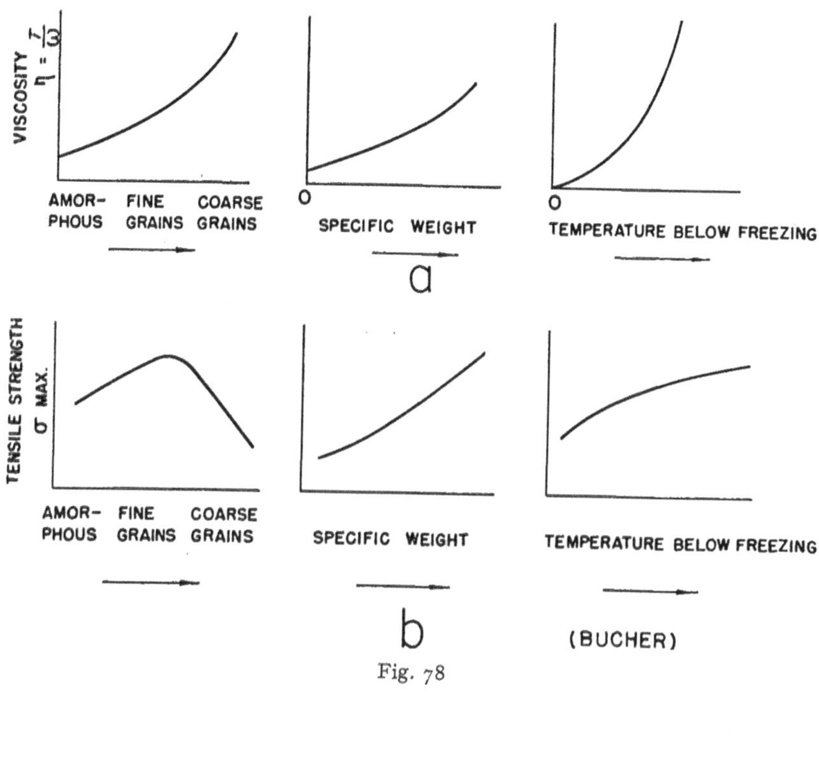

Fig. 78

ture leads to an increase in viscosity (Figure 78a). The tensile strength of snow behaves in a somewhat different way: compression increases the tensile strength; an increase in grain size or temperature, however, affects the strength in the way shown in Figure 78b.[78]

It may then be concluded that vehicles having higher pressures will develop larger propelling forces than those which exercise low pressures. This conclusion is rather contrary to the commonly accepted concepts of the snow-going vehicle, whose performance is supposed to be improved when the ground pressure is very low, or whose "flotation" is very high.

The mechanical characteristics of snow are more affected by changes in structure due to compression than those of soils. According to the experiments performed by Kondratiev, Krahelskij, and Shakov,[77] the change in snow density γ with reference to the pressure applied varies in accordance with the line shown in Figure 79a. The relationship of density and temperature follows the curves shown in Figure 79b.

Figure 79 indicates the existence of practical limits of snow compaction at two ranges of temperature, i.e., it indicates that for these temperatures, pressures higher than approximately 0.4 kg/cm² (or 5.6

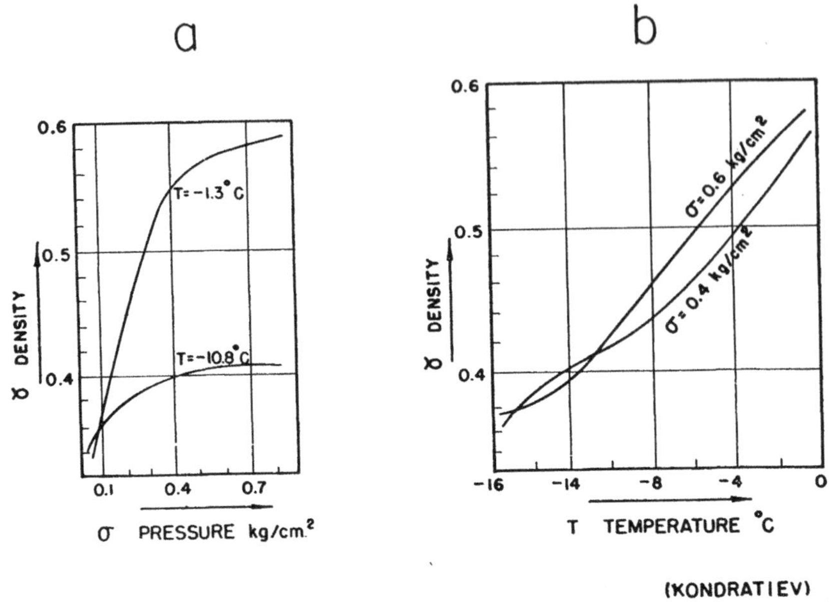

Fig. 79

lb/sq in.) do not change the snow structure as far as γ is concerned. Since vehicles usually exercise pressures in the vicinity of 5.6 lb/sq in., it may be contended that there exists within given limits of temperature a certain uniformity of snow density which is to be related to vehicle performance. Of course, an investigation of a larger number of temperatures and snow types than those explored by Kondratiev is required in order to generalize this conclusion.

On the basis of available experience, it is known that the granular, so-called "sugar snow" which at low temperatures does not compact at any practical pressures will not be quite in line with this conclusion. On the other hand, however, it is known that sugar snow behaves almost like dry sand, and as such may be subject to the same assumption which characterizes sand: $\tau = \sigma \tan \phi$. It does not compact unless shaken by vibrations.

Kondratiev and his colleagues propose the following formula for determining snow density γ under pressure p:

$$\gamma = \gamma_o + \frac{K_1 p\,(K_2 - t)}{p + K_3}\,, \qquad (147)$$

where t is the temperature (°C), γ_o is the original density, p is the compression pressure measured in kg/cm², and K_1, K_2, and K_3 are constants.

When working on snow compaction for airstrips, the Russian authors found the following values of K for $\gamma_o = 0.18$:

$$K_1 = 0.0038\,, \quad K_2 = 96\,, \quad K_3 = 0.08\,.$$

These values were determined by compacting snow with a disk whose area was equal to 100 sq cm. Under the above-described conditions, the depth of compaction reached 15 to 18 cm. It is evident that the coefficients K would be different for other sizes and forms of the compacting device.

Frictional Properties of Snow

Snow (ice) is the only solid that is slippery by its very nature, as was truly noticed by Dobrowolski.[79] This peculiar characteristic is due to the fact that ice is not a homogeneous substance, but a three-phase system in which solid, liquid, and gaseous (vapor) states coexist at a thermodynamic equilibrium corresponding to a given pressure and temperature. Any change of pressure or temperature causes an immediate shift in the phase composition, thus involving fluctuations in the structure of the matter. Since snow is an aggregate of ice crystals and air, the factors involving the phase change are particularly active. An increase in pressure and temperature between a sliding body and snow will invariably increase the amount of liquid phase, thereby providing the rubbing surfaces with a lubricant which considerably reduces the friction coefficient. This is why a ski or a skate slides easily when applied to "solidified water."

This reasoning is generally accepted in the explanation of the low coefficient of friction between ice and a sliding solid. Haefeli [76] investigated the problem, with interest focused upon the slow-sliding masses of snow which result in avalanches. It is thought that his work is of a rather general nature and does not relate directly to the problems of an oversnow transport. However, a discussion of the properties of the water film, which is credited with the lubrication of sliding surfaces, is of particular interest since it elucidates the nature of the quantities involved. The extremely low coefficients of friction of the order of 0.0005 and the thickness of the water film as discussed by Haefeli may be related in the following way: According to Newton's equation, the medium thickness s of such a film may be determined from the formula

$$\mu\sigma = \eta \frac{v}{s} \text{ (g/cm}^2\text{)}$$

and

$$s = \eta \frac{v}{\mu\sigma} \text{ (cm)}, \tag{148}$$

where η is the water viscosity for 0° C, equal to 1.83×10^{-5}; v is the speed of sliding, equal to 4 mm/min = 0.0067 cm/sec; and σ is the unit load, equal to 20 gr/cm². Hence,

$$s = \frac{1.83 \times 10^{-5} \times 0.0067}{0.0005 \times 20} = 1.2 \times 10^{-5} \text{ cm.}$$

This example gives an idea of the amount of water which may be involved in the lubrication of snow surface. It is obvious that the conditions of friction are thus very sensitive to the slightest changes in temperature.

The relationship between speed and friction belongs to important problems of snow mechanics. Tests made to determine the friction developed between snow and glass indicate that an increase in speed causes a steady increase in μ comparable to that characterizing any lubricant. For $v = 0$, μ is either zero or so small that it cannot be detected. For temperatures below 0° C (see Figure 80a), the variation of μ with speed v

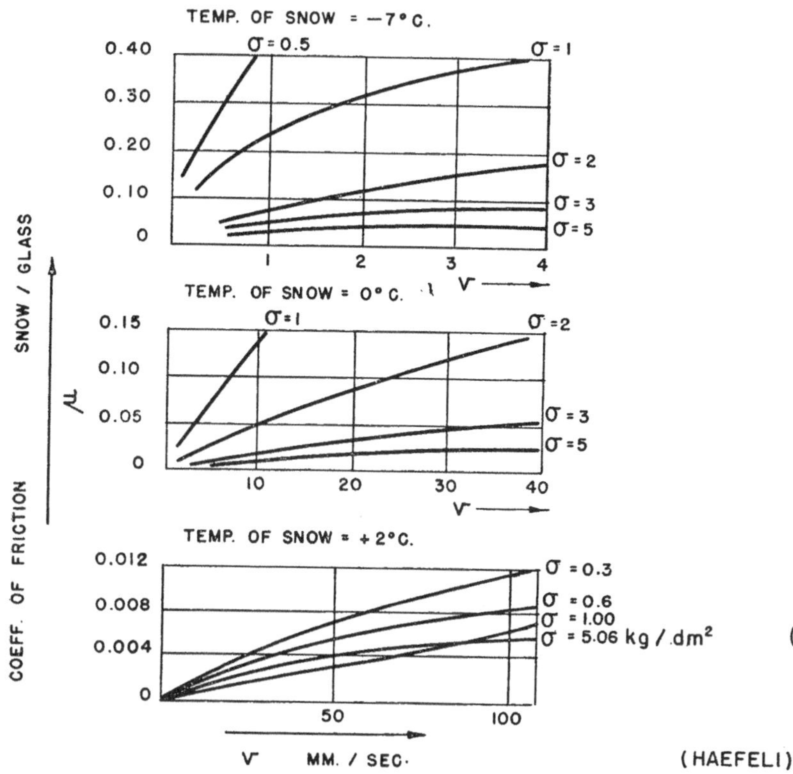

Fig. 80

for σ larger than 1 kg/dm² appears to be of the same character as the change in the stress and strain in a plastic flow. For other temperatures (Figure 80b and c), a more or less straight-line increase in μ with speed v is indicated. A general relationship between speed and friction has not been established as yet, although Haefeli has proposed some equations referring to very slow speed and temperatures below 0° C.[76]

The variation of μ with pressure σ is shown in Figure 81. The tests performed show that in all cases μ increases with speed, with the exception shown in Figure 81c, where minimum μ is observed in the vicinity of $\sigma = 1$ kg/dm². No special explanation of the phenomenon is offered. It may be noted that the basic equation of "friction" by Coulomb, $\tau = c + \sigma \tan \phi$, does not hold true in this case.

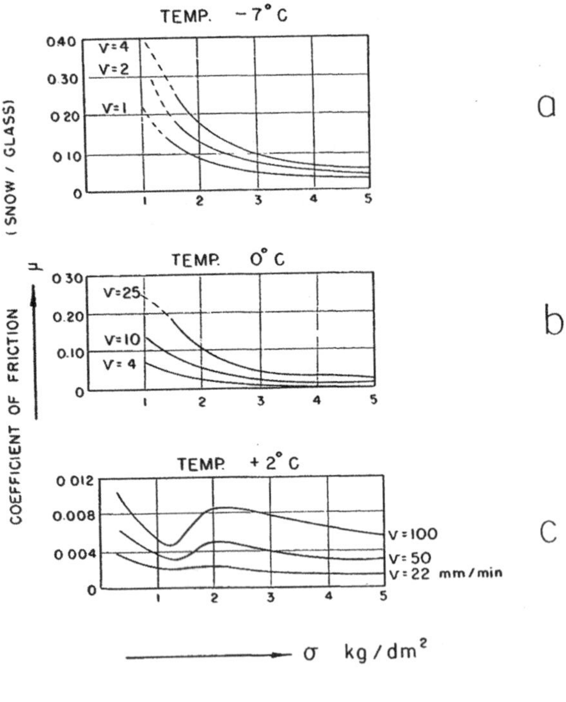

Fig. 81

Some tests to determine the friction between ice and steel were made by Croce [108] (also see Reference 109). The results showed again that μ increases with speed and seems to tend toward a definite limit which is not surpassed at speeds higher than a certain maximum. This limit is reached more quickly when the load is high.

An increase in load leads, in general, to a reduction in friction. Gorubunov [110] related friction to the density of snow and quotes the following values of μ for snow and wood:

Snow Density γ	0.1	0.15	0.24	0.37	0.44	0.5	0.55
Friction Coefficient μ	0.1	0.09	0.08	0.06	0.045	0.03	0.02

The mechanism of snow friction is not very clearly understood and there is no over-all theory which would attempt to solve the problem quantitatively. Haefeli suggested that the mechanics of sliding is not far apart from that of lubrication in general.[76] In such a case, the mathematical methods introduced in the theory of lubrication, and recently perfected by the rapid development of hydrodynamics,[111] would be helpful in a quantitative study of friction. However, the thermodynamic aspect of the phase system of sliding snow cannot be neglected; it presents considerable, if not insurmountable, difficulties in any rational treatment of the problem. It may also be possible that the electrical properties of snow [103] may play an important role. The emerging importance of over-snow transport in the Arctic opens this vast field for investigation.

The nature of the friction between snow and other solids was qualitatively explored first by Bowden, Tabor, and Hughes; [112] the results of their work may be stressed as follows. Sliding contact usually takes place in a few points, the area of which is much smaller than the ski area. In cold weather, it is possible that the real area of a sliding body which is in contact with ice crystals is as small as 1/1000 of the total "contact" area, if low pressure is considered. In such a case, the heat generated through friction would have to be abundant to produce enough water for lubrication, though the water film would not be visible. This invisible water may be detected at the coefficient of friction of the order of $\mu = 0.03$. Klein [113] estimates that for a settled snow of medium hardness, the real contact area amounts to 20% of the total ski area at loadings up to 200 lb/sq ft and 50% and 500 lb/sq ft. His estimates were based on the observed area of water droplets which could have been seen through a glass window located in the bottom of a ski and do not represent the true contact area between ice crystals and the sliding surface.

Experiments have shown that when the load is increased, the area of the true contact, and thus that of the water film, has to increase in order to adjust to the higher pressure, and that the friction coefficient is independent of load and area within a certain range.

In the final phase of load increase, when the whole surface carries the load through a water film, practically perfect slider-bearing conditions may be assumed, and the assessment of the friction coefficient, in accordance with the theory of lubrication, offers quite reasonable though approximate values.

In low temperatures, when the heat of friction produces insufficient film, the friction increases and, as Bowden pointed out, reaches values at extremely low temperatures having the same order of magnitude as those of other rubbing solids (Figure 82a). This occurrence suggests that the heat conductivity of the material from which a ski is made plays an important role in reducing the friction coefficient. Solids having low heat conductivity, such as ice, will not dissipate the friction energy which would in turn melt snow and produce water film. High-conductivity materials, such as brass, act in the opposite direction, as was experimentally demonstrated by Bowden and Hughes (Figure 82b). Their investigation also pointed out that a large amount of water increases the friction. The above phenomenon also was observed by Klein, who suggested that friction increases are due to surface tension. However, since an approximate evaluation of the quantities involved is not available, the plausibility of this explanation cannot be checked. To sum up, there is no theory which would quantitatively explain the frictional properties of ice.

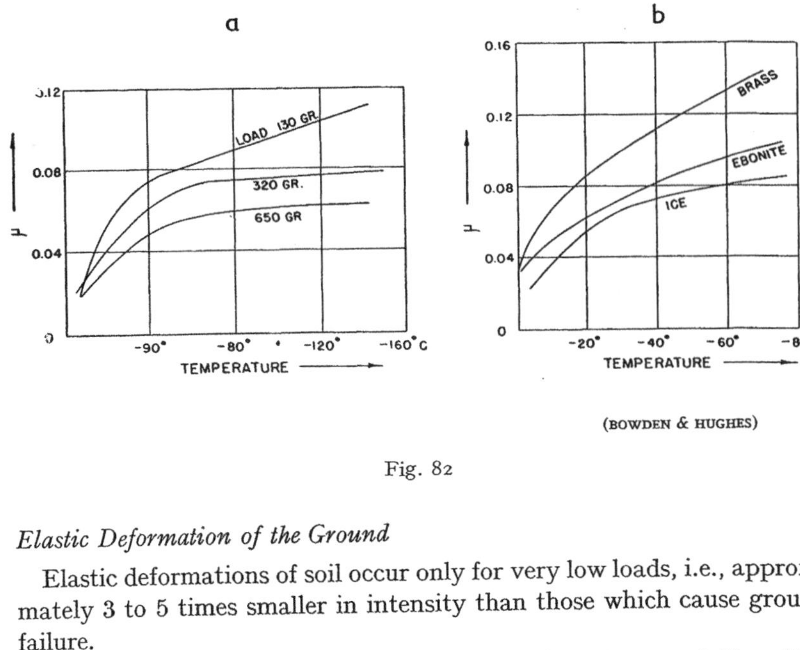

Fig. 82

Elastic Deformation of the Ground

Elastic deformations of soil occur only for very low loads, i.e., approximately 3 to 5 times smaller in intensity than those which cause ground failure.

Although such small ground deformations do not appreciably affect vehicle performance, some knowledge of the mechanism of the elastic strain may be useful in the evaluation of other problems, such as the theory of small-scale tests.

The strain pattern is implicit in the previously discussed properties of the stress function Φ and in Hooke's law, and may be determined accordingly.[56] By assuming that the soil is loaded with a point load W, and is perfectly elastic throughout its semi-infinite, homogeneous mass, Boussinesq determined the following values for the vertical sinkage Δz and horizontal displacement Δr of soil mass located at depth z and distance r from load W (Figure 83a):[54]

$$\Delta z = \frac{W}{2\pi r} \frac{1 + \nu}{E} [2(1 - \nu) + \cos^2 \varphi] \sin \varphi$$

$$\Delta r = \frac{W}{2\pi r} \frac{1 + \nu}{E} [-(1 - 2\nu) + \cos \varphi + \cos^2 \varphi] \sin \varphi \tan \varphi/2 \, .$$

The deflection of the surface ($\varphi = 90°$)

$$\Delta z_s = \frac{W}{\pi r} \frac{1 - \nu^2}{E} \tag{149}$$

$$\Delta r_s = - \frac{W}{2\pi r} \frac{1 - \nu - 2\nu^2}{E} \, . \tag{150}$$

From formulas (149) and (150), the settlement of the edges of a circular or rectangular flexible load can be determined by the integration of the elementary loads acting upon the given surface. For a rectangular area $(l) \times (w)$, the element $dx\, dy$ supports the load dW under the action of a uniform pressure p (Figure 83b). According to equation (149),

$$d(\Delta z_s) = \frac{dW}{\pi r} \frac{1 - \nu^2}{E} \, .$$

Since $r = \sqrt{x^2 + (w - y)^2}$ and $dW = p\, dx\, dy$,

$$\Delta z_s = \int_0^l \int_0^w \frac{\dfrac{1 - \nu^2}{E}}{\pi \sqrt{x^2 + (w - y)^2}} p \, dx \, dy \, .$$

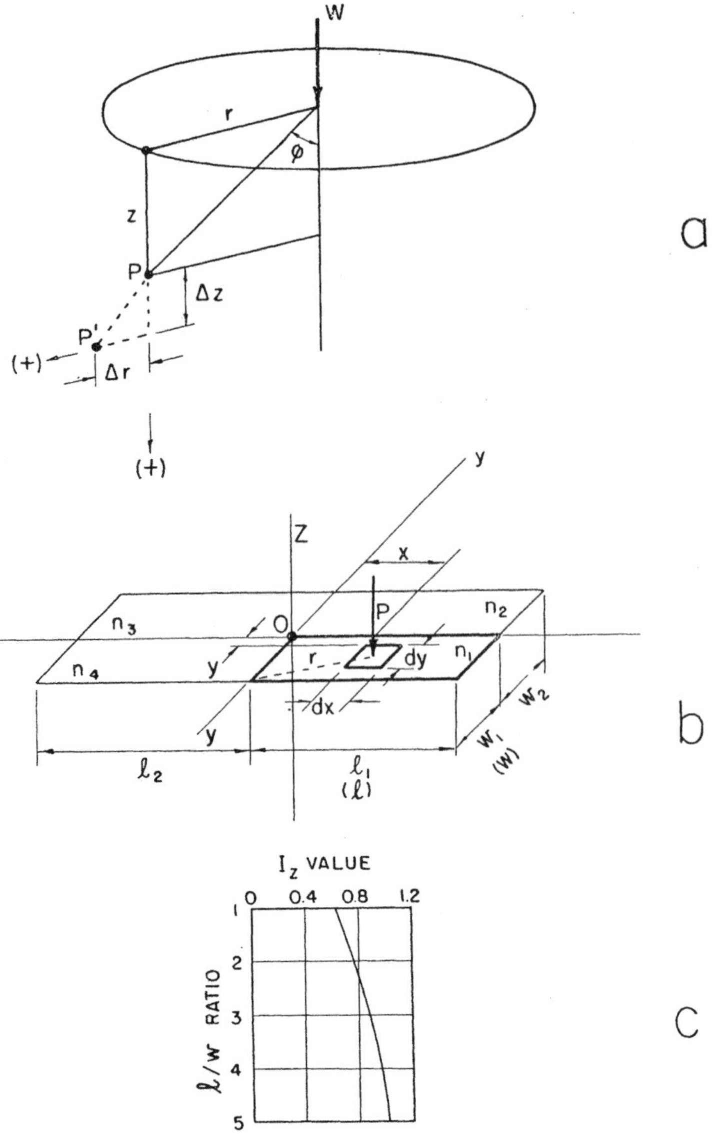
W
r
φ
z
P
Δz
P'
(+)
Δr
(+)
a
y
Z
x
P
n_2
n_3
O
x
y
r
dy
n_1
n_4
dx
w_2
y
ℓ_2
ℓ_1
(ℓ)
w_1
(w)
b
I_z VALUE
0 0.4 0.8 1.2
1
2
3
4
5
ℓ/w RATIO
c

Fig. 83

If $w - y = \zeta$ and $-dy = d\zeta$ are substituted in the above equation, then the first integration with reference to dy gives

$$\int_0^w - \frac{d\zeta}{\sqrt{x^2 + \zeta}} = \left| - \log \left[(w - y) + \sqrt{x^2 + (w - y)^2} \right] \right|_0^w$$

$$= - \log x + \log (w + \sqrt{x^2 + w^2}) .$$

The second integration with reference to dx requires the solution of two integers in order to determine Δz_s:

$$\Delta z_s = \frac{p}{\pi} \frac{1 - \nu^2}{E} \int_0^l \log (w + \sqrt{x^2 + w^2}) \, dx - \int_0^l \log x \, dx .$$

The second integer immediately gives

$$\int_0^l \log x \, dx = \left| x (\log x - 1) \right|_0^l = - l \log l + l .$$

In order to find the first one, denote $\log (w + \sqrt{x^2 + w^2}) = u$. Then

$$du = \frac{\dfrac{x \, dx}{\sqrt{x^2 + w^2}}}{w + \sqrt{x^2 + w^2}} = \frac{x \, dx}{(w + \sqrt{x^2 + w^2}) \sqrt{x^2 + w^2}} .$$

Denote further $x = v$ and $dx = dv$. Since $\int u \, dv = uv - \int v \, du$,

$$\int \log (w + \sqrt{x^2 + w^2}) \, dx = x \log (w + \sqrt{x^2 + w^2}) -$$

$$- \int \frac{x^2 \, dx}{(w + \sqrt{x^2 + w^2}) \sqrt{x^2 + w^2}} .$$

By denoting $w + \sqrt{x^2 + w^2} = \zeta$, etc., it will be found that the second part of the above integral

$$\int \frac{x^2 \, dx}{(w + \sqrt{x^2 + w^2}) \sqrt{x^2 + w^2}} = \int \frac{\sqrt{\zeta^2 - 2w\zeta}}{\zeta} \, d\zeta$$

$$= \sqrt{\zeta^2 - 2w\zeta} - w \log (\sqrt{\zeta^2 - 2w\zeta} + \zeta - w)$$

$$= x - w \log (x + \sqrt{x^2 + w^2}) .$$

Thus the first integral under consideration is

$$\int_0^l \log (w + \sqrt{x^2 + w^2})\, dx$$

$$= \left| x \log (w + \sqrt{w^2 + x^2}) - x + w \log (x + \sqrt{x^2 + w^2}) \right|_0^l$$

$$= l \log (w + \sqrt{l^2 + w^2}) - l + w \log (l + \sqrt{l^2 + w^2}) - w \log w \,.$$

And finally,

$$\Delta z_s = l \log (w + \sqrt{l^2 + w^2}) - l + w \log (l + \sqrt{l^2 + w^2}) -$$
$$- w \log w - l \log l + l \,. \qquad (151)$$

If the ratio l/w is denoted by n, then, after simplifying, formula (151) may be written in the following form:

$$\Delta z_s = \frac{pw}{\pi} \frac{1 - \nu^2}{E} \left[n \log \frac{1 + \sqrt{n^2 + 1}}{n} + \log (n + \sqrt{n^2 + 1}) \right] . \qquad (152)$$

Denote

$$\frac{1}{\pi} \left[n \log \frac{1 + \sqrt{n^2 + 1}}{n} + \log (n + \sqrt{n^2 + 1}) \right] = I_z \,. \qquad (153)$$

Then

$$\Delta z_s = \frac{1 - \nu^2}{E} p\, w\, I_z \,.$$

If the area under consideration is composed of four rectangles $l_1 \times w_1$, $l_1 \times w_2$, $l_2 \times w_1$, and $l_2 \times w_2$ (Figure 83b), the deflection at the beginning of the coordinates (point P) will be obtained by superimposing the deflections caused separately by particular areas:

$$\Delta z_s = \frac{1 - \nu^2}{E} p\, (w_1 I_z' + w_2 I_z'' + w_3 I_z''' + w_4 I_z'''') \,, \qquad (154)$$

where the values I_z', I_z'', I_z''', and I_z'''' are those determined by equation (153) for $n_1 = l_1/w_1$, $n_2 = l_1/w_2$, $n_3 = l_2/w_2$, and $n_4 = l_2/w_1$, respectively. The values of $I_z = f(n)$ are quoted from Terzaghi in Figure 83c.

The above example of computing an elastic settlement under a rectangular flexible area uniformly loaded with a given pressure indicates the complexity of the problem. The mathematical difficulties in such computations are enormous and it is no wonder that the settlement, for instance, of an elliptic area loaded with a uniform pressure has not yet been computed by automotive engineers, though it might be directly referred to the sinkage of a flexible pneumatic tire.

Effect of Size of Loaded Area on the Elastic Settlement of the Ground

Consider a square plate so that $l = w$ (Figure 83b), and a circular plate having radius r. In following a procedure similar to that outlined in the preceding section, it will be found that the settlement of the center of the rectangular and circular areas will be

$$\Delta z_{\square}' = 2.24\, p\, w \frac{1 - \surd^2}{E} \tag{155}$$

$$\Delta z_{\circ}' = 2\, p\, r \frac{1 - \surd^2}{E}, \tag{156}$$

respectively. The elastic deformation of the soil at the corners of the rectangular area and at the circumference of the circular area will accordingly be

$$\Delta z_{\square}'' = 0.5\, \Delta z_{\square}' \tag{157}$$

$$\Delta z_{\circ}'' = 0.6\, \Delta z_{\circ}'. \tag{158}$$

It will be seen then that the elastic settlement increases with the dimensions of the loaded area.[114] Thus, small-scale loading tests are greatly handicapped and cannot provide a simple relationship between the sinkage of areas of varying sizes.

If it is therefore assumed that the elastic sinkage Δz is proportional to pressure p, i.e., $p = k\, \Delta z$, then it must be realized that the coefficient of proportionality k is not only a function of soil properties $(E \surd)$ but also of the dimensions and form of the loading surface. Experiments also show that the settlement decreases with increasing unit load, thus adding more complexity to the phenomena involved.

Nonelastic Settlement of Soils and Snow

The transition from elastic to nonelastic (plastic) deformation does not take place abruptly, but gradually spreads from a small nucleus located in the vicinity of the loading surface over the whole soil mass.[74]

The computation of this intermediate state of the stress-strain pattern appears tedious. It might be pursued by means of the relaxation method.[115] The work by Glover and Cornwell suggests a method for investigating the stability of granular masses which are divided into alternate plastic and elastic zones by a group of planes.[116] It appears, however, that only approximate results may be achieved by any similar method.

The main difficulty in the evaluation of the nonelastic settlement of soils and snow lies in the fact that the sinkage of the loaded area is due to two separate factors: (1) the compression of soil volume, which may be elastic, or not, and (2) the sidewise flow which leads to the lateral and upward displacement of the soil particles located in the vicinity of the mass undergoing compression.

Experience shows that the smaller the loaded area, the stronger is the effect of the lateral displacement of the ground. Volume compression is significant only for large areas. Experiments by Kögler,[117] performed in sand and clay with circular disks having diameters from 8 to 100 cm, have reflected the effect of the discussed factors and indicate that the effect of the lateral displacement increases with the decrease in area-perimeter ratios. The results are shown in Figure 84a and b. It should be noted that a similar observation was made by Housel, who assumed that the settlement takes place because of the volume compression and shear along the perimeter area.[118] Terzaghi showed, however, that the perimeter-area factor may be considered as a purely empirical coefficient without any physical meaning.[54]

An experimental attempt to evaluate the settlement of bare and spudded rectangular plates was made recently by Pollitt and Lewis.[119] In a search for the explanation of the sinkage of motor vehicles, the above workers performed a series of experiments in sand and clay on square plates ranging from 4×4 to 8×8 in. The results obtained lead to the conclusion that when the pressure is close to 8 psi, sinkage is proportional to the plate size. It was also found that the presence of spuds and "cut-away" areas tends to reduce sinkage. Thus, "open" tracks appear to be both lighter and better load carriers than "closed" tracks.

In view of the short duration of loadings applied to vehicles, sinkage as a function of time appears to be of importance; the investigations by Pollitt and Lewis show an empirical approach to this problem. By assuming that the resistance R to the sinkage is due to the load factor q, depth

factor λ, and speed of sinkage factor ε, the British investigators proposed the equation

$$R = a\,(q + \lambda z + \varepsilon\, v^2)\,, \tag{159}$$

where a is the area of the plate, z the sinkage, and v the speed of the sinkage. By measuring and computing related values, the following empirical figures were proposed for the investigated soils ($c = 0.8 - 0.85$ psi, $\phi = 7.7° - 10.2°$, $\gamma = 73 - 99$ lb/cu ft, moisture content $= 25-50\%$):

Plate Size, in.	q	λ	ε
4 × 4	5.56	0.636	0.745
6 × 6	4.41	0.607	0.745
8 × 8	3.75	0.520	0.745

The curves of settlement as related to time are shown in Figure 84c.

General information on the theories of ground settlement is described in Reference 120. That information, however, reveals that much remains to be done in order to produce at least a qualitative picture of the sinkage of vehicles at various loads, speeds, repeated loads, etc.

The frequently prevailing methods are based on the assumption of the variable modulus of elasticity E of soil, which may be determined from a nonconfined compression test. In such a test, the obtained stress-strain curve is replaced, within the range of working loads σ' and σ'', by a straight line (Figure 85a), thus yielding the value of $E = \tan \alpha$.

It may be marginally noted that in the case of laterally confined compression, when a reduction in the void ratio may be measured as an index of the compression, the modulus of elasticity is expressed in terms of pressure p and void ratio ε.

Assume that a unit soil mass consists of a prism having height a of a solid and height b of voids so that the unit volume of the prism under investigation is $a + b = l$ (Figure 85b). During the process of compression, a remains constant whereas b changes, and $dl = db$. The specific change of length is therefore

$$\frac{dl}{l} = \frac{db}{a + b}\,.$$

The definition of the void ratio is [63]

$$\varepsilon = \frac{b}{a}\,.$$

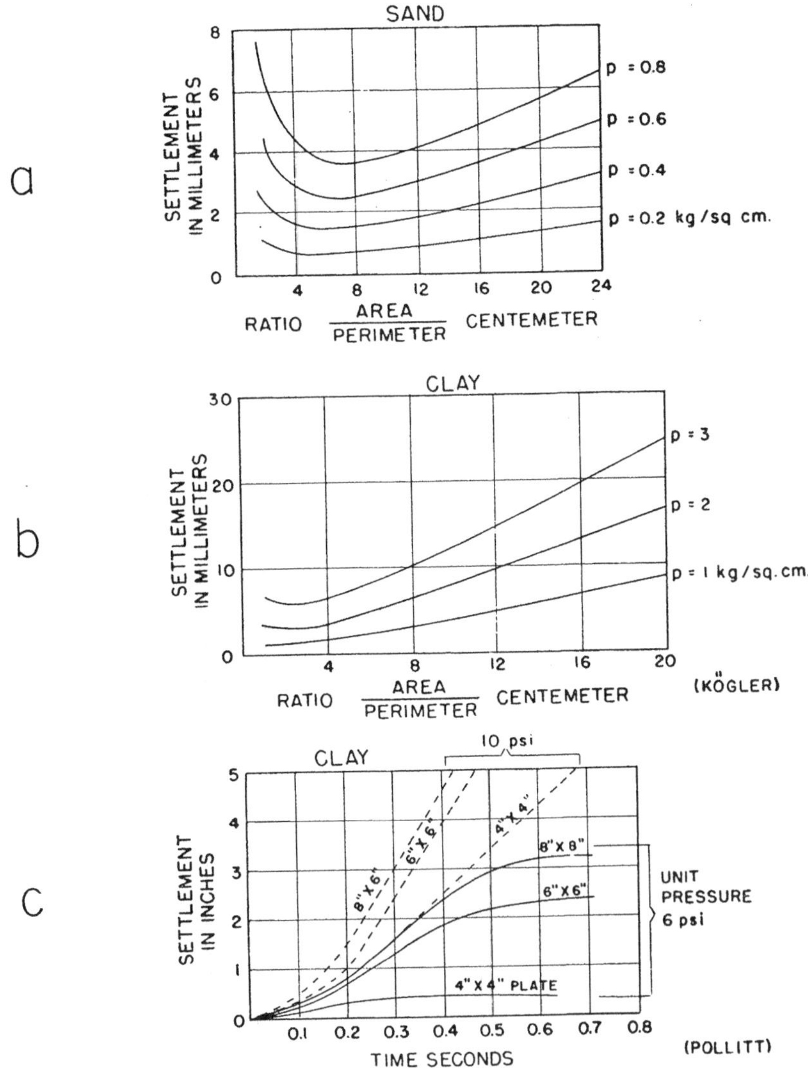
a
SAND
SETTLEMENT IN MILLIMETERS
p = 0.8
p = 0.6
p = 0.4
p = 0.2 kg/sq cm.
RATIO AREA/PERIMETER CENTEMETER
b
CLAY
SETTLEMENT IN MILLIMETERS
p = 3
p = 2
p = 1 kg/sq.cm.
RATIO AREA/PERIMETER CENTEMETER
(KÖGLER)
c
CLAY
10 psi
SETTLEMENT IN INCHES
8"X6"
6"X6"
4"X4"
8"X8"
6"X6"
4"X4" PLATE
UNIT PRESSURE 6 psi
TIME SECONDS
(POLLITT)

Fig. 84

Then, since $db = a\,d\varepsilon$, the unit deformation of soil may be written as follows:

$$\frac{dl}{l} = \frac{a\,d\varepsilon}{a + b} = \frac{d\varepsilon}{1 + \frac{b}{a}} = \frac{d\varepsilon}{1 + \varepsilon}. \tag{160}$$

The definition of the modulus of elasticity is: $E = -\,dp/(dl/l)$. When the value dl/l from equation (160) is substituted into this equation, it will be seen that

$$E = -\frac{dp}{d\varepsilon}(1 + \varepsilon). \tag{161}$$

The value $dp/d\varepsilon$ in equation (161) may be replaced by empirical data obtained from the confined compression test,[82, 121] and the settlement of the surface of the ground layer having thickness h may be theoretically determined if the vertical stress σ_z is known:

$$\Delta z_s = \int_0^h \frac{\sigma_z\,dz}{E}.$$

The complexity of such computations, however, appears prohibitive. Further, the confined compression in which no lateral soil yielding occurs appears to have little direct application in the case of vehicles. Nevertheless, a study of the strength of undisturbed vs remolded soils, in which the void ratio is one of the primary changes, may be helped by the introduction of the above-outlined relations among ε, p, and E.

Simplified computations based on the assumption that the ratio f between the vertical stress σ_z and the corresponding strain e_z changes with depth z in accordance with the straight-line equation $f = f_o + az$, where a is a constant factor, give an interesting picture of the relation between a unit load and the radius of the circular area required to produce the same settlement in an ideally elastic mass $[f = f_o = E/(1 - \surd^2)]$, clay, and sand (Figure 85c).[54]

Figure 85 clearly indicates that the stress-strain relationship of real soils cannot be easily expressed. Computations of the loads and sinkage related to irregular loading areas, such as wheels, tracks, etc., present considerable difficulties. For this reason, only the most simplified assumptions have been accepted, the least accurate of which is the one

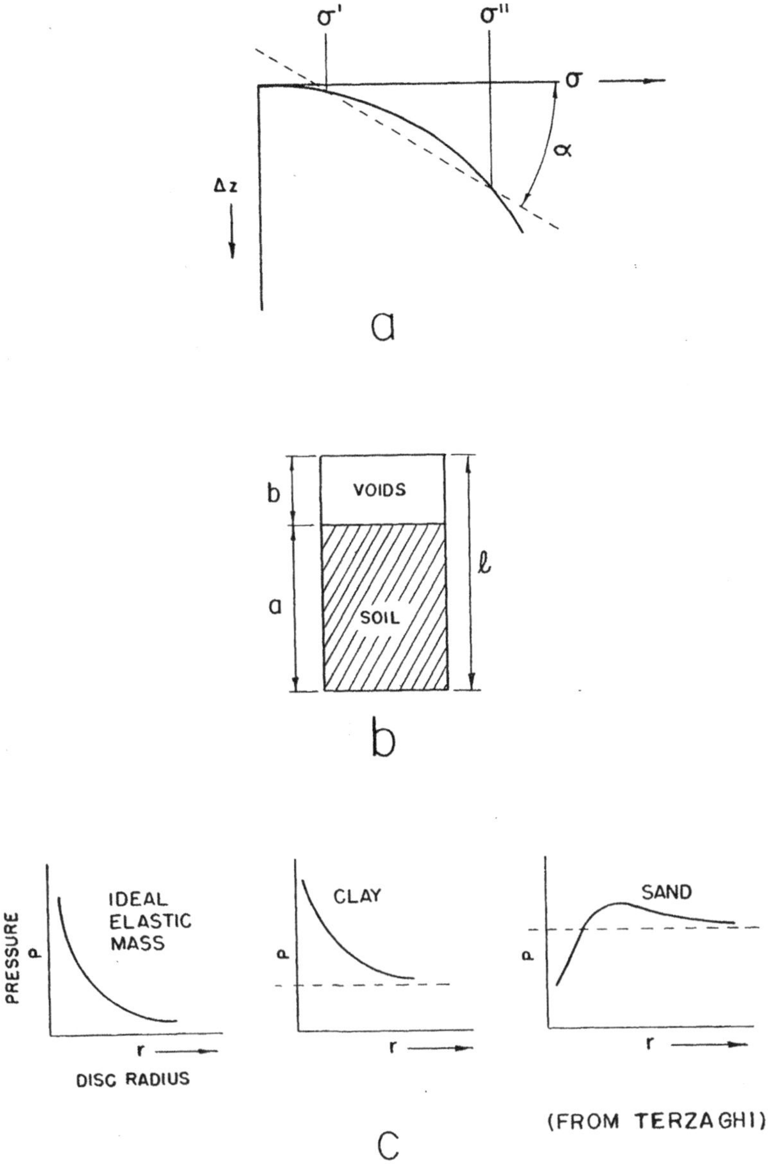
σ'
σ''
σ
α
Δz
a
b
VOIDS
ℓ
a
SOIL
b
PRESSURE
p
IDEAL
ELASTIC
MASS
r
DISC RADIUS
CLAY
p
r
SAND
p
r
(FROM TERZAGHI)
C

Fig. 85

which postulates a straight-line relationship between sinkage z and pressure p, as if soil were elastic:

$$p = kz\,, \tag{162}$$

where k is not called, however, the modulus of elasticity, but is merely considered as a coefficient of proportionality, also sometimes called the "bedding number." It may be interesting to note that both the criticism of equation (162) (see Reference 121) and its acceptance [123] have been recorded. Under these conditions, only a clear understanding of the limitations of the discussed formula may justify its use or rejection.

A more general form of equation (162) was proposed by Letoshnev, Goriatchkin, and others,[37] who assumed the relationship in the following form:

$$p = kz^n\,, \tag{163}$$

where n is the exponent depending on the type of soils, and which, as experiments by Bernstein indicate, is equal to 1/2 in average conditions.[124] Experiments by Klein [113] suggest indirectly that the same value of n may be proposed for average snow. A similar functional relationship between p and z was assumed for snow compaction by Kondratiev and others.[77] The problem, however, is that k is a function of the size of the loading area and cannot be considered as a "modulus" of deformation.

To sum up, it may be concluded that, for practical reasons, only very approximate assumptions may be made in order to evaluate the nonelastic deformation of soils and snow. The discrepancy between these assumptions and real conditions, as may be seen from the previous discussions, indicates the amount of work which is to be done in order to obtain a more satisfactory solution of such an important problem of vehicle mobility as such sinkage vs load in a given terrain.

VI. THE MECHANICS OF A WHEEL

In the previous chapter it was shown how man-made locomotion had been based on wheels and how the wheel had become the foremost element of practically every vehicle. In this section some problems of the wheel itself will be discussed and its movement will be analyzed in conjunction with the type of road upon which it operates.

There are four cases of the application of a wheel, as far as the mechanical properties of the materials involved in the construction of both the wheel and the road are concerned: the application of (1) a rigid wheel to a rigid surface, (2) a rigid wheel to an elastic or plastic surface, (3) an elastic wheel to a rigid surface, and (4) an elastic wheel to an elastic or plastic surface.

The first case refers theoretically to an ideally rigid wheel which runs on a perfectly rigid surface, where no deformations of material occur, and is the ideal case of locomotion in which no resistance is encountered; it has no practical meaning unless related to the more practical case of a steel wheel and the railroad.

The second case, which is related to the movement of a rigid wheel upon an elastic or plastic surface, may be referred to the action of a steel wheel on a rubber track or soft soil. An elastic wheel applied to a rigid surface, as mentioned in case (3), will be seen in the action of an automobile tire which runs on a concrete pavement, whereas the fourth case, comprising an elastic wheel rolling on an elastic or plastic bed, corresponds to an elastic tire applied to a plastic or elastic soil.

Rigid Wheel on a Rigid Surface

The case of a railway wheel was considered generally in Chapter I in conjunction with some problems of animal locomotion. It was mentioned that the movement resistance of such wheels, including the air resistance of the whole train, varies with speed in the same way as that shown by the border line of Gabrielli and von Kármán, which was determined for all types of land, sea, and air vehicles (Figure 32). It may now be suggested that the Gabrielli-Kármán line, which determines the minimum power required to drive any vehicle at a given speed, expresses the limiting value of the rolling resistance of steel wheels applied to a steel track. In fact, this type of locomotion requires the minimum propulsion power in all man-made land vehicles.

It is logical to assume that if air resistance is neglected, the rolling resistance of railway wheels will depend entirely upon the stress-strain relation of the surfaces in contact. Thus, a stress-strain pattern involved in rolling would determine the resistance of wheel motion. It should be possible, therefore, to relate this resistance, as taken per unit weight, to the ultimate strength of materials used, which would lead to a dimensional interpretation of equation (37) such as that proposed by Gabrielli and von Kármán.[3, 125]

These questions, which are interesting from a general angle, certainly deserve further attention, although the discussed relation may be difficult to demonstrate because the stress-strain picture between a wheel and the rail is rather complex.[126, 127] Since, from the automotive point of view, the questions related to the rolling of a rigid wheel upon an elastic and/or plastic medium, or to the problem of an elastic wheel in the same conditions, are of more immediate interest, the mechanics of a rigid wheel on a rigid surface will not be further considered.

Single Rigid Wheel on a Soft Ground

The problem of a rigid wheel is the oldest one. Many investigators, such as Grandvoinet,[33] Gerstner,[128] Coulomb,[33] Morin,[129] and Reynolds,[17] were preoccupied with it a hundred years ago, at which time the related phenomenon was investigated with the purpose of determining the rolling resistance of a rigid wheel acting upon various surfaces. In more modern times, Bernstein,[124] Shultz,[37] Goriatchkin,[37] Letoshnev,[128] Kuhne,[35] Meyer,[130] Swiezawski,[131] and many others have worked on similar questions.

The solutions which resulted from this work were based on more or less arbitrary assumptions. The variable properties of soils have been expressed by certain empirical coefficients whose meaning and significance were not quite based on any physical facts. Although some of the results obtained have a definite practical value, their theoretical generalization is impossible because of the lack of systematic studies of the stress-strain pattern of soils under the action of a wheel and the lack of the utilization in this respect of modern methods applied in elasticity and plasticity problems. A novel approach made in this field by Nuttall, who dealt with the problem of the wheel by means of dimensional analysis,[132] opened wide fields for investigation. In this approach, the tedious and uncertain analytical study of the role and effects of the various factors involved is replaced by a synthetic insight into the summary effect of apparently important factors, without investigating their separate mechanisms.

Such work, however, is barely in its first stage. The necessity of further developing the research along the discussed lines appears to be imperative, for it is impossible to imagine an understanding of the wheel-soil relationship unless it is analyzed by means of methods developed by mechanics.

Since most of the early investigators were interested in moving farm machinery or animal-drawn carts which, of course, did not have pneumatic tires, their theories refer to rigid wheels running on a soft medium. It is interesting to note that the introduction of tires in agriculture did stimulate innumerable fact-finding tests which have a definite value as far as the elucidation of the nature of the problem is concerned (References 133 to 150) but which did not produce any general theory of the relationship between soil and wheel. This, in consequence, handicapped the development of off-road vehicles, particularly the military ones whose requirements, as was mentioned before, are more far-reaching than those of farm implements. A fair picture of the existing theoretical knowledge of the problem of a rigid wheel applied to soil may be obtained by following the work by Bernstein [124] and Letoshnev.[128]

Assume that the pressure p underneath a sinking wheel is an exponential function of certain soil properties n and of the depth of sinkage z in accordance with equation (163). Then

$$p = kz^n ,$$

where k is the coefficient of proportionality. Accordingly, the work spent per unit of area in compressing the soil under the sinking wheel to depth z_0 is

$$L = \int_0^{z_0} p\, dz = \int_0^{z_0} kz^n\, dz = k\frac{z_0^{n+1}}{n+1} .$$

Various investigators assumed various values for n. Gerstner, Shultz, Goriatchkin, and Grandvoinet used $n = 1$ and hence $L_1 = kz_0^2/2$. Bernstein assumed $n = \frac{1}{2}$ and thus $L_2 = 2kz_0^{3/2}/3$. If it is assumed that $n = 0$, then the work of compression will be $L_3 = kz_0$.

Assume that the ground compressed by the circumference of a wheel moves only vertically. Under these circumstances, it may be admitted that the work L per unit area, expanded for the compression soil, is equal to the rolling resistance R which is counteracted by the pulling force located at the center of the wheel. If the distance rolled upon is l and the wheel width is b, then the work of traction is Rl and the work of soil compression is Lbl. Since both quantities must be equal,

$$Rl = Lbl \, .$$

Accordingly, the movement resistance is

$$R = Lb \, .$$

As previously discussed, in a general case, when $L = kz_o^{n+1}/(n+1)$,

$$R = kb \frac{z_o^{n+1}}{n+1} \, . \tag{164}$$

If the various values of n which were proposed by the various investigators are assumed,

$$n = 1 \, ; \qquad R_1 = \tfrac{1}{2} kbz_o^2 \tag{165}$$

$$n = \tfrac{1}{2} \, ; \qquad R_2 = \tfrac{2}{3} kbz_o^{3/2} \tag{166}$$

$$n = 0 \, ; \qquad R_3 = kbz_o \, . \tag{167}$$

The above equations express the rolling resistance as a function of coefficient k, wheel width b, and sinkage z. In order to obtain an analytical solution of the relationship between load and wheel dimensions, consider the equilibrium of forces acting upon a wheel (Figure 86).

A wheel which sinks to depth z_o is acted upon by a load W and a pulling force equal to the movement resistance R. Assume that N is the reaction of the soil equal to the sum of the elementary reactions dN of the ground resistance against rolling. These reactions are assumed to be perpendicular to the circumference of the wheel. Then, according to the denotations shown in Figure 86, the following equations may be written:

$$R - \int_0^{\theta_o} dN \sin \theta = 0 \tag{168}$$

$$-W + \int_0^{\theta_o} dN \cos \theta = 0 \, . \tag{169}$$

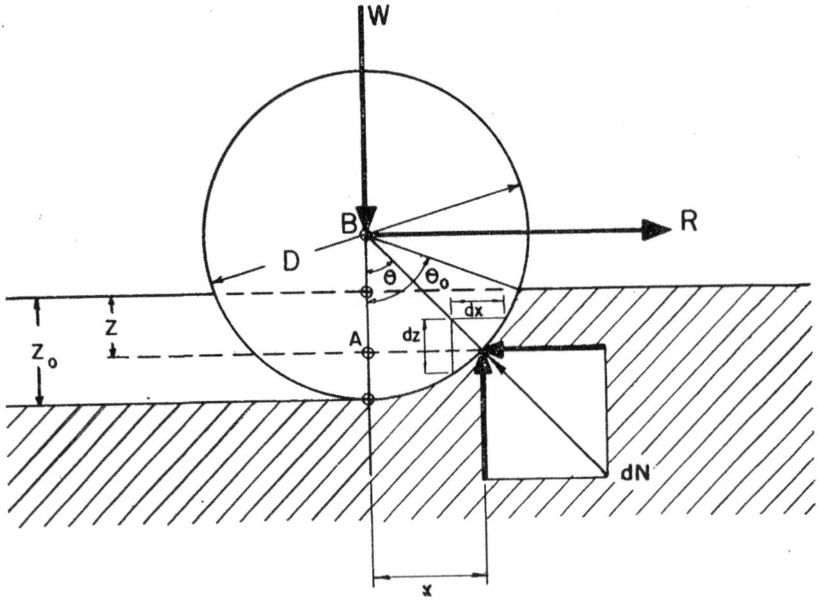

Fig. 86

The elementary reaction $dN \cos \theta = -pb\, dx$ and $dN \sin \theta = pb\, dz$. Accordingly, equations (168) and (169) may be written as follows:

$$R = \int_0^{z_0} pb\, dz \tag{170}$$

$$W = -\int_0^{z_0} pb\, dx\,. \tag{171}$$

Equation (170) has already been solved in the form of equations (165), (166), and (167). Now, equation (171) also may be solved when assuming, for example, that $p = kz^n$. Then,

$$W = -\int_0^{z_0} bkz^n\, dx\,. \tag{172}$$

From the geometrical configuration of dimensions shown in Figure 86, it results that

$$\overline{AB} = (D/2) - (z_o - z)$$

$$x^2 = (D/2)^2 - (\overline{AB})^2 .$$

Hence,

$$x^2 = [D - (z_o - z)]\,(z_o - z)$$

or, approximately,

$$x^2 = D(z_o - z) \qquad (173)$$

and

$$2x\,dx = -D\,dz\,. \qquad (174)$$

By combining equations (173), (174), and (172), the following may be obtained:

$$W = bk\int_0^{z_o} z^n \cdot \frac{\sqrt{D}\,dz}{2\sqrt{z_o - z}}\,.$$

Denote $z_o - z = t^2$. Accordingly, $dz = -2t\,dt$ and

$$W = bk\sqrt{D}\int_0^{t=\sqrt{z_o}} (z_o - t^2)^n\,dt\,.$$

By expanding $(z_o - t^2)^n$ into a series and taking only the first two members of that series $(z_o{}^n - nz_o{}^{n-1}\,t^2 + \ldots)$, it will be found that

$$W = bk\sqrt{D}\int_0^{\sqrt{z_o}} (z_o{}^n - nz_o{}^{n-1}\,t^2)\,dt$$

or

$$W = \frac{bk\sqrt{Dz_o}}{3}\,z_o{}^n\,(3 - n)\,. \qquad (175)$$

This equation expresses the relationship among load W, diameter D, width b, and depth of sinkage z_o. If it is assumed again that $n = 1, \frac{1}{2}$, and 0, then

$$n = 1\,; \qquad W_1 = \tfrac{2}{3}\,bkz_o\sqrt{Dz_o} \qquad (176)$$

$$n = \tfrac{1}{2}\,; \qquad W_2 = \tfrac{5}{6}\,bkz_o^{\frac{1}{2}}\sqrt{Dz_o} \qquad (177)$$

$$n = 0\,; \qquad W_3 = bk\sqrt{Dz_o}\,. \qquad (178)$$

The relationship between the rolling resistance, load, and dimensions of wheels may be obtained by eliminating z_0 from equations (164) and (175). In the case when the stress of compression increases proportionally to the depth ($n = 1$),

$$0.633\, W_1^4 = bkD^2R_1^3 \,. \tag{179}$$

If ground properties are such that the n value is equal to $\frac{1}{2}$, then

$$0.590\, W_2^6 = (bk)^2D^3R_2^4 \,. \tag{180}$$

When the ground stress does not depend on the depth of sinkage z ($n = 0$), then

$$W_3^2 = bkDR_3 \,. \tag{181}$$

Solving equations (179), (180), and (181) gives

$$R_1 = \frac{0.86}{\sqrt[3]{bk}} \frac{W_1^{4/3}}{D^{2/3}} \tag{182}$$

$$R_2 = \frac{0.876}{\sqrt{k'b}} \frac{W_2^{3/2}}{D^{3/4}} \tag{183}$$

$$R_3 = \frac{1}{k} \frac{W_3^2}{bD} \,. \tag{184}$$

Equation (182) is identical with that deduced by Gerstner; equation (183) has the same form as the formula by Bernstein. The coefficient k, however, does not depend on soil type only. It also depends on wheel width b and was proposed in the following form:

$$k' = 2a' + a''b \,, \tag{185}$$

where a' and a'' are empirical constants related to the shear along the edges of a wheel root and to the compression of the area which equals the area of the root in a given type of soil. Thus, the idea of the perimeter-area ratio discussed in the preceding chapter in conjunction with the work by Kögler,[117] Housel,[118] and Terzaghi [54] was considered in a somewhat different form by Bernstein approximately 40 years ago. Some further work developed by the Russian school along these lines will be discussed later.

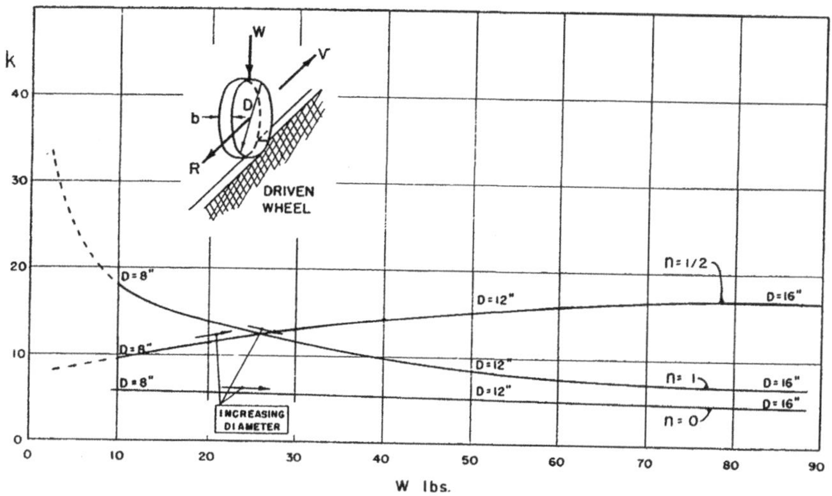

Fig. 87

A demonstration of the practical value of formulas (182), (183), and (184) may be considered debatable. Rational tests which would check the relationships among wheel dimensions, load, rolling resistance, and soil properties have not yet been made. The available data are seldom complete, or comparable, and a full answer to the question raised is lacking. If, however, the figures for R, W, D, and b considered in Nuttall's tests,[132] which seem to offer the only data so far obtained in fully controlled conditions, are fed into equations (182), (183), and (184), then the degree of constancy of k for a given type of soil (n) may be considered as a measure of the accuracy of the discussed formula. The results of such a computation are shown in Figure 87. The lines plotted in this graph refer to mean values computed for various wheel diameters D ranging from 16 in. to 8 in. They indicate that for large diameters, all the formulas seem to give a reasonable accuracy, since k is fairly stable. It is evident, however, that equation (184), which is based on $n = 0$, is most accurate within the range of investigated diameters D, loads W, and soil type. However, the most frequently accepted value is $n = \frac{1}{2}$. Perhaps soils *in situ* conform better with this figure than the soil used by Nuttall (Davidson Loam) in the laboratory bin.

Generally speaking, it would appear that the equation $p = kz^n$ is reasonably correct and that both k and n do express approximately those soil properties that affect the rolling resistance of a wheel at given load and dimensions. It should be remembered, however, that the validity of these equations is acceptable within a narrow range of sinkage values z, which was assumed when simplifying the integration. For higher sinkage, a wide variation of k is to be expected with the variation of wheel dimensions.

Rigid Wheels in Tandem on a Soft Ground

Since sinkage obviously affects rolling resistance, it must be assumed that the way in which wheels are arranged in an undercarriage affects the total rolling resistance of a vehicle. It would be interesting, therefore, to investigate the movement resistance of an ordinary four-wheel carriage, or, what it amounts to, of two rigid wheels in tandem.

According to previous considerations, the following may be the approximate answer to the problem: Assume that the loads acting on the front and rear axles are W' and W'', respectively. The radii and sinkages of the front and rear wheels are denoted by r_1, z_1, and r_2, $z_1 + z_2$, respectively (Figure 88). By following Bernstein's assumption that $p = k'\sqrt{z}$, i.e., that $n = \frac{1}{2}$ and $k' = 2a' + a''b$, it may be admitted that the unit load underneath the particular wheels is:

$$\text{front wheels: } p' = k'\sqrt{z_1}$$

$$\text{rear wheels: } p'' = k'\sqrt{z_1 + z_2}\,.$$

These unit loads will prevail when the rear wheel follows the same rut as the first wheel. According to the denotations, Figure 86, and equation (169),

$$W = \int_0^{\theta_o} dN \cos\theta\,.$$

On the other hand, $dN\cos\theta = pb\,ds\cos\theta$; since, in Bernstein's assumption, $p = k'\sqrt{z}$, the total wheel load W is

$$W = \int_0^{\theta_o} bk'\sqrt{z}\cos\theta\,ds\,.$$

If it is recalled that $ds = r\,d\theta$ and $z = r(\cos\theta - \cos\theta_o)$, then

$$W = \int_0^{\theta_o} bk'r\sqrt{r(\cos\theta - \cos\theta_o)}\cos\theta\,d\theta\,.$$

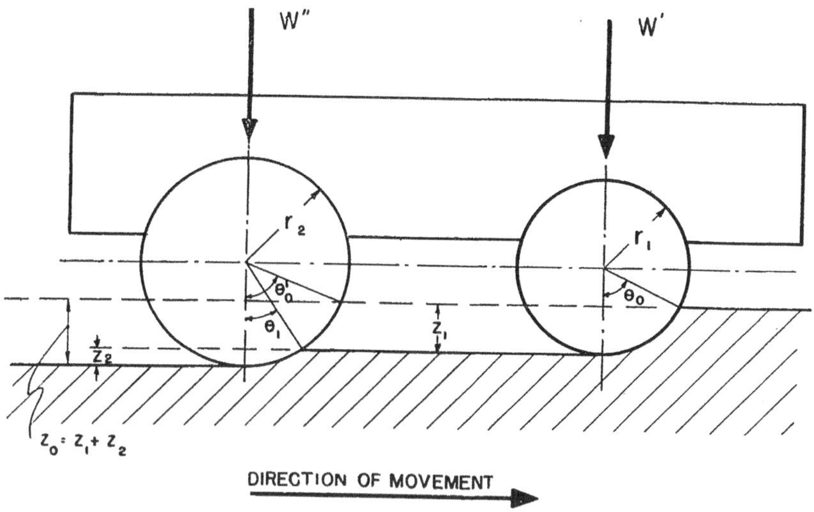

Fig. 88

When solving this equation, consider a small sinkage. Thus, a very small value of θ will result. Accordingly, $\cos\theta\, d\theta \cong d(\sin\theta) \cong d\theta$. By writing a series expansion for $\cos\theta$ and $\cos\theta_o$, from which only the first two members are taken ($\cos\theta = 1 - (\theta^2/2) + \ldots$), and by substituting these simplified values in the above equation, the following will be obtained:

$$W = bk' \frac{r^{3/2}}{\sqrt{2}} \int_0^{\theta_o} (\sqrt{\theta_o^2 - \theta^2})\, d\theta\,, \tag{186}$$

$$W = bk' \frac{r^{3/2}}{\sqrt{2}} \frac{1}{2} \left| \theta_o^2 \sin^{-1} \frac{\theta}{\theta_o} + \theta\sqrt{\theta_o^2 - \theta^2} \right|_0^{\theta_o}$$

or, finally,

$$W = \frac{\pi b k' r^{3/2}}{4\sqrt{2}} \theta_o^2\,.$$

But for the first wheel of the tandem,

$$z_1 = r_1 - r_1 \cos\theta_o\,.$$

After expanding $\cos\theta_o$ into a series and taking the first two terms, it will be found that

$$z_1 = r_1 - r_1\left(1 - \frac{\theta_o^2}{2} + \ldots\right)$$

and

$$\theta_o^2 = 2z_1/r_1 \,.$$

Thus, the load of the front wheel is

$$W' = \frac{\pi b k' z_1 \sqrt{r_1}}{2\sqrt{2}} \,. \tag{187}$$

A similar relationship for the rear wheel may be obtained from formula (186) if θ_1 is substituted for θ_o in the limit of the integral, θ'_o for θ_o under the square root, and r_2 for r in the numerator:

$$W'' = \frac{bk' r_2^{3/2}}{\sqrt{2}} \int_0^{\theta_1} (\sqrt{\theta'^2_o - \theta^2})\, d\theta \,.$$

Since $\sqrt{\theta'^2_o - \theta^2} = \theta'_o\left(1 - \tfrac{1}{2}\dfrac{\theta^2}{\theta'^2_o} - \ldots\right)$, etc.,

$$\int_0^{\theta_1} (\sqrt{\theta'^2_o - \theta})\, d\theta = \theta'_o\theta_1 - \tfrac{1}{6}\frac{\theta_1^3}{\theta'_o} = \theta'_o\theta_1\left(1 - \tfrac{1}{6}\frac{\theta_1^2}{\theta'^2_o}\right).$$

The expression $\theta_1^2/6\theta'^2_o$ is very small and may be neglected. Thus,

$$W'' = \frac{bk' r_2^{3/2}}{\sqrt{2}} \theta'_o \theta_1 \,. \tag{188}$$

If, as was previously deduced for the first wheel, it is assumed that $z_o = r_2 - r_2 \cos\theta'_o \cong r_2\theta'^2_o/2$, $z_2 \cong r_2\theta_1^2/2$, and hence

$$\theta'_o = \sqrt{2z_o/r_2}$$
$$\theta_1 = \sqrt{2z_2/r_2}\,,$$

it will be found that

$$W'' = bk' \sqrt{2r_2} \sqrt{z_o z_2} \,. \tag{189}$$

From equations (187) and (189), z_1 and z_2 can be determined as a function of the loads W' and W'' and the wheel radii r_1 and r_2. By solving these equations with reference to z_1 and z_2 and recalling that $z_o = z_1 + z_2$, the following formula will be obtained:

$$z_o = \frac{2\sqrt{2}\,W'}{b\pi k'\sqrt{r_1}} + \frac{(W'')^2}{2b^2 r_2 z_o (k')^2}\,.$$

The solution of this quadratic equation gives

$$z_o = \frac{1}{bk'}\left(\frac{W'\sqrt{2}}{\pi\sqrt{r_1}} + \sqrt{\frac{2W'^2}{\pi^2 r_1} + \frac{W''^2}{2r_2}}\right). \tag{190}$$

If the relationship between the rolling resistance R and sinkage z is assumed to be expressed by formula (166), which incorporates the value of $n = \frac{1}{2}$, then

$$R = \tfrac{2}{3}\, bk' z_o^{3/2}\,.$$

By substituting the value z_o from equation (190) into this formula, it will be found that

$$R = \frac{2}{3}\,\frac{1}{\sqrt{bk'}}\left(\frac{W'\sqrt{2}}{\pi\sqrt{r_1}} + \sqrt{\frac{2W'^2}{\pi^2 r_1} + \frac{W''^2}{2r_2}}\right)^{3/2} \tag{191}$$

or, since $r_1 = D_1/2$ and $r_2 = D_2/2$,

$$R = \frac{2}{3}\,\frac{1}{\sqrt{bk'}}\left(\frac{2W'}{\pi\sqrt{D_1}} + \sqrt{\frac{4W'^2}{\pi^2 D_1} + \frac{W''^2}{D_2}}\right)^{3/2}. \tag{192}$$

Formula (192) is valid for two wheels in tandem loaded with loads W' and W''. It is also valid for a four-wheel trailer with axles loaded in the same way. In this case, loads W' and W'' would mean axle loads.

Assume that the trailer load is W and $W'/W'' = a$. Then,

$$W = W''(1 + a)$$

$$W' = \frac{a}{a+1}\,W$$

$$W'' = \frac{1}{a+1}\,W\,.$$

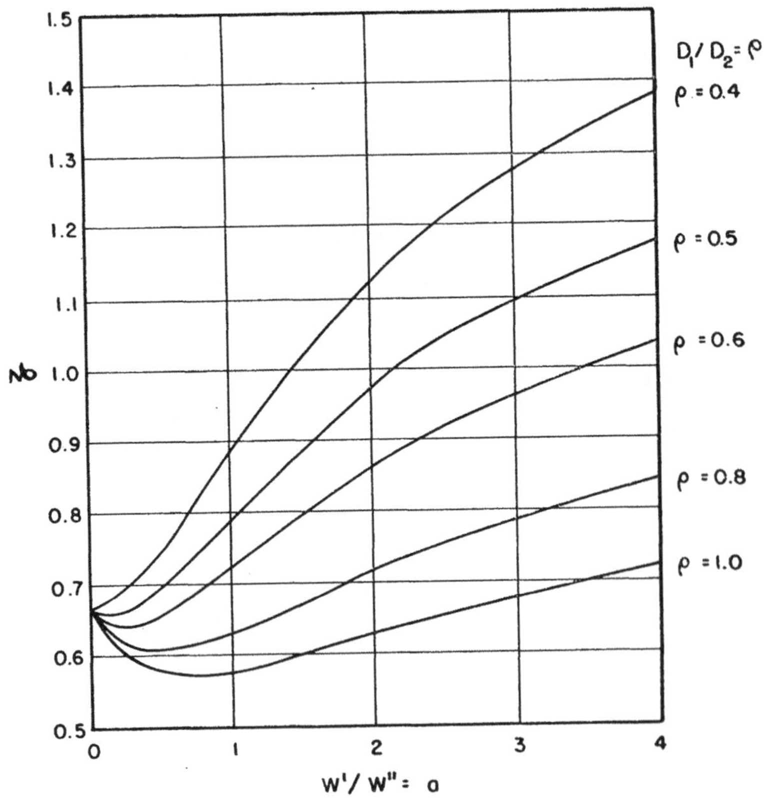

Fig. 89

When these values are substituted into formula (192),

$$R = \tfrac{2}{3} \frac{1}{(a+1)^{3/2}} \frac{W^{3/2}}{\sqrt{bk'}} \left(\frac{2a}{\pi \sqrt{D_1}} + \sqrt{\frac{4a^2}{\pi^2 D_1} + \frac{1}{D_2}} \right)^{3/2} . \tag{193}$$

If D_1/D_2 is denoted by ϱ, or if $D_1 = \varrho D_2$, then equation (193) yields

$$R = \tfrac{2}{3} \frac{1}{(a+1)^{3/2}} \frac{W^{3/2}}{D_2{}^{3/4} \sqrt{bk'}} \left(\frac{2a}{\pi \sqrt{\varrho}} + \sqrt{1 + \frac{4a^2}{\pi^2 \varrho}} \right)^{3/2} . \tag{194}$$

In order to simplify this equation, the following denotation may be made:

$$\zeta = \frac{2}{3(a+1)^{3/2}} \left(\frac{2a}{\pi \sqrt{\varrho}} + \sqrt{1 + \frac{4a^2}{\pi^2 \varrho}} \right)^{3/2} . \tag{195}$$

Then

$$R = \frac{\zeta}{\sqrt{bk'}} \frac{W^{3/2}}{D_2^{3/4}} . \tag{196}$$

The coefficient ζ depends entirely on the relative diameters of the trailer wheels and load distribution. It will be seen that the general form of equation (196), which refers to two wheels in tandem or to a four-wheel trailer, is identical with the form of equation (183) previously deduced for a single wheel or for a two-wheel trailer. The coefficient ζ indicates the extent to which two wheels in tandem affect the total rolling resistance of the towed vehicle.

In order to ease the investigation of the problem within the limits of the assumptions made, various values of ϱ and a were substituted in equation (196). The results are shown in Figure 89.

Comparison between Tandem and Dual Wheels

For tandem wheels (or a two-axle trailer) whose center of gravity is so located that a uniform load distribution over the wheels is obtained ($a = 1$), and whose diameters are equal ($\varrho = 1$), the value ζ of equation (195) is

$$\zeta = \frac{1}{3\sqrt{2}} \left(\frac{2}{\pi} + \sqrt{1 + \frac{4}{\pi^2}} \right)^{3/2} = 0.580 . \tag{197}$$

This value used in conjunction with equations (196) and (183) leads to the determination of the relative merits of a tandem wheel arrangement versus a dual wheel arrangement. Formula (196) is valid in both cases if the appropriate value of k' is adopted. It is obvious that for tandem wheels [see equation (185)],

$$k'_t = 2a' + a''b ,$$

and for dual wheels,

$$k'_a = 4a' + 2a''b .$$

Thus, according to equations (196) and (197) for tandem arrangements and equation (183) for dual (single) wheel arrangements, with a uniform distribution of load on both wheels,

$$\frac{R_t}{R_d} = \frac{\dfrac{0.580\, W^{3/2}}{\sqrt{(bk')_t}\, D^{3/4}}}{\dfrac{0.876\, W^{3/2}}{\sqrt{(2bk')_d}\, D^{3/4}}} = 0.662 \sqrt{\frac{K_{2b}}{K_{1b}}} \tag{198}$$

if $(bk')_t$ is denoted by K_{1b} and $(2bk')_d$ by K_{2b}.

The experimental work performed by Russian investigators [128] leads to the empirical values of K_b which are plotted on Figure 90 with reference to the width b of the wheel rim and as a function of soil type. It will be noted that K_b increases in the case of practically all soils, the exception being, however, the gravel road on which k' shows a rapid decrease.

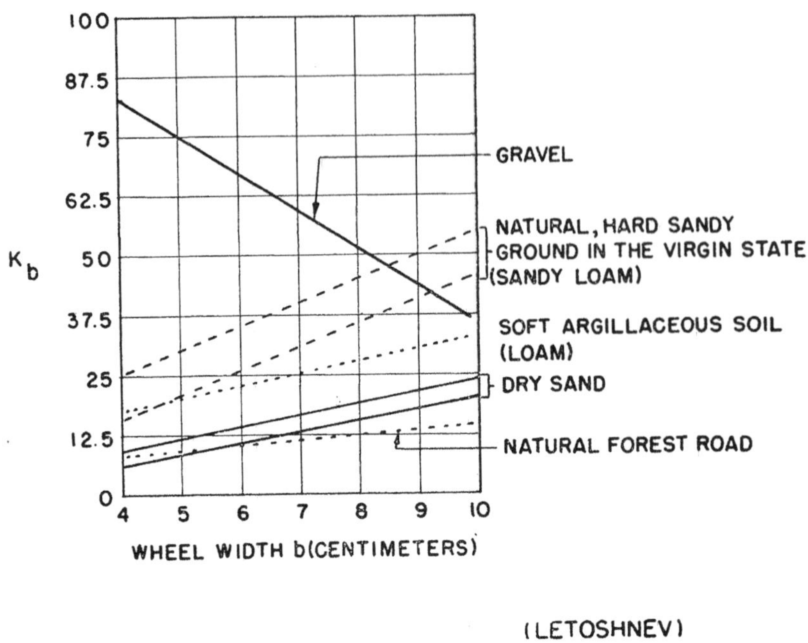

Fig. 90

Take a hard, natural sandy road. For a tandem arrangement and wheel width $b = 4$ cm, $K_{1b} = 25$. In the same conditions for a dual wheel arrangement, i.e., for $2b = 8$ cm, $K_{2b} = 45$. Hence, equation (198) will give

$$\frac{R_t}{R_d} = 0.662 \sqrt{\frac{45}{25}} = 0.888 .$$

The same ratio for dry sand will be ($K_{2b} = 15$; $K_{1b} = 6$)

$$\frac{R_t}{R_d} = 0.662 \sqrt{\frac{15}{6}} = 1.047 .$$

In other words, the dual arrangement of wheels will produce 12% higher and 5% lower rolling resistance than the tandem arrangement of the same wheels in a hard sandy ground and in loose sand, respectively.

It is worthwhile to note that McKibben and Davidson [151] investigated a similar problem. They found that in loose sand the tandem wheels encounter 23% less movement resistance than a dual configuration. The experiments were performed in such a way that the 6 × 16.00 tires were separated in the dual arrangement by 8 in., which may account for the lack of agreement, with reference to the trend in loose sand, between the experimental and calculated results.

Effect of Weight Distribution and Wheel Dimensions upon the Rolling Resistance of Carriages in Soft Ground

The effect of types of weight distribution upon the rolling resistance of vehicles may be stressed on the basis of equation (196) and Figure 89 as follows:

(a) Overloading the front axle increases the rolling resistance. The increase is more intensive when the diameter of the front wheels is smaller than that of the rear wheels. In general, it appears to be more profitable to allow more sinkage at the rear than in the front.

(b) When the dimensions of all the wheels are the same, the load distribution affects the rolling resistance to a smaller degree than when the diameter of the front wheels is different from that of the rear wheels. The most convenient distribution is when the load is spread uniformly on the front and rear.

(c) Rolling resistance increases with an increase in the difference between the diameters of the front and rear wheels. For example, for

$$a = 4 \quad \text{and } \varrho = 1\,, \qquad \zeta = 0.724$$
$$a = 0.25 \text{ and } \varrho = 1\,, \qquad \zeta = 0.604$$
$$a = 1 \quad \text{and } \varrho = 1\,, \qquad \zeta = 0.580$$

but for

$$a = 4 \quad \text{and } \varrho = 0.4\,, \qquad \zeta = 1.390$$
$$a = 0.25 \text{ and } \varrho = 0.4\,, \qquad \zeta = 0.694$$
$$a = 1 \quad \text{and } \varrho = 0.4\,, \qquad \zeta = 0.891\,.$$

Thus, a four-fold overloading of the front axle ($a = 4$) of a trailer having identical wheel diameters ($\varrho = 1$) would increase the rolling resistance by

$$\frac{0.724 - 0.580}{0.580}\,100 = 25\%\,.$$

But a four-fold overloading of the front, where the wheels are much smaller than in the rear ($\varrho = 0.4$), would increase the rolling resistance by

$$\frac{1.390 - 0.891}{0.891}\,100 = 56\%\,.$$

Accordingly, for rear wheels larger than the front ones ($\varrho = 0.4$), it is more profitable to overload the rear axle ($a = 0.25$). In such a case, the rolling resistance may be reduced by

$$\frac{0.891 - 0.694}{0.891}\,100 = 22\%$$

if the rear-axle load is four times higher than that at the front axle.[128]

Practical lessons which were learned the hard way in recent years also lead to similar conclusions. However, if these lessons had been supported by a reliable theory, then such ideas as building a vehicle with front dual tires, or an airplane with tires of enormous width, would not have been conceived.

The need for an over-all theory may be illustrated by the fact that the previous conclusions are based on Bernstein's assumption of $n = \frac{1}{2}$ [equation (183)]. If it is assumed that Gerstner's $n = 1$ is valid [equation (182)], which may be quite plausible as shown in Figure 87, then the conclusions regarding the effect of load upon the wheels of a four-wheel carriage would be different. It would be found that overloading the front wheels may lead to an economy in rolling resistance. The effect of the wheel diameters would be identical, however, with that determined for $n = \frac{1}{2}$.[128]

It may be agreed, however, that Gerstner's assumption refers to unusual soil properties and that Bernstein's assumption is more acceptable since it apparently refers to a more common type of soil encountered. Wheel behavior in less common types of soil or snow remains to be investigated. In any case, it may be stressed again that McKibben and Davidson's experiments [151] support the general conclusion which may be drawn by means of Bernstein's equation, i.e., by assuming $n = \frac{1}{2}$.

Elastic Wheel on a Hard Surface; Geometry of the Contact Area

An unloaded, uniformly shaped pneumatic tire (Figure 91a) is subjected to tension σ acting in the meridian direction. Since the state of stresses is symmetrical with reference to the center, the shearing forces τ vanish and the tire assumes a constant curvature $r = b/2$ (Figure 91b). Thus, the determination of tire form is a matter of geometry.[152] Accordingly,

$$h = \frac{b}{2}\left[1 + \sqrt{1 - \left(\frac{b_r}{b}\right)^2}\right]. \tag{199}$$

When a tire is resting on a rigid surface under a load (Figure 91b), then in view of the absence of shearing forces the curvature r_1 of the tire is constant and the determination of its shape again is a matter of geometry. Accordingly, when following Figure 91b,

$$h = f + r_1(1 + \cos\delta) \tag{200}$$

and

$$b_r = b' + 2r_1 \sin\delta\,. \tag{201}$$

Since the deflection does not cause the length u of the circumference of the tire cross section to change,

$$u = b' + 2r_1(\pi - \delta) = b\left[\pi - \sin^{-1}\left(\frac{b_r}{b}\right)\right]. \tag{202}$$

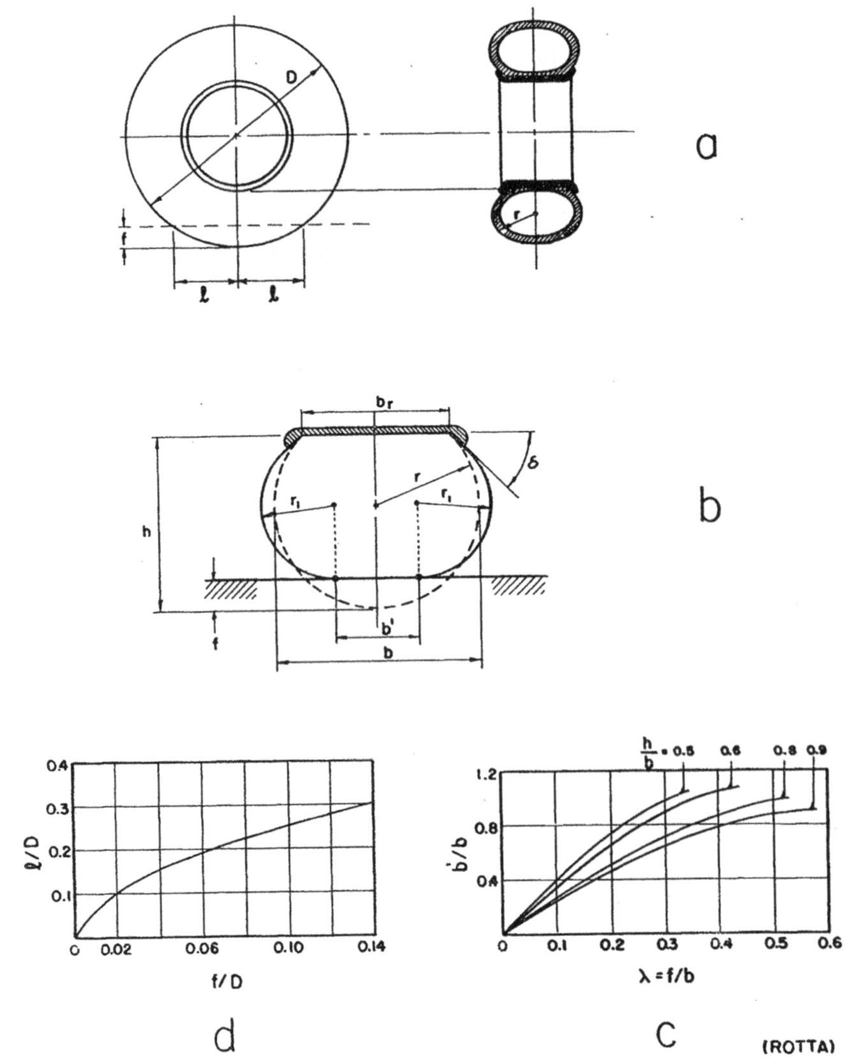
D
f
l
l
r
a
b_r
h
r
r_1
r_1
δ
f
b'
b
b
0.4
0.3
0.2
0.1
l/D
0
0.02
0.06
0.10
0.14
f/D
d
h/b = 0.5
0.6
0.8
0.9
1.2
0.8
0.4
b'/b
0
0.1
0.2
0.3
0.4
0.5
0.6
λ = f/b
c
(ROTTA)

Fig. 91

Denote $b_r/b = \psi$, $f/b = \lambda$, and $2r_1/b = \varrho$. Then, from equations (201) and (202),

$$\varrho = \frac{\pi - \sin^{-1}\psi - \psi}{\pi - \delta - \sin\delta}. \tag{203}$$

Combining equations (203), (200), and (199) gives

$$\lambda = \frac{1}{2}\left[1 + \sqrt{1 - \psi^2} - \frac{\pi - \sin^{-1}\psi - \psi}{\pi - \delta - \sin\delta}(1 + \cos\delta)\right]. \tag{204}$$

And finally, from (201),

$$\frac{b'}{b} = \frac{\psi(\pi - \delta) - (\pi - \sin^{-1}\psi)\sin\delta}{\pi - \delta - \sin\delta}. \tag{205}$$

A graphic solution of some of the above relations is shown in Figure 91c. The length $2l$ of the contact area (Figure 91a) of a tire deflected by f is shown in Figure 91d. Thus, assuming that the ground contact area may be determined as an area of a rectangle having its dimensions $2l \times b'$ reduced by 15%, it will be found that

$$A = 1.7\, l\, b'. \tag{206}$$

Pressure Distribution of Elastic Tires on a Hard Surface

The pressure distribution in the case of an ideally elastic tire and rigid surface would be uniform and equal to the pressure of inflation. However, the presence of tire treads and the stiffness of the carcass change the picture. The problem was investigated by Martin,[153] who based his work on the experimental measuring of pressure distribution by means of a special apparatus.

Figure 92a shows a solid rubber tire loaded up to $W = 3700$ lb, and resting upon a hard pavement. Assume that the local tire pressure is proportional to the rubber deflection z. Then the ground pressure is

$$p = c_r z,$$

where c_r is the elastic constant of the rubber. Experimental verification showed the practical value of this assumption. Accordingly, the maximum pressure in the center would be

$$p_{\max} = c_r f.$$

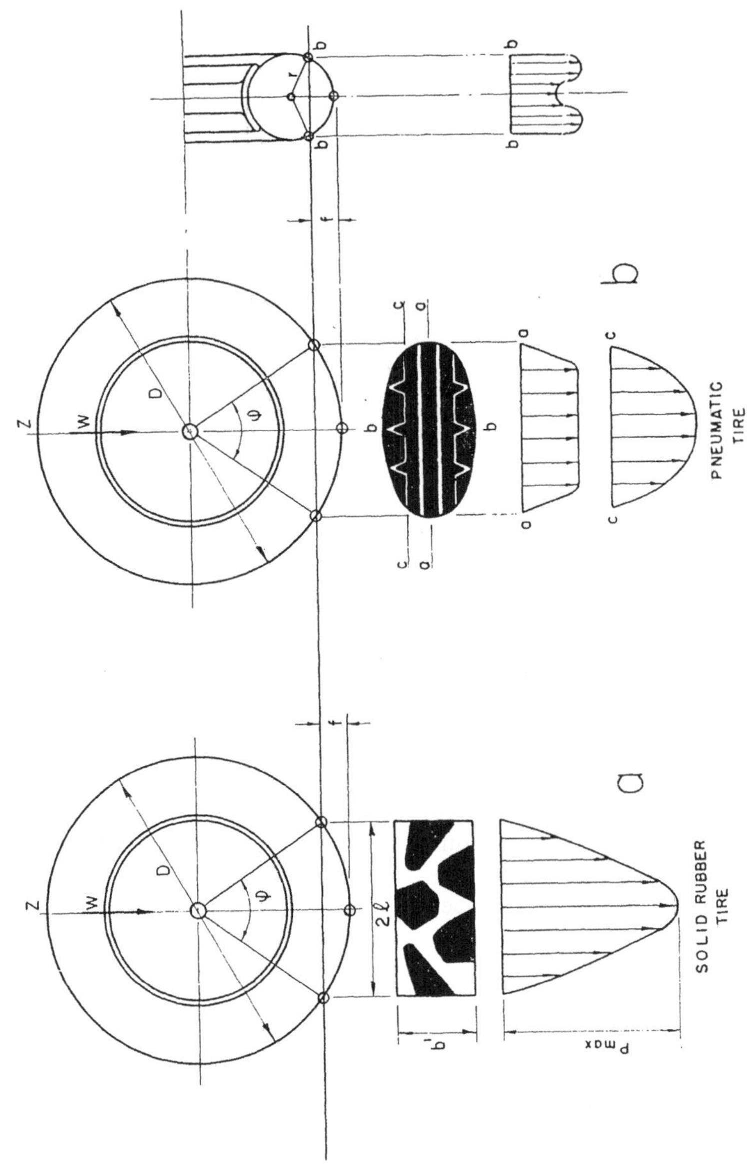

(MARTIN)

Fig. 92

An average pressure per unit width of the tire would be obtained by dividing the unit load by the length of the contact area. This gives

$$p_{\max} = c_r \frac{D\,(\varphi - \sin\varphi)}{8 \sin(\varphi/2)} . \tag{207}$$

By assuming that the carrying surface of the tire is determined by the tread design and is smaller than $2l \times b'$ (Figure 92a), and by introducing the ratio

$$a = \frac{\text{real area of contact}}{\text{area of } 2l \times b'} ,$$

the following formula for $p_{\max}$ as a function of the wheel load W and maximum deflection f was proposed:

$$p_{\max} = \frac{0.53\,W}{ab' \sqrt{f\,D/2}} . \tag{208}$$

For pneumatic tires, the problem appears to be more complex since the pressure distribution depends not only on the inflation but also on the stiffness of the tire carcass.

On the basis of the measurements of pressure distribution in the various sections a–a, b–b, and c–c shown in Figure 92b, the following semi-empirical equation may be considered:

$$W = (p_i + p_c) \frac{f^2}{f+1} \sqrt{2Dr - 2f\,[(D/2) + r]} , \tag{209}$$

where W (kg) is the tire load, f is the deflection in centimeters, $D/2$ and r the tire radii in centimeters, p_i the inflation pressure in kg/cm^2 and p_c the mean vertical pressure of the carcass alone. Experiments performed with airplane tires showed a good relationship between W and f, as defined in the above formula.

Besides the vertical pressure distribution of the tire, the distribution of the horizontal tensions also is of importance since it is mainly responsible for the gripping power of the tire rolling resistance and for the tire wear. Martin [153] considered this case in a general way. Other studies on this subject may be found in papers by Wedemeyer [154] and Schmidt.[155]

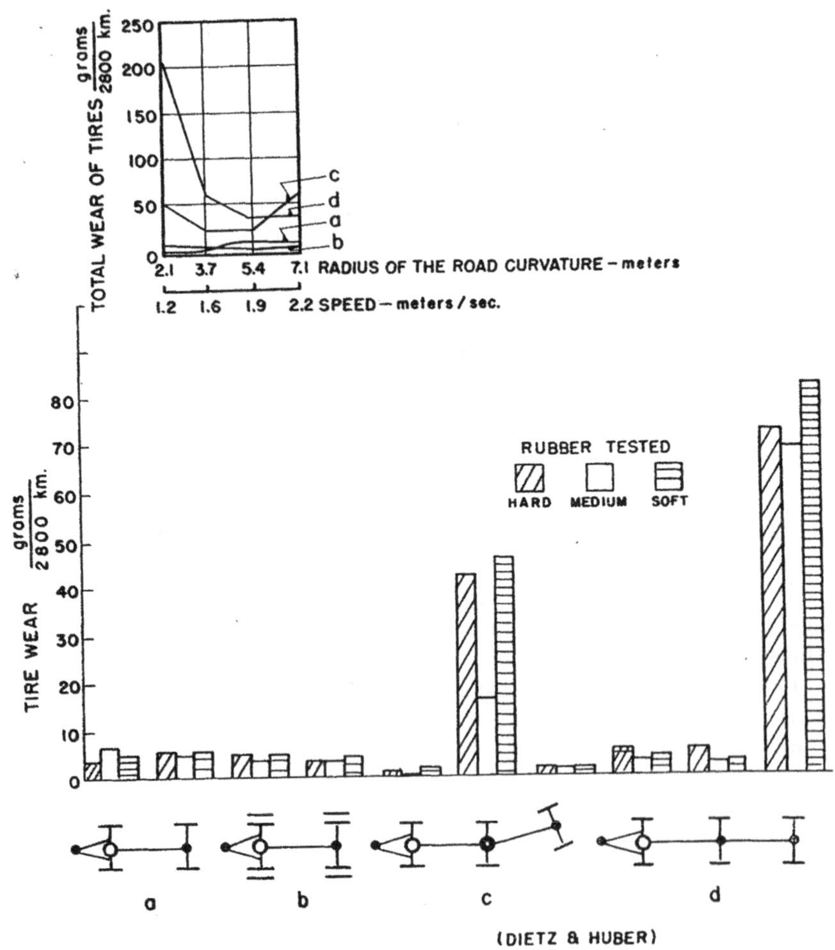

Fig. 93

It should be borne in mind, however, that the tangential pressure distribution is to be related not only to the tire load in static conditions but also to the over-all arrangement of the wheel system which affects the general character of the wheel load. Investigations by Dietz and Huber [156] of scale models of trailers whose tires were arranged as shown in Figure 93 indirectly throw some light on this problem. It is not surprising to find out that a six-wheel trailer of type *d* wears out the tires most quickly. The same can be applied to six-wheelers having identical wheel arrangements. If it is assumed that the tire wear, as shown in Figure 93, is proportional to the resultant tangential stress, then the graph shown in this figure may be quantitatively interpreted in terms of the tire loads which prevail at given wheel arrangements. A method for the determination of dynamic tire loads will be found in Reference 157. The dynamic loads acting upon the road due to the wheel impact will be discussed later.

Rolling Resistance of Elastic Tires on Hard Surface

The distribution of load and the mechanical and geometrical characteristics of the tire and its speed affect the rolling resistance on a hard road. A rigorous study of how all the relevent factors may be correlated with the resistance of a tire to motion is not known. Kamm presented a semi-empirical formula which expresses the rolling resistance R in kilograms as a function of inflation, load, and speed in the following way:

$$R = 5.1 + \frac{5.5 + 18W}{p} + \frac{8.5 + 6W\,(v/100)^2}{p}, \qquad (210)$$

where W is the wheel load in tons, p is the tire pressure in kg/sq cm, and v is the speed in km/hr. Kluge and Haas [158] found experimentally that the formula gives good results for conventional passenger-car tires having a pressure $p = 20 \div 50$ psi ($p = 1.4 \div 3.5$ kg/cm^2) and speeds up to approximately 95 mph (150 km/hr). However, a study of unconventional types of tires and higher speeds is required it the validity of this formula is to be extended.

Other attempts to determine the rolling resistance of a car in terms of inflation pressure of tires and speed are described in Reference 159. In particular, some information on the nature of the tire slip, which produces the main components of motion resistance, will be found in the work by Reynolds, [17] Swiezawski,[131] Gough and Jones,[157] and de Carbon.[160]

Elastic Tire on a Soft Ground: General

Although cross-country transport performed by means of wheeled vehicles relies entirely on pneumatic tires, the mechanics of the interrelation between soil and tire is practically unknown. Undoubtedly, the complexity of the phenomena involved is responsible for this situation.[17] The vital necessity of gaining more knowledge in this field, however, cannot excuse the lack of a more general solution of the discussed subject. The present state of knowledge may be based only on qualitative conclusions drawn from the previously analyzed problem of soil and snow mechanics.

As was previously shown, the pressure distribution in the ground depends on the mechanical properties of the soil, on the form and size of the loading area, on the degree of its elasticity, and on the type of pressure distribution applied to the loading surface.

If it is assumed that an elastic-tire deformation on soft ground is of the same nature as that on hard ground, which may be true only in a first approximation, then the geometry of the contact area would be roughly similar to the geometry of the tire previously described.

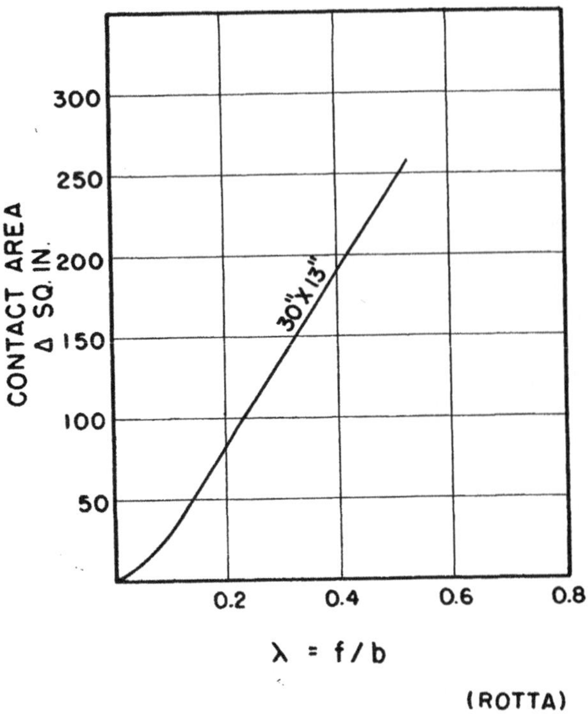

Fig. 94

Consider a 30-in. × 13-in. pneumatic tire. If, for instance, the ground contact area Δ is computed by means of equation (206) for various $\lambda = f/b$ and for $h/b = 0.75$, then it will be seen that Δ increases very rapidly with increasing f/b (see Figure 94).

From equations (93) and (94), it follows that the vertical soil stress σ_z decreases proportionally with a decrease in p_o. Thus, a decrease in the pressure p_o due to an increase in the loading surface Δ by tire deflation acts most favorably upon the soil stress and explains the advantages of "low-pressure" tires on soft ground. The numerical relations between σ, loading area, and pressure may be computed in accordance with the previously discussed data.

The advantages of elastic tires in a soft-ground operation depend, however, on the degree of elasticity of the ground contact area. The latter, if too elastic, may be considered to be disadvantageous when compared to rigid loading surfaces. Take a circular area as a reference for a qualitative comparison: the computation by Kögler (Figure 63d) shows that a rigid loading area exercises a peak pressure which is almost 50% smaller than that exercised by an elastic load area of the same dimensions under the same unit pressure. Hence, a pneumatic tire with strong side walls and a rigidly supported central portion will do better than a very flexible pneumatic tire. The "rigidity" of the considered tire, however, should be such as to allow the necessary deflection which is required to produce the given area Δ.

The type of load applied affects the ground pressure considerably, as was shown by equations (94) and (93). It will be seen that a conic type of load distribution [equation (94)] produces higher σ_z than a uniform distribution [equation (93)]. Thus, tires with a flat tread rim may be considered better than those having a tread whose center is higher than the side portion. Incidentally, a pneumatic tire designed for a hard surface often does not satisfy this requirement, and this invariably makes the conditions of tire operation worse in a soft ground. If the pressure is sufficiently low, however, this effect may, to some extent, be offset by the rigidity of the side wall of the tire. The walls will then support the protruding central portion of the tire tread on both its ends, letting it deflect freely in the center so that, in the final analysis, the sides of the tire will be more loaded and the pressure distribution will not be of a conical type, but will be similar to that shown in the right-hand side of Figure 92b. This figure illustrates the complexity of pressure distribution in general and suggests that without a rational study of pressure distribution and without an analysis of the form of tire deflection in various types of soils, the true merits and disadvantages of available tire types will remain unknown.

The problems of a pneumatic tire applied to a soft ground have never been considered beyond the purely empirical approach. Although innumerable data are available,[133–150] an over-all theory is lacking. Generally speaking, it should be possible to relate the properties of an elastic wheel acting upon a soft ground to those of a rigid wheel in the same conditions. McKibben has shown [38] experimentally that a pneumatic tire only forms a flatter and larger loading surface than a steel wheel of the same diameter and has a curvature larger than the wheel radius. It might be possible, therefore, to replace the flat elastic tire surface by the rigid arc of a larger wheel (Figure 95). Under these conditions, the problem would be reduced to determining the diameter D_1 of the imaginary rigid wheel which would replace the diameter D of the elastic wheel.

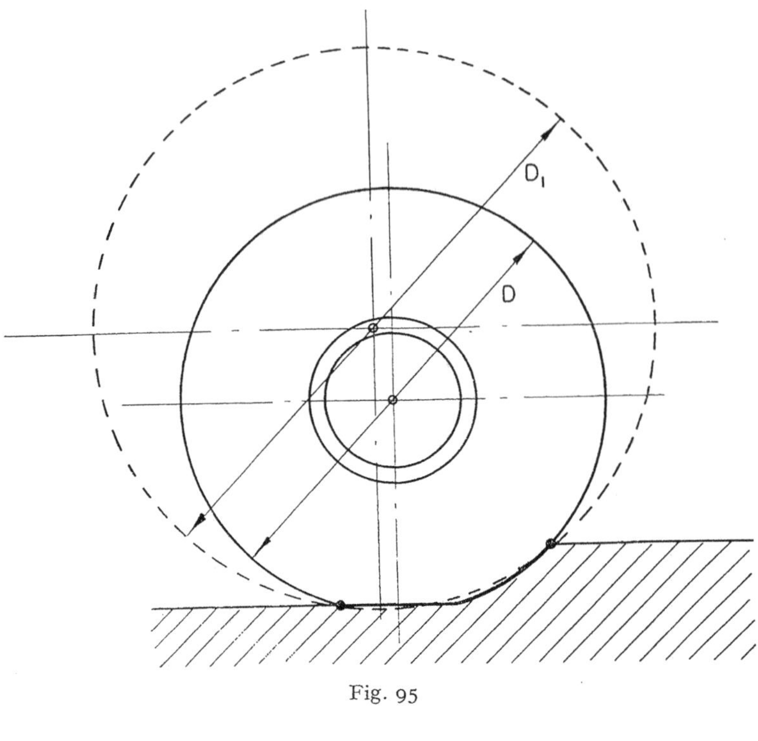

Fig. 95

If a successful solution of the problem can be based on such an assumption, then it may be agreed that the methods of Bernstein, Letoshnev, and Goriatchkin, as previously discussed, are also applicable, with proper corrections, to pneumatic tires. Preliminary findings in this field encourage further work in this direction.

It is worthwhile to note that the form of rigid wheels remains unaffected, whether they are driven or driving elements. This cannot be said about elastic wheels.[152] The forms of the loading surface of driving and freely rolling wheels may be so much different that the differentiation between these two types of wheels appears of paramount importance. Unfortunately, very little is known in this respect, and thus the theoretical evaluation of the behavior of a propelling wheel versus a towed wheel is practically impossible.

Rational Wheel Form

The general effect of wheel form upon rolling resistance was investigated by Grandvoinet more than a century ago. The results of his studies, quoted by Goriatchkin, are reproduced in Figure 96. In this figure, a percentage increase or decrease of rolling resistance is referred to the increase in wheel load W and wheel dimensions D and b. Since no soil type was specified, the graph has only a general qualitative value. Similar information may be found in the test results reported by other investigators (References 38, 39, and 133–136), who supplied a number of figures referring to particular conditions, mainly selected from agricultural experience. Such an approach is not sufficient today because it does not refer to an unlimited number of possible combinations of various factors conceivable in an industrial and military application of a wheel.

A more quantitative and general study of wheel form as expressed by the D/b ratio may be performed by means of Bernstein's or Gerstner's formulas, equations (183) and (182). When rolling resistance R is plotted as a function of wheel width b for a constant load W (Figure 97), it is seen that Bernstein's values give a somewhat steeper curve for small wheel diameters than Gerstner's figures. However, for small diameters, both curves indicate a rather sharp reduction in R with an increase in b. For large diameters, both curve types flatten to such an extent that no practical difference in R exists between wide and narrow wheels. Accordingly, it may be concluded that both equations indicate the strongest effect of width increase in reducing the rolling resistance to be only within the range of small diameters. For given constant loads, large-

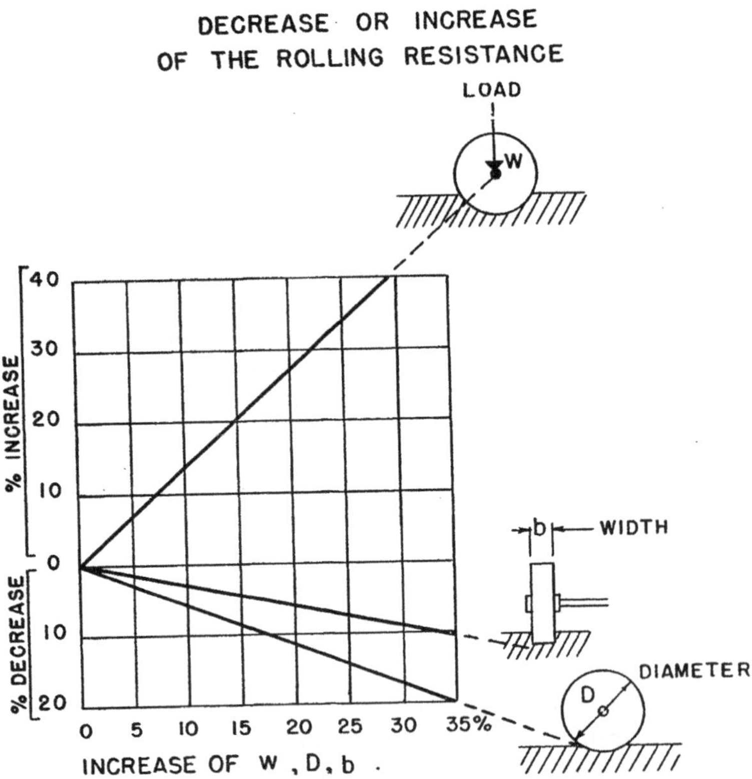

(GRANDVOINET QUOTED FROM GORIATCHKIN)

Fig. 96

diameter wheels are rather irresponsive to the attempts made to reduce the rolling resistance by widening the rims.

In general, equations (182) and (183) and Grandvoinet's graph (Figure 96) indicate that the rolling resistance R decreases more rapidly with an increase in diameter than with an increase in width, a fact long known. On the basis of the present experience, it can again be stressed that a very large and narrow wheel, having the largest possible D/b aspect, may be a more economical means of transport under given conditions than a wheel with a small D/b ratio.

A qualitative support of this conclusion can be sought in the analysis of the stress distribution in soil mass. For an elliptical elastic loading

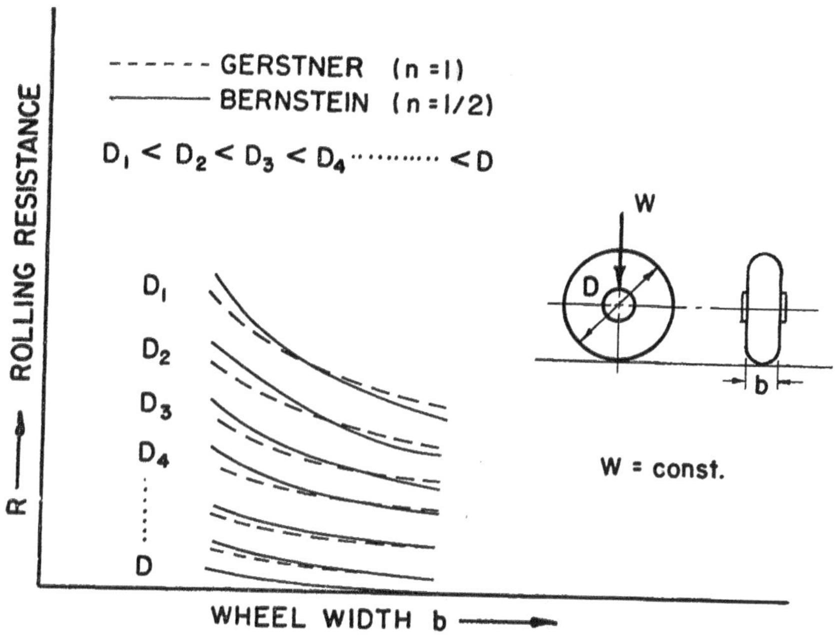

Fig. 97

area, the vertical stress σ of a pneumatic tire at depth z may be computed by means of formula (95). A graphical solution of the integral included in this formula will show that the same ground contact area (200 sq in.) and the same wheel load produce different soil pressures, depending on the form of that area. If the form of the ground contact area is related to a given wheel form (Figure 98), then it will be seen that, at the same pressure p_o, narrow, large-diameter wheels produce less stresses in soil than the so-called "high-flotation" tires which are very wide in comparison to their diameter. In the considered case, the difference in σ_z is as much as 40%.

This conclusion appears to contradict the usually accepted trends. It seems quite doubtful, however, whether these trends have been based on any rational assumption. The history of the application of a tire to a soft ground indicates that practically no special tires were developed for that purpose as far as their general form is concerned. The basic form of a pneumatic tire was frozen on the highway since the main concerns were to reduce the shock and vibrations of the wheel, provide

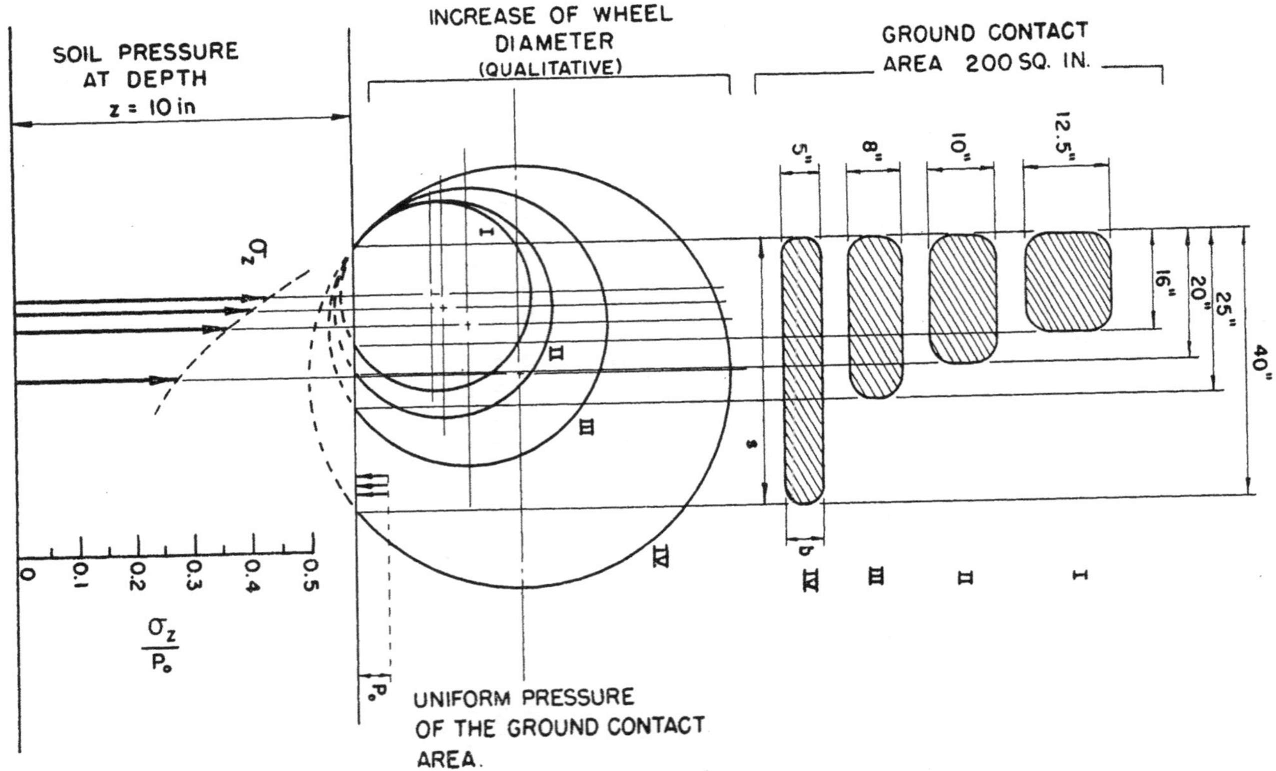
SOIL PRESSURE
AT DEPTH
z = 10 in
INCREASE OF WHEEL
DIAMETER
(QUALITATIVE)
GROUND CONTACT
AREA 200 SQ. IN.
5"
8"
10"
12.5"
16"
20"
25"
40"
s
b
I
II
III
IV
σz
0
0.1
0.2
0.3
0.4
0.5
σz / Po
Po
UNIFORM PRESSURE
OF THE GROUND CONTACT
AREA.

Fig. 98

adequate stability of the ride at high speeds, and increase the life of the tire. This led to the establishment of certain standard forms of air volume and carcass. As has been shown, the form of this volume plays a far more important role in soft ground than on a highway, and a careful analysis of the prevailing concepts appears inevitable if more information on a high-performance, pneumatic, soft-ground tire is to be gained.

It should be noted that the considerations referred to in this section are based on formulas which are valid when the elastic state of stresses is assumed. For wheels which impart plastic deformation to soils, the problem grows in complexity, and at present the only promising hope of determining a rational D/b form aspect seems to be in the realm of dimensional analysis as performed by Nuttall.[132] Further details on this subject will be discussed in Chapter XI, which deals with this analysis.

Another important argument in favor of a large-diameter, narrow wheel, which is related to the problem of slippage and wheel spinning, will be analyzed in detail in Chapter VII when the nature of slip and the relationship between a track and the wheel are discussed.

Bearing Capacity of Wheels and Flotation

The problem of the load-bearing capacity of wheels, or the so-called "flotation," may be elucidated qualitatively when analyzing formula (141) or (142), which were developed for strip loads. The proposed approach, however, is suitable only when the ground contact area in question may be considered as a strip, i.e., when the diameter D of a wheel is large in comparison with its width b. In such a case, the wheel form, which gives ground contact areas IV and possibly III (in Figure 98), would correspond to the assumed conditions, whereas the wheel forms related to areas II and I might be analyzed in the first approximation by a formula developed, for instance, for a rectangular or, perhaps, a circular load.

The structure of such formulae is identical to that of formula (141); they differ only in numerical coefficients, which are of an empirical nature because of the difficulties involved in the solution of the three-dimensional problem. For a circular area, the safe load which may be carried by a disc having a diameter $2r$ is:[54]

$$W_t^\circ = \pi r^2 \, (1.3c \, N_c + qN_q + 0.6\gamma r N_\gamma) \, . \qquad (211)$$

In this equation, W_t° represents the total load which is supported by the circular area. Since formula (141) refers to a load per unit of the length s of the strip load (Figure 98), it has to be multiplied by s in order to give the value of the total load $W_t^{\square}$. If, according to the previously used denotations for a wheel, it is assumed that $2l = b$, then

$$W_t^{\square} = sb\,(cN_c + qN_q + 0.5b\gamma N_\gamma)\,. \tag{212}$$

To sum up, equation (211) expresses the bearing capacity or "flotation" of a tire whose diameter is small enough compared to its width in order to produce a circular contact area with the ground, whereas equation (212) determines the load-carrying capacity of a large-diameter, narrow wheel, which gives an approximately rectangular contact area whose length s is larger than its width b. Both equations refer to those safe loads which do not cause soil failure by a plastic flow and which theoretically do not produce considerable sinkage. In order to compare the merits of both wheels in keeping the tire "afloat," assume that the ground contact areas of both tires are equal, i.e., that

$$\pi r^2 = bs\,.$$

For the sake of simplicity, consider that there is no surcharge ($q = 0$). Then, for a purely cohesive soil ($\phi = 0$), like saturated clay or wet snow,

$$W_t^{\circ} = 1.3\pi r^2 c N_c \tag{213}$$

$$W_t^{\square} = sbcN_c \tag{214}$$

and

$$\frac{W_t^{\circ}}{W_t^{\square}} = \frac{1.3\pi r^2}{sb}\,.$$

Since, as was previously stated, $\pi r^2 = bs$,

$$\frac{W_t^{\circ}}{W_t^{\square}} = 1.3\,. \tag{215}$$

In other words, in a cohesive soil, small-diameter, wide tires will have 1.3 times higher load-bearing capacity than a narrow, large-diameter tire of the same ground contact area.

Next, consider a purely frictional soil, like dry sand ($c = 0$). From equations (211) and (212),

$$W_t^{\circ} = 0.6\pi\gamma r^3 N_\gamma \tag{216}$$

$$W_t^{\square} = 0.5\gamma s b^2 N_\gamma\,, \tag{217}$$

and when it is assumed, as previously, that $\pi r^2 = bs$ and $r = \sqrt{bs/\pi}$,

$$\frac{W_t^\circ}{W_t^\square} = \frac{0.6\pi r^3}{0.5sb^2} = 0.675\sqrt{s/b}\,. \tag{218}$$

Formula (218) refers to a strip load in which, according to the original assumptions, $s/b >> 1$. It will be seen, however, that in order to produce values of W_t° higher than $W_t^\square$, $\sqrt{s/b}$ must be at least larger than 1.47, or $s/b > 2.16$. Since the tires actually used seldom have proportions which would satisfy this condition, it may be concluded that, in frictional soils, the large-diameter tire at $s/b < 2.16$ is more favorable than a tire with a circular bearing area of the same size produced by a small diameter.

Since most soils show both frictional and cohesive properties, the evaluation of the problem would depend on the proportion of c and ϕ. In a general case, it may be expected that the difference between both discussed types of tires tends to vanish, with rather small-diameter, wide tires being favored if high-bearing capacity, or "flotation," is to be obtained. The real problem, however, lies in the questions of whether or not the high flotation requirement can always be satisfied and whether or not the described tendency is favorable in all conditions.

It is evident that in a very large number of cases the soil is so weak that it cannot support wheels unless they have prohibitive sizes in comparison to the load they carry.[161] In such cases, until the hard pan is reached, sinkage will unavoidably occur and, in consequence, will lead to a considerable increase in the rolling resistance which is directly proportional to the tire width. In the final analysis, then, the advantages of "high flotation" may not only be completely cancelled but they may also become detrimental, as has been actually demonstrated in daily experience. In the discussed type of soil, sinkage will take place up to the time when the firm base is able to support the load. The deeper the wheel sinks, the firmer the grip and the supporting power become. But also, at the same time, locomotion becomes more difficult. Under these circumstances, a direct answer as to what degree of "flotation" would be the optimum does not exist. Equations (182), (183), and (184), however, suggest that in all conditions, a large-diameter, narrow wheel would be more advantageous than a small, wide tire, since it would produce the same rolling resistance on the hard pan but a smaller drag in the soft layer.

If it is agreed that the Ford Model T was a country-road vehicle superior to modern automobiles, then at least a part of the explanation of this paradox may be found in the form of the tires used at that time. It may also be recalled that the ancient carts and nineteenth-century wagons did not roll on wheels having the form of modern pneumatic tires.

In conclusion, then, the problem of the load-bearing capacity or "flotation" of wheels and tires cannot be considered without taking into account the problems of rolling resistance, mechanical properties of soil, and the problem of soil stratification.

As experiments with tracks indicate, there is a critical value beyond which "flotation," which is tacitly associated with the idea of "low ground pressure" and large ground contact area, does not contribute to vehicle performance, but becomes a factor of high resistances to motion. A quantitative study of this problem is essential if the basic questions on the form of cross-country tires are to be solved.

Tractive Effort; Spuds of Rigid Wheels, and Tire Tread

Theoretical considerations and experience indicate that the spuds or grousers of rigid wheels, or ribs of elastic tire treads clog with the soil once they are in contact with the ground, and that they play an insignificant and secondary role in developing the tractive effort above a definite maximum, provided the ground is homogeneous. The main physical effect of grousers would be, strictly speaking, the increase of the diameter of the wheel from D to D', where $D' = D + 2h$, according to the denotations shown in Figure 99a. For a wide spacing of spuds, the grouser effect will not take place in the simple form described above. A similar situation will occur in the case of a narrow wheel, in which the lateral flow of soil may take place. In any case, however, the commonly accepted desirability of the "self-cleaning" of treads, which is supposed to provide a better traction, loses, in the light of the above remarks, its original meaning and suggests a study of the tractive effort and spud action from another angle.

If the soil is not homogeneous, then the role of any grouser is similar to that of a cutter which helps the wheel to dig through soft, unstable layers and to reach the hard strata in which a sufficient tractive effort may be developed.[161] From that moment on, however, the grouser action does not differ much from the action described above.

In the first approximation, the net force of traction T_n of a propelling wheel may be considered to be the difference between the gross tractive

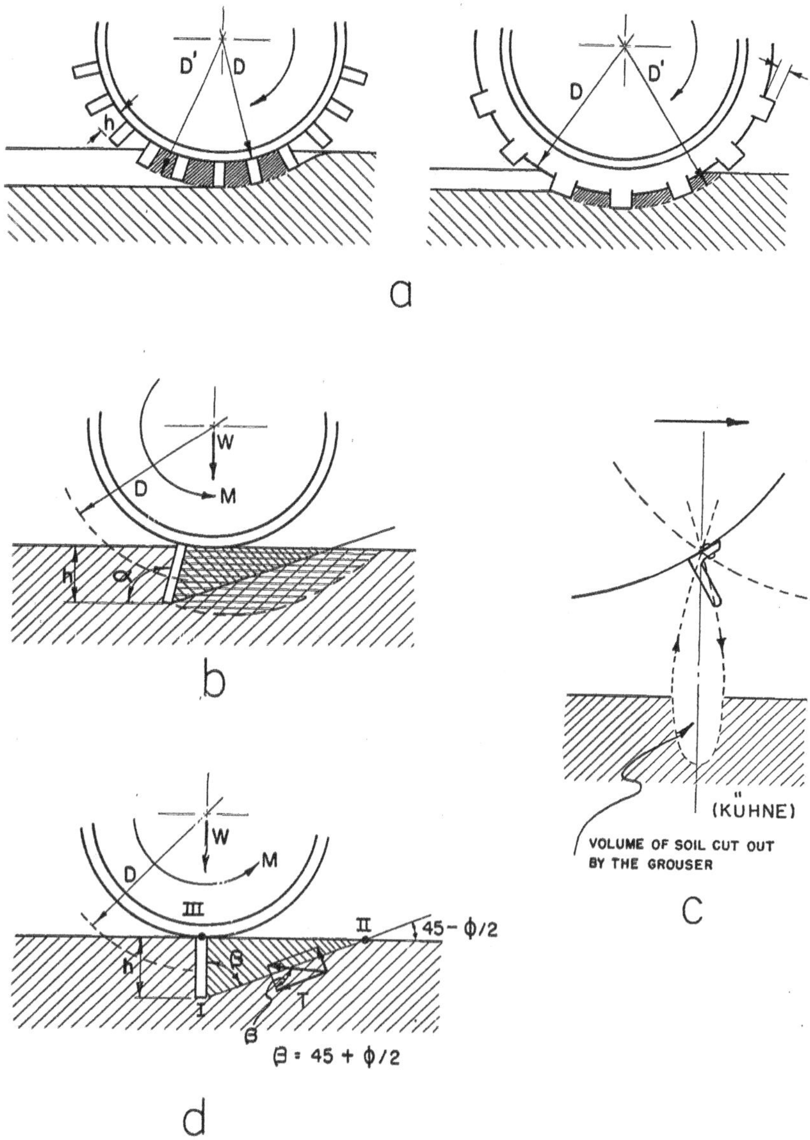

D'
D
h
D
D'
h
a
W
D
M
h
α
b
(KÜHNE)
VOLUME OF SOIL CUT OUT
BY THE GROUSER
C
W
D
M
III
II
45 − ϕ/2
h
β
I
T
β
β = 45 + ϕ/2
d

Fig. 99

effort T_g available in the soil and the rolling resistance of the wheel R:

$$T_n = T_g - R \,. \tag{219}$$

It should be borne in mind, however, that R is not only a function of load W and wheel dimensions D and b, which produce a given ground contact area Δ as Bernstein and the others showed, but also is a function of the horizontal load T_g. If it is assumed that the soil properties are constant,

$$T_n = T_g - f(W, T_g, \Delta) \,.$$

The difficulty in theoretically or experimentally separating T_g from the function $f(W, T_g, \Delta)$ is the source of most of the misunderstanding in the evaluation of the performance of a driving wheel. In the light of the present limited knowledge in this field, the following analysis may be ventured. In purely frictional soils ($c = 0$), wheel dimensions will play a secondary role, as far as T_g is concerned, since the shearing resistance of soil τ, which supplies the gross traction, is expressed by equation (124),

$$\tau = \sigma \tan \phi \,,$$

and does not depend on the form of the sheared area. It depends only on the normal stress σ and the coefficient of friction $\tan \phi$. Thus, T_g will be solely a function of weight W and friction $\tan \phi$, and equation (219) may be written in the following form:

$$T_n = W \tan \phi - f'(W, \phi, \Delta) \,. \tag{220}$$

In the case of purely cohesive soils ($\phi = 0$), the shearing strength of soil is expressed by equation (125):

$$\tau = c \,.$$

Accordingly, the gross tractive effort T_g does not depend on the vehicle weight W and is equal to the ground contact area Δ times the coefficient of cohesion c. Rolling resistance R, however, as Bernstein showed, is related to wheel load W through the mechanism of sinkage, and equation (219) will take the following form:

$$T_n = \Delta c - f''(W, c, \Delta) \,. \tag{221}$$

Formulas (220) and (221) clearly indicate that in frictional soils, weight (W) is the decisive factor in developing the tractive effort, whereas in cohesive soils, the dimensions of the wheel (Δ) are essential. It thus becomes obvious that a soil in which both cohesion and friction exist will behave in a different way, depending on the preponderance of one of these factors. This is why any test in which only the load and geometry of a wheel are considered cannot lead to a general answer, unless the mechanical properties of the soil c and ϕ are specified. For an average type of soil including both c and ϕ,

$$T_n = (W \tan \phi + \Delta c) - f'''(W, \phi, c, \Delta) . \qquad (222)$$

An attempt to evaluate the problem quantitatively was made by Nuttall and the writer [162] in a discussion on the study of the effect of lug height on wheel performance by Reed and Shields.[163] Assume that the solution of equations (220) and (221) or (222) is obtained with the help of Bernstein's equation (183). Then

$$f'(W, \phi, \Delta) = \frac{0.876}{\sqrt{k'_1 b}} \frac{W^{3/2}}{D^{3/4}}$$

$$f''(W, c, \Delta) = \frac{0.876}{\sqrt{k'_2 b}} \frac{W^{3/2}}{D^{3/4}}$$

$$f'''(W, \phi, c, \Delta) = \frac{0.876}{\sqrt{k'_3 b}} \frac{W^{3/2}}{D^{3/4}} ,$$

where k'_1, k'_2, and k'_3 reflect the ground properties. If load W and wheel dimensions D and b are known, then the sinkage z_o may be determined directly from equation (177), and this will lead to the evaluation of Δ in the case of rigid wheels. In the case of pneumatic tires, Δ may be assessed by means of formula (206). Thus, T_n may be determined in accordance with equation (222) for a given ϕ and c.

As an example, take a dry sand having $c = 0$ and $\phi = 35°$. The question is what will be the net tractive effort of a rigid, moderately spudded wheel loaded up to $W = 1000$ lb and having the following dimensions: $D = 40$ in. $\cong 100$ cm; $b = 4$ in. $\cong 10$ cm. For such a wheel, according to the table on Figure 90, $K_b = k'_1 b = 25$. Since this value is in kilograms and centimeters, the given W and D values have to be multiplied by 2 in order to obtain the answer in pounds; hence,

$$f'_1(W, \phi, \Delta) = 2 \frac{0.876}{\sqrt{25}} \frac{500^{3/2}}{100^{3/4}} = 124 \text{ lb} .$$

Assume, now, that the grouser height is increased by 2 in. so that the wheel diameter $D = 40$ in. $= 100$ cm is increased to $D_1 = 44$ in. $= 110$ cm. Under these circumstances,

$$f'_2(W, \phi, \Delta) = 2 \frac{0.876}{\sqrt{25}} \frac{500^{3/2}}{110^{3/4}} = 115 \text{ lb} .$$

According to the original assumptions, the gross traction T_g will be the same in both cases and will be equal approximately to

$$W \tan \phi = 1000 \times 0.7 = 700 \text{ lb}.$$

Hence, according to equation (220), the net tractive effort produced by the shallow-spudded wheel will be

$$T'_n = 700 - 124 = 576 \text{ lb},$$

and for the higher-spudded wheel,

$$T''_n = 700 - 115 = 585 \text{ lb}.$$

The insignificance of this increase of traction in homogeneous soils has been observed in actual experiments. The proposed explanation of the relative inefficiency of high spuds thus appears to be confirmed and indicates the validity of previous conclusions regarding their true role.

It will be interesting to note that the attempt made by Stewart, Weiss, and Magill to solve analytically the problems of the wheel in an assumed soil type leads to an equation which is based on a concept similar to that previously discussed. By assuming that the diameter of the grousered wheel merely increases from D to D' (Figure 99a) and that the wheel develops friction of soil on soil, Weiss and his collaborators deduced the following formula for the gross tractive effort:

$$T_g = (W + R) \tan \phi + cbs ,$$

where s is the length of the wheel arc in contact with the ground.[164] This formula may be presented in the following form, assuming that $bs = \Delta$ and that $T_n = T_g - R$:

$$T_n = (W \tan \phi + \Delta c) - R(1 - \tan \phi) . \qquad (223)$$

Equation (223) is identical in structure with equation (222), and when only frictional soil is assumed ($c = 0$), it will give

$$T_n = W \tan \phi - R(1 - \tan \phi) .$$

If R is determined by Bernstein's formula (183), this equation will produce slightly higher values than equation (220) since it assumes that a part of the rolling resistance ($R \tan \phi$) contributes to the net traction T_n.

In the case of the previously discussed wheel, according to equation (223),

$$T'_n = 700 - 124\,(1 - 0.7) = 663 \text{ lb}$$

$$T''_n = 700 - 115\,(1 - 0.7) = 665 \text{ lb}\,,$$

which compares favorably with the figures of 576 and 585 obtained from equation (220). Thus, the discussed attempt to solve the problem analytically indicates the same order of magnitude of the forces involved. Equation (223), if applied to purely cohesive soils ($\phi = 0$), will take the following form:

$$T_n = \Delta c - R\,,$$

which is identical to equation (221). It should be noted that Stewart, Weiss, and Magill have developed an analytical solution for the rolling resistance which leads to a value of R having the same order of magnitude as that determined by formula (182), (183), or (184).

It may then be concluded that the approximate methods of wheel evaluation produce results that are in fairly good agreement. However, further refinement of these methods is necessary to take into account the important effects of the factors which were eliminated in the original oversimplified assumptions.

If the grousers are spaced far enough apart so that each may act individually without clogging the soil, then the problem is somewhat different. Rathje [165] and Kanafojski [166] have determined the pattern of shear of soil by a wheel lug (Figure 99b). This pattern is practically identical to that shown in Figures 69 and 70. The passive earth pressure acting upon the blade may thus be determined in accordance with formula (136) or (138) when the wheel is wide enough as compared to the depth h of the lug. The low values of shear obtained in this way are usually assumed to be augmented by the frictional force $W \tan \mu$ between a wheel and soil. In the case of the application of equation (138), the driving torque M may be assumed to be approximately equal to

$$M \cong \frac{D}{2}\left(\tfrac{1}{2}\,\gamma b h^2 \frac{K_p}{\sin \alpha \cos \mu} + W \tan \mu\right). \tag{224}$$

The inaccuracy of formula (224) results mainly from the lack of consideration of the vertical loads applied to the wedge under shear and of the changing value of M during the passage of the lug through the soil.

Another approximate determination of the forces developed between the soil and a lugged wheel was proposed by Halkinov,[167] who also did not consider the additional rolling resistance due to the excavation of the soil by the lug. The excavation forms a cavity having trochoids as walls (Figure 99c) and may be the source of sizable resistance [35] which lowers the net traction.

The method by Halkinov considers the grouser to be in a vertical position and a straight-line shear along the line to be sloped to the horizon at $45 - (\phi/2)$ (Figure 99d). This assumption, which was made originally by Coulomb, leads to the conclusion that $\beta = 45 + (\phi/2)$. If the shearing stress of soil is τ_o, and the shearing force along the I-II surface is T_1, then

$$T_1 = \frac{hb}{\cos\beta}\tau_o ,$$

where b is the grouser width. Another shearing force T_2 is developed along two triangular areas I-II-III $= (h^2 \tan\beta)/2$ located on both sides of the grouser:

$$T_2 = \frac{2h^2 \tan\beta}{2}\tau_o .$$

In order to obtain the total force T, T_1 and T_2 must be added:

$$T = T_1 + T_2 = \tau_o\left(\frac{hb}{\cos\beta} + h^2 \tan\beta\right).$$

Accordingly, the driving force T_o supplied by the lug would be

$$T_o = \frac{T}{\sin\beta} = \tau_o\left(\frac{hb}{\cos\beta\,\sin\beta} + \frac{h^2}{\cos\beta}\right). \qquad (225)$$

The somewhat arbitrary assumptions made in deducing this formula were apparently corrected by the experimental determination of the unspecified value of τ_o, which is quoted to vary between 10 and 35 psi for average agricultural soil.

The total value of the traction moment M developed by the grousered wheel would be

$$M = \frac{D}{2}(T_o + W \tan \mu) , \tag{226}$$

where μ is the coefficient of friction between soil and steel.

An attempt has been made in the preceding paragraph to explain the action of grousers and tire treads in soft homogeneous soil. On the basis of the material reviewed, it was shown that this action is rather insignificant as far as the tractive effort, or grip between the tire and soil, is concerned. The conclusion reached appears to be in agreement with experiments and indicates that any change in the traction of pneumatic tires, for instance, is chiefly a secondary effect of tread design and is not caused by an improved interaction of tread and soil.

The tread design directly affects the stiffness of the tire carcass and thus influences the size and form of the loading area as well as the character of load distribution. These factors, as was shown in the chapter dealing with soil mechanics, are of primary importance in regard to wheel performance. Thus, it is not the tread itself, but rather the stiffness and related geometrical properties of a tire that result from the tread design which are responsible for a rather unpredictable behavior of various patterns. Systematic investigations which would elucidate this problem are unknown; only experimental studies in limited test conditions have been made so far.

Tire Tread and Slippery Surface

Notwithstanding how inefficient the tread design may be in improving traction in soft ground, the effectiveness of the tread pattern in securing a firm grip between a tire and a slippery hard surface is paramount. Although this fact has been recognized since the early stages of the automobile, rational studies in this subject have not been carried out, and again only an empirical approach may be recorded. Since empirics cannot embrace the infinite number of factors involved, the existing solutions cannot be properly evaluated and generalized.

An example of a theoretical approach to this question and of the methods available is shown below with the purpose of elucidating the nature of the discussed problem.

Assume that a given bar of the tire tread is in contact with the pavement covered with a viscous fluid h inches thick (Figure 100a). In a two-

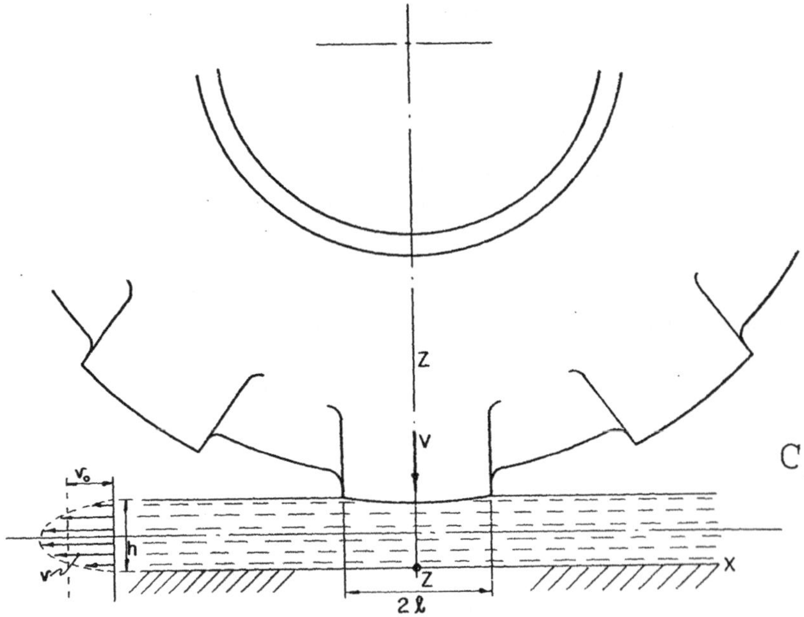

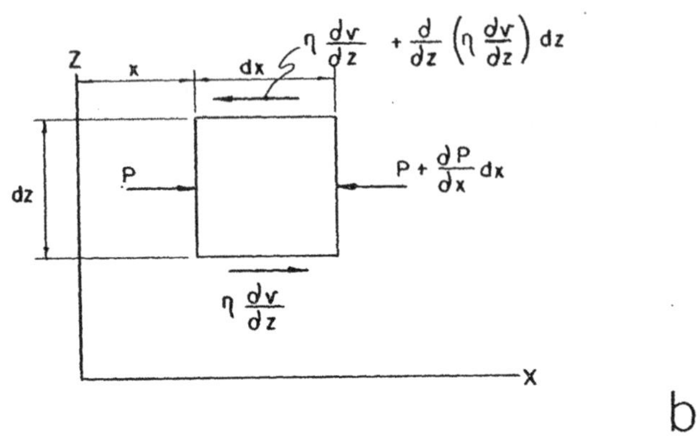

Fig. 100

dimensional case, the equilibrium of the forces acting upon an elementary prism (Figure 100b) in the direction of the xx axis will be as follows:

$$-\eta \frac{\partial v}{\partial z} dx - \eta \frac{\partial^2 v}{\partial z^2} dz\, dx + \eta \frac{\partial v}{\partial z} dx + p\, dz - p\, dz - \frac{\partial p}{\partial x} dx\, dz = 0$$

or, when simplified,

$$\frac{\partial p}{\partial x} + \eta \frac{\partial^2 v}{\partial z^2} = 0 , \tag{227}$$

where p = unit load acting upon a given area, v = speed of flow of the fluid particle, and η = viscosity of the fluid covering the road.[168]

Integrating equation (227) gives

$$\frac{\partial v}{\partial z} = -\frac{1}{\eta} \frac{\partial p}{\partial x} z + c_1$$

and

$$v = -\left(\frac{1}{2\eta} \frac{\partial p}{\partial x} z^2 + c_1 z + c_2\right).$$

Since $v = 0$ for $z = 0$ and for $z = h$, then $c_2 = 0$ and

$$\frac{1}{2\eta} \frac{\partial p}{\partial x} h^2 + c_1 h = 0 .$$

Thus

$$c_1 = -\frac{1}{2\eta} \frac{\partial p}{\partial x} h .$$

And finally,

$$v = \frac{z}{2\eta} \frac{\partial p}{\partial x} (h - z) . \tag{228}$$

The average speed of fluid particles v_o may be found from the following equation (Figure 100a):

$$v_o = \frac{1}{h} \int_0^h v \, dz = \frac{1}{2\eta} \frac{\partial p}{\partial x} \frac{1}{h} \int_0^h (zh - z^2) \, dz = \frac{1}{\eta} \frac{\partial p}{\partial x} \frac{h^2}{12} . \tag{229}$$

Assume that the pressure acting upon the fluid causes a reduction of h by dh. Thus the volume of fluid squeezed out between 0 and x is $x\,dh$. For an incompressible fluid, this volume equals $h\,dx$. Hence,

$$x \, dh = h \, dx$$

or

$$\frac{dh}{dt} x = \frac{dx}{dt} h .$$

Since $dx/dt = v_o$, the speed of the "sinkage" of the tire is

$$\frac{dh}{dt} = \frac{h}{x} v_o \,. \tag{230}$$

When equations (230) and (229) are combined, the following will be obtained:

$$\frac{\partial p}{\partial x} = \eta \frac{dh}{dt} \frac{12x}{h^3} \,.$$

Integrating this formula gives

$$p = \eta \frac{dh}{dt} \frac{6x^2}{h^3} \,.$$

Since the load V per unit width of tread must be supported by average pressure p,

$$\tfrac{1}{2}V = \int_0^l p \, dx = 2\eta \frac{dh}{dt} \frac{l^3}{h^3} \,.$$

Thus the speed of the "sinkage" of the tire due to the squeezing out of the "fluid" which covers the pavement is

$$\frac{dh}{dt} = \frac{Vh^3}{4\eta l^3} \,. \tag{231}$$

It will be seen that the speed with which a tire reaches a firm grip with the road by forcing the lubricating fluid out of the contact area depends on the viscosity of the lubricating film. The larger the viscosity, the slower the speed with which the tire may reach a dry surface. Fresh rain gives a film of clean water with low viscosity: it is relatively easy to squeeze it out. Highways covered with water for a lengthy period, during which water mixes with dirt, are covered with high-viscosity moisture which gives slow dh/dt and, hence, a high probability of "viscous friction." At high speeds of locomotion, when the time of the contact of a tread bar with the road, as shown in Figure 100a, is very short, the fluid may not be squeezed out and instead of a dry adhesion between the tread and the road, a sliding coefficient of friction, due to the presence of a fluid, will prevail. Thus the adhesion upon wet surfaces depends on the speed of locomotion, a case which practically does not exist in the case of dry roads.

It will be seen from formula (231) that the load V improves the traction as it increases dh/dt. However, the tire-tread length l is of paramount importance. The finer the tread, the shorter the time required for reaching conditions of "dry friction" through the squeezing of the fluid. For a wide tread, the speed of tire "sinkage" is so low that only the slow speed of locomotion can ensure safe driving. At high speeds of tire rotation, skidding and sliding may occur.[168]

It may be noticed that, in general, it is necessary to provide such tread forms as will secure unobstructed channels for the flow of fluid during the process of squeezing the lubricating film enclosed between the tread and the road. The considered example appears to have a broad meaning, for the method applied might also be useful if soil or snow is assumed to be a viscous mass and if the conditions of flow under the action of a wheel in such a mass are investigated.

VII. CRAWLERS OF TRACK-LAYING VEHICLES

Tracked vehicles are basically wheeled ones. Their only peculiar feature is the portable rail which is laid down in the front of the carriage and picked up in the rear as the vehicle progresses—a process which was so admirably described in the Edgeworth patent almost two hundred years ago.[25]

Because of this quality, the wheel does not act directly upon the ground, but transfers the load to the soil through the intermediary of track plates. Although the form of the ground contact area in this case and in the case of a simple wheel is radically different, the laws of soil mechanics remain unchanged and may be usefully applied to many problems. It was Micklethwait who first called attention to this fact.[88]

Sinkage and Pressure Distribution Prevailing Along the Ground Contact Area as a Result of Track Tension, Suspension Design, and Soil Properties

There seems to be nothing more misleading in regard to vehicle performance than the idea of the so-called "mean ground pressure," defined as the result of dividing the load by the projection of the ground contact area. Although this idea may serve well when comparing the performance of vehicles with rigid loading surfaces, it is of little use when dealing with flexible loading areas such as the tracks of high-speed vehicles.

In order to illustrate in a general way the origin of these difficulties, assume that the vehicle is at rest. In this case, only the vertical forces acting upon the ground need to be considered. Further assume that the track is perfectly flexible and made of a continuous strip. Also, the diameter of the wheels is supposed to be small enough so that the wheels will not interfere with the natural deflection of the track, at least in the middle of the track span. The problem is to determine the shape of the deflected track portion located between two adjacent bogie wheels and to derive a formula expressing the pressure distribution. Assuming that the xx and zz axes are traced as shown in Figure 101, and that the ground

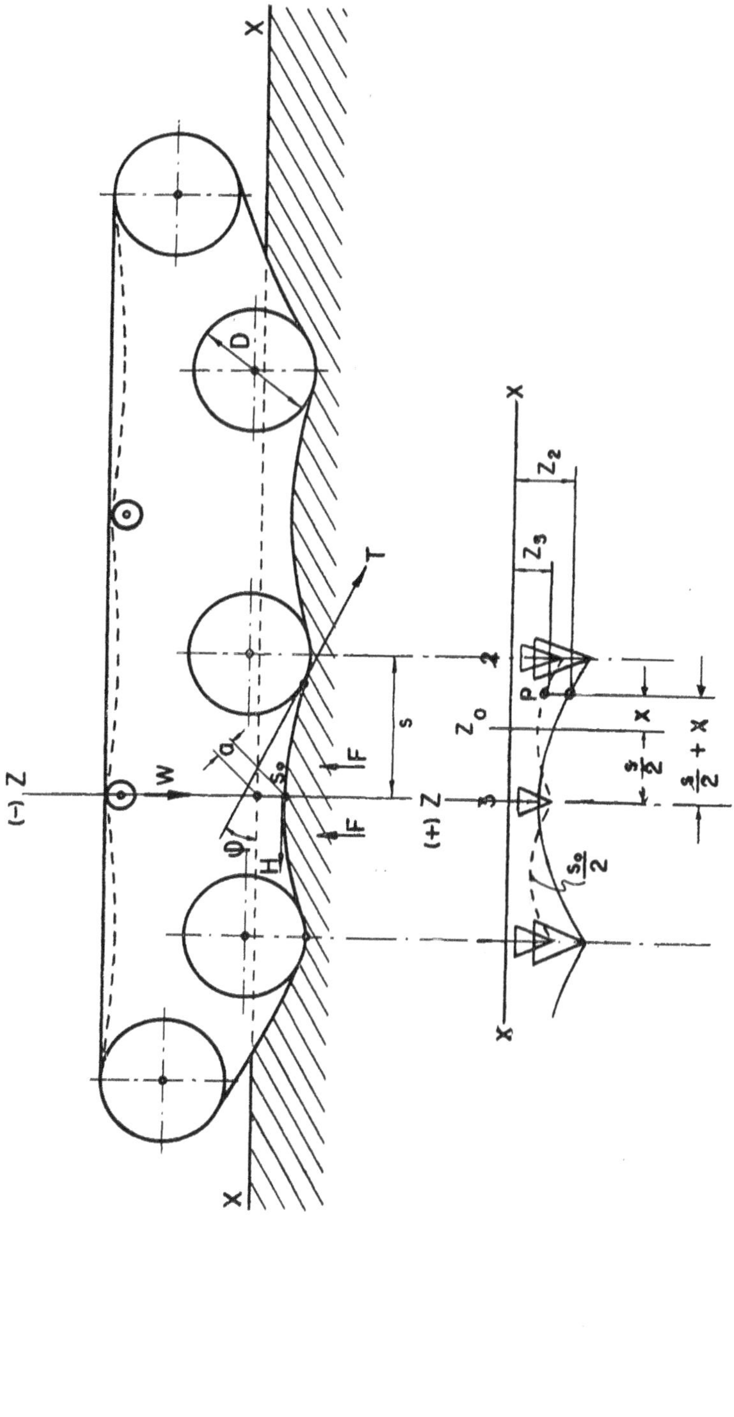

Fig. 101

pressure p is proportional to the depth of sinkage z in accordance with equation (162),

$$p = kz .$$

The following condition of equilibrium may be written if the denotations shown in Figure 101 are followed:

$$H - T \cos \varphi = 0$$

$$F - T \sin \varphi = 0 ,$$

where H and T are tension forces acting upon the track and F is the ground reaction. From the above equations, it follows that

$$\frac{F}{H} = \tan \varphi ,$$

but since

$$\tan \varphi = \frac{dz}{dx} ,$$

then

$$F = H \tan \varphi = H \frac{dz}{dx} . \tag{232}$$

On the other hand,

$$F = 2 \int pl \, dx = 2 \int kzl \, dx , \tag{233}$$

where $2l$ is the width of the track. When equations (232) and (233) are combined,

$$H \frac{dz}{dx} = 2 \int kzl \, dx$$

or

$$H \frac{d^2z}{dx^2} = 2kzl .$$

Solving the above differential equation gives

$$z = Ae^{x\sqrt{2kl/H}} + Be^{-x\sqrt{2kl/H}} .$$

Differentiating this equation gives

$$\frac{dz}{dx} = \sqrt{\frac{2kl}{H}}\,(Ae^{x\sqrt{2kl/H}} - Be^{-x\sqrt{2kl/H}})\,.$$

For $x = 0$, $dz/dx = 0$ and $A = B$; hence,

$$z = A\left[e^{x\sqrt{2kl/H}} + e^{-x\sqrt{2kl/H}}\right].$$

For $x = 0$, $z = a$ and $A = a/2$. Accordingly, the final equation for the track shape is

$$z = \frac{a}{2}\left[e^{x\sqrt{2kl/H}} + e^{-x\sqrt{2kl/H}}\right] = a\cosh\left(\sqrt{\frac{2kl}{H}}\,x\right). \qquad (234)$$

Determine now the track load W as a function of track dimensions, sag, track tension, and soil properties. According to the denotations of Figure 101,

$$\frac{W}{2} = 2\int_0^s klz\,dx$$

or, when substituting the value of z from equation (234),

$$\frac{W}{2} = 2\int_0^s kla\cosh\left(\sqrt{\frac{2kl}{H}}\,x\right)dx\,.$$

Integrating this equation gives

$$W = 2a\,\sqrt{2klH}\,\sinh\left(\sqrt{\frac{2kl}{H}}\,s\right). \qquad (235)$$

In order to determine the length s_o of the track as a function of x and z, the following may be written:

$$s_o = \int_0^s \sqrt{1 + \left(\frac{dz}{dx}\right)^2}\,dx\,.$$

But when equation (234) is differentiated, it will be found that

$$\frac{dz}{dx} = a\sqrt{\frac{2kl}{H}}\sinh\left(\sqrt{\frac{2kl}{H}}\,x\right).$$

Hence,

$$s_o = \int_0^s \left[1 + \frac{2a^2kl}{H}\sinh^2\left(\sqrt{\frac{2kl}{H}}\,x\right)\right]^{\frac{1}{2}} dx\,,$$

or, after this function is developed into a series,

$$s_o = \int_0^s \left[1 + \tfrac{1}{2}\,\frac{2a^2kl}{H}\sinh^2\left(\sqrt{\frac{2kl}{H}}\,x\right) - \ldots\right] dx\,.$$

If only the central portion of the track is considered (x is small as compared to s), then it may be assumed that $\sinh(\sqrt{2kl/H}x) \cong \sqrt{2kl/H}x$. This assumption appears justifiable because the wheels always have definite diameters and take the load at the outer points of the track span. Thus the load distribution at these points is not related to the track deflection, but is governed by other rules. A study of the problem by assuming rigorous values of $\sinh(\sqrt{2kl/H}x)$ would then have little practical meaning and would lead to involved computations. Therefore, a simplified form of sinh is adopted:

$$s_o = \int_0^s \left[1 + \frac{a^2}{2}\left(\frac{2kl}{H}\right)^2 x^2\right] dx\,.$$

Integrating this equation gives

$$s_o = s + \frac{a^2s^3}{6}\left(\frac{2kl}{H}\right)^2. \tag{236}$$

Equation (235) also may be developed into a series corresponding to $\sinh(\sqrt{2kl/H}x)$. When only the first two terms are taken, it will be found that

$$W = 4akls + \frac{4ak^2l^2s^3}{3H}. \tag{237}$$

From equation (236),

$$\frac{2kl}{H} = \sqrt{\frac{6(s_o - s)}{a^2s^3}}. \tag{238}$$

By substituting this value into equation (237), the following will be obtained:

$$W = 4akls + \frac{2kls^3}{3}\sqrt{\frac{6(s_o - s)}{s^3}}$$

and

$$a = \frac{W}{4kls} - 0.41\sqrt{s(s_o - s)}\,. \tag{239}$$

The considered vehicle may serve as an instrument in determining the "elasticity" of the soil as expressed by the coefficient k. The latter may be found from equation (239):

$$k = \frac{W}{4ls\,[a + 0.41\sqrt{s(s_o - s)}]}\,. \tag{240}$$

It should be borne in mind, however, that equation (240) is only approximate and that if a better approximation is needed, a more accurate value of $\sinh(\sqrt{2kl/Hx})$ should be considered when determining s_o. Also, the wheels should be replaced by knife-edge supports (Figure 101). The discussed formula might be useful in determining k values for tracks.

Equation (234) for the track shape may be presented in an approximate form in which the value $\sqrt{2kl/H}$ is substituted for by the value taken from equation (238):

$$z = a\cosh\left(\sqrt[4]{\frac{6(s_o - s)}{a^2 s^3}}\,x\right).$$

When cosh is developed into a series, the final equation of the central portion of the track deflection may be written as follows:

$$z = a\left[1 + \frac{x^2}{2!}\sqrt{\frac{6(s_o - s)}{a^2 s^3}} + \frac{x^4}{4!}\left(\frac{6(s_o - s)}{a^2 s^3}\right) + \dots\right]. \tag{241}$$

The value a is determined from equation (239). Thus the ground pressure at any point x may now be found from the equation $p = kz$, or from a general form of this equation as expressed by formula (163): $p = kz^n$. Accordingly, the pressure at any point will be known as a function of load W, dimensions s_o, l, s, and soil constant k. It will be seen from equation (241) that the sinkage z depends on wheel spacing $2s$ and track slack $(s_o - s)$.

Equation (239) discloses an interesting relationship between the track settlement a and soil coefficient k. If k increases, or in other words, if the soil becomes harder, then a decreases. In a soft soil, when k is small, the value of a increases, which is in agreement with observations. The distance $a = 0$ when

$$\frac{W}{4kls} = 0.41\sqrt{s(s_o - s)}\,.$$

If this condition is satisfied, then the central part of the track located at the beginning of the coordinates does not carry any load at all. In general, this takes place when

$$k \geqslant \frac{W}{1.64ls\sqrt{s(s_o - s)}}\,.$$

It may be interesting to note that when the track is ideally tight, or when $s_o = s$,

$$z = a\,.$$

Then the whole track sinks to depth a and the process is characterized by the assumably uniform load distribution of the rigid surface.

Figure 102 shows an example of the pressure distribution for three types of ground defined by $k = 5$, 7.6, and 10. It will be seen that for the track dimensions and loads denoted on this drawing, the load in the track center reaches a value of approximately 4 psi in case "a," zero in case "b," and is negative in case "c." In a softer ground, there would be a greater load on the middle portion of the track. It is interesting to note that in case "c" the load is not carried on approximately

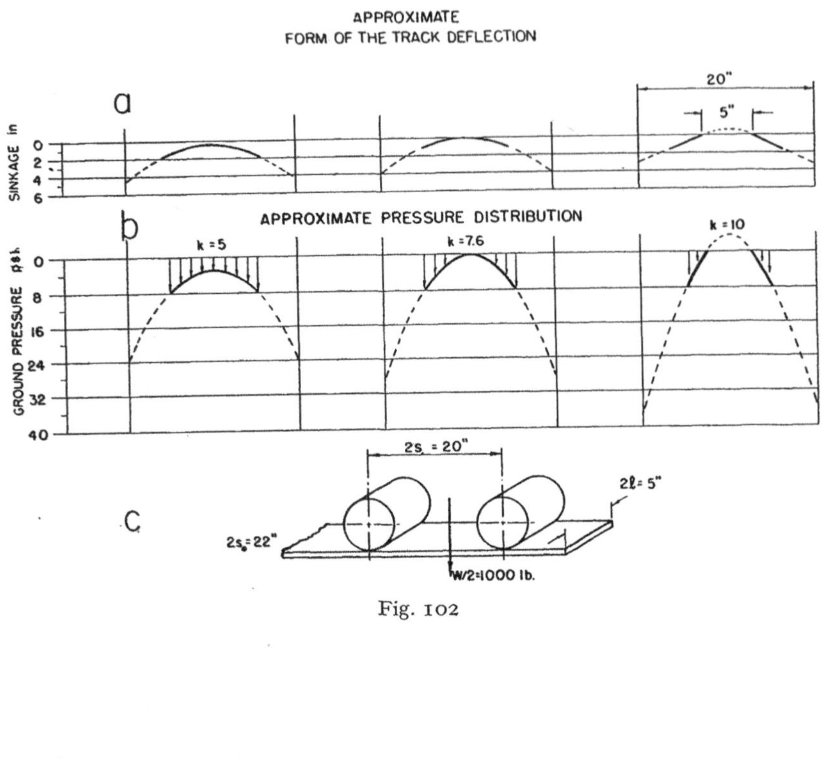

Fig. 102

5 in. of the track; therefore, $5/20 = 25\%$ of the track length is not "used" if the ground "hardness" reaches $k = 10$.

The above considerations illustrate the inadequacy of the "mean-ground-pressure" concept established in the early twenties and show why the attempts to evaluate vehicle performance in terms of this concept have failed.

Track Sinkage and the Number of Supporting Wheels

In the analysis referred to in the preceding section, and also in Figure 102, it was found that when the ground reaches a certain "hardness," not all of the "ground contact area" of the track carries the weight of the vehicle. Starting with the soil determined by $k = 7.6$, the central track portion of the considered track becomes unloaded to such an extent that when the soil reaches $k = 10$, the whole middle track length of approximately 5 in. is free from vertical loads. This suggests an immediate conclusion that, for instance, the additional wheel 3 located between the previously considered wheels 1 and 2 (Figure 101) would increase the bearing power of the track and thus reduce the sinkage z. If equations (239) and (241) are combined, and if only the first two members of the series in equation (241) are considered, the following formula for the track line will be obtained:

$$z = \frac{W}{4kls} - 0.41s\sqrt{\frac{s_o}{s} - 1} + 1.23\,\frac{x^2}{s}\sqrt{\frac{s_o}{s} - 1}\,. \tag{242}$$

In order to illustrate the effect of the proposed increase of wheel number upon sinkage z, assume that the considered track length s_o, which does not change if extra wheels are added, will amount to $s_o/2$ when supported by wheels 3 and 2. The beginning 0 of the coordinates will also be shifted by s_o.

In the case of three wheels, according to the original assumptions, s_o/s will remain constant but s will be divided in half. The value of x in the previous combination of two wheels (Figure 101) will be equal to $(s/2) + x$. Thus the comparative track sinkage which would occur at point P of the soil mass under the action of two and three wheels would be

$$\left.\begin{aligned} z_2 &= \frac{W}{4kls} - 0.41s\sqrt{\frac{s_o}{s} - 1} + 1.23\,\frac{[(s/2) + x]^2}{s}\sqrt{\frac{s_o}{s} - 1} \\ z_3 &= \frac{W}{4kls} - 0.21s\sqrt{\frac{s_o}{s} - 1} + 2.46\,\frac{x^2}{s}\sqrt{\frac{s_o}{s} - 1}\,. \end{aligned}\right\} \tag{243}$$

Equations (243) indicate that z_2 will be larger than z_3, which is in agreement with observations. These equations also show a general method for the numerical evaluation of the sinkage z in a given type of soil (k) as a function of the number of bogies (n), track slack (s_o/s), load (W), and track dimensions ($2l$, s_o).

In the particular case discussed above, if it is assumed that the unit ground pressure $W/4ls = 10$ psi, $k = 5$, $s = 10$ in., and $\sqrt{(s_o/s - 1} = 0.5$, then for $x = 5$ in.,

$$z_2 = 6.15 \text{ in.}$$

$$z_3 = 4.05 \text{ in.}$$

Thus, in given conditions, an increase in the number of wheels from n to $(2n - 1)$ will result in a reduction of sinkage by $100\,(6.15 - 4.05)/6.15 = 34\%$, which is quite a sizable improvement. Equations (243) are based on the assumption that the track is perfectly flexible. Rigid tracks would somewhat reduce the necessity of having a large number of wheels since the tracks themselves would increase the load uniformity.

On the basis of the foregoing remarks, the analysis of the suspension design of contemporary tracked vehicles appears to indicate a certain degree of nonuniformity in the assessment of the right number of wheels. In particular, the German practice of using overlapping wheels during the last war seems to have exaggerated the need for a large number of track wheels. Practically the same terrain performance could have been obtained with a smaller number of wheels and without the cost and weight of the overlapping system, assuming that other requirements, for instance the life of rubber tires, could have been satisfied.

Sinkage and Load Distribution under the Wheels Supporting a Track

Assume a flexible track, loose enough in order to wrap freely around the wheels, as shown in Figure 103. Such a condition is not exaggerated and may be frequently encountered.

If the vehicle is at rest, then the pressure distribution below the wheel which supports the track may be computed on the same basis as in the case of the track: $p = kz$. However, the procedure by Voellmy[123] illustrates that other assumptions are possible. He found that, in frictional soils, a close agreement exists between the measured and computed pressures if the latter is based on the assumption

$$P_r = kzy_r\,, \tag{244}$$

where P_r is the radial soil load per unit length of the cylinder and y_r is the radial deformation caused by the virtual displacement z_o of the cylinder (Figure 103). According to the denotations made on this figure, the following equations may be derived:

$$z = \frac{D}{2}(\cos\theta - \cos\theta_o)$$

$$y_r = z_o \cos\theta .$$

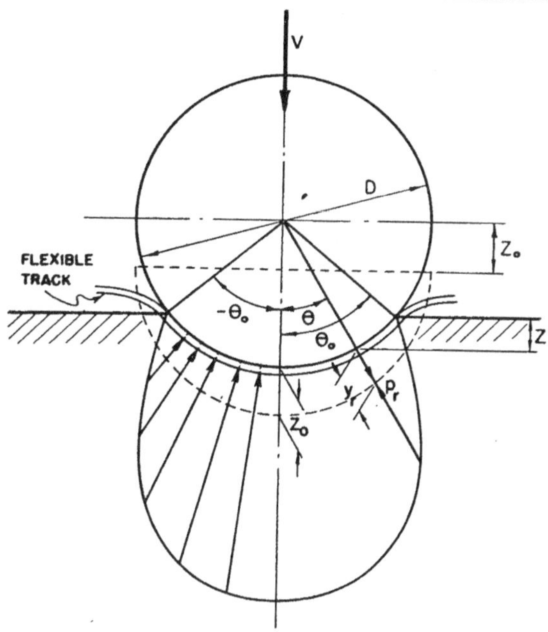

Fig. 103

Substituting these values into equation (244) gives

$$P_r = \frac{kD}{2} z_o (\cos\theta - \cos\theta_o)\cos\theta . \qquad (245)$$

On the other hand, the cylinder load V must be balanced by the vertical pressure:

$$V = \int_{-\theta_o}^{+\theta_o} p_r \frac{D}{2}\cos\theta \, d\theta .$$

This equation and equation (245) will give

$$V = \frac{kD^2}{4} z_o \int_{-\theta_o}^{+\theta_o} (\cos\theta - \cos\theta_o)\cos^2\theta \, d\theta$$

and

$$V = \frac{kz_o D^2}{12}(3\sin\theta_o + \sin^3\theta_o - 3\theta_o\cos\theta_o) .$$

Hence, the cylinder sinkage z_o becomes

$$z_o = \frac{12V}{kD^2(3\sin\theta_o + \sin^3\theta_o - 3\theta_o\cos\theta_o)} . \qquad (246)$$

The radial pressure may be determined from equations (245) and (246):

$$p_r = \frac{6V(\cos\theta - \cos\theta_o)\cos\theta}{D(3\sin\theta_o + \sin^3\theta_o - 3\theta_o\cos\theta_o)} . \qquad (247)$$

The form of load distribution along the circumference of the cylinder is shown in Figure 103.

The pressure distribution under wheels which move on a flexible loose track may be determined qualitatively when based on the previously discussed equation (163). For Bernstein's condition ($n = \frac{1}{2}$), this equation becomes $p = k\sqrt{z}$. Assuming that the normal load per unit length of the cylinder is $p_r = p/\cos\theta$, it will be seen that (Figure 104)

$$z_o = \frac{D}{2}(1 - \cos\theta_o)$$

and

$$\cos\theta_o = 1 - \frac{2z_o}{D} .$$

From equation (177), according to Bernstein,

$$W = \tfrac{5}{6}\, bkz_o\sqrt{D} .$$

Hence,

$$z_o = \frac{6W}{5bk\sqrt{D}} .$$

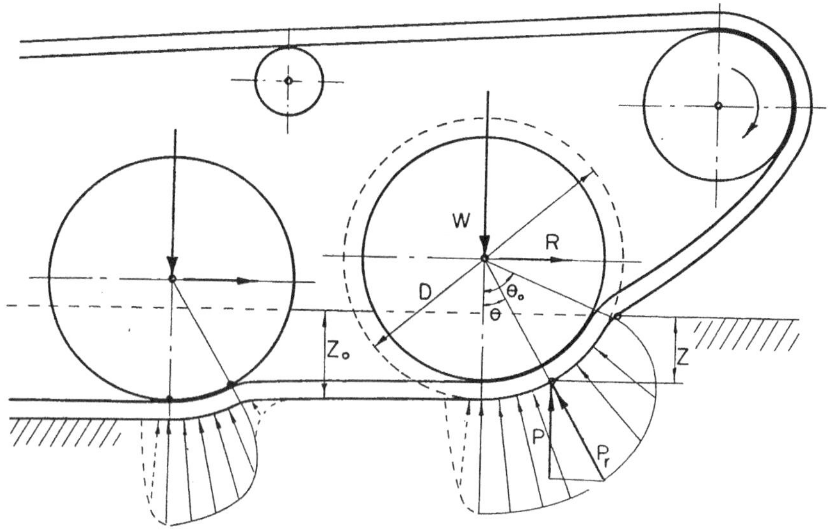

Fig. 104

Combining this equation with the equation $\cos\theta_o = 1 - (2z_o/D)$ gives

$$\cos\theta_o = 1 - \frac{12W}{5kbD\sqrt{D}}.$$

Since $z = (D/2)(\cos\theta - \cos\theta_o)$,

$$z = \frac{D}{2}\left(\cos\theta - 1 + \frac{12W}{5bkD\sqrt{D}}\right) \tag{248}$$

and

$$p_r = \frac{p}{\cos\theta} = \frac{k\sqrt{z}}{\cos\theta} = \frac{k\sqrt{\dfrac{D}{2}\left(\cos\theta + \dfrac{12W}{5bkD\sqrt{D}} - 1\right)}}{\cos\theta}$$

or

$$p_r = \frac{k\sqrt{\dfrac{D}{2}\,(5bkD\sqrt{D}\cos\theta - 5bkD\sqrt{D} + 12W)}}{\sqrt{5bkD\sqrt{D}}\cos\theta}. \tag{249}$$

The pressure distribution balanced by the resistance to motion R and corresponding to equation (249) is shown in Figure 104. An abrupt vanishing of the load below the vertical diameter of the wheel obviously will not take place, and there will be some elastic load effect as shown by the dashed line.

The quoted examples of computations of the pressure underneath wheels which support a track indicate to what extent the results depend on the accepted assumptions. The necessity of checking these assumptions by experiments becomes paramount. The reviewed methods illustrate the nature of the problems involved.

In general, it will be seen that the longitudinal distribution of the load between the track and soil is far from being uniform and varies not only as a function of track-support design, track flexibility, etc., but also as a function of the mechanical soil properties which have been defined by a rather arbitrary and ambiguous coefficient k.

Deflection and Transverse Load Distribution of the Track Plates

The transverse distribution of the pressure along the tracks is subject to similar considerations. Assume an elastic track plate of length $2l$ to which two loads W are applied by supporting wheels located at a distance a from the ends of the track (Figure 105a). A force acting on an element dx of the plate is $p\,dx$ if p is the pressure exercised on the ground. A cross section of the plate located at a distance x (where $x < a$) is subjected to the shearing force $dS = p\,dx$ and to the bending moment M. The latter may be determined in the following way.

Consider the cross section *m-m* of the beam under bending which is subjected to forces W, W_1, $W_2 \ldots W_n$ (Figure 105b). The bending moment which prevails at this section is

$$M = W_A x - \sum_0^x W(x - l) .$$

If this equation is differentiated with respect to x, then

$$\frac{dM}{dx} = W_A - \sum_0^x W .$$

But $W_A - \Sigma W$ is the shearing force S in the cross section *m-m*; hence,

$$\frac{dM}{dx} = S$$

and

$$\frac{d^2M}{dx^2} = \frac{dS}{dx} .$$

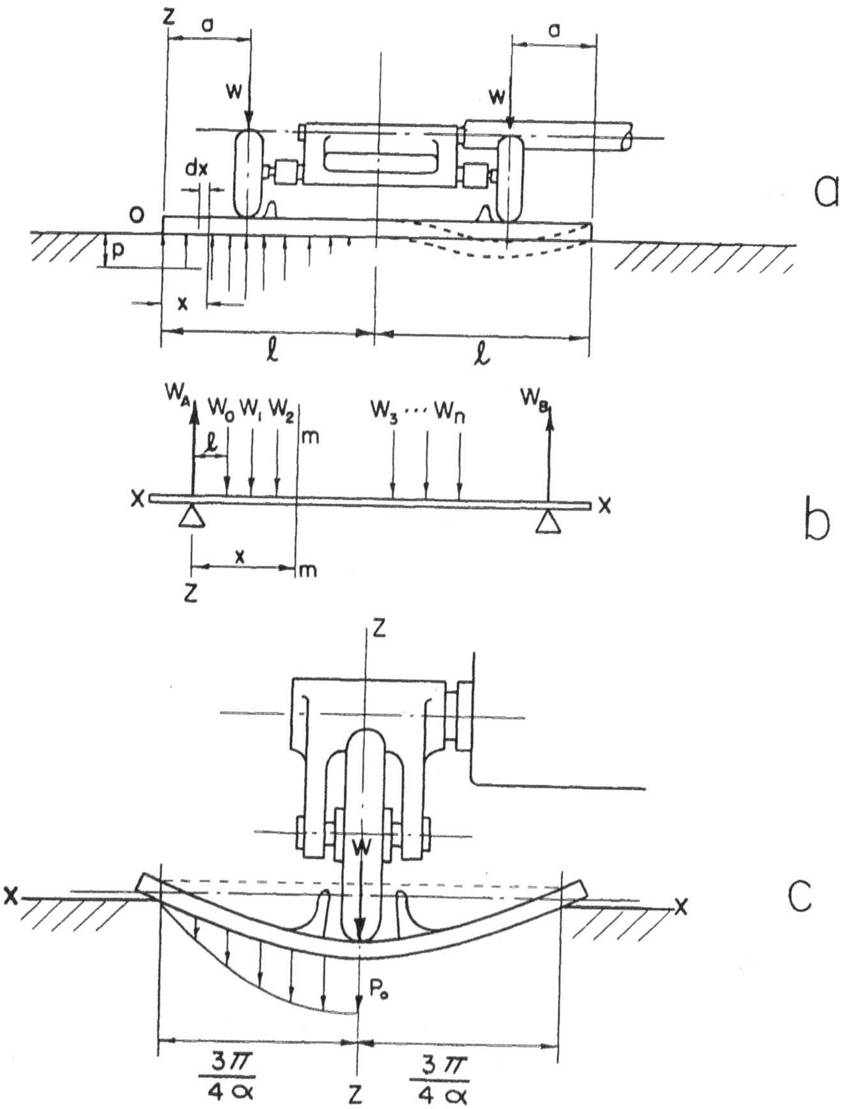

Z
a
a
W
W
dx
O
p
x
ℓ
ℓ
a
W_A
W_0 W_1 W_2
m
W_3 ··· W_n
W_B
ℓ
X
X
x
m
Z
b
Z
W
X
X
P_o
3π/4α
3π/4α
Z
c

Fig. 105

Since $dS/dx = p$,

$$\frac{d^2M}{dx^2} = p\,. \tag{250}$$

Under the action of the moment M, the beam will bend so that its centerline will take the form of an elastic curve whose equation in the xx and zz coordinates (Figure 105b) is: [169]

$$EI\frac{d^2z}{dx^2} = -M\,, \tag{251}$$

where E is the modulus of elasticity and I the moment of inertia. Differentiating this equation with reference to x gives

$$-\frac{d^2M}{dx^2} = EI\frac{d^4z}{dx^4}\,. \tag{252}$$

When this result is combined with equation (250), it will be found that

$$-EI\frac{d^4z}{dx^4} = p\,. \tag{253}$$

Now again an assumption is to be made with regard to the relationship between the sinkage z and the corresponding pressure p exercised on the ground. In a general case, as was previously discussed, $p = kz^n$. In order to simplify our considerations, however, take $n = 1$. Accordingly, $p = kz$. Thus, from equation (253),

$$EI\frac{d^4z}{dx^4} = -kz\,.$$

The general solution of this equation is: [170]

$$z = C_1e^{ax}\cos ax + C_2e^{ax}\sin ax + C_3e^{-ax}\cos ax + C_4e^{-ax}\sin ax\,, \tag{254}$$

where $a = \sqrt[4]{k/4IE}$. In order to determine the constants C_1, C_2, C_3, and C_4, the limiting conditions must be formulated. First consider the external portion of the beam denoted by length a. Equation (252) gives

$$\frac{dM}{dx} = -EI\frac{d^3z}{dx^3}\,.$$

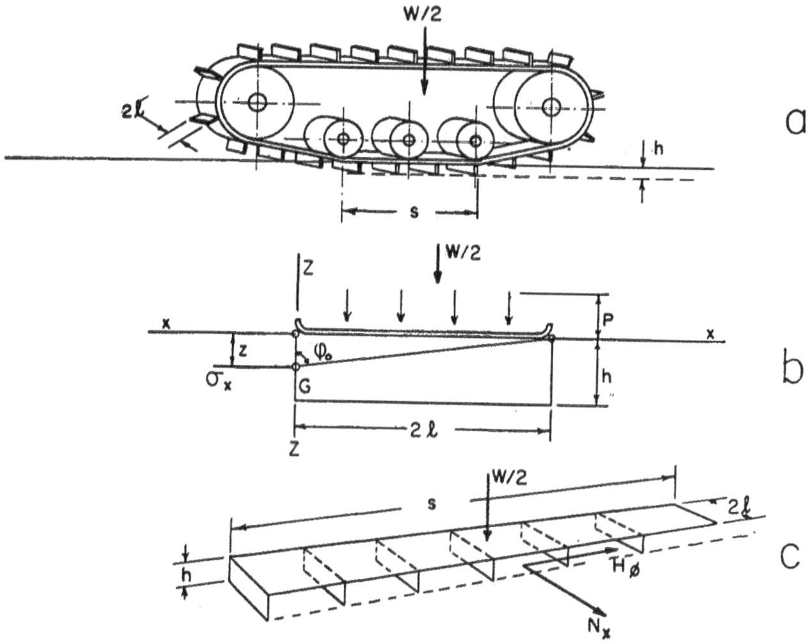

Fig. 106

tals of soil mechanics discussed in Chapter V suggest a basic reservation of the "ground-pressure" formula as far as the evaluation of the bearing capacity of tracks is concerned.

An analytical approach to this problem which could take into consideration all the elements involved, such as pressure distribution, flexibility of the loading area, its form and dimensions, as well as the mechanical properties of soil, is beyond the reach of a practical solution. However, a study of a properly simplified picture of the track-soil relationship may shed considerable light on some concepts which have at least a fundamental character.

When discussing the mechanical properties of soil which affect vehicle design and performance, it was suggested that particularly two of these properties, friction ϕ and cohesion c, are of essential significance, notwithstanding whether they reflect the real physical meaning of friction and cohesion, or are merely coefficients of an equation which describes the shearing strength of the ground at a given moment. The following considerations will further illustrate this point.

A track represents an ideal strip load if its width $2l$ is small as compared to the length s of the ground contact area (Figure 106). Assume that there is no horizontal force. Then the bearing capacity of such a strip load may be directly determined by means of equation (141), if the failure takes place because of general shear, or by equation (142), if it is caused by local shear failure. Since the above-mentioned equations refer to loads per unit of track length s, the critical vehicle weight W which may be supported by the two tracks can be evaluated by the equation

$$W = 4sl(cN_c + qN_q + \gamma l N_\gamma) \tag{262}$$

in the case of general shear, or by

$$W = 4sl\left(\tfrac{2}{3}\, cN'_c + qN'_q + \gamma l N'_\gamma\right) \tag{263}$$

if the failure takes place because of local shear.

Now, only the values of ϕ and c, as well as the specific gravity of soil γ, have to be known in order to choose the coefficients N_c, N_q, and N_γ from the table on Figure 73b, and to determine the safe vehicle weight W which may be supported in a given soil by two tracks having ground contact areas s inches long and $2l$ inches wide.

Suppose there is no initial sinkage so that the surcharge load q may be assumed to be zero. Also consider the general shear failure only. Then, in a purely frictional soil, like dry sand ($c = 0$), the safe vehicle weight W will be determined by

$$W_\phi = 4sl^2\gamma N_\gamma k_s\,, \tag{264}$$

where k_s is an arbitrary safety factor. In purely cohesive soils, like saturated clay or wet snow ($\phi = 0$),

$$W_c = 4slcN_c k_s\,. \tag{265}$$

By dividing equations (264) and (265), assuming that, in accordance with Figure 73b, $N_\gamma = 50$ ($\phi = 35°$) for sand, and $N_c = 5$ ($\phi = 0$) for plastic clay, it will be found that

$$\frac{W_\phi}{W_c} = \frac{10\gamma}{c}\, l\,.$$

If it is further assumed that, for instance, $\gamma = 0.06$ lb/cu in. and $c = 2$ psi, then it will be seen that the frictional soil will support more load than the selected cohesive soil, the ratio being

$$\frac{W_\phi}{W_c} = 0.3\, l\,.$$

This equation stresses the importance of track width l in the case of frictional soils, and the importance of cohesion c in plastic soils. As formula (264) indicates, the bearing capacity of a track in frictional soils increases with the square power of l, whereas in the case of cohesive soils it increases with the first power of track width [equation (265)]. Thus, for sandy soils, the measure of the bearing capacity is the ground contact area times half the track width, $2(2ls)l$, whereas in plastic clayish soils this characteristic is expressed merely by the contact area, $2(2ls)$. It is logical, therefore, to conclude that in order to increase the bearing capacity, or "flotation," of a track, it is more advisable to increase the track width $2l$ than to extend the lengths s of the ground contact area. A mere increase in the $2ls$ area, with the s/l ratio constant, is not an effective measure when attempting to reach a higher "flotation" in frictional soils.

The overlooking of this fact has been the cause of a basic misunderstanding in the interpretation of vehicle performance. "Flotation," defined as the bearing capacity of a given track contact area, does not depend on the size of this area alone, as is implied in the concept of "mean ground pressure," but also on the form of the area as defined by the s/l ratio. The more frictional the soil, the greater the effect of the form of the ground contact area. Since most of the soils are, for practical purposes, frictional, the s/l factor cannot be neglected.

Although the above considerations refer to a homogeneous semi-infinite medium, the effect of soil stratification should not change the basic conclusions. If, however, a relatively thin plastic soil layer located on a hard pan is considered, then the problem may be stated in a somewhat different way. For instance, in such a case, flotation may not be a desired factor. If the ground clearance is large enough, it may be hoped that the vehicle will reach the hard bottom as quickly as possible in order to be able to develop a sufficient grip which would overcome the resistance to motion. Thus, the vehicle should develop the pressure needed to squeeze out the soft mass located between the tracks and the hard substratum. An evaluation of the forces involved in such a case may be performed by means of equation (119) if the assumed condition of plasticity checks with the given type of mud, or snow. In such a case, for track width $2l$ and length s,

$$W = 4s \int_0^l \sigma_z \, dx = \frac{4sk_o}{h} \int_0^l x \, dx$$

and

$$W = \frac{2k_o s l^2}{h}, \tag{266}$$

where h is the half thickness of the soft soil layer (Figure 64) and k_o is a constant equal to the yielding stress of the mass in compression, divided by $\sqrt{3}$ [see equation (111)].

It will be seen, in this case again, that the width of the track l, which enters into equation (266) in a second power, plays an important role. It is again more economical to reduce the track width than to reduce the length in order to obtain a better squeezing-out effect of the mud, for instance.

When considering the "flotation" by means of formula (262), it was assumed that the bearing capacity of two tracks equals the doubled capacity of a single track. This assumption, however, is valid only when the track spacing is large enough to clear the soil masses under failure. According to Figures 72 and 73, the soil masses in question extend sidewise through the radial shear and Rankine zones (Figure 107). If these zones overlap to a large extent, the interference of stresses and strains causes the bearing capacity of two tracks to be different from the doubled capacity of a single track. The solution of such a case, if possible at all, would certainly be very complex. Similar phenomena were observed and investigated by Buchholz in conjunction with his studies of anchor plates,[171] and by Rathje, who worked on the resistance of sand against shear by using various cutter forms.[165] Their work may illustrate the problem reviewed. In most cases, however, soil fails because of local shear, and the question of track spacing appears to be less significant.

The Main Tractive Effort

The force that may be developed in the soil with the purpose of propelling a vehicle under given conditions will be called the tractive effort H. Such a force is substantially larger than the so-called draw-bar pull H_a defined as the net towing capacity of the vehicle. The difference

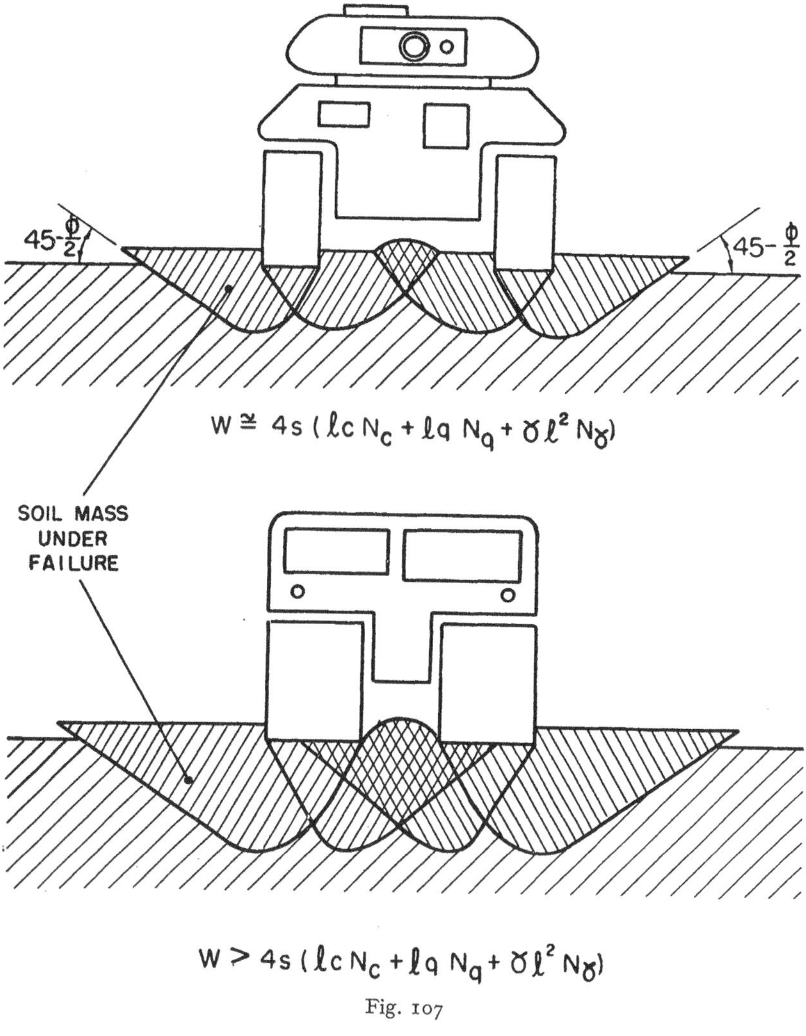

Fig. 107

between H and H_a is the external rolling resistance R caused by the interaction of the soil and the track.

The occurrence of tractive effort is made possible because of the existence of the shearing strength of the soil. It was shown in Chapter V that the generally accepted theory of soil shear is based on Coulomb-

Mohr's condition, which postulates the following relationship between shearing strength τ, normal pressure σ, angle of friction ϕ, and cohesion c [equation (123)]:

$$\tau = c + \sigma \tan \phi .$$

Experimental evidence of the shear phenomena which take place in the soil as a result of vehicle motion at the point close to a stall indicates that the main shear occurring at the ground contact area is only a large-scale replica of the shearing process produced in the laboratories by means of the so-called "shear box" under the conditions expressed by equation (123).

Figure 108 shows this analogy and directly implies that the horizontal tractive forces H and vehicle weight W which act upon the ground due to vehicle action may be related by equation (123):

$$H_m = 2\Delta c + W \tan \phi , \tag{267}$$

where Δ is the approximate area of the shear, roughly equal to the "ground contact area" between the track and the ground.

Assume that the soil is purely frictional ($c = 0$). Under these conditions, the maximum tractive effort which may be developed by any tracked vehicle is

$$H_m = W \tan \phi . \tag{268}$$

It will be seen that such a tractive effort depends only on the vehicle weight W. The heavier the tractor, the more pull it develops. The dimensions and design details of the track, as well as those of the suspension, are basically irrelevant. Thus, in sand or sandy soils, vehicle weight will remain a decisive factor in the development of the propelling forces. These forces cannot be larger than $W \tan \phi$. Since ϕ in sand is practically constant and is approximately equal to 35°, the limiting value of H is $W \tan 35° = 0.7W$. This is why the tractive effort per unit of vehicle weight (H/W) is always constant in sand, and never exceeds the value of approximately 70% of the vehicle weight. The size and form of the tracks and of the ground contact area do not affect the main tractive effort as long as the area is large enough to provide the necessary bearing capacity which may be evaluated by means of equation (263). Thus, the criterion of track design for sandy soils is the "flotation" determined by this equation.

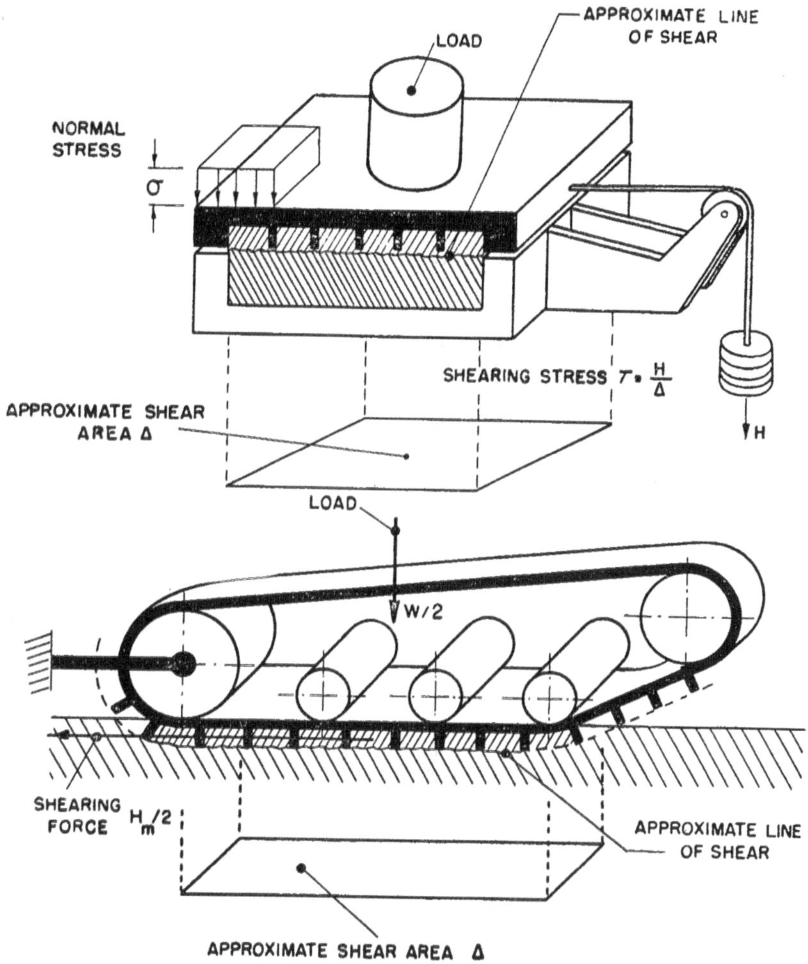

Fig. 108

Consider next a purely cohesive soil in which $\phi = 0$. In this case, equation (267) takes the following form:

$$H_m = 2\Delta c \,. \tag{269}$$

Now, the conditions of traction are entirely different. Vehicle weight does not enter into the formula. The main tractive effort basically depends on the size of the ground contact area Δ. The larger the tracks, the larger the tractive effort. Tracks designed for operation in plastic soils in accordance with equation (265) will provide only sufficient "flotation," and may have to be made much larger in order to provide sufficient traction. Thus, the criterion of track design in this case is track size: the larger the size, the better the traction; vehicle weight is irrelevant. These basic conclusions, although only recently formulated, have been indicated by daily experience for a long time.

The shear area Δ remains constant when the dimensions and form of the track spud design change. The location of Δ in the soil mass depends on whether the grousers are high or low. However, the location does not affect the tractive effort. The latter depends solely on the form and size of Δ, which do not change in given conditions. Thus, the grouser design and size do not fundamentally affect the maximum tractive effort—a paradox learned after many years of trials. There is, however, a secondary effect of the spud which, for practical purposes, cannot be neglected. This effect will be discussed in the next section.

In the case of nonhomogeneous soils, the grouser action is somewhat different. The track spud, like the deep, sharp tread of a pneumatic tire applied to similar soil, may act as a cutter which helps the wheel to remove the weak surface layer and to reach a firm stratum which has enough shearing strength to provide the necessary tractive effort. From now on, however, the role of the spud becomes secondary again.

In conclusion, it may be stressed that two factors are predominant in securing large tractive efforts: track load and track size. The more frictional the soil, the more dominant the load factor. The effect of track size increases only with increasing ground cohesion. Other details of design are of secondary nature. Since average soil is a combination of both frictional and cohesive ingredients, their relative preponderance is the only rational clue in establishing the main design concept in which track load and track size are in desired balance.

The Action of Spuds and the Total Tractive Effort

As was mentioned in the preceding section, the main tractive effort developed either by frictional or cohesive forces is to be related to the shear area located beneath the tracks, often identified as the ground contact area.

It can easily be seen (Figure 106c), however, that there are two other areas of shear per each track, which are located along the sides of the tracks. If the depth of the spuds is denoted by h, then the four additional areas will amount to $4hs$, and cannot be neglected in the estimates of additional tractive effort.

Assume that the soil cohesion is c. The increase in traction due to the shear along the described areas will be

$$H_c = 4hsc\,. \tag{270}$$

A similar increase caused by the friction $\tan \phi$ between the walls of the prism $2lhs$ sliding with the track and the rest of the soil which is stationary may be evaluated in the following way. Assume that the pressure p acting upon the bearing area of the track (Figure 106b) "radiates" in conformity with the general principles of soil mechanics discussed in Chapter V. Accordingly, it brings into existence a normal pressure σ_x acting at an arbitrary point G located at the wall of the investigated soil prism.

Since the track may be considered as a strip load, σ_x may be computed by means of equation (73). In this equation only, the limits of integration must be changed since point G is located in such a way that $\varphi_1 = 0$ and $\varphi_2 = \varphi_o$ (see Figures 57d and 57e, and then compare with Figure 106b). Accordingly, assuming that $V = p$,

$$\sigma_x = \frac{2p}{\pi}\int_0^{\varphi_o} \sin^2\varphi\, d\varphi = \frac{p}{\pi}(\varphi_o - \sin\varphi_o \cos\varphi_o)\,.$$

On the other hand,

$$\varphi_o = \cot^{-1}\left(\frac{z}{2l}\right);\ \sin\varphi_o = \frac{2l}{\sqrt{z^2+4l^2}}\,;\ \cos\varphi_o = \frac{z}{\sqrt{z^2+4l^2}}\,.$$

Thus, the total force N_x acting vertically upon the unit lateral area $s \times h$ will equal

$$N_x = \int_0^h \sigma_x\, dz = \frac{p}{\pi}\int_0^h \left[\cot^{-1}\left(\frac{z}{2l}\right) - \frac{2lz}{z^2+4l^2}\right] dz\,.$$

The first integral may be solved by denoting $z/2l = u$; then,

$$2l\int_0^{h/2l} \cot^{-1} u\, du = h\cot^{-1}\left(\frac{h}{2l}\right) + l\log\left(1+\frac{h^2}{4l^2}\right).$$

The second integral will be solved directly:

$$-\int_0^h \frac{\frac{z}{2l}}{1+\frac{z^2}{4l^2}}\,dz = -l\log\left(1+\frac{h^2}{4l^2}\right).$$

Thus, finally, when adding both integrals, and multiplying by p/π,

$$N_x = \frac{p}{\pi}\left[h\cot^{-1}\left(\frac{h}{2l}\right)\right].$$

The total frictional force H_ϕ caused by the normal forces N_x and acting upon the four lateral surfaces of shear of two tracks is

$$H_\phi = 4sN_x\tan\phi\,.$$

If it is assumed that $p = W/4ls$, the last two equations will yield

$$H_\phi = 0.64\,W\tan\phi\left[\frac{h}{2l}\cot^{-1}\left(\frac{h}{2l}\right)\right]. \tag{271}$$

By adding equations (267), (270), and (271), the total maximum tractive effort developed by two tracks fitted with spuds h inches high and having a ground contact area equal to $2(2ls)$ inches, subjected to load W, will be obtained for a given soil defined by c, ϕ, and γ:

$$H = H_m + H_c + H_\phi\,.$$

Since $\Delta = 2ls$,

$$H = 4lsc + W\tan\phi + 4hsc + 0.64\,W\tan\phi\left[\frac{h}{2l}\cot^{-1}\left(\frac{h}{2l}\right)\right]$$

or

$$H = 4lsc\left(1+\frac{h}{l}\right) + W\tan\phi\left\{1 + 0.64\left[\frac{h}{2l}\cot^{-1}\left(\frac{h}{2l}\right)\right]\right\}. \tag{272}$$

In order to investigate more closely the effect of grouser height h upon the tractive effort, assume that the soil is cohesionless ($c = 0$). Then,

$$H = W\tan\phi\left\{1 + 0.64\left[\frac{h}{2l}\cot^{-1}\left(\frac{h}{2l}\right)\right]\right\}. \tag{273}$$

In the case of frictionless soil, equation (272) takes the form

$$H = 4sc\,(h + l)\,. \tag{274}$$

Equations (273) and (274) are plotted in dimensionless form in Figure 109. The effect of grouser height upon the tractive effort is shown as a function of the $h/2l$ ratio. This ratio, rather than the absolute value of the grouser height, is responsible for the increase of the tractive effort in frictional soil, as may be deduced from equation (273).

Figure 109 shows that, in frictional soils, the action of spuds is practically negligible, irrespective of spud height. Since, for practical reasons, the ratio of spud height to track width is seldom larger than 0.2, the grouser action in dry sand may be considered small. It will be seen, however, that, in cohesive soils, the total tractive effort is radically increased by the application of spuds. For $h/2l = \frac{1}{2}$, i.e., for the grouser height equal to the half-track width, the traction is doubled. It is unfortunate that such a track cannot be made mechanically sound for practical reasons. Nevertheless, even short lugs contribute sizably to the performance of the vehicle if the soil is sticky.

Most soils combine frictional and cohesive factors in varying degrees. Thus, a deep lug which is effective in one case may be entirely idling in another. A mere change of weather, for instance, may increase the spud effectiveness. Rain usually increases the cohesive properties of soil, and thus makes a deep-spudded track work better than it did in dry weather. Figure 109 indicates that the effect of grousers cannot be evaluated without specifying the mechanical properties of the soil, and that there is no over-all scale which could be expressed in terms of grouser shape and size in order to state the relative merits of various track designs.

It will be interesting to note that in all the previous considerations, the spacing of the grouser plates was not mentioned. Although this factor appears to have a secondary role in track efficiency, its study cannot be neglected. In principle, spuds may be assumed to be a source of stresses which also "radiate" toward the sides of the soil prism sheared by the track, and which further contribute to the frictional increase in tractive effort. In such an assumption, the number of spuds, and thus the spacing, would enter into the final equation, the basic form of which would be similar to that of equation (271).

Experimental and theoretical work would be needed, however, to ascertain the degree of the practicability of such an approach. So far, only formula (272) has been checked experimentally: the values obtained are about 10% smaller than those actually measured in the field.

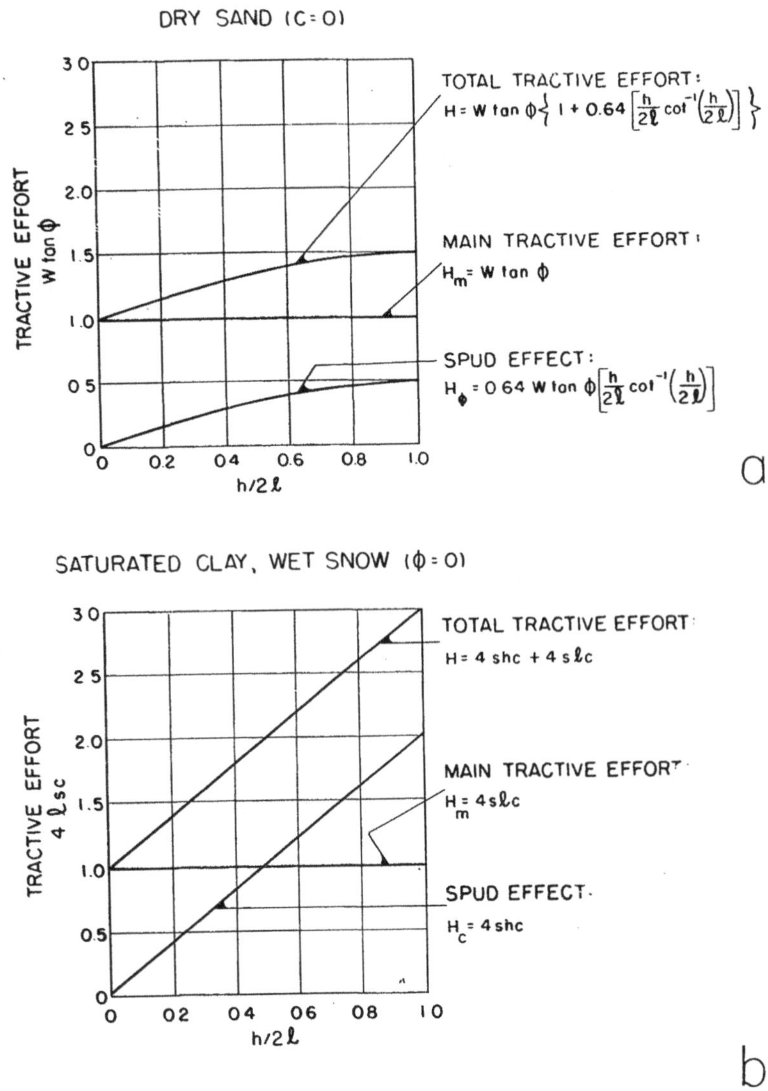

Fig. 109

Track Slip and Tractive Effort

The analogy between the soil shear produced by laboratory apparatus and the shear caused by a track, as shown in Figure 108, requires some elucidation in order to be more general. If the upper half of the shear box, or the blocked track, is moved with reference to the lower half of the soil mass, then the maximum shearing force will not be developed instantaneously upon the start of the relative motion, as is known from daily experience. The porosity or looseness of the soil will require some degree of compaction which results in a slip being produced before the full shearing stress is developed. During the described shearing process, all the grousers will encounter the same resistance and will slip to the same extent, assuming that the ground is homogeneous.

If the relative motion of the soil under shear is plotted as a function of the corresponding stress, then three basic curves may be determined, the shape of which will depend on the three types of soil strength involved. Figure 110 shows the general character of the discussed lines. Curve A refers to loose frictional or plastic soils. The shearing strength τ_a of such a mass is reached after the initial period of compaction, which takes place at a distance j_a. The stress remains practically on the same level, irrespective of any further slip d. Soil B is of different structure; it is not loose but forms a more or less solid coherent mass. Dry settled clays or firm silts in the undisturbed state may belong to this category. A similar type of "soil" is represented by frozen snow, which, in low temperatures, more closely resembles a spongy cohesive mass than a loose granular structure. This type of ground reaches the required degree of compression (j_b) rather quickly, and then shears off rapidly—a process in which the original structure of the mass is destroyed. The new structure does not have the strength of the original mass and the τ_b value decreases rapidly. The third type of soil C has intermediate properties: upon reaching the required compaction at the slip distance j_c, which corresponds to the maximum value of τ_c, the value of the shearing strength also decreases, but not as low and as rapidly as in case B. Most types of snow in moderate climates and average agricultural soils belong to this category.

The described phenomena take place when the whole surface of a body is in uninterrupted contact with the soil during the process of shear. This condition may be assumed to be satisfied if the shear is performed in accordance with the scheme shown in Figure 108, i.e., when the track is blocked before it is pulled with reference to the ground.

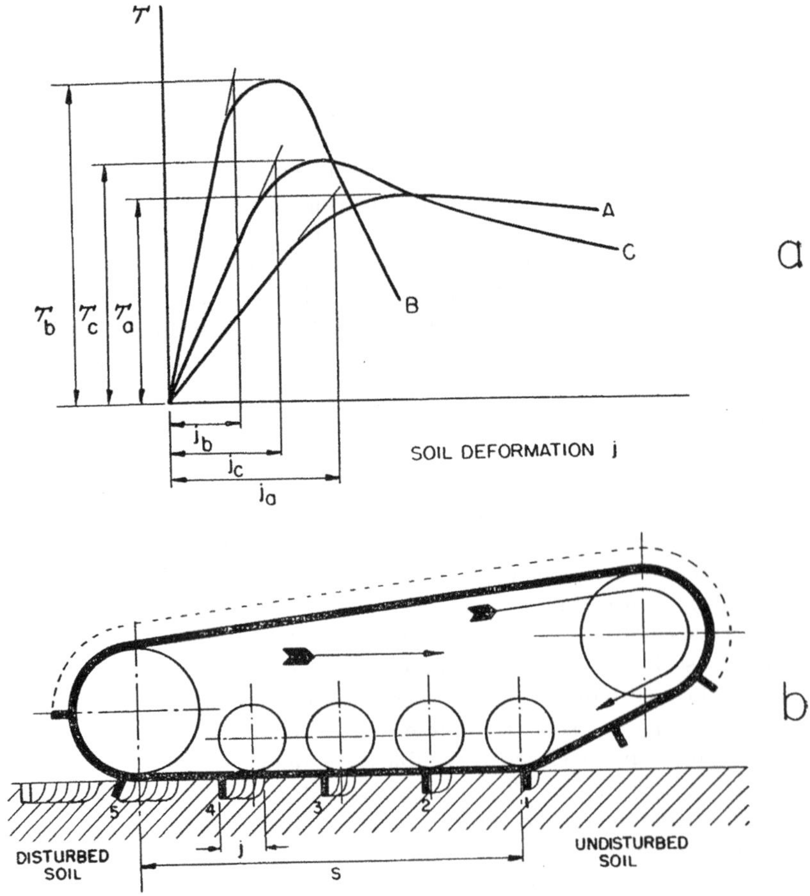

Fig. 110

In the case of a vehicle in motion, however, the situation is somewhat different. Figure 110b shows the frontal portion of the track adjacent to spud 1 coming into contact with the undisturbed soil. At this point, it cannot develop the same shearing force as the other grousers 2, 3, 4, and 5 because they have already applied loads to the ground for a certain period of time when passing along the ground contact area. It is only after the frontal portion of the track has moved along this area in a relative motion with reference to the vehicle that the gradual compression of the soil or slipping on the soil surface may produce the tractive effort which would balance the resistance to motion. The maximum slip produced in such a way will be reached before section 1 of the track leaves the ground at the rear of the vehicle. Since the track cannot stretch, the amount of horizontal soil distortion or slip j which occurs beneath the particular track sections 1, 2, 3, 4, and 5 accumulates consecutively along the ground contact area and reaches its maximum value at a constant rate (Figure 110b).

It will thus be seen that a stalled track, as shown in Figure 108, does not reproduce quite accurately all the effects that result from the traction of a crawler in motion. This fact, although stressed a long time ago,[33] has seldom been appreciated.

The relationship between the tractive effort and slip resulting from the foregoing analysis may be described in the following way. It is an interesting coincidence that the form of the soil-shear curves shown in Figure 110a is identical to the form of the curves pertaining to the displacement (x) and natural frequency time ($\omega_n t$) of an aperiodic vibration. The general equation of such a motion is

$$x = A_1 e^{(-b+\sqrt{b^2-1})\omega_n t} + A_2 e^{(-b-\sqrt{b^2-1})\omega_n t} ,$$

where b is the coefficient of damping larger than unity.[172] In order to write a corresponding equation in terms of soil deformation j, which will replace $\omega_n t$, and in terms of soil stress τ, which will substitute the displacement x, introduce the factor K_1, from which the scale of j with reference to τ can be determined. This coefficient depends on the degree of soil looseness, or compactness, and reflects the amount of slip j required to obtain the corresponding value of τ. Consider further that the type of curve can be determined by the coefficient K_2 which replaces the value of damping b, and which predetermines whether the shear curve is of the A, B, or C type (Figure 110a). Finally, to evaluate the constants A_1 and A_2, the coefficient K_3 is introduced, which at this moment is defined as the slope of the tangent to the discussed curve at the beginning of the coordinates. According to these assumptions,

$$\tau = A_1 e^{(-K_2+\sqrt{K_2^2-1})K_1 j} + A_2 e^{(-K_2-\sqrt{K_2^2-1})K_1 j} . \qquad (275)$$

The value K_3, as was mentioned above, is to be included in the constants A_1 and A_2. These constants may be determined in the following way: for slip $j = 0$, $\tau = 0$; hence,

$$0 = A_1 + A_2 . \qquad (276)$$

Also for $\tau = 0$ and $j = 0$, $d\tau/dj = K_3$, and

$$\frac{d\tau}{dj} = A_1 e^{(-K_2+\sqrt{K_2^2-1})K_1 j}\left(-K_2+\sqrt{K_2^2-1}\right)K_1 +$$

$$+ A_2 e^{(-K_2-\sqrt{K_2^2-1})K_1 j}\left(-K_2-\sqrt{K_2^2-1}\right)K_1 = K_3$$

or

$$K_1(A_1 - A_2)\sqrt{K_2^2-1} - K_1K_2(A_1+A_2) = K_3 . \qquad (277)$$

From equations (276) and (277),

$$A_1 = \frac{K_3}{2K_1\sqrt{K_2^2-1}}$$

$$A_2 = -\frac{K_3}{2K_1\sqrt{K_2^2-1}} .$$

After substituting these values into equation (275), it will be found that

$$\tau = \frac{K_3}{2K_1\sqrt{K_2^2-1}}\left[e^{(-K_2+\sqrt{K_2^2-1})K_1 j} - e^{(-K_2-\sqrt{K_2^2-1})K_1 j}\right] . \qquad (278)$$

The meaning of the coefficient K_1 reflects the degree of soil compactness, in accordance with the original assumption. Its value may be determined, for instance, by trial and error when plotting the τ (j) curve. It is believed that an average type of soil will be expressed by $K_1 = 0.3$. A "brittle" or highly coherent soil may have a value of K_1 close to unity. K_2, which determines the fundamental character of the shear curve, presumably reaches a value of 2 for an average soil and approaches unity for "brittle" or coherent soils, always being larger than unity. Loose dry sands, or supersaturated ideally plastic clays may have a value of K_2 equal to approximately 3. The discussed coefficient (K_2) is dimensionless and its value may be established empirically when the stress-strain curve of a shear test is studied.

Since, in equation (278), the portion enclosed by the brackets is dimensionless, the value of $K_3/2K_1\sqrt{K_2^2-1}$ must have the dimension of lb/sq in. in order to express τ in the same units. The value of K_3 may be defined in the following way. From equation (123), which was derived from the Coulomb-Mohr hypothesis of soil failure, it follows that $\tau_{max} = c + \sigma \tan \phi$. Thus, the maximum value of the coefficient enclosed by the brackets of equation (278) and multiplied by $K_3/2K_1\sqrt{K_2^2-1}$ is equal to $c + \sigma \tan \phi$. Hence,

$$K_3 = \frac{2K_1\sqrt{K_2^2-1}\,(c+\sigma\tan\phi)}{\left[e^{(-K_2+\sqrt{K_2^2-1})K_1 j} - e^{(-K_2-\sqrt{K_2^2-1})K_1 j}\right]_{\max}}. \tag{279}$$

Now equation (278) may be written in a simplified form if the maximum value of the function enclosed in the brackets of the denominater of equation (279) is denoted by $y_{\max}$:

$$\tau = \frac{c+\sigma\tan\phi}{y_{\max}}\left[e^{(-K_2+\sqrt{K_2^2-1})K_1 j} - e^{(-K_2-\sqrt{K_2^2-1})K_1 j}\right]. \tag{280}$$

Thus, the constant K_3 has been replaced by a function of the stress prevailing at the moment of shear failure. In such a way, equation (280) correlates the shearing stress τ with the soil properties c and ϕ and slip j for a given normal stress σ. This equation is empirical and can be determined for each type of soil by finding the appropriate K_1, K_2, ϕ, and c values for an assumed σ.

As an illustration, take a firm "brittle" soil which may be represented by a hard, well-settled, compact, and undisturbed silt. In this case the following K values may be envisaged as general soil characteristics: $K_1 = 1$, $K_2 = 1.1$. Then, for these values,

$$y_{\max} = \left[e^{(-K_2+\sqrt{K_2^2-1})K_1 j} - e^{(-K_2-\sqrt{K_2^2-1})K_1 j}\right]_{\max} = 0.31\,.$$

If the shear test has shown, for instance, that $c = 6.5$ psi and $\phi = 10°$, then for $\sigma = 3$ psi, according to equation (280),

$$\tau_{\text{silt}} = \frac{6.5 + 3\tan 10°}{0.31}\left[e^{-0.65j} - e^{-1.55j}\right]. \tag{281}$$

In the case of a loose sandy loam with a small amount of moisture content, the following K values may represent the stress-strain curve of a shear test: $K_1 = 0.3$, $K_2 = 2$. If, for instance, the shear test discloses that $c = 0.2$ and $\phi = 33°$, then, for $\sigma = 3$ psi, equation (280) may be written analogically in the following form:

$$\tau_{\text{loam}} = \frac{0.2 + 3\tan 33°}{0.76}\left[e^{-0.08j} - e^{-1.11j}\right]. \tag{282}$$

Both curves τ_{silt} and τ_{loam} are plotted in Figure 111. They appear to

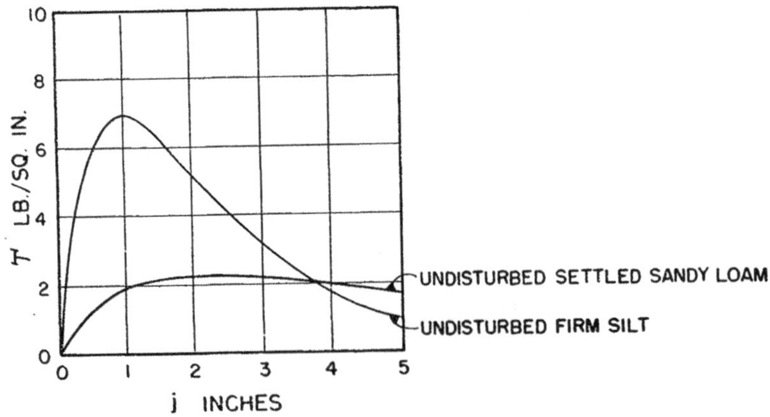

Fig. 111

present an acceptable stress-strain relationship which may be encountered in average conditions. Although equation (280) does not include all the complexities of the shear-strain relationship of various soils, it does attempt to express this relation and may serve as a useful tool in further investigation of the slip-tractive effort relationship of a track moving in a given soil.

The discussed equation, if written in dimensionless form, also may be helpful in attempts to classify soil types in accordance with the form of their $\tau(j)$ curve:

$$\frac{\tau}{c + \sigma \tan \phi} = \frac{e^{(-K_2 + \sqrt{K_2^2 - 1})K_1 j} - e^{(-K_2 - \sqrt{K_2^2 - 1})K_1 j}}{y_{max}}. \qquad (283)$$

By following the previously discussed pattern of slip distribution along the ground contact area, it may be agreed that the amount of soil distortion j which equals zero at the beginning and reaches its maximum value (j_m) at the end of that area (Figure 110b) can be represented by a triangle (Figure 112). On the other hand, the slip i_o of a vehicle is expressed by the conventional formula

$$i_o = 1 - \frac{v_a}{v_t},$$

where v_a and v_t are the actual and theoretical speeds of locomotion, respectively. The slip and the amount of maximum distortion of the soil

j_m may be related as follows. The distance of shear j_m is equal to the speed of slip v_j times the time t in which it occurs:

$$j_m = v_j t .$$

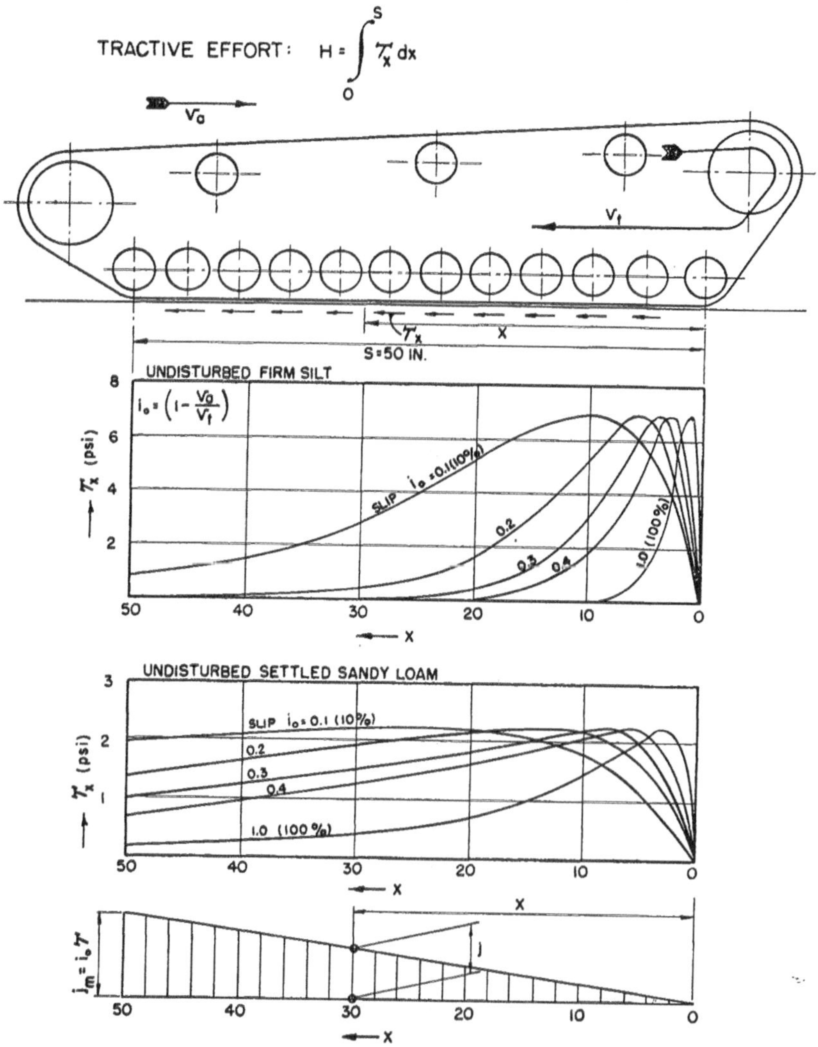

Fig. 112

Since, however, $v_j = v_t - v_a$,

$$j_m = (v_t - v_a)\, t \, .$$

On the other hand, $t = s/v_t$, where s is the distance along which j_m has occurred. Hence,

$$j_m = s\left(1 - \frac{v_a}{v_t}\right) = i_o s \, . \tag{284}$$

The amount of soil distortion which takes place at any point located at a distance x from the front of the ground contact area (Figure 111a) will be

$$j = j_m \frac{x}{s} \, . \tag{285}$$

From equations (284) and (285), it follows that

$$j = i_o x \, . \tag{286}$$

Since the local tangential stress, or the local tractive effort per unit of track area, is determined by equations (280) and (286), the total tractive effort developed by one track will be

$$\tfrac{1}{2}H = 2l \int_0^s \tau \, dx \tag{287}$$

or

$$H = 4l \int_0^s \frac{c + \sigma \tan \phi}{y_{\max}} \left[e^{(-K_2 + \sqrt{K_2^2 - 1}) K_1 i_o x} - e^{(-K_2 - \sqrt{K_2^2 - 1}) K_1 i_o x} \right] dx$$

and

$$H = \frac{4sl\,(c + \sigma \tan \phi)}{K_1 i_o s \; y_{\max}} \left[- \frac{1}{-K_2 + \sqrt{K_2^2 - 1}} + \frac{e^{(-K_2 + \sqrt{K_2^2 - 1}) K_1 i_o s}}{-K_2 + \sqrt{K_2^2 - 1}} + \right.$$

$$\left. + \frac{1}{-K_2 - \sqrt{K_2^2 - 1}} - \frac{e^{(-K_2 - \sqrt{K_2^2 - 1}) K_1 i_o s}}{-K_2 - \sqrt{K_2^2 - 1}} \right] . \tag{288}$$

Equation (288) expresses the tractive effort in terms of soil properties c, ϕ, track dimensions $2l$ and s, unit track load $\sigma = W/2\,(2ls)$, and slip i_o for a given type of soil defined by the constants K_1 and K_2.

The tractive effort per unit track area $4ls$ was computed in accordance with equation (288) for the previously specified load and silt and is plotted in Figure 113. This figure shows the changes of $H/4ls$ with the distortion of soil $j_m = i_o s$, i.e., with vehicle slip i_o.

Since $j = i_o x$, it results from equation (280) that the tractive forces τ_x per unit of the track area located at a distance x from the front of the ground contact surface (Figure 112) equal

$$\tau_x = \frac{c + \sigma \tan \phi}{y_{\max}} \left[e^{(-K_2 + \sqrt{K_2^2 - 1}) K_1 i_o x} - e^{(-K_2 - \sqrt{K_2^2 - 1}) K_1 i_o x} \right]. \quad (289)$$

Thus, it will be seen that the distribution of the elements of the tractive effort along the ground contact area is not even and depends on the amount of slip as well as on the properties of the soil.

Curves showing the distribution of τ_x for the previously specified undisturbed soils which may belong, as it was proposed, to the categories of firm silts and settled sandy soil are plotted in Figure 112, in accordance with equation (289).

It may appear surprising to find that in firm soils (undisturbed firm silt), which do not display much strength after breaking, only the frontal portion of the track produces tractive effort, whereas the rear contributes very little at small slip ($i_o = 10\%$), and practically nothing at higher slip.

The situation is somewhat different in soils which do not lose so much of their original strength after the distortion is extended beyond the $(j)_{opt}$ value (Figure 113). Undisturbed, settled sandy loam referred to in Figure 112 indicates that the whole second half of the track develops a uniform pull for 10% slip. However, even in this case, at 100% slip, almost the total tractive effort is concentrated in the front of the track, with the rear portion practically idling.

The most uniform distribution of the tractive effort for loose plastic or frictional soil masses will be obtained when their strength is not reduced after the distortion determined by $(j)_{opt}$ is reached. For an average track length s, such soils (dry sand, supersaturated clay) produce the maximum pull H at 20—40% of slip. An unscaled curve of this type is shown by the dashed line in Figure 113 with the purpose of illustrating the general shape of a function which may represent the discussed type of soil. This shape is presumably determined by $K_1 \cong 0.2$ and K_2 ranging from 3 to 4.

Figure 113 leads to an important conclusion regarding the length s of the ground contact area. In this graph, the value $H/4ls$ is plotted as a

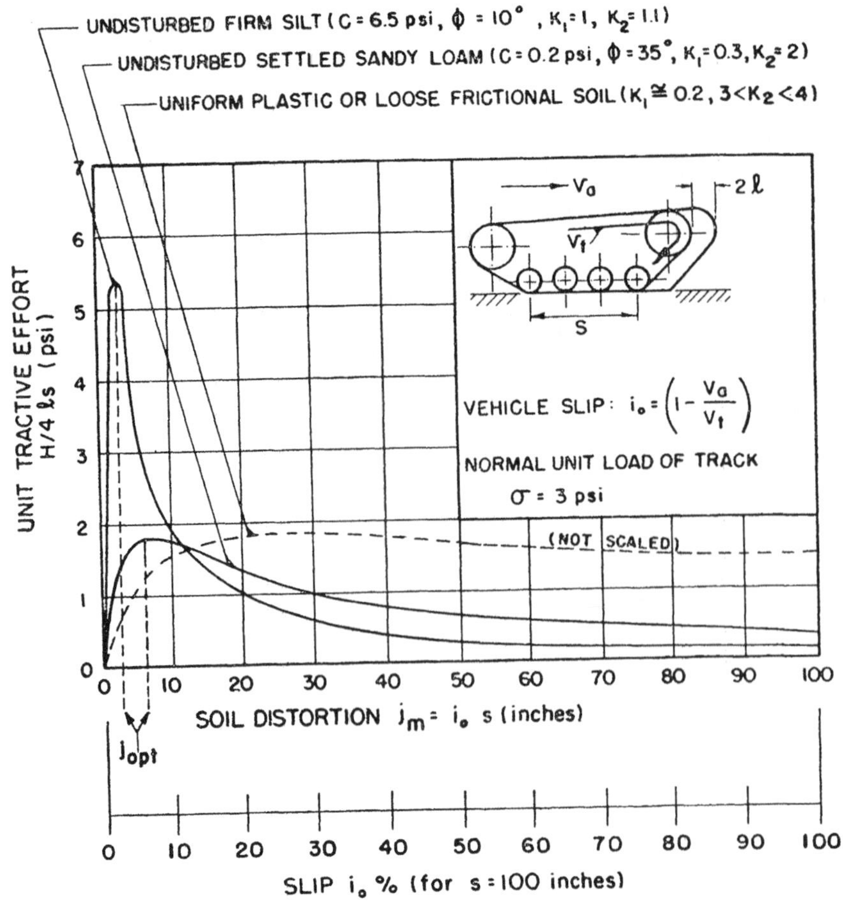

Fig. 113

function of $j_m = i_o s$. If, for instance, a constant s equal to 100 in. is assumed, the j_m scale may be written in terms of slip: $i_o = j_m/100$. Such a scale is traced below the j_m abscissa and indicates that the maximum traction will be developed in the previously considered soils at the following approximate slips:

$$s = 100 \text{ in.} \begin{cases} \text{silt: } & i_o s = 2 \quad \text{or} \quad i_o = 2\% \\ \text{loam: } & i_o s = 5 \quad \text{or} \quad i_o = 5\% \,. \end{cases}$$

If, however, the track length s is only 50 in., then the maximum traction will be obtained at different slips:

$$s = 50 \text{ in.} \quad \begin{cases} \text{silt:} \quad i_o = (2/50)100 = 4\% \\ \text{loam:} \, i_o = (5/50)100 = 10\% \, . \end{cases}$$

Thus the absolute value of the track length is responsible for the amount of slippage, all other variables being equal. In general, then, it may be concluded that a short, stubby vehicle will slip more than a long one. Since slip is a source of power loss, a long, narrow track should be more economical than a short, wide track. The concept of a sleek vehicle derived from previous morphological studies (Chapter IV) may be considered as being strengthened by this evidence.

Track Sinkage and Track Bogging Due to Slip

The sinkage of vehicles due to soil deformation is a source of loss of power and traction, and may lead to stalling. This explains why a study of this problem is important. Sinkage may be due to elastic and/or plastic soil deformations. An elastic deformation of soil, as discussed in Chapter V, is rather insignificant and is not dangerous from a vehicle point of view. It is important only in laboratory tests performed with scale models. This type of work, however, is in its early stages, as will be seen in Chapter XI; so, for the time being, only the nonelastic settlement of a vehicle will be considered.

It was mentioned before that satisfactory methods for the evaluation of sinkage due to a plastic flow of soil are practically nonexistent and that only estimates based on approximate assumptions are available. When the wheel-and-track problem was considered, equation (163) was accepted as the basis for computing the sinkage z under the pressure p:

$$p = kz^n \, .$$

The amount of wheel sinkage z_o was determined by equations (176), (177), and (178), in which various values of n were used, or by equation (246), in which a slightly modified relationship between p and z was assumed [see equation (244)]. The sinkage z_2 of the second wheel behind the first one in a tandem arrangement also was determined in a similar way by equation (189). Track sinkage in static conditions was expressed by the equation of the track line, formula (241), and the magnitude of the sinkage of wheels supporting loose tracks was determined by equation

(248). The deflection of elastic track plates due to soil yielding, also based on equation (163), was expressed by equation (254).

However, the sinkage of a vehicle does not occur only because of the elastic deformation or plastic flow of soil. The slip of the track and the consequent soil distortion described in the preceding section also may lead to a sinkage whose nature is different from that of the ground settlement under the action of a vertical load. The mechanism of the sinkage due to slip is similar to the mechanical removal of soil by blades or scoops.

The "digging-in" phenomenon of a vehicle due to the removal of soil from under its tracks has been recognized. Figure 112 illustrates the way in which some correlation of the tractive effort, soil properties, and amount of sinkage caused by a slip may be obtained. The bottom sketch shows how soil distortion, which may now be identified with soil removal, increases from zero to j_{max} during the time when a given part of the track remains in contact with the ground. Accordingly, it may be expected that the vehicle sinkage at the rear will be larger than that at the front, since the removal j of the soil reaches a maximum at the end of the ground contact area and is zero at the beginning. Maximum sinkage z (Figure 114) will then occur at the end point and may be evaluated in the following way. If the soil deformation caused by the track of a moving vehicle is observed in an imaginary cross section behind a glass plate, then in the case of a smooth track, the vertical lines of a grid attached to the soil will be deformed, as shown in Figure 115a. If the track is provided with spuds having grousers h_g high, then the picture of soil deformation will be as shown in Figure 115b.

According to equation (284), the maximum shift of soil j_m which takes place at the rear of the vehicle is $j_m = i_o s$. The volume of soil thus shifted or "ejected" from beneath the tracks, which are $2l$ in. wide, will be $2lhi_o s$, if h is assumed as the mean depth at which the slippage extends into the soil mass (Figure 115a and b). Since the soil removed at the rear is accumulated gradually along the ground contact area as j increases linearly, it may be expected that the volume ejected by the particular track portions also increases like j and may be represented by a triangular prism $2l$ in. wide, z in. high, and s in. long, as shown in Figure 115c. Accordingly,

$$2lhi_o s = \tfrac{1}{2}(2lzs)$$

and

$$z = 2hi_o \,. \tag{290}$$

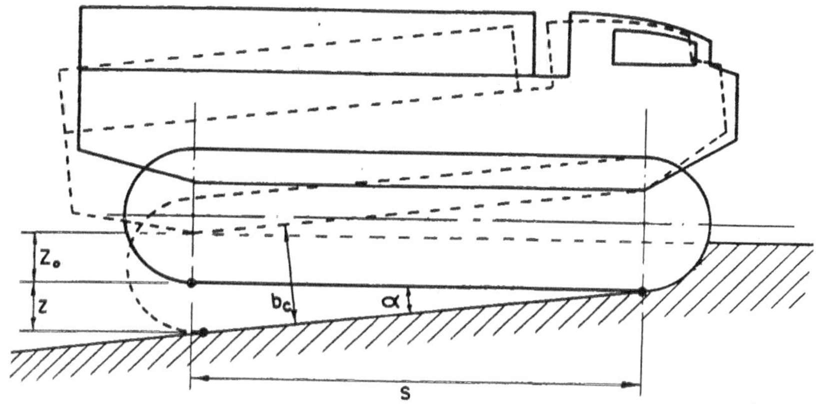

Fig. 114

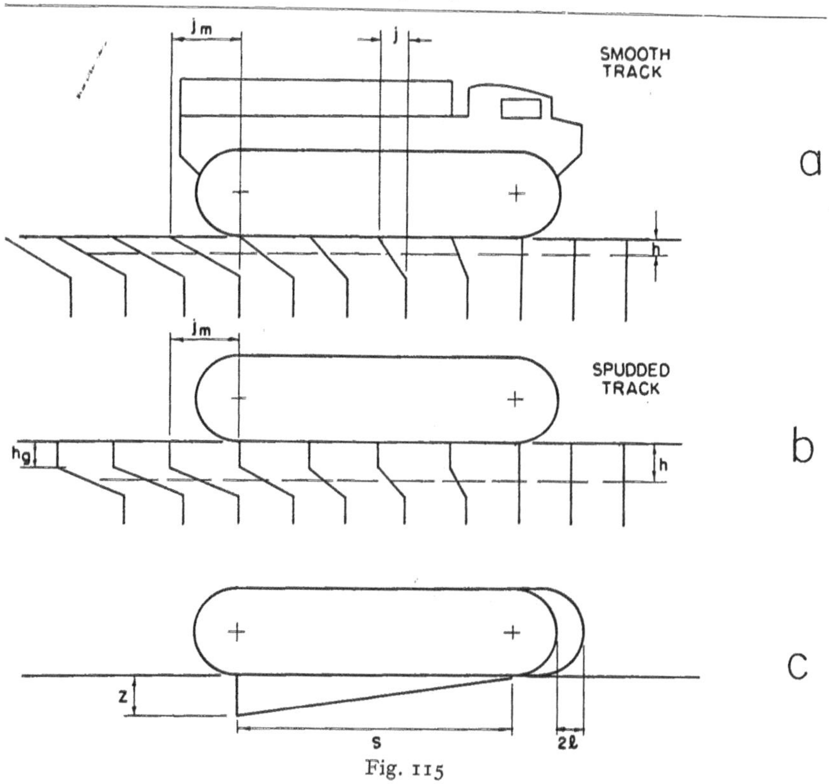

Fig. 115

Thus the sinkage z at the rear of the ground contact area would be proportional to the slip i_o and to the mean thickness h of the "boundary layer" of the soil which moves along with the slipping track.

It is evident that such a layer is much thinner in the case of smooth tracks than in the case of heavily spudded ones. Formula (290) thus indicates that a smooth track will "dig in" to a lesser extent than a grousered track, which is in agreement with observations.

A general picture of a vehicle that is sunken because of slip is shown in Figure 114. If it is assumed that the initial vehicle sinkage is z_o and that the load is evenly distributed along the tracks which form a rigid continuous platform, as, for instance, in the case of agricultural tractors, the necessary belly clearance b_c may be roughly evaluated as being equal to $z_o + z$, or in accordance with equation (290),

$$b_c = z_o + 2hi_o \,. \tag{291}$$

Equation (291) is valid if the angle α is not too large. The value of this angle may be estimated from the function

$$\alpha = \tan^{-1}\left(\frac{z}{s}\right) = \tan^{-1}\left(\frac{2hi_o}{s}\right). \tag{292}$$

For a slip close to 100% ($i_o \sim 1$), and for an average value of $h/s = 0.05$, $\alpha = 0.1$ radian, which represents an angle between 5° and 6°. Such is the order of magnitude of the tipping of a vehicle due to slip close to 100%. This angle immediately levels out after 100% slip is reached. At this point, the j value does not increase gradually from zero to j_m during the passage of the track along the ground contact area. Since the vehicle is immobilized, j_m is not constant at any point. Thus the soil is removed uniformly by the whole track length s at the rate of $2lhs$ per each track revolution, and the vehicle sinks horizontally, if the uniform ground pressure is not disturbed by, for instance, the drawbar pull.

The actual rearing of a tracked vehicle when it bogs down under the action of high slip has to be distinguished from the considered ideal case. Such a tipping may be caused by an uneven load distribution over the ground contact area (tail-heavy vehicle) or by the moments of external forces.

Motion Resistance Due to Compaction of Soil, Bulldozing, and Soil Drag

The power of soil necessary to support the locomotion of a vehicle can then develop only when a specific deformation of the ground takes

place. Such a deformation consumes a definite amount of energy which is wasted in the form of the external resistance to motion.

It may be recalled that the previously discussed rolling resistance of a wheel [equation (164)] was based on the assumption that the source of the resistance is a vertical compaction of the soil. The assumption that there is only a vertical motion of the soil particles when compacted is probably the crudest oversimplification of the problem of a wheel, although the theoretical results obtained on this basis have some practical value (cf. discussion pertaining to Figure 87). Fortunately, in the case of a track, this assumption appears to be quite justified and may be considered to be a fair basis on which an explanation of the nature of the main external resistance of crawlers to motion may be made.

In order to evaluate this type of resistance, which is encountered when a vehicle crosses a granular or plastic medium, consider again equation (163). The work of soil compaction per unit of track area (Figure 116a) performed at depth z_o under the pressure $p = kz^n$ will amount to

$$L = \int_0^{z_o} p\, dz = \int_0^{z_o} kz^n\, dz = k\,\frac{z_o^{n+1}}{n+1}\,.$$

The work done at a distance s per track width $2l$ and per one track is

$$L_t = 2lsk\,\frac{z_o^{n+1}}{n+1}\,. \tag{293}$$

This work may be considered as being performed by the force R_c which resists the motion at the same distance s; hence,

$$R_c = 2lk\,\frac{z_o^{n+1}}{n+1}\,. \tag{294}$$

If, as in the case of a wheel, various n values are assumed, then for

$$n = 1, \quad R_{c_1} = lkz_o^2 \tag{295}$$

$$n = \tfrac{1}{2}, \quad R_{c_2} = \tfrac{4}{3}lkz_o^{3/2} \tag{296}$$

$$n = 0, \quad R_{c_3} = 2lkz_o\,. \tag{297}$$

It is obvious that equations (295), (296), and (297) are identical with equations (165), (166), and (167) when considering that the track width $2l$ replaces the wheel width b.

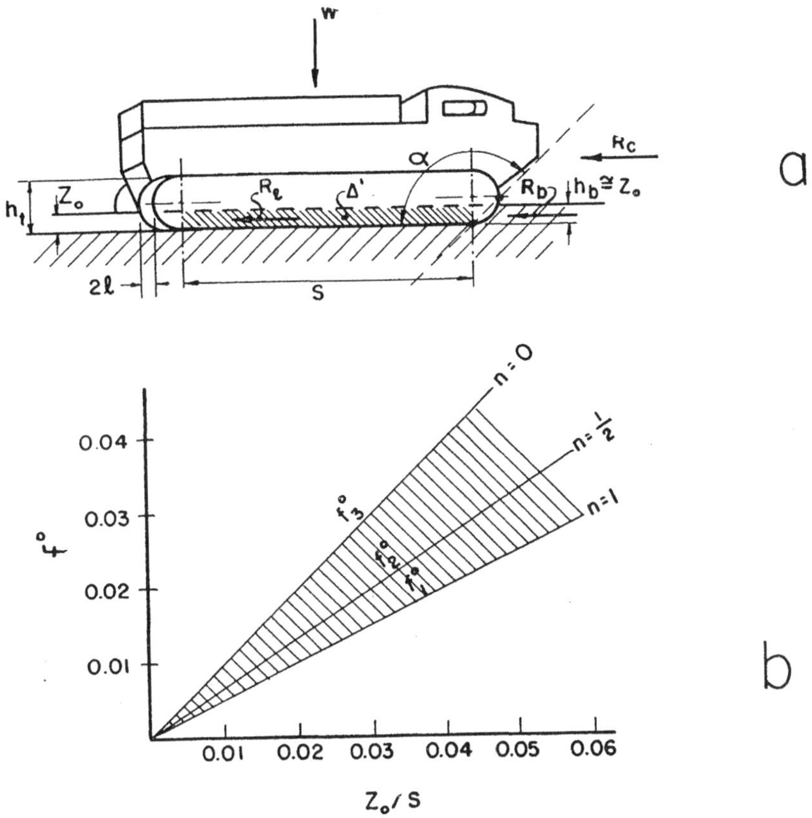

Fig. 116

Since the ground pressure p is exercised by the vehicle weight $W/2$ spread over an area of $2ls$,

$$\tfrac{1}{2}W = \int_0^s 2lp\,dx = 2lk \int_0^s z_o{}^n\,dx = 2lkz_o{}^n s\,, \tag{298}$$

and the load per one track which causes sinkage z_o is

for
$$n = 1, \quad \tfrac{1}{2}W_1 = 2lskz_o \tag{299}$$
$$n = \tfrac{1}{2}, \quad \tfrac{1}{2}W_2 = 2lskz_o{}^{\frac{1}{2}} \tag{300}$$
$$n = 0, \quad \tfrac{1}{2}W_3 = 2lsk\,. \tag{301}$$

The external motion resistance f^o, due to soil compaction per unit of vehicle weight, may be obtained by dividing equations (295), (296), and (297) by equations (299), (300), and (301), respectively:
for

$$n = 1\,, \quad f^o{}_1 = \tfrac{1}{2}\frac{z_o}{s} \tag{302}$$

$$n = \tfrac{1}{2}\,, \quad f^o{}_2 = \tfrac{2}{3}\frac{z_o}{s} \tag{303}$$

$$n = 0\,, \quad f^o{}_3 = \frac{z_o}{s}\,. \tag{304}$$

Since z_o and s are not independent variables, it is assumed that the unit resistance of motion due to soil compaction has to be evaluated in terms of the z_o/s ratio. Sinkage alone does not define f^o.

The variation of f^o with reference to z_o/s and n is shown in Figure 116b. Experimental evidence is not in disagreement with this graph and indirectly corroborates the rather narrow limits within which the compaction resistance may vary.

This resistance, however, is not the only one to be considered, particularly when the sinkage z_o is very large in comparison to the height h_t of the tracked gear (Figure 116a). Under these circumstances, there is an additional drag caused by the bulldozing effect of the frontal portion of the track submerged in the soil to depth h_b, usually equal to the sinkage z_o, and by the drag of the soil mass which penetrates above the track. This soil is pulled along by the track and the suspension, producing a resistance which is due to the shear developed between the moving and stationary part of the soil over an area Δ'.

In a first approximation, the value of the bulldozing force R_b may be assumed to be equal to the horizontal projection of the passive earth pressure $P_p{}^f$ [see equation (137) and Figure 70], particularly if the tracks are wide enough with reference to z_o. The angle between R_b and $P_p{}^f$ is $\alpha + \mu - 90$; hence, $R_b = P_p{}^f \sin(\alpha + \mu)$ per unit of track width.

If it is assumed that there is no surcharge ($q = 0$) and that $\mu = \phi$, then equation (137) multiplied by track width $2l$ and $\sin(\alpha + \phi)$ will give the following total bulldozing force per one track:

$$R_b = \frac{l \sin(\alpha + \phi)}{\sin\alpha \cos\phi}\,(2z_o c K_c + \gamma z_o{}^2 K_\gamma)\,. \tag{305}$$

The lateral drag R_l caused by the shearing of the soil along the Δ' area (Figure 116a) is usually significant in highly cohesive soils. It may be readily estimated as

$$R_l = k_r z_o s c \tag{306}$$

per one track, where k_r may vary from 2 to 4, depending on the width of the belly. In the case of dense muds, R_l reaches rather high values.

The Rational Form of a Tracked Vehicle

The problem of track form may be reduced to the problem of the form of the ground contact area in the same way as it was done when discussing a wheel. Assume that there are two tracks having contact-area dimensions as shown in Figure 117. One track is wide and short, the other is narrow and long. The load $W/2$ and the ground pressure are equal in both cases; hence, $2l_a s_a = 2l_b s_b$. Which form is more rational?

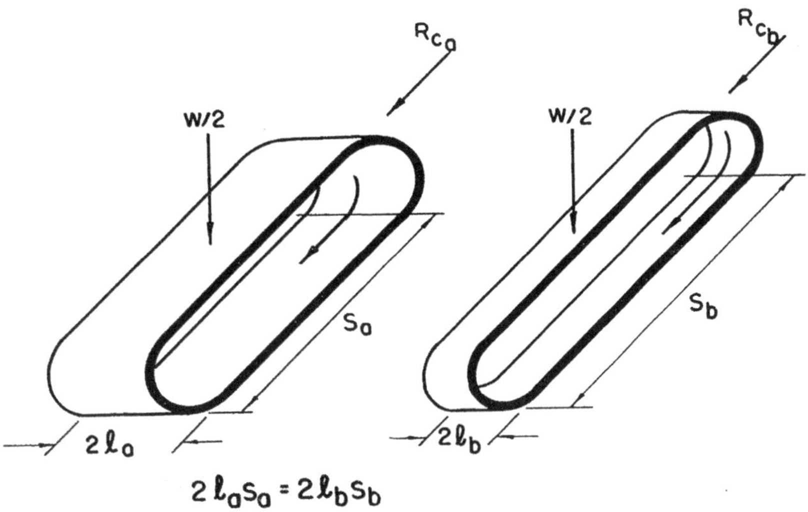

Fig. 117

If the problem is approached from the point of view of external motion resistance, then formula (294) combined with (298) gives

$$R_c = \frac{2lk}{n+1}\left(\sqrt[n]{\frac{W}{4lsk}}\right)^{n+1} = \frac{2lk}{n+1}\left(\sqrt[n]{\frac{p}{k}}\right)^{n+1}, \tag{307}$$

and since the ground pressure $p = W/4l_a s_a = W/4l_b s_b =$ constant,

$$\frac{R_{ca}}{R_{cb}} = \frac{l_a}{l_b}.$$

According to the original assumptions, $l_a > l_b$; hence, $R_{ca} > R_{cb}$. In other words, the motion resistance due to the soil compaction of a short, wide track will be greater than that of a long, narrow track. Thus the answer to the original question regarding the track form appears to be simple: the narrower the track, the lower the resistance due to soil compaction. A similar conclusion was reached by Little,[173] who considered the problem by assuming $p = kz$. It will be seen that his conclusion may be referred to the more general case of $p = kz^n$.

However, if the "flotation" or bearing capacity of tracks is investigated in frictional soils, then the results will be entirely different. Equation (262) indicates that the wider the track compared to its length, the better the "flotation," since l enters into formula (262) in a second power whereas s is in the first power.

The same problem is faced as in the case of a wheel. It appears justifiable, therefore, to ask which is more important as far as vehicle performance is concerned—the flotation defined by equation (262), or a low R_c value computed from equation (307). Unfortunately, the answer is not known, for equation (307) is based on an arbitrary assumption of $p = kz^n$ which cannot be expressed in terms of c, ϕ, and γ, the values which belong to the main structure of equation (262).

It is certain, however, that in the case of nonhomogeneous, stratified soils, such as mud resting on a hard pan, or snow covering a hard ground, the top layer might be so weak that any amount of bearing capacity practically available will not suffice. If the ground clearance b_c (Figure 114) is large enough, the vehicle will sink until it reaches the hard bottom upon which it can move. Under these conditions, since R_b is directly proportional to l, narrow, widely spaced tracks will produce less bulldozing drag R_b than wide tracks narrowly spaced, as may be evaluated from equation (305). Similarly, it may be expected that the capability of a narrow track to pull along the dirt, which causes the lateral drag R_l, will certainly decrease, for the height z_o of the dragged soil mass will not enter into equation (306) as a value equal to the sinkage. Undoubtedly, a much smaller value would be more appropriate in such a case because the dirt would flow more freely across a narrow-tracked gear and would not be pulled along to the same extent as in the case of a wide envelope of the track shell. These observations clearly indicate that at some points there is a conflict between "high flotation" and good performance.

Experimental evidence confirms this conclusion and indicates that an excess of "flotation" beyond a certain limit is detrimental. Thus, a narrow, long track may often be more efficient than an exceedingly wide, short track.

The most important advantage of a narrow, long track, however, may be deduced from the discussion related to Figure 113; the longer the track, the smaller the slip, all other conditions being equal.

Since slip is the source of a large internal resistance and thus of sizable power and economy losses, the advantages of a narrow, long-tracked vehicle seem to be uncontested. Reducing the slip through the adoption of an elongated form appears to be the most direct way to improve the economy of tracked vehicles. Such a concept is in line with the conclusions reached in Chapter IV when the morphology of motor vehicles was discussed. The only obstacle which at present makes it impossible to improve the situation is the existing concept of the steering mechanism and its consequences.

The Similarity and Difference Between a Track and a Wheel

If a survey were made of the vehicle types which are used in similar conditions and for similar purposes, it would undoubtedly disclose a certain lack of uniformity in the assessment of the merits of particular types. In the most extreme case, it might not be difficult to find, for instance, that both tracked and wheeled vehicles are used in practically identical missions without there having been determined the suitability of a given type for fulfilling a given mission. Such a study would inevitably lead to a comparison of the wheel and the track, and to the determination of the extent to which a wheel may replace a track, or vice versa.

The problem of the track versus the wheel is not new. Its perplexity, however, has never been generally solved in a rational way, despite the successful conversion of agricultural crawlers into rubber-tired tractors. Even today, the planning of policies in the development of vehicle types is based more on empirical knowledge acquired in the past than on a rigorous and quantitative evaluation of the differences or similarities between the particular types. Such a procedure cannot lead to satisfactory results, particularly if an unorthodox development is undertaken.

Although the solution of the problem is as far away as it has ever been, the comments and conclusions made in the preceding chapters show the nature of the relationship between the two considered types of running gear, and at least state the problem.

Consider equations (165), (166), and (167). If R is divided by respective W values defined by equations (176), (177), and (178), then the rolling resistance of a wheel per unit load is
for

$$n = 1\,,\quad f^{o}_{1} = \tfrac{3}{4}\sqrt{\frac{z_o}{D}} \tag{308}$$

$$n = \tfrac{1}{2}\,,\quad f^{o}_{2} = \tfrac{4}{5}\sqrt{\frac{z_o}{D}} \tag{309}$$

$$n = 0\,,\quad f^{o}_{3} = \sqrt{\frac{z_o}{D}}\,. \tag{310}$$

Similar equations for a track are given in formulas (302), (303), and (304). It will be seen that the unit resistance to motion due to soil compaction, which incidentally is in most cases the main component of the total resistance, can be expressed in the same terms for both the track and the wheel. The variable in the case of a wheel is $\sqrt{z_o/D}$, and in the case of a track, z_o/s. It should be remembered that wheel sinkage and track sinkage are of different magnitudes.

In the case of Gerstner's soil ($n = 1$), the relationship between R and W for a wheel may be expressed by equation (182):

$$R_w = \frac{0.86}{\sqrt[3]{kb}}\,\frac{W^{4/3}}{D^{2/3}}\,.$$

A similar equation for a track may be obtained by combining formulas (295) and (299):

$$R_t = lk\left(\frac{W}{2lsk}\right)^2.$$

If it is assumed that the load W is the same as in the case of the wheel, then by dividing R_w by R_t,

$$\frac{R_w}{R_t} = 1.72\,\frac{2lk_w}{\sqrt[3]{k_t b}}\,\frac{s^2}{D^{2/3}W^{2/3}}\,. \tag{311}$$

Equation (311) expresses the difference between a track and a wheel as far as the motion resistance due to soil compaction is concerned. It contains, however, too many independent variables. In order to eliminate the weight, an additional condition is assumed: for instance, the weight W should be such as to cause a wheel sinkage $z_o = D/3$. Such a limit is reasonable because a wheel which sinks more than this amount might be unrealistic. Accordingly, from equation (176),

$$W = \tfrac{2}{3}\, bkz_o \sqrt{Dz_o} = 0.127\, bkD^2 \,.$$

If the load in equation (311) is substituted into the above formula, then

$$\frac{R_w}{R_t} = 1.47\, \frac{2lk_w}{k_t b} \left(\frac{s}{D}\right)^2 . \qquad (312)$$

It is further assumed that the ratio of widths of the track and of the given wheel, as well as of the related k values, is $\psi = 2lk_w/k_t b$. Thus, the ratio of the respective motion resistances will be as follows:

$$\frac{R_w}{R_t} = 1.47\, \psi \left(\frac{s}{D}\right)^2 . \qquad (313)$$

Equation (313) is plotted in Figure 118, assuming that $k_w = k_t$. It can be seen that in a soft ground, the resistance to motion of a track is always lower than that of a comparable wheel, a fact known since the invention of the track. The interesting point of equation (313), however, is that it gives numerical values of the relationship between the sizes of a wheel and a track as a function of the resistance to motion under assumed sinkage-load conditions. If, for instance, it is necessary to find the diameter D of a wheel which would offer the same resistance as a track having the length of ground contact area equal to s, then from equation (313),

$$D = s\sqrt{1.47\, \psi}\,.$$

This function, which is plotted in Figure 119, gives the order of magnitude of the dimensions of similar tracks and wheels which encounter the same external motion resistance due to soil compaction.

The foregoing remarks refer to towed wheels and tracks when only resistance to motion is of interest. In the case of propelling gears, an all-important factor is the tractive effort. Thus, the next problem of paramount interest is to know how the traction of wheeled and tracked

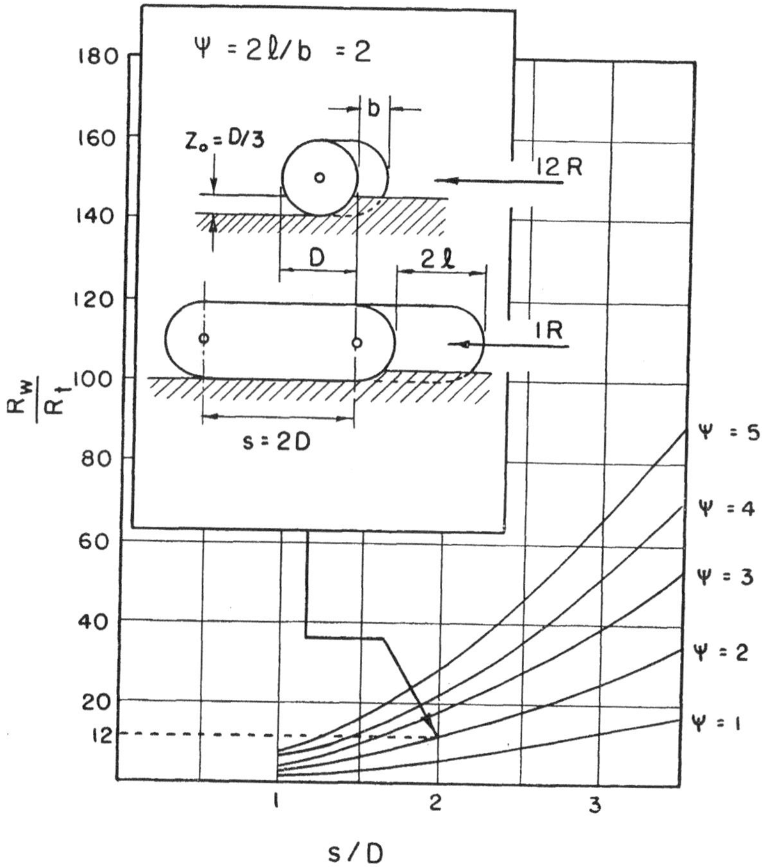

Fig. 118

vehicles compare. An answer to this question, which would only elucidate the nature of the problem involved, may be found in the interpretation of equation (288).

Equation (288) is plotted in Figure 113, with the tractive effort H of two tracks related to the ground contact area $\Delta = 4ls$. Assume an equivalent elastic wheel, having width $b = 2l$ and length of ground contact s_w. A pair of these wheels corresponding to two tracks would have a contact area $\Delta = 2bs_w$.

Assume that the number of pairs of wheels used in an imaginary sub-

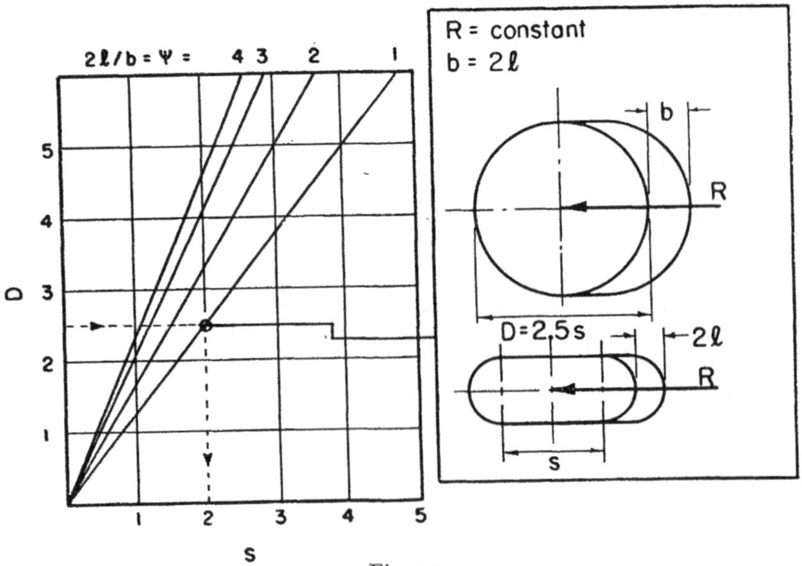

Fig. 119

stitution of a track is n_o, so that the total ground contact area of n_o wheels is equal to that of the two tracks:

$$4ls = 2n_o bs_w .$$

Hence, the load W may be the same in both cases if the same "ground pressure" is desired, and the scale of ordinates $H/4ls = H/2n_o bs_w$ of Figure 113 would not change basically (particularly in cohesive soils), notwithstanding whether it is applied to wheels or to an "equivalent" tracked vehicle. However, if n_o wheels are considered, the track abscissa $i_o s$ will represent the $i_o s_w$ scale of the wheel. For example, consider the ground contact area of a track to be δ times longer than that of a wheel ($s = \delta s_w$). Then it will be seen that the same unit tractive effort which may be developed by the track with, say, 10% slip will be developed by the wheel with $10\delta\%$ slip.

Since the length of the ground contact area determines the slip, it may be concluded that wheels having shorter contact areas will slip to a considerably larger extent than tracks, with all other conditions being equal.

It would be quite possible to make a cylinder or a number of wheels wide enough to provide a ground contact area equal to that of given tracks. If it is assumed that, in the case of tracks and comparable wheels having the same width, both the vertical and traction loads are equal, it will be seen immediately that the number of times the wheel slips more than the track depends on the number of times the ground contact area of the track is longer than that of the wheel. In order to produce equal slips, the wheel diameter should be increased to such a size so as to produce an s_w length equal to s (Figure 120). Figure 120 also suggests that since the nonsupporting portions of a slack track destroy the continuity of the length of the ground contact area and cause this length to be similar to a series of contact areas s_w' of individual wheels, the slip of such a track will increase in accordance with the previous considerations. This leads to the conclusion that vehicles with slack track and widely spaced supporting wheels resemble simple wheeled vehicles and will produce more slip than those vehicles which provide a continuous support for the track. Experimental evidence confirms this conclusion.

It should be stressed again that there is only as much difference between the track and the wheel as there is geometrical difference in the form of the respective ground contact areas. Even in the case of rigid wheels, when the curvature of the rim enters into the picture, this conclusion appears valid if the sinkage is not too high. A wheel with an acceptable diameter can provide enough flotation if it is wide enough, or if the number of wheels is sufficiently large. It will not, however, produce the tractive effort comparable to that of a track without excessive slips which, in practical conditions, lead to the wasteful and downbogging spin, unless the diameter of the wheel is large enough in order to produce a sufficiently long ground contact area. Thus it is possible and advisable to replace tracks by wheels in all towed vehicles; tracked trailers do not appear to have any reason for existence. It is impossible, however, to substitute wheels for tracks in the case of propelling vehicles. A sufficient number of wheels may produce enough "flotation" but never enough traction, unless their diameter is prohibitive. Because of this fact, a track may be considered as a wheel whose form has been rationalized by an extensive increase in the theoretical diameter. Unfortunately, however, the central bearing located economically and conveniently in the hub of the wheel had to be replaced by a heavy, costly, and uneconomical system of rollers which forms the suspension of the tracked vehicle. Nevertheless, as experience shows, even this penalty is worthwhile in order to overcome the inherent disadvantages of a wheel operating in soft ground: the slip.

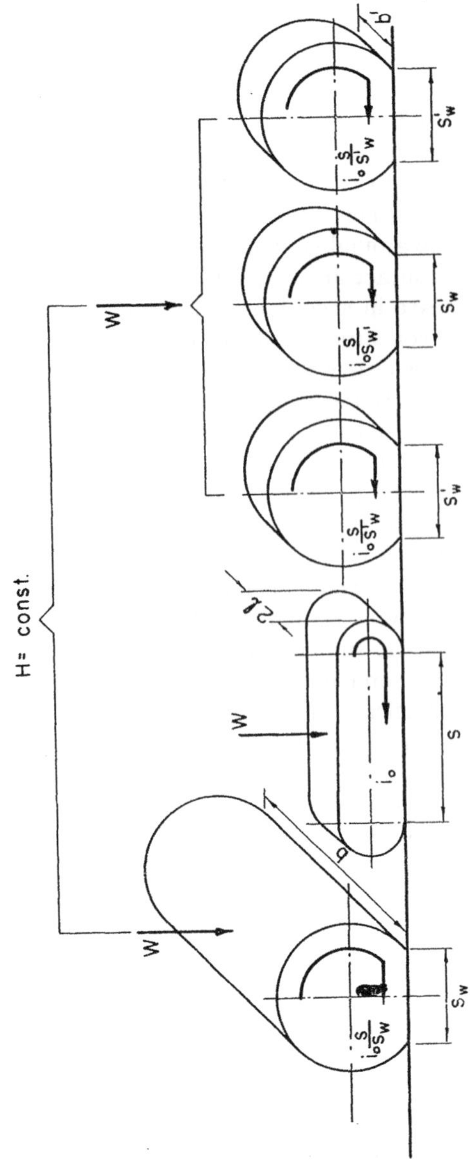

$$P_{,} = \frac{W}{s_w b} = \frac{W}{2\ell s} = \frac{W}{\Sigma s'_w b'}$$

Fig. 120

VIII. SKIS, SLEIGHS, AND TOBOGGANS

Although the use of skis, sleighs, and toboggans is not as widespread as the use of wheels or tracks, the difficulties encountered when this type of transportation is required fully justify an analysis. The importance of this analysis is further stressed by the fact that locomotion in areas where snow covers the earth's surface for a considerable part of the year or where heavy equipment must be transported over swampy terrain belongs to unformulated and unsolved problems.

The Distribution of Pressure Beneath the Ski and the Location of the Ski Axle

The problem of pressure distribution under a ski is identical with that in the case of plates and tracks, which were considered in Chapter V and in Chapter VII. It was shown in these chapters that the pressure which prevails along the ground contact area depends on the mechanical properties of the soil and on the elasticity of the loading surface. Thus, the method previously discussed refers directly to the solution of the distribution of unit loads underneath a ski or toboggan, and equation (254) combined with equation (163) gives the distribution of pressure if the ski is supported at two points (Figure 105a). In the case of one-point support, the method used in conjunction with the problem related to Figure 105c yields the answer.

In order to illustrate the problem further, the following case may be of interest. A ski whose length is $2s$ rests in the snow and is supported at one point located in the center (Figure 121). The cross section of the ski is rectangular and has a constant width; however, the height increases toward the center so that the modulus of rigidity is always proportional to the bending moment caused by snow reactions. The problem is to determine the distribution of pressure along the snow contact area.[170]

The equation of the elastic curve of the ski is as follows [see equation (251)]:

$$\frac{d^2z}{dx^2} = -\frac{M}{EI} = -c',$$

where c' is a constant which determines the assumed relationship between M and EI. When integrating twice,

$$z = -c' \frac{x^2}{2} + C_1 x + C_2 .$$

For $x = s$, $dz/dx = 0$ and hence $C_1 = c's$.

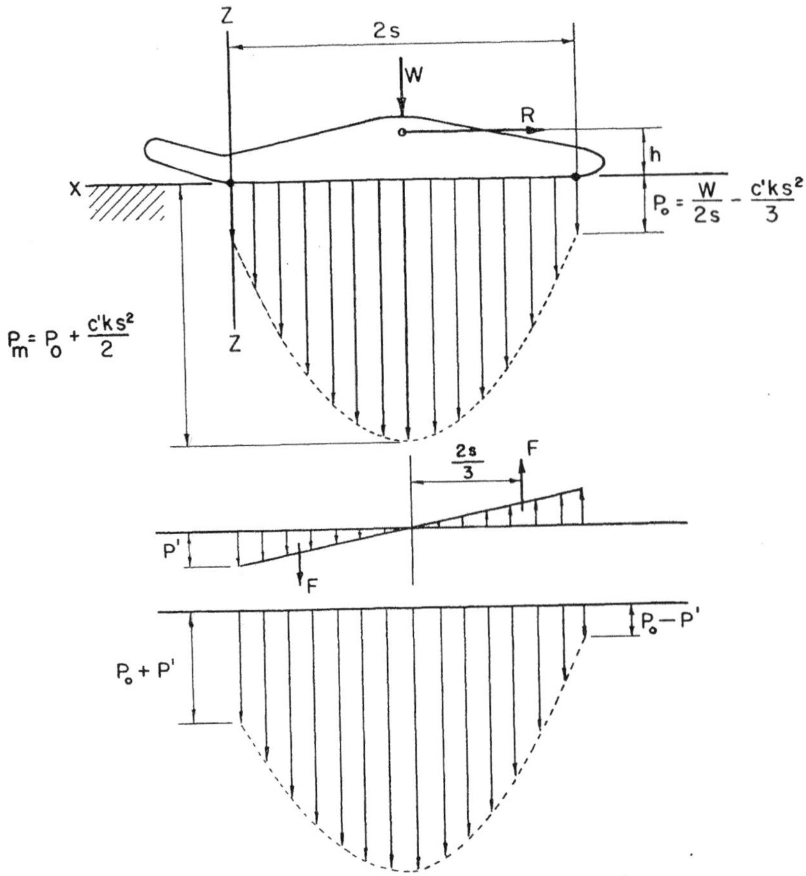

Fig. 121

It is assumed that, in accordance with equation (163), for $n = 1$, $p = kz$. Then

$$p = kC_2 + \frac{c'k}{2}(2sx - x^2) . \tag{314}$$

Since this is an equation of a parabola, the pressure distribution will be shaped by that curve. The pressure at the ends of the ski ($x = 0$ and $x = 2s$) is

$$p_o = kC_2 , \tag{315}$$

whereas in the middle ($x = s$),

$$p_m = kC_2 + \frac{c'ks^2}{2} . \tag{316}$$

By combining equations (314), (315), and (316), the following may be written:

$$p = p_o + \frac{p_m - p_o}{s^2}(2sx - x^2) ,$$

but

$$\int_0^s p\, dx = \frac{W}{2}$$

and

$$\frac{W}{2} = \int_0^s \left[p_o + \frac{p_m - p_o}{s^2}(2sx - x^2)\right] dx = \frac{p_o + 2p_m}{3} s$$

or

$$p_o = \frac{3W}{2s} - 2p_m .$$

From equations (315) and (316),

$$p_m = p_o + \frac{c'ks^2}{2} . \tag{317}$$

The last two equations yield

$$p_o = \frac{W}{2s} - \frac{c'ks^2}{3} . \tag{318}$$

Finally, from equations (318) and (315),

$$C_2 = \frac{W}{2sk} - \frac{c's^2}{3}$$

and

$$p = \frac{W}{2s} - \frac{c'ks^2}{3} + \frac{c'k}{2}(2sx - x^2)\,. \tag{319}$$

The regularity of the parabolic load distribution, as shown in equation (319), will be disturbed by the resistance force R applied to the hinge of the ski, which creates an overturning moment Rh (Figure 121). If it is assumed that the ski is rigid, which may be correct since the R forces are much smaller in comparison with the W loads, it will be found approximately that

$$F = \tfrac{1}{2}p's\,.$$

The equilibrium of moments requires that

$$4F\frac{s}{3} = Rh\,.$$

When these two equations are combined,

$$p' = \frac{3Rh}{2s^2}\,.$$

Thus, the load at the front and rear of the ski will be respectively

$$\left.\begin{aligned} p_f &= p_o + p' = \frac{W}{2s} - \frac{c'ks^2}{3} + \frac{3Rh}{2s^2} \\ p_r &= p_o - p' = \frac{W}{2s} - \frac{c'ks^2}{3} - \frac{3Rh}{2s^2}\,. \end{aligned}\right\} \tag{320}$$

In order to offset the Rh moment, the axle is not usually located in the center of the ski, but is displaced toward the rear. For high speeds which are encountered at aircraft or propeller-driven skis, such an arrangement also trims the running surface, which thus creates a component of lift, similar to that of an airplane wing. In such a case, Russian authors recommend that the ski axle at the rear end be located at a distance equal to 0.4 of the ski length.[174] Canadian work does not give a definite answer as to the optimum location of this point.[113] More detailed specifications for aircraft skis, apparently based on empirical results, may be found in References 175, 176, and 177. Generally speaking, if the R value is known, the optimum position of the axle which would provide a uniform load distribution or trimming effect may be determined by means of the method used in deducing equations (320). In most cases, however, the load distribution of an elastic ski will never be uniform and, particularly in the case of highly flexible runners, the difference between p_m and p_o may be very large.

Load-Bearing Capacity

The problem of the bearing capacity of a ski applied to snow appears to be somewhat different from that of a wide toboggan which is used, for instance, for hauling heavy loads over a swampy plastic ground. In the first case, the main task is to provide through sinkage a considerable initial compression required to produce a sufficiently strong supporting mass of snow. In the case of plastic soil, either the absolute minimum of sinkage or maximum buoyancy is exclusively required.

First consider the case of snow. As Figure 74 indicates, the shearing strength changes with load, so that the tangent to the curve of $\tau = f(\sigma)$ at a given σ will determine the corresponding imaginary c and ϕ. If, therefore, the function $\tau = f(\sigma)$ is known experimentally and if the unit ski load σ is predetermined, then the appropriate value of c and ϕ may be used in equation (262), which will give the order of magnitude of the safe load.

Experience indicates, within practical limits, that the lower the unit load σ, the larger the angle ϕ and the smaller the cohesion c. This leads to the conclusion, which may be considered to be in agreement with experiment, that for low unit loads, the bearing capacity increases faster with ski width $2l$ than with its length s, since the value of ϕ, and hence N_γ, is relatively large and the expression $\gamma N_\gamma l^2$, in which l is squared, plays an extensive role. For higher loads, however, which make the $\tau = f(\sigma)$ curve level off, ϕ becomes so small and c so large that N_γ may be neglected. Then the bearing capacity is merely proportional to the ski surface $2ls$, irrespective of its form. In other words, if the loads applied to sleds or skis are high, ϕ may be considered to be negligible and the load-bearing capacity, in practical cases, can be considered to be proportional to the area of the ski. Thus, a long, narrow runner would carry the same safe load as a short, wide one, without causing general or local shear failure.

The situation is different in the case of sleighs used for civil engineering or military purposes to move heavy loads across soft, plastic soil masses. The bearing capacity of such a mass may not be easily expressed in terms of c, and in terms of equation (265), in which $2l$ and s mean the width and length of the ski.

In most cases, if the condition of plasticity expressed by equation (111) is satisfied, equation (265) would be more appropriate in estimating the safe load W. The correlation of this equation with experiments made in slushy, wet, plastic snow resting on a hard stratum or supersaturated clay under similar conditions appears promising. In such circumstances, the width $2l$ of the ski is a predominant factor since it enters into equation (265) as a square power, whereas length s is in the first power. This may explain why the engineers use a toboggan-type gear whenever a soft, plastic marsh ground is encountered in hauling heavy equipment.

To sum up, it may be concluded again that the bearing capacity of a ski depends on the type of ground negotiated. The importance of the form of the ground contact area may fluctuate in accordance with the relations established by soil and snow mechanics and discussed in conjunction with tracks in the preceding chapter.

Sliding Resistance Due to Snow Compaction

The movement of a ski is hampered by snow compaction in the same way as that of a track. Assuming the validity of equation (163), it will be found that, in accordance with equation (294), the considered resistance of the sliding of a ski is

$$R_c = 2lk \frac{z_o^{n+1}}{n+1},$$

where $2l$ is the width of the ski and z_o is the sinkage. The coefficient of "friction" due to this resistance is identical with that expressed by equations (302), (303), and (304):
for

$$n = 1, \quad \mu_c = \tfrac{1}{2}\frac{z_o}{s} \tag{321}$$

$$n = \tfrac{1}{2}, \quad \mu_c = \tfrac{2}{3}\frac{z_o}{s} \tag{322}$$

$$n = 0, \quad \mu_c = \frac{z_o}{s}, \tag{323}$$

It may be deduced from equation (326) that the maximum capillary drag coefficient is constant if, under the assumed conditions, circumstances permit the generation of an increasing number of droplets of decreasing diameter to support the load. Under these conditions, the maximum capillary drag would be of the order of approximately 10% of the unit load of the ski. For $p_o = 200$ lb/sq ft, the drag in question can amount to $0.096 \times 200 = 19.2$ lb/sq ft. The order of magnitude of the size of the droplet may be determined in these conditions from equation (324), assuming that for 32° C, $\tau_s = 0.0052$ lb/ft:

$$d = \frac{4 \times 0.0052n'}{200} = 0.0012n' \text{ in.}$$

Thus, when a capillary drag exists, the presence of water may hardly be seen. The water droplets observed and counted by Klein [113] and Eriksson [180] through a glass window placed at the ski bottom probably belong to water films whose curvature r was so large that the capillary forces were entirely negligible.

However, the size and number of droplets and, thus, the capillary drag do not depend on load p_o alone. The water content primarily depends on the temperature and on the frictional conditions of the ski and snow surfaces which, through the generation of heat, may produce water droplets in a certain volume at certain places. Hence, it is impossible, at this stage, to see a clear picture of the part played by the capillary forces, unless more experimental information is obtained. If any theory of ski sliding is to be proposed, the invisible droplets, which lubricate the ski and at the same time create capillary tension, will have to be detected, measured, and counted, their distribution along the contact area determined, and the chemistry as well as the physics of the engaged surfaces examined in the light of available information.[179,181]

Viscous Drag

As was mentioned in the preceding section, it seems that the water droplets observed by Klein and Eriksson on the bottom of a ski could produce only an insignificant capillary force. Their viscous behavior, however, may not be negligible. Newton's equation [equation (148)], which was discussed in Chapter V in conjunction with the work by Haefeli on snow friction, [76] may be used again as a tool in a qualitative assessment of the problem.

According to equation (148), the coefficient of viscous friction μ_v is

$$\mu_v = \eta \frac{v}{p_w d} \cdot$$

From equation (324), the film thickness $d = 2r$ which would support the load through surface tension equals

$$d = \frac{4\tau_s}{p_w} \cdot$$

Hence,

$$\mu_v = \eta \frac{v}{4\tau_s} \cdot \tag{327}$$

The surface tension τ_s of water at 32° F is 70 g/sec², and the viscosity η at that temperature is 1.83×10^{-2} poises. Accordingly, from equation (327),

$$\mu_v = 6.5 \times 10^{-5}\, v\,. \tag{328}$$

For a speed of 60 mph = 2650 cm/sec, the viscous drag will amount to

$$\mu_v = 6.5 \times 2670 \times 10^{-5} = 0.173\,.$$

For 10 mph, which is the maximum speed in which Klein performed his experiments of aircraft skis, the viscous drag would amount to $\mu_v = 0.029$. Figure 123 indicates that at the average speed of landing or taking off, the viscous drag alone at 60 mph may be twice as high as the total sliding resistance of a bakelite ski at 70 mph. Since it is not known how the sliding resistance changes with the material at high speeds, it is difficult to say whether the advantages of bakelite, for instance, are as distinct at high speeds as they are at the low speeds investigated by Klein. Thus, the question of material for an aircraft ski remains rather open, and the behavior of a ski landing gear remains rather unknown.

The complexity of the problem may be seen in the fact that the process of lubrication is entirely hypothetical because the water generation is not uniformly distributed along the whole contact area. Since the melting of snow requires a certain amount of heat, the snow particles located at the rear of a ski may be surrounded with more water than those located in the front, as they were exposed to the process of friction for a longer period of time. Accordingly, the distribution of viscous forces will be more intensive in the rear. At the front of the ski, a practically

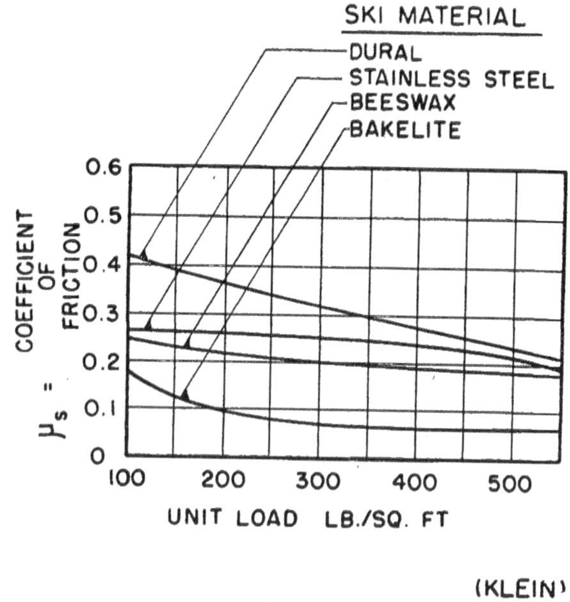

Fig. 123

solid friction is to be expected; the abrasion of the ski, which at this point is always quite sizable, confirms the hypothesis. It is possible that the distribution of frictional and hydrodynamic forces starts, in certain conditions, with solid friction in the front of the ski, surface tension in the middle, and a viscous friction in the rear. In such a case, the physics and chemistry of the interacting surfaces offer the only key to the improvement of the sliding performance of the ski. Incidentally, from the work by Klein,[113] it may be deduced that much more can be accomplished by changing the chemical and physical structure of the runner than by increasing the unit load with the purpose of generating more water and thus producing a better lubrication (Figure 123).

The problem is certainly not simple and may be further complicated by the existence of the so-called boundary lubrication which may appear because of the presence of water vapor.[182] This type of lubrication is produced by exceedingly thin layers, perhaps the monomolecular films, which are able to reduce friction very considerably. The mechanism of this type of lubrication is not known, and as far as is known, has not been considered in the problem of the ski.

It is obvious that not only the material of the runner but also the outside temperature will affect the hydrodynamic effects. This was particularly stressed by Erikson[180] in regard to the viscous friction which depends entirely on speed. The effects of speed upon the sliding of a ski are different below and above freezing temperature. At low

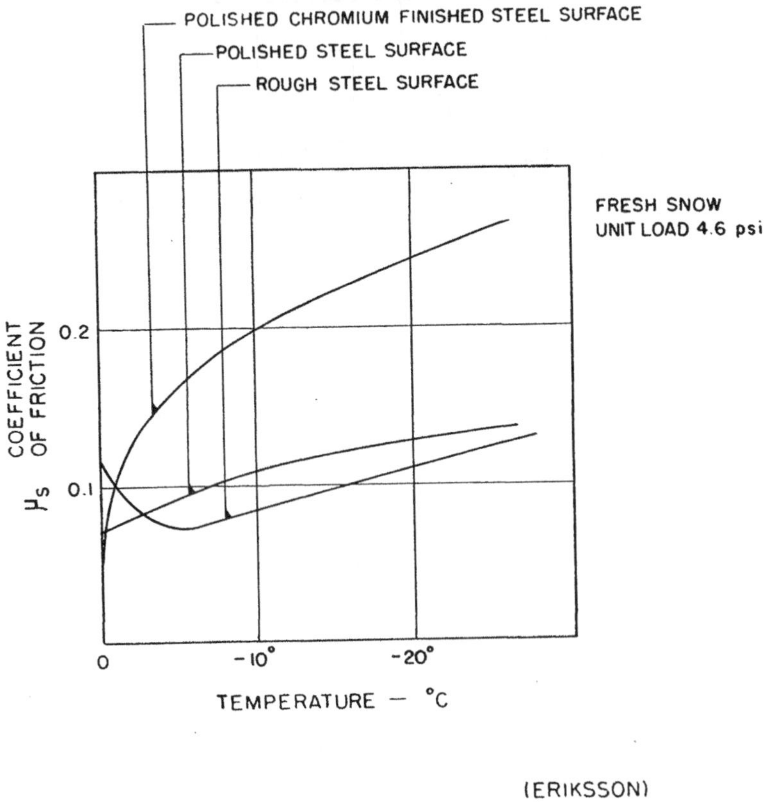

Fig. 124

temperatures, when initially solid friction may even occur,[112] increasing speed produces an increasing volume of water and thus replaces gradually the solid friction with the viscous one, thus reducing the sliding resistance. Close to 30° F, however, water is already present in snow in sizable quantities, and the viscous friction is the only one which may occur. In such a case, any increase in speed will produce an increase in the viscous drag, and thus in the motion resistance in general. If the thickness of the water film is larger than that assumed in equation (324), then the viscous forces should be investigated by means of the methods established in the theory of lubrication.[111]

Decreasing the dimensions of the runner at a constant load may not have the same effects as keeping the dimensions constant at an increased load, because the size of the contact area is not proportional to the number of contact points between the snow and the ski. Under these circumstances, a smaller ski may be expected to contact more points per unit surface than a large ski. This would reduce the local friction, produce less water, and, consequently, lead to a higher sliding friction.[180] Such an occurrence is particularly probable when the snow surface is composed of hard, large, irregular ice crystals. It will thus be seen, then, that the concept of unit pressure alone cannot explain the relations among load, ski size, and hydrodynamic forces.

The finish and the hardness of the surface of a ski also have a considerable effect upon the viscous action of the lubricating film. According to Eriksson, hard steel runners produce a greater melting effect than soft ones, particularly at temperatures below freezing. The smoother the surface, the smaller the friction at temperatures close to 32° F.

At temperatures below freezing, however, a fine surface more rapidly increases the number of contacts with the melted snow and reduces the pressure, which may not be sufficient to produce more lubricating films. Then the viscous friction, perhaps increased by a capillary drag, is not as low as it would have been in the case of a runner with a rough surface.[180] The discussed effect of the physics of surfaces is illustrated in Figure 124, reproduced from Eriksson's work.

Dynamic Resistance of Sliding

Besides the resistance of compaction and the hydrodynamic forces which oppose the sliding motion of a ski, there are forces of dynamic nature which may be several times larger than all the other resistances taken together.

The origin of these forces may be seen in the trim of the ski, which results either from accidental load distribution or from the deliberate shifting of the axle toward the rear of the ski. The latter practice has been adopted in both aircraft skis and propeller-driven sleds.[174] The planing area thus established may produce a very high drag and a certain amount of lift which, at speeds developed in the vicinity of 50 mph, play an important role in ski behavior.

A qualitative analysis of the nature of such a dynamic lift and drag may be performed in the following way. Consider a planing surface $2l$ in. wide and tilted to the horizon at an angle α (Figure 125). The speed with which the ski "sinks" into the soil mass is $v \tan \alpha$, if v is the horizontal speed of travel; the lifting force F will then be expressed in a general way by the equation of momentum,

$$F = \frac{d(mv)}{dt} \tan \alpha ,$$

where v is considered to be constant, whereas the mass m of the displaced snow is variable. Accordingly, the lift is

$$F = v \frac{dm}{dt} \tan \alpha . \tag{329}$$

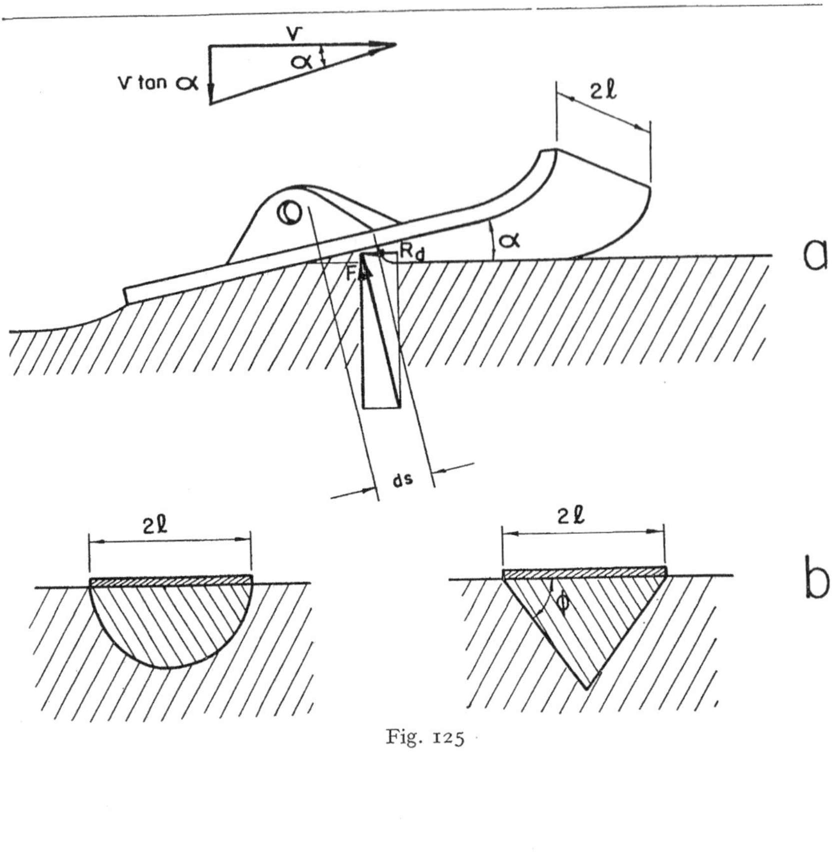

Fig. 125

In the case of a fluid, the virtual mass of the displaced medium is assumed to be a half cylinder having a diameter of $2l$ (Figure 125b). Probably the same assumption might be valid in the case of plastic masses which generally may be considered as being similar to viscous fluids. In the case of frictional soils or snow, however, the triangular shape of the virtual mass appears to be more appropriate. The work by Rathje [165] indicates that this shape is practically round, although the investigations by the writer show that it is clearly a triangular form.[90] Assuming the latter case, it may be appropriate to take the value ϕ of the base angle (Figure 125b) as being equal to the angle of friction of the virtual mass, which would be in accordance with Terzaghi's assumption discussed in Chapter V. In the case of snow, the ϕ value should be referred to the prevailing dynamic force since the shape of the $\tau = f(\sigma)$ curve cannot be based on constant values of ϕ and c. An investigation performed with frictional and cohesive masses would be needed in order to determine the behavior of snow mass under a gliding plane.

If it is assumed in a qualitative evaluation of the problem that the virtual mass has a semi-cylindrical shape (Figure 125b), then its mass per element ds located at the very front of the constant area is $\gamma \pi l^2 \, ds/2$, where γ is the mass density. Since $ds = v \, dt/\cos \alpha$, the virtual mass

$$dm = \frac{\gamma \pi l^2 \, v \, dt}{2 \cos \alpha} . \tag{330}$$

Combining equations (329) and (330) yields

$$F = \frac{\gamma \pi l^2 \, v^2 \sin \alpha}{2 \cos^2 \alpha} , \tag{331}$$

and the dynamic drag R_d which is equal to $F \tan \alpha$ becomes

$$R_d = \frac{\gamma \pi l^2 \, v^2 \sin^2 \alpha}{2 \cos^3 \alpha} . \tag{332}$$

The coefficient of dynamic resistance (drag/lift ratio) obviously is

$$\mu_d = \tan \alpha . \tag{333}$$

An investigation which could produce a practical solution of the dynamic forces acting upon a ski may be helped by the methods developed in the experimental investigation of planing surfaces by Epstein.[183] The methods used in a theoretical approach to the problem of surfaces planing in a heavy fluid may be studied in the monograph by Murray,[184] which also contains a bibliography of the subject. Similar work in the field of surfaces planing on snow is not known, and the important question of the motion resistance of a ski at high speeds does not seem to have been approached yet. Under these conditions, it is impossible to evaluate fully the effect of the previously discussed phenomena such as sliding resistance due to snow compaction, capillary drag, and viscous friction, upon the total resistance of a ski at high speeds. Thus, the principles of design of an aircraft ski, or propeller-driven sleds, are not theoretically established.

Total Sliding Resistance

The total sliding resistance of a ski μ_s will be the sum of the particular resistances considered in this chapter. Thus, the sliding friction coefficient is equal to the sum of the quantities expressed by equations (321)–(323), (326), (328), and (333). Accordingly,

$$\mu_s = \mu_c + \mu_\tau + \mu_v + \mu_d \,. \tag{334}$$

The uncertainties discussed in conjunction with the hydrodynamic coefficients μ_τ and μ_v and the lack of information concerning the value of the dynamic coefficient μ_d make it difficult to evaluate quantitatively the role of particular phenomena in the formation of μ_s. This is why it is felt that Klein's conclusions formulated in respect to aircraft skis [185] should be limited instead to slow-running sleighs which have been investigated by other authors.[186]

The Russian approach to this problem appears to be more primitive since it deals with a simplified structure of the components of the total sliding friction μ_s. According to Juvenatiev,[174] the motion resistance of a ski is composed of: (a) the main frictional force acting at the bottom of the ski, (b) snow compression, (c) snow throwing in the front, and (d) the frictional force acting on the sides of the ski.

The inaccuracy of this scheme is obvious, although it has to be credited with the consideration of dynamic resistance, which has not been investigated in any previous work.

The necessity of further research may be illustrated by the controversial statements found in the literature. Vietchinkin, for instance, maintains that polishing the surfaces of the runners reduces the drag.[174] Eriksson, as was previously mentioned, does not entirely share this opinion. Klein recommends high unit loads: 500 lb/sq ft or more. The Russian school suggests 100 lb/sq in. Eriksson finds that the harder the surface of the runner, the lower the sliding resistance. Klein proposes the use of bakelite. Undoubtedly, each investigator is right within the limits of his experience. More general conclusions, however, necessitate more research.

Definition of Snow Conditions

The physics and chemistry of the surface of a ski are not the only factors which affect the sliding resistance. Snow conditions also play an

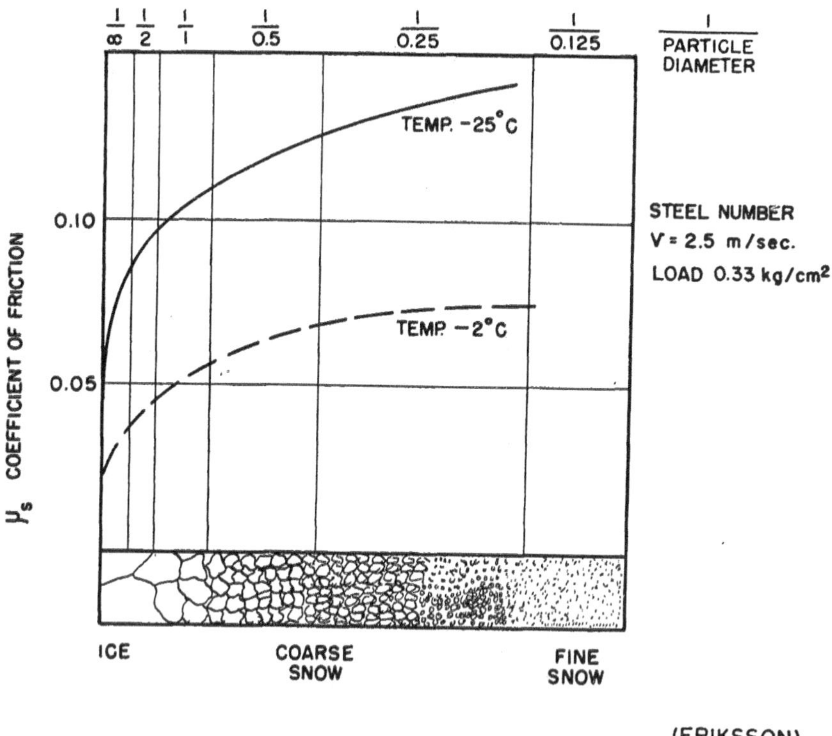

Fig. 126

equivalent role in the whole picture. However, the present scheme of describing these conditions is entirely inadequate. In some works, the undefined term of "hardness" specified in terms of "lb/sq in."[174, 185] is exclusively used, which makes any attempt to analyze related phenomena impossible. The methods applied in determining the moisture content in the field are not accurate enough, and much more sensitive colorimetric equipment is needed in order to record the small contents of water in snow in the vicinity of the melting point.[108]

The reference to snow structure is based on qualitative definitions such as "slightly wet dendrific structure," "wind-toughened surface," etc., which supply only inadequate descriptive data for the development of any physicochemical theory of sliding. An attempt to correlate μ_s with the size of snow crystals may be found in Swedish work; [180] however, the findings, although important and in agreement with the opinion expressed in Reference 185, are limited to a few particular cases investigated (Figure 126). Measurements of friction ϕ and cohesion c have been performed in a

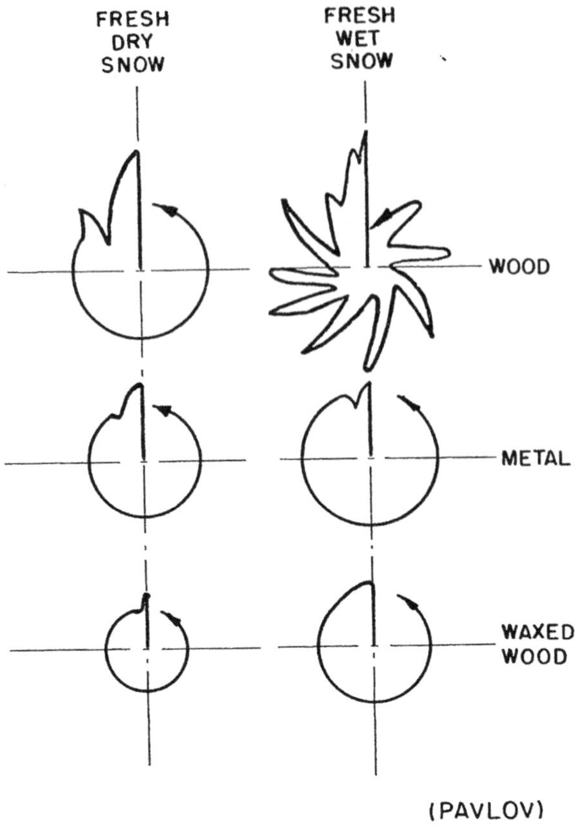

Fig. 127

preliminary way only and for the purpose of exploring the concept itself. So far, c and ϕ have not been used in the evaluation of the movement resistance of a ski.

The only measurement which seems to have attracted the attention of many investigators and which relates undoubtedly to the frictional properties of snow cover is the density γ of snow. Since snow never is uniform throughout its depth, the distribution of density of particular layers appears of paramount importance in the evaluation of sliding properties; the work by Abe, Defant, and Shepelevsky should be mentioned in this connection.[96] As equation (332) indicates, an accurate knowledge of γ is essential in the assessment of the dynamic resistance to motion.

The changes in snow conditions caused by a repeated crossing of sleds are only qualitatively known. As was shown in Chapter V, the problem relates to snow compaction and has not been fully explored. Korunov evalutes the decrease in μ_s by 0.005 to 0.0055 for each sled which follows a preceding one.[96] How far this improvement goes is not known. Some information in this respect may be given by Figure 79.

The freezing of skis to the bed after short periods of rest raises another problem related to snow conditions. Pavlov investigated the values of adhesion in fresh dry and fresh wet snow when using small-size sleds, and measured the force needed to start and then to sustain the motion.[96] The result of his experiments is shown in Figure 127. The effect of snow type is evident, and the jerking character of the pull of wooden sleds in wet snow appears surprising. The necessity of snow description which would yield a clue in respect to all these phenomena appears paramount.

The Form of a Ski

The effect of the form of a ski upon its movement resistance has the same significance as that of a wheel or track. The relationship between the form and performance of a ski is, however, more involved than in the case of other running gears, because of the complicated nature of the force of resistance which may build up from frictional, hydrodynamic, and dynamic components.

Take the form of the ground contact area. This area produces the major portion of the coefficient of sliding resistance μ_s, and is responsible for the generation of frictional, hydraulic, and compaction forces. The frictional forces occur mainly on the front portion of the ski, and as may be assumed, gradually undergo a change from capillary to viscous tensions. The latter generate at the rear portion of the contact area as a result of an increasing production of capillary droplets which subsequently fuse into lubricating films of water. In this process, both the load and the number of points of contact play an important role. Since, however, the occurrence of the relevant physical processes along the contact area is not known, it is difficult to judge which should be the general form of a ski for a given load in order to provide minimum frictional and capillary drag and maximum viscous lubrication. The problem is further complicated by the dynamic disturbance of the static load, as discussed previously, and requires a special study before any answer to the question of the rational form of a ski may be given.

The coefficient of the capillary drag at its peak may not depend on load or area [see equation (326)]; hence, the portion of the ski adjacent to the front where this drag mainly spreads does not directly postulate the form problem. This problem, however, may be formulated immediately when considering the portion of the ski directly behind the bow. It must be shaped in such a way so that high local pressures are produced in order to melt as large quantities of snow as possible and to lubricate the rest of the ski. What the relevant shape should be is an open question. The rear portion, which presumably provides the main sliding surface where viscous friction reduces the drag, also may be independent for practical purposes of load and loading area, as long as the film structure is based on capillary tension [equation (328)]. Thus, the hydrodynamic effects do not yield any special clue as to the form of the ski. In the case where the abundance of lubricating water makes equation (328) invalid, the sliding of the ski and thus its form possibly may be assessed by means of the methods developed in the theories of lubrication.[111] Preliminary analysis in this direction appears promising. At this stage, only a general statement may be made to the effect that the longer the contact area, the longer the time in which an adequate water quantity may be produced. Thus, particularly in cold climates, the ski should be narrow and long.

A clear indication with regard to the form of a ski may be deduced from the compaction resistance. Equations (321), (322), and (323) suggest that the greater the length, the smaller the drag due to snow compaction. These conclusions have been confirmed by all the experimenters. However, if the bearing capacity of a ski or toboggan and not the hydrodynamic drag is considered in a frictional snow, or soil mass, then, according to equation (262), an increase in the width $2l$ of the ski will more effectively increase the safe load since it enters into the formula as a second power. A similar conclusion will be drawn from equation (265) in the case of soft plastic masses. Under these circumstances, a wide, short ski or toboggan is more rational than a long, narrow runner, as was mentioned before.

As shown in equation (332), the dynamic force of resistance due to the planing of a ski increases with an increase in its width. This would suggest again a narrow, long ski, if there were no other effects which, like in the case of a planing boat, might, in the final analysis, reduce the drag. The problem is open for investigation. Broadly speaking, it is not known why a ski should not be, for instance, triangular in shape.

The form of the bottom of a ski apparently does not affect the sliding friction.[185] If, however, an aircraft gear or a fast-moving propeller sled is considered, then the analogy with the wedge-shaped bottom of a seaplane may have some relation to the problem of impact loads. Consider that the ski bottom is triangular in shape, as shown in Figure 128, and that it

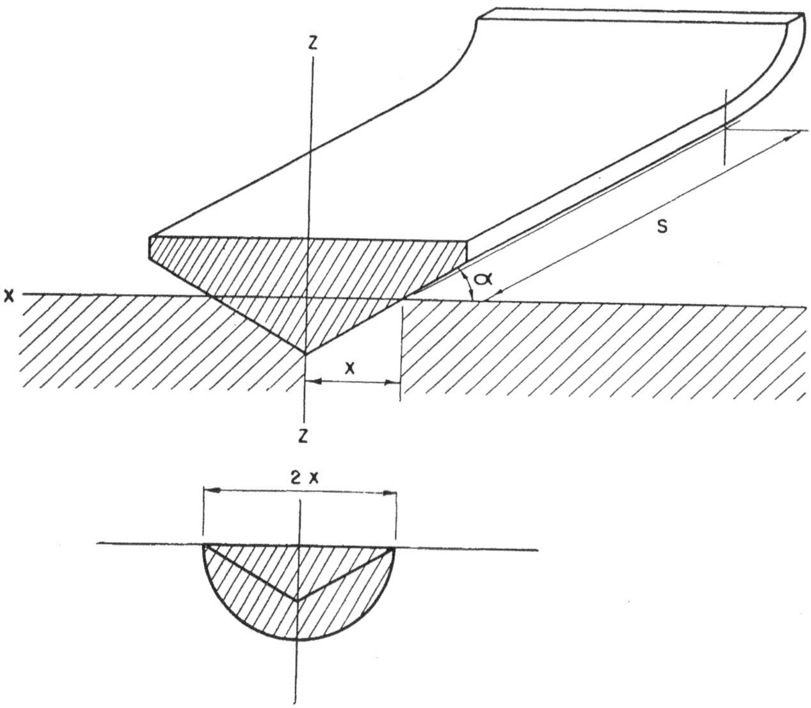

Fig. 128

carries V load per inch of length s. The dynamic loads of such a body may be determined if it is assumed that the momentum at time t, when the speed of impact is v, equals the original momentum when the downward speed of the surface was v_o.[187] Since, in this case, the inertia of the original mass V/g also is increased by the previously assumed virtual mass, which is equal to the volume of a half cylinder having a diameter $2x$ (the buoyancy of the wedge may be neglected for small x values), the equation of the momentum will be as follows:

$$\frac{V}{g} v + \tfrac{1}{2} \pi x^2 \gamma v = \frac{V}{g} v_o , \tag{335}$$

where γ, as previously, is the mass density. Since $v = (dx/dt) \tan \alpha$, equation (335) yields

$$\frac{dx}{dt} = \frac{v_o \cot \alpha}{1 + \dfrac{\gamma \pi x^2 g}{2V}}$$

and

$$\frac{d^2x}{dt^2} = -\frac{\gamma \pi x g}{V} \frac{v_o^2 \cot^2 \alpha}{\left(1 + \dfrac{\gamma \pi x^2 g}{2V}\right)^3} .$$

But $dz/dt = (dx/dt) \tan \alpha$; hence the impact force $F = -\dfrac{V}{g}\dfrac{d^2z}{dt^2}$ equals

$$-\frac{V}{g}\frac{d^2z}{dt^2} = \frac{v_o^2 \cot \alpha}{\left(1 + \dfrac{\gamma \pi x^2 g}{2V}\right)^3} \gamma \pi x . \tag{336}$$

The pressure p per unit of length s will be obtained when dividing equation (336) by $2x$:

$$p = \frac{\gamma v_o^2}{2} \frac{\pi \cot \alpha}{\left(1 + \dfrac{\gamma \pi x^2 g}{2V}\right)^3} .$$

This pressure reaches its maximum in the middle of the ski ($x = 0$) at the moment of the first contact and equals

$$p_{max} = \frac{\gamma v_o^2}{2} \pi \cot \alpha . \tag{337}$$

It will be seen that the effect of the angle α in reducing the maximum impact pressure is very sizable and changes directly as cotangent α. For a flat ski bottom ($\alpha = 0$), the pressure cannot be determined from equation (337) because it reaches infinity. In such a case, the compressibility of snow should be considered.

A wedge-shaped bottom may increase the rigidity of the ski and may make it more difficult for it to break away when frozen to the bed. The problem of the impact loads, however, which may be critical in the cases of heavy gear, cannot be neglected, and the study of the form of the ski bottom, as related to these loads is important; the problem of the strength of a ski, even in the case of sport gear, cannot be oversimplified[188] and presents a complex picture in itself, as was shown by Vietoris.[189] Also, the dynamic loads which may act upon landing gears are by no means simple.[190] Thus, the alleviation of impact forces should be approached not only from the point of view of the suspension design but also from the point of view of the form of the ski, particularly of its bottom.

The form of the bow, although it affects ski performance, has not yet been explored. Only few experimental data are available.[113, 174] Attempts based on the assumption that snow or soil particles should slide downward in a compacting motion instead of being pushed ahead of the ski lead to complex mathematical solutions.[191] Moreover, assumptions related to the stress-strain pattern of such a process are not fully justified. The study of the methods developed by Rathje,[165] and more recently in the theory of plasticity in conjunction with the cutting of metals, should be helpful in this respect.[59, 192]

Perhaps, in the case of the planing of a ski, the shape of its bow is not essential. However, in the case of slow-running toboggans designed for the hauling of heavy equipment, the form of the front portion is of paramount importance, as was stressed in the work by Weiss.[191]

IX. MECHANICS OF A MOTOR VEHICLE

In the three preceding chapters, questions pertaining to the mechanics of a wheel, a track, and a ski were considered with the purpose of formulating the problems rather than attempting to solve them. In the present chapter, the mechanics of a motor vehicle as a whole will be investigated in a similar way. The subject, however, appears to be so wide that only a few topics may be considered at this time. In discussing these topics, the problems related to off-the-road vehicles will be emphasized.

The Distribution of Loads under Wheeled Vehicles

In the previous discussion of a wheel or a track, the loads acting upon these elements were assumed to be available. In reality, however, their values are not readily given but depend on the general configuration of the vehicle. Furthermore, they are not constant, but change according to the conditions of locomotion.

Consider a wheeled tractor towing a trailer up a hill sloped at an angle α (Figure 129a). The wheel loads W_1 and W_2 act ahead of the vertical through the axle by the coefficient of rolling resistance. If the vehicle is moving with uniform velocity, the wheels are in equilibrium and the resultant of W_1 and R_1 must pass through the axle.

From the equations of equilibrium applied to the tractor as a unit, W_1 and W_2 may be defined as follows:

$$W_1 = \frac{W(b \cos \alpha - h_1 \sin \alpha) - F_h h_2 + r(R_1 + R_2)}{a + b} \tag{338}$$

$$W_2 = \frac{W(a \cos \alpha + h_1 \sin \alpha) + F_h h_2 - r(R_1 + R_2)}{a + b} . \tag{339}$$

Adding (338) and (339) gives

$$W_1 + W_2 = W \cos \alpha , \tag{340}$$

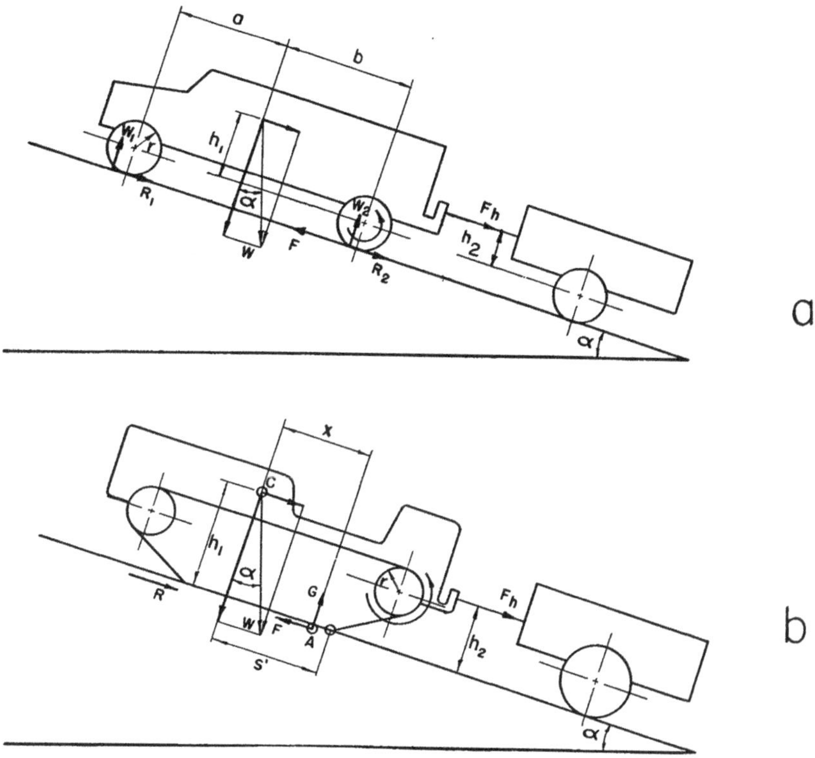

Fig. 129

as it should for vertical equilibrium, and equilibrium of the forces up the slope gives

$$F = R_1 + R_2 + F_h + W \sin \alpha \,. \tag{341}$$

Finally, under the condition that the resultant of W_2 and R_2 passes through the axle, the driving torque M_D equals

$$M_D = Fr \,. \tag{342}$$

The unknown loads W_1, W_2 and resistance $R = R_1 + R_2$ may be determined from equations (341) and (342), and, for instance, from Bernstein's formula

$$R_1 + R_2 = \frac{\zeta}{\sqrt{bk'}} \, \frac{W^{3/2}}{(2r)^{3/4}} \,. \tag{343}$$

It will be seen that the load varies with the type of ground (k), location of the center of gravity (a, b, h_1), height of the drawbar (h_2), wheel dimensions (b and r), and with the configuration of the soil surface (α). Acceleration forces were omitted for the sake of simplicity.

Equations (338), (339), and (343) indicate that the distribution of loads which takes place on a hard pavement is different than that which occurs on a soft ground. The commonly applied assumption which defines the rolling resistance R as a function of the wheel load W by means of the formula

$$R = f^o W ,$$

in which various empirical values for f^o are given for various terrain types, is only approximately correct, for not only the soil qualities but also the wheel dimensions and a more complex relationship between R and W [see equations (196)] affect the distribution of loads at any moment.

The efficiency and economy of multi-axle propulsion depend on whether or not the driving torque M surpasses the traction capacity of a given axle, which is closely related to the W_2 load determined by equation (339). If more axles are driven, then their torques must depend on respective loads, and the theoretical torque distribution among the axles should be adjusted accordingly. The problem of such a distribution was discussed by Strohhäcker.[193]

Equation (338) shows that the unloading of the front axle may, in some cases, reach proportions in which the driving of this axle becomes entirely ineffective, and the question of whether the rear wheels or all wheels should drive the tractor comes into the picture. An analysis of such a problem will be found in Reference 194.

In uniform motion, the performance and stability of a tractor-trailer unit, as shown in Figure 129a, depend on the degree to which the front wheel is unloaded and the rear uploaded. If the rear axle is the driving one, as assumed in the discussed figure, then when overloaded, it may not produce enough net traction T_n required for the propulsion of the vehicle, as was determined by equation (219). If both axles are driving, then the overloading of the rear and the underloading of the front also would produce the same final effect, which also may lead to the stalling of the locomotion. The vehicle will be entirely unstable when $W_1 = 0$. The corresponding critical trailer pull F_h will be obtained from equation (338) when W_1 is assumed to be zero:

$$F_h \leqslant \frac{W(b \cos \alpha - h_1 \sin \alpha) - r(R_1 + R_2)}{h_2} . \tag{344}$$

The dynamic change of wheel loads of two- and three-axle tractors when climbing or descending a hill was computed by Meyer and was presented in a series of graphs which were produced in Reference 195. This particular problem also was investigated by Strohhäcker, who computed data for the determination of the maximum tractive effort which may be developed in dynamic conditions by the propelling axles of heavy highway tractors.[196, 197]

The effect and role of suspension in the dynamic distribution of loads were considered in an exhaustive work by Jante,[198] whereas Steeds investigated the load distribution of rigid six wheelers as a result of torque reactions.[199] The recent work on agricultural tractors by Barger, Carleton, McKibben, and Bainer also considers the load distribution in vehicle motion.[200] A comparative evaluation of various maximum torques, accelerations, slope performances, etc., depending on whether front, rear, or all axles are driven, was given by Goldbeck.[201] All the above-quoted analyses refer to a hard surface and apply only qualitatively to the problem of soft-ground crossing, since they do not consider soil properties as expressed by equation (196).

Load Distribution under Tracked Vehicles

Phenomena similar to those which result in shifting of loads from the front toward the rear of a vehicle may be observed in the case of tracked vehicles. Figure 129b shows such a vehicle; the denotations are similar to those marked in Figure 129a. In this case, the soil reactions G and F, which balance the vehicle weight W and motion resistance R, respectively, are located at a distance x with reference to the center of gravity C. The equation of moments with reference to point A, called the "center of the ground pressure," is

$$-Wx \cos \alpha + Wh_1 \sin \alpha + F_h h_2 = 0 ,$$

and therefore,

$$x = \frac{Wh_1 \sin \alpha + F_h h_2}{W \cos \alpha} . \tag{345}$$

If x is larger than s', then the vehicle will overturn. The overloading of the rear portion of a vehicle due to the shifting of the center of the ground pressure toward the rear also deteriorates the track performance, as may be deduced from all previous considerations. In order to remedy this situation, a shifting of the center of gravity C toward the front is necessary. The evaluation of the required shift may be performed in the same way as in the case of wheeled vehicles.

In order to determine the load acting upon each wheel which supports a loose track, take a rigid suspension which has only three equidistant wheels acting as track supporters (see Figure 130). The equations of equilibrium yield

$$W_1 + W_2 + W_3 - W = 0 \tag{346}$$

$$W\frac{l}{2} - W_2 l - 2W_3 l = 0 . \tag{347}$$

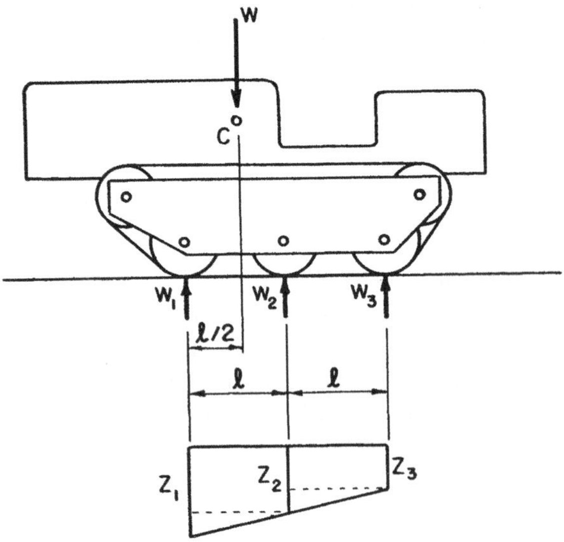

Fig. 130

In accordance with equation (163), assume that the ground reaction is

$$\left.\begin{aligned} P_1 &= \frac{W_1}{\Delta} = kz_1{}^n \\ P_2 &= \frac{W_2}{\Delta} = kz_2{}^n \\ P_3 &= \frac{W_3}{\Delta} = kz_3{}^n , \end{aligned}\right\} \tag{348}$$

where Δ is the area of the ground contact of each wheel. This area is not necessarily constant for all the wheels. If, however, this condition is accepted, which may be true for a small sinkage z and hard ground, then in the case of Bernstein's $n = \frac{1}{2}$,

$$\left.\begin{aligned} z_1 &= \left(\frac{W_1}{\Delta k}\right)^2 \\ z_2 &= \left(\frac{W_2}{\Delta k}\right)^2 \\ z_3 &= \left(\frac{W_3}{\Delta k}\right)^2 . \end{aligned}\right\} \tag{349}$$

Since the suspension is rigid, the sinkage z will occur in such a way that the centers of the wheels will remain on a straight line. This condition may be expressed by the equation

$$z_1 - z_2 = z_2 - z_3$$

or

$$z_3 = 2z_2 - z_1 . \tag{350}$$

By substituting into equation (350) the values of equations (349),

$$W_3^2 = 2W_2^2 - W_1^2 . \tag{351}$$

The solution of equations (346), (347), and (351) yields the following values for the three unknown loads:

$$W_1 = \left(\frac{11}{12} - \frac{\sqrt{19}}{12}\right) W = 0.556W$$

$$W_2 = \left(-\frac{1}{3} + \frac{\sqrt{19}}{6}\right) W = 0.390W$$

$$W_3 = \left(\frac{5}{12} - \frac{\sqrt{19}}{12}\right) W = 0.054W .$$

The above equations show to what extent the assumed location of the center of gravity in given conditions disturbs the uniformity of the distribution of pressure.

A similar method may be applied in the evaluation of loads in the case of a spring suspension as shown schematically in Figure 131a. If any of such suspension structures is replaced by an equivalent imaginary spring, then the scheme of a vehicle with spring-wheel supports may be presented as shown in Figure 131b. By denoting the deflections of particular springs by $f_1, f_2, f_3 \ldots f_n$, the following n equations will be obtained:

$$\left.\begin{aligned} W_1 &= c_s f_1 \\ W_2 &= c_s f_2 \\ W_3 &= c_s f_3 \\ &\cdots\cdots\cdot \\ W_n &= c_s f_n, \end{aligned}\right\} \quad (352)$$

where c_s is the spring constant, i.e., a load per deflection of one inch.

The deflection of springs also may be expressed by the following equations:

$$\left.\begin{aligned} f_2 &= f_1 + l_2 \tan\delta \\ f_3 &= f_1 + l_3 \tan\delta \\ &\cdots\cdots\cdots\cdots \\ f_n &= f_1 + l_n \tan\delta\,, \end{aligned}\right\} \quad (353)$$

which will be $(n - 1)$ in number. Two more equations may be established when considering the equilibrium of forces and moments:

$$\left.\begin{aligned} W_1 + W_2 + W_3 + \ldots W_n &= W \\ W_2 l_2 + W_3 l_3 + \ldots W_n l_n &= W l_o\,. \end{aligned}\right\} \quad (354)$$

Thus, the total number of equations will be $n + n - 1 + 2 = 2n + 1$. Since there are $2n + 1$ unknowns (n forces W, n deflections f) and one unknown δ of the angle of tilt, the problem may be considered as being solved by equations (352), (353), and (354).

A similar problem of load distribution may be encountered with a lever-type suspension, a few examples of which are shown in Figure 132a. In order to determine the load acting upon each bogie wheel, the equations of equilibrium may be written in the same way as was done previously. Denote the distances as shown in Figure 132b and reverse the problem: instead of finding the load distribution of the suspension, determine the dimension of levers so that the load distribution will take

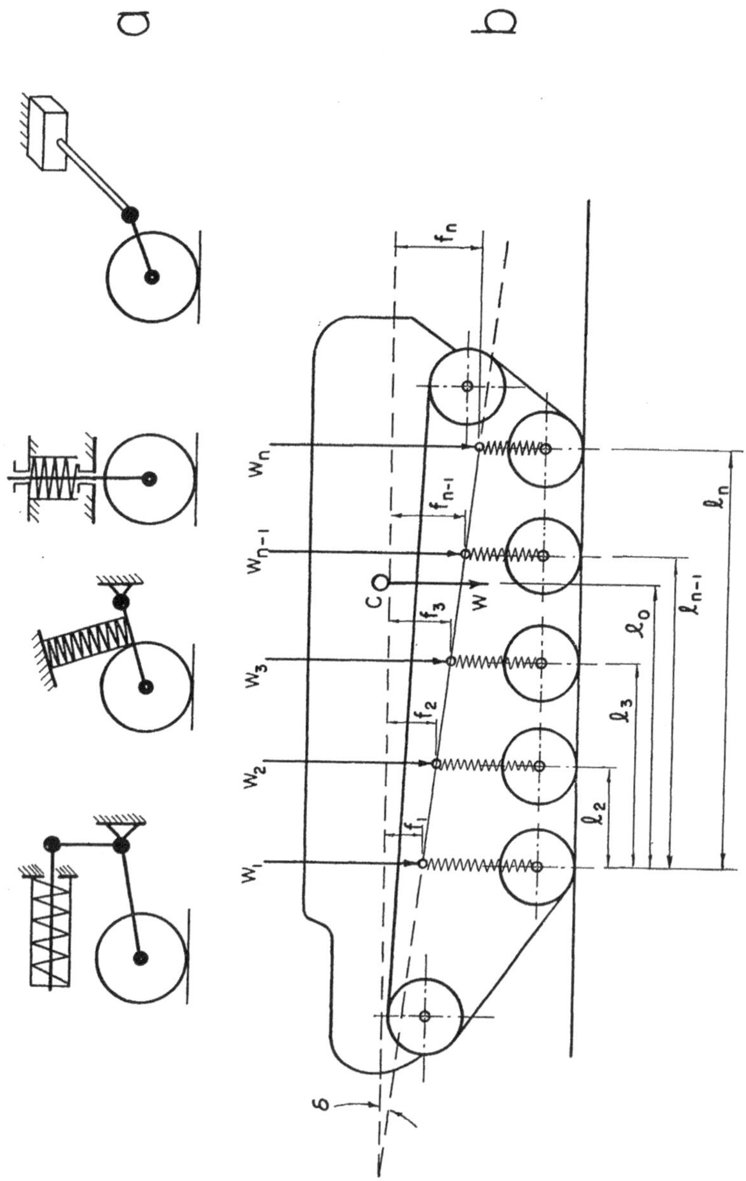

Fig. 131

place in accordance with an arbitrarily assumed scheme. This scheme is shown in the lower portion of Figure 132b and is assumed to be parabolic. Hence, it has to conform with the following equation:

$$W_x = W_1 + Cl_x^{\frac{1}{2}} ,$$

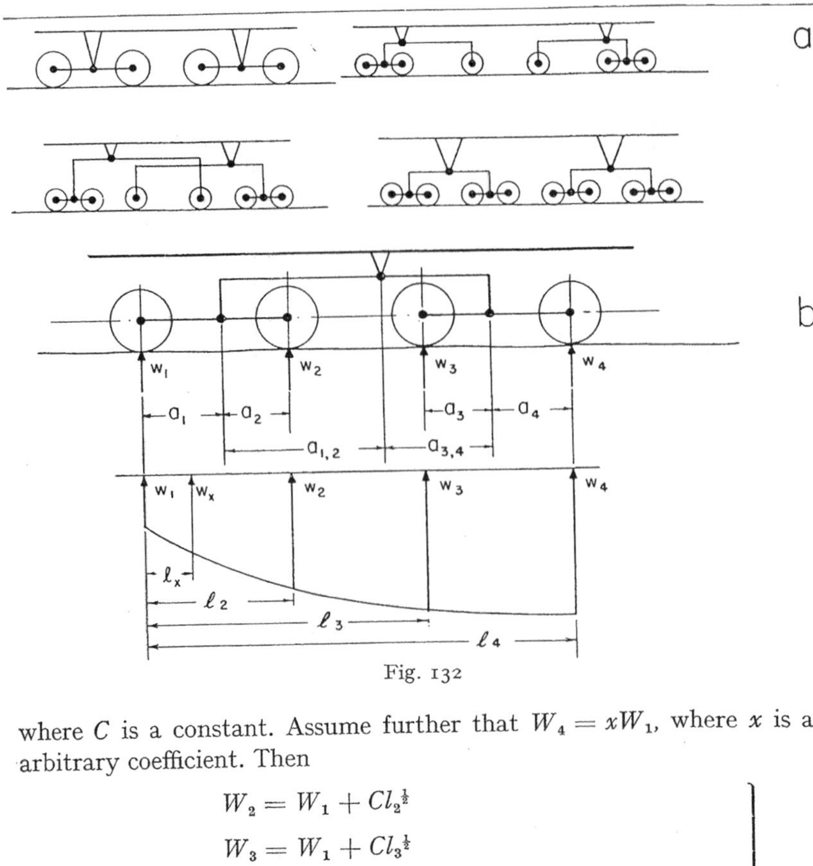

Fig. 132

where C is a constant. Assume further that $W_4 = xW_1$, where x is an arbitrary coefficient. Then

$$\left.\begin{aligned} W_2 &= W_1 + Cl_2^{\frac{1}{2}} \\ W_3 &= W_1 + Cl_3^{\frac{1}{2}} \\ W_4 &= W_1 + Cl_4^{\frac{1}{2}} \\ W_4 &= xW_1 \\ \text{also} \quad W_1 + W_2 + W_3 + W_4 &= W \,. \end{aligned}\right\} \quad (355)$$

From these five equations, the five unknowns W_1, W_2, W_3, W_4, and C

may be determined. In order to determine the dimensions of the levers, the following conditions may be established:

$$\left.\begin{aligned}\frac{a_1}{a_2}&=\frac{W_2}{W_1}\\ \frac{a_3}{a_4}&=\frac{W_4}{W_3}\\ \frac{a_{12}}{a_{34}}&=\frac{W_4+W_3}{W_2+W_1}\,.\end{aligned}\right\}\qquad(356)$$

When dealing with a design problem as considered above, usually some over-all dimensions of the suspension are given, as determined by other considerations. Thus, the over-all length of the levers may be assumed as being known and denoted by the letters a, b, and c. In this case,

$$\left.\begin{aligned}a_1+a_2&=a\\ a_3+a_4&=c\\ a_{12}+a_{34}&=a_2+a_3+b\end{aligned}\right\}\qquad(357)$$

and the six unknowns a_1, a_2, a_3, a_4, a_{12}, and a_{34} may be determined from the six equations (356) and (357).

In the case when a combination of spring and lever suspension is contemplated, as shown, for instance, in Figure 133, the distribution of the loads also may be analyzed in a similar way. Assume that the most advantageous load distribution will be such in which

$$W_3 : W_1 = W_3 : W_6 = n'\,.$$

Find the loads acting on the wheels, the spring deflections, the length of spring levers, and the location of the vehicle center of gravity if the total spring length s is given. From Figure 133, the following equations may be derived:

$$\left.\begin{aligned}\frac{W_2-W_1}{W_3-W_1}&=\frac{l_2}{l_3}\\ \frac{W_3-W_6}{W_4-W_6}&=\frac{l_6-l_3}{l_6-l_4}\\ \frac{W_4-W_6}{W_5-W_6}&=\frac{l_6-l_4}{l_6-l_5}\,.\end{aligned}\right\}\qquad(358)$$

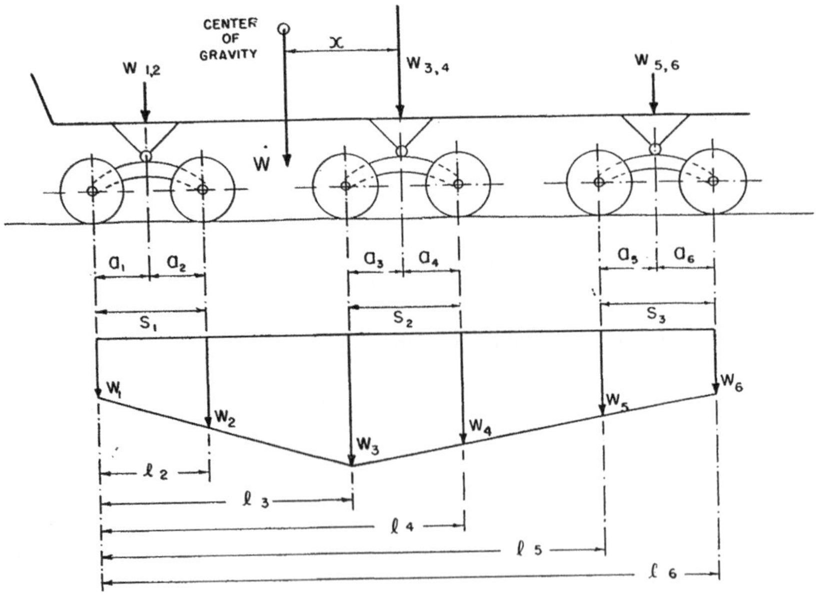

Fig. 133

According to the original assumption,

$$\left.\begin{aligned} \frac{W_3}{W_1} &= n' \\ \frac{W_3}{W_6} &= n' . \end{aligned}\right\} \quad (359)$$

From the equation of equilibrium,

$$W_1 + W_2 + W_3 + W_4 + W_5 + W_6 = W . \quad (360)$$

Thus, six unknown forces W may be determined from the above six equations (358) — (360).

A force W exercised by a spring equals

$$W = c_s f ,$$

where c_s is the spring constant and f the deflection. Accordingly, the following equations may be written

$$\left.\begin{aligned} W_1 + W_2 &= W_{12} = c_s f_{12} \\ W_3 + W_4 &= W_{34} = c_s f_{34} \\ W_5 + W_6 &= W_{56} = c_s f_{56}\,. \end{aligned}\right\} \quad (361)$$

Since the W values and c_s are known, f_{12}, f_{34}, and f_{56} may be determined from equations (361).

The dimensions a of levers will be determined from the following formulas:

$$\left.\begin{aligned} a_1 &= s_1 \frac{W_2}{W_{12}} \\ a_2 &= s_1 - a_1 \\ a_3 &= s_2 \frac{W_4}{W_{34}} \\ a_4 &= s_2 - a_3 \\ a_5 &= s_3 \frac{W_6}{W_{56}} \\ a_6 &= s_3 - a_5\,. \end{aligned}\right\} \quad (362)$$

Finally, the sought-for location x of the center of gravity of the vehicle may be determined from the equation of moments:

$$x = \frac{W_{12}(a_2 - l_4 - l_2 + a_3) - W_{56}(a_4 + l_5 - l_4 + a_5)}{W}\,. \quad (363)$$

Thus, the load distribution in static and dynamic conditions may be fully controlled right at the design stage. The improper distribution of load may lead to a critical inefficiency of a vehicle. This has been demonstrated experimentally when testing nose- and tail-heavy agricultural tractors fitted with rigid spring suspensions.[50] The results of these tests are shown in Figure 134, in which the net tractive effort is plotted as a function of the slip of the track in a sandy soil. It will be seen that a nose-heavy sprung vehicle (d) gives the best results. It should be noted that a tail-heavy, rigidly suspended tractor (a) is particularly poor since it produces approximately five times less traction than the best vehicle tested. The results shown in Figure 134 have been corroborated in daily experience. Although the load distribution has been proven to be one of the most important factors affecting vehicle performance, a theoretical study and evaluation of the problem appears to have been often neglected.

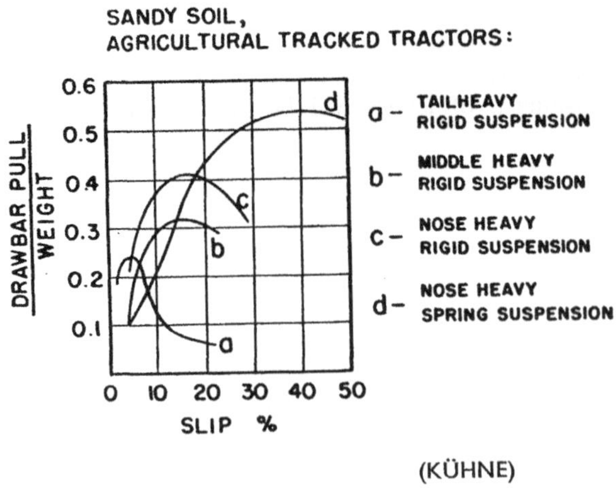

Fig. 134

Dynamic Loads under Vehicle Action

The loads considered in the previous section were of a static nature, i.e., based on the assumption that the vehicle movement is very slow. It is obvious that such an assumption is partially valid only, since soil reactions may be increased considerably at higher speeds of locomotion.

The action of soil under dynamic loads has not been thoroughly explored. Casagrande,[202] in a study of soil behavior under a shock similar to that caused by an explosion, arrived at the conclusion that the shearing resistance of soil increases under dynamic loads. The extent to which a similar phenomenon would exist in the case of loads due to vehicle motion is not known, and the entire field seems to be open for investigation.

The magnitude of dynamic loads caused by a pneumatic tire moving on a bumpy road was explored recently by Klöppel and Moppert,[203] who presented a general method by means of which the dynamic loads acting between the ground and a wheel may be evaluated. In this case, the road was assumed to be a rigid surface, which does not apply rigorously to a quantitative exploration of the phenomena related to riding over soft humps.

In an analysis of dynamic loads, it is essential to assume a certain form of ground unevenness which can be fed into the equations of motion and is representative of the real configuration of the road. Kamm [204] assumed in his studies a vertical step-like unevenness. This resembles the case of shocks caused by railway wheels with a flattened portion of the circumference, as studied by Popp,[205] and does not reflect the true shape of the terrain or road bumps. Timoshenko [206] and Lehr [207] assumed a sinusoidal form of surface unevenness, which, as Wedemeyer showed,[208] corresponds for all practical purposes to the practical shape of the ground wave. The equation of such a type of wave may be derived directly from Figure 135a, and reads as follows:

$$x_1 = \frac{h}{2}\left(1 - \cos\frac{2\pi vt}{l}\right).$$

Since $2\pi v/l = \omega_g$, where ω_g is the frequency of the ground wave,

$$x_1 = \frac{h}{2}(1 - \cos\omega_g t)\,. \tag{364}$$

Following Klöppel and his associate, assume that the static load W_3 represents the portion of vehicle weight which rests upon the investigated axle. The symbol c_s is the spring constant, whereas W_2 is the dead weight of the axle and adjacent suspension portion. The symbol c_r is the elastic constant of the tire, and W_1 is the weight of the tire portion which is under deflection (Figure 135b).

According to the definition of the ground wave [equation (364)],

$$x_1 = \frac{h}{2}(1 - \cos\omega_g t)\,,$$

and the equations of motion of weights W_2 and W_3 (W_1 remains on the ground) become

$$\frac{W_2}{g}\frac{d^2x_2}{dt^2} + c_r(x_2 - x_1) + c_s(x_2 - x_3) = 0$$

$$\frac{W_3}{g}\frac{d^2x_3}{dt^2} + c_s(x_3 - x_2) = 0\,.$$

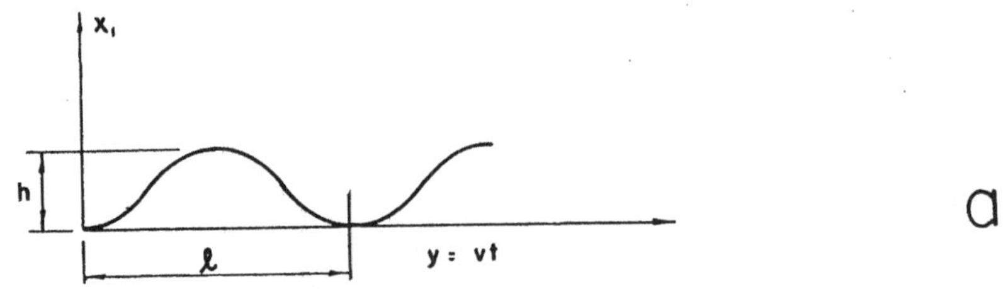

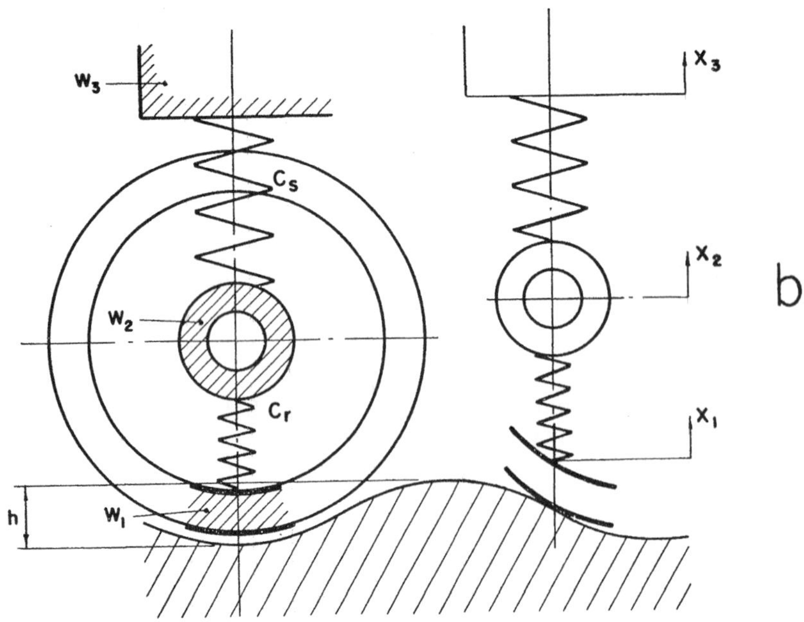

Fig. 135

Or after rearrangement,

$$\left.\begin{aligned} \frac{W_2}{g}\frac{d^2x_2}{dt^2} + (c_r + c_s)\,x_2 - c_s x_3 - c_r x_1 = 0 \\ \frac{W_3}{g}\frac{d^2x_3}{dt^2} + c_s x_3 - c_s x_2 = 0\,. \end{aligned}\right\} \quad (365)$$

If weight W_3 is immobilized and $x_1 = 0$, then the natural frequency ω_{12} with which W_2 will oscillate may be directly obtained from the following equation:

$$\frac{W_2}{g}\frac{d^2x_2}{dt^2} + (c_r + c_s)\, x_2 = 0\,,$$

and, in accordance with the theory of vibration,[206] its squared value equals

$$\omega_{12}{}^2 = -\frac{(c_r + c_s)g}{W_2}\,. \tag{366}$$

If W_2 is held still, then the natural frequency ω_{23} of the oscillations of W_3 is determined by the following equation:

$$\frac{W_3}{g}\frac{d^2x_3}{dt^2} + c_s x_3 = 0$$

and equals

$$\omega_{23}{}^2 = -\frac{c_s g}{W_3}\,. \tag{367}$$

By combining equations (365), (366), and (367),

$$\left.\begin{aligned} &\frac{d^2x_2}{dt^2} - \omega_{12}{}^2x_2 + \frac{c_s}{c_r + c_s}\omega_{12}{}^2x_3 = -(1 - \cos\omega_g t)\,\frac{c_r\omega_{12}{}^2h}{2(c_r + c_s)} \\ &\frac{d^2x_3}{dt^2} - \omega_{23}{}^2x_3 + \omega_{23}{}^2x_2 = 0\,. \end{aligned}\right\} \tag{368}$$

The solution of equations (368), assuming that for initial conditions $x_2(0) = 0$, $(dx_2/dt)(0) = 0$, $x_3(0) = 0$, and $(dx_3/dt)(0) = 0$, is

$$\left.\begin{aligned} x_2 &= A_2\cos\omega_1 t + B_2\cos\omega_2 t + C_2 + D_2\cos\omega_g t \\ x_3 &= A_3\cos\omega_1 t + B_3\cos\omega_2 t + C_3 + D_3\cos\omega_g t\,, \end{aligned}\right\} \tag{369}$$

where

$$
\left.\begin{aligned}
A_2 &= -\frac{h}{2}\frac{\omega_2^2}{\omega_2^2-\omega_1^2} + D_3\frac{(\omega_g^2+\omega_{23}^2)(\omega_g^2-\omega_2^2)}{\omega_{23}^2(\omega_2^2-\omega_1^2)}\\
B_2 &= +\frac{h}{2}\frac{\omega_1^2}{\omega_2^2-\omega_1^2} - D_3\frac{(\omega_g^2+\omega_{23}^2)(\omega_g^2-\omega_1^2)}{\omega_{23}^2(\omega_2^2-\omega_1^2)}\\
A_3 &= -\frac{h}{2}\frac{\omega_2^2}{\omega_2^2-\omega_1^2} + D_3\frac{\omega_g^2-\omega_2^2}{\omega_2^2-\omega_1^2}\\
B_3 &= +\frac{h}{2}\frac{\omega_1^2}{\omega_2^2-\omega_1^2} - D_3\frac{\omega_g^2-\omega_1^2}{\omega_2^2-\omega_1^2}\\
C_2 &= C_3 = \frac{h}{2}\\
D_2 &= \frac{\omega_{23}^2+\omega_g^2}{\omega_{23}^2}D_3\\
D_3 &= -\frac{h}{2}\frac{c_r}{c_r+c_s}\frac{\omega_{12}^2\,\omega_{23}^2}{(\omega_g^2+\omega_1^2)(\omega_g^2+\omega_2^2)}\\
&= -\frac{h}{2}\frac{\omega_1^2\,\omega_2^2}{(\omega_g^2+\omega_1^2)(\omega_g^2+\omega_2^2)}\,.
\end{aligned}\right\} \quad (370)
$$

The frequencies ω_1 and ω_2 may be determined in the usual way: the homogeneous equations, derived from equations (368) with $\omega_g = 0$, are converted to

$$
\left.\begin{aligned}
\frac{d^4x_3}{dt^4} - (\omega_{12}^2+\omega_{23}^2)\frac{d^2x_3}{dt^2} + \omega_{12}^2\omega_{23}^2x_3\frac{c_r}{c_r+c_s} &= 0\\
\frac{d^4x_2}{dt^4} - (\omega_{12}^2+\omega_{23}^2)\frac{d^2x_2}{dt^2} + \omega_{12}^2\omega_{23}^2x_2\frac{c_r}{c_r+c_s} &= 0
\end{aligned}\right\} \quad (371)
$$

and then ω_1^2 and ω_2^2 are the roots of their characteristic equation:

$$
\omega^4 - \omega^2(\omega_{12}^2+\omega_{23}^2) + \omega_{12}^2\omega_{23}^2\frac{c_r}{c_r+c_s} = 0\,. \qquad (372)
$$

This equation yields two ω^2 values: ω_1^2 and ω_2^2.

Since the constants $A_2, B_2, A_3, B_3, C_2, C_3, D_2, D_3$ are determined by equations (370) and the frequencies ω_1 and ω_2 are given by equation

(372), the values of x_2 and x_3 may be computed by means of equations (369). The dynamic load F_{dyn} acting between the tire and the ground is then

$$F_{dyn} = c_r(x_1 - x_2) ,$$

where x_1 is determined by equation (364). The total load acting upon the ground is $F = F_{stat} + F_{dyn}$. Since $F_{stat} \simeq W_2 + W_3$,

$$F = W_2 + W_3 + c_r(x_1 - x_2) . \tag{373}$$

Klöppel and Moppert computed dynamic loads for a truck wheel riding over a sinusoidal wave 1 cm high and 1 m long. Detailed results are shown in Figure 136. They found that in order to double the total shock force F in comparison with the static load, at least 4.5 waves of the assumed form ($h = 1$ cm, $l = 100$ cm) have to be encountered.

It will be seen from Figure 136 that for $h = 1$ cm, the dynamic load is

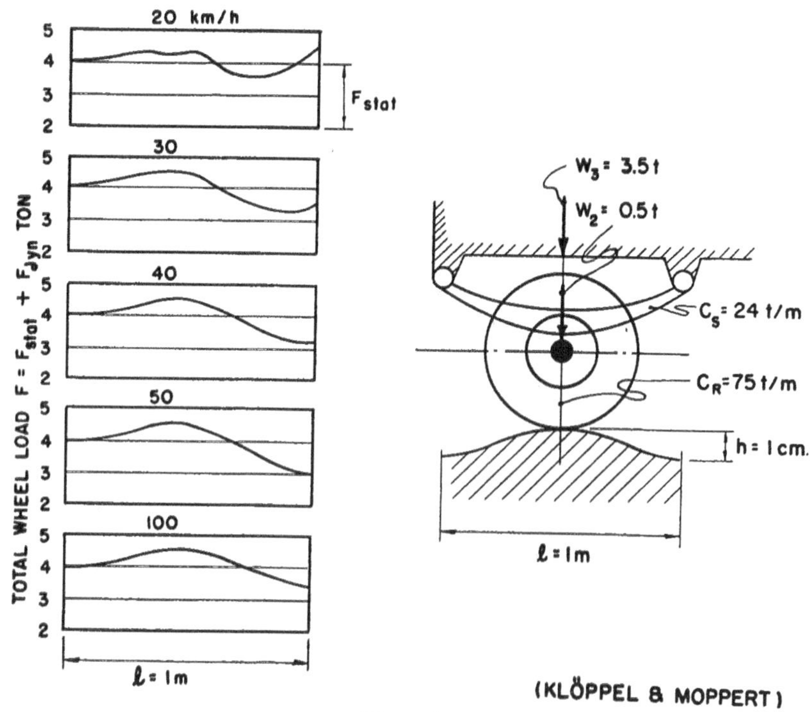

Fig. 136

approximately 20% higher than the static load. Since F_{dyn} is directly proportional to h, the doubling of the height of the wave up to 2 cm will increase F_{dyn} up to 40% over F_{stat}. The increase of the wave length l reduces F_{dyn}. Highway research indicates that l is usually between 40 and 80 in., whereas an average h value is 0.5 in. The problem of road waves has been carefully investigated because it relates directly to both the traffic safety and road damage caused by dynamic loads. More information on this subject will be found in References 209, 210, and 211.

It should be noted that a similar problem in the realm of cross-country locomotion has not been explored, although studies by Lehr, who formulated the fundamentals of springing cross-country vehicles, indicate the necessity of some basic knowledge in this respect.[212] A preliminary investigation by the writer also indicates that "terrain waves" of $h = 0.4$ ft and $l = 6$ ft are not uncommon, and that a certain regularity in this field definitely exists. Thus, equation (364) may serve the evaluation of terrain unevenness as well.

An example computed by Klöppel indicates the strong effect of dynamic forces, which may increase by 40% above the static load at a "wave" height of approximately 1 in. This effect takes place when the wheel is provided with a pneumatic tire. It may be shown that dynamic soil loads increase rapidly with an increase in c_r, i.e., when the tire becomes harder. A limiting case of a rigid wheel was investigated by Lehr.[207] As Den Hartog pointed out, it is evident from the study of vibrations that the main role of a pneumatic is to reduce the impact forces upon the road.[213] The springing role of the tire in regard to vehicle mass appears to be secondary. This conclusion indicates the necessity of a certain reassessment of tire forms and of a quantitative re-evaluation of their role when applied to cross-country vehicles, which move with different speed and in different ground-surface conditions than road vehicles. Discussions performed in Chapter VI also have indicated the necessity of such an investigation because of static tire loads.

Some Criteria of Vehicle Performance

It is customary to qualify vehicle performance either in terms of the "mean ground pressure," which was mentioned before, and/or in terms of the maximum drawbar pull per unit of vehicle weight. Misleading or contradictory conclusions which result from the application of these criteria in the assessment of various types of vehicles have been the source of considerable confusion, making some clarification of the problem imperative.

It was shown in Chapters VI and VII that the real distribution of loads beneath tracks and wheels is very complex and that it depends not only on the mechanical and geometrical properties of the loading area, but also on the mechanical properties of the soil. A picture which would reflect the effect of all the variables involved would be so complex that it would obscure the true image of the phenomena more than elucidate the relevant problems.

It appears advisable, therefore, not to introduce rigorous solutions at this stage, even if it were possible, but to accept simplifying assumptions in a further study of the criteria of vehicle performance. However, more refinement in the ideas of the mean ground pressure and of the unit tractive effort or drawbar pull would be needed in order to obtain results which are more satisfactory than those obtained heretofore. Such a refinement might be readily made by considering the mechanical properties of soil, c, ϕ, and γ, when the unit loads which may be supported by a given ground are investigated.

Assume that the evaluation of vehicle performance is contemplated in a type of soil that provides enough bearing capacity to keep the vehicle afloat right on the surface, or with a negligible sinkage. Thus, the effect of surcharge, which must be considered in the evaluation of the safe load W, may be neglected [see equations (262) and (263)]. Accordingly, the safe unit ground pressure may be determined as follows, if equation (262) is considered to be the basis of evaluation:

$$p = \frac{W}{4sl} = cN_c + \gamma l N_\gamma \,. \tag{374}$$

Equation (374) indicates that two vehicles with the same nominal ground pressure will behave in different ways in different soils, or in different atmospheric conditions when drainage, evaporation, or precipitation will cause changes in the temporary values of c, ϕ, N_c, and N_γ, because the l dimension may be different in both cases.

In the case of a stratified plastic soil, such as that depicted in Figure 64b, equation (265) may be the basis for the determination of the order of magnitude of a safe ground pressure:

$$p = \frac{W}{4sl} = \frac{kl}{2h} \,. \tag{375}$$

It also shows the effect of track width $2l$.

If the soil mass is ideally plastic and semi-infinite, the magnitude of the safe ground pressure may be expressed by equation (374) when assuming that $\phi = 0$ and, hence, $N_\gamma = 0$:

$$p = \frac{W}{4sl} = cN_c\,. \tag{376}$$

The "ground pressure" in such soil conditions gives the performance measure which is valid for any track irrespective of its form.

Equation (375) indicates that for increasing h, p decreases, and that the flow or failure of soil will take place at infinitely small loads if $h \to \infty$. This is in accordance with original assumptions of the plastic flow which, in the considered case, was obstructed by frictional forces exercised by the surfaces of both plates squeezing the mass (Figure 64b). For $h = \infty$, the forces acting upon the lower plate vanish. Thus, when the thick layers of a plastic mass which rest upon a rigid surface are considered, a better qualitative evaluation of the flow would be given by equation (376).

The above considerations illustrate the complexity of the problem and indicate the amount of effort required for a better clarification of the true meaning of the safe ground pressures. However, even now it appears evident that this meaning cannot be divorced from the mechanical structure and geometrical configuration of the soil or snow masses involved. Some soil types and particularly most snow types favor a "high-pressure" vehicle rather than a high-flotation vehicle, as was discussed in Chapter V. A specification of the ground pressure as a criterion of vehicle mobility without a definition of the soil types which are supposed to be negotiated and without considering the form of the loading areas does not actually specify very much.

It should be stressed that any such specification has particularly little meaning in the case of frictional soils, when c is zero or very small. For these soils, equation (374) yields the following value for the "safe" p pressure:

$$p = \frac{W}{4sl} = \gamma l N_\gamma\,, \tag{377}$$

which, as it will be seen, depends entirely on the width $2l$ of the track. Thus, for frictional soils, the ground contact area $4sl$ alone does not represent the criterion of the mean ground pressure at all. The form of this area, defined by the ratio s/l, also has to be specified.

This may be illustrated in an example of two vehicles, A and B, having the same weight and ground contact areas Δ, but different track dimensions: $2l_A \times s_A$ and $2l_B \times s_B$ ($\Delta = 4l_A s_A = 4l_B s_B$), respectively. When the vehicles considered are tested in dry sandy soils, their performance will not be the same. If $l_A/s_A > l_B/s_B$, then, obviously, vehicle A will be better than B since its "flotation," as determined by the safe pressure p expressed by equation (377), will be more satisfactory. In cohesive soils, however, the performance of both vehicles will be identical. Experiments fully confirm this conclusion.

Similar conclusions in regard to the meaning of the "ground pressure" may be deduced for a wheeled vehicle if equation (211) or (212) is considered. From the point of view of these equations, both the track and the wheel are identical.

In order to investigate the effect of soil properties upon the maximum tractive effort-weight, or drawbar pull-weight ratios of a tracked vehicle, consider equation (272):

$$H = 4lsc\left(1 + \frac{h}{l}\right) + W \tan\phi \left\{1 + 0.64\left[\frac{h}{2l}\cot^{-1}\left(\frac{h}{2l}\right)\right]\right\}.$$

If it is assumed that the sinkage is negligible, then from equation (262),

$$W = 4sl\,(cN_c + \gamma l N_\gamma)\,.$$

Dividing H by W as expressed by the above equations gives

$$\frac{H}{W} = \frac{c\left(1 + \frac{h}{l}\right)}{cN_c + \gamma l N_\gamma} + \left\{1 + 0.64\left[\frac{h}{2l}\cot^{-1}\left(\frac{h}{2l}\right)\right]\right\}\tan\phi\,. \qquad (378)$$

The value of H/W corresponds to what is called the "coefficient of adhesion" μ_a responsible for the tractive forces developed between the soil and the loading area. Various tentative figures based on experimental data have been quoted without stating the track type and only for types of soils specified in terms like "country road," "sandy soil," "arable ground," "heavy terrain," etc.[33] In contrast, equation (378) defines the form of the relationship between H and W as a function of the mechanical properties of soil, c, ϕ, γ, and track dimensions h/l.

It results from equation (378) that for dry sand ($c = 0$),

$$\mu_a = \left\{1 + 0.64 \left[\frac{h}{2l} \cot^{-1}\left(\frac{h}{2l}\right)\right]\right\} \tan\phi \,, \tag{379}$$

and for plastic soils ($\phi = 0$),

$$\mu_a = \frac{1 + \dfrac{h}{l}}{N_c} \,. \tag{380}$$

Experience shows that equations (379) and (380) check closely, and a comparison between the pull per ton of weight of two vehicles embraces not only the problem of soil in which the vehicles are assessed, but also the track dimensions as specified by the h/l ratio.

In the case of a driving wheel, the relationship between H and W for low sinkage and at the very moment preceding motion may be evaluated in a similar way. Assume that the contact area of a smooth elastic wheel is, for example, circular and has a radius r, and that the load per wheel is W/n_o, where n_o is the number of wheels. Then, according to equation (211), the safe load is

$$W = \pi r^2 n_o (1.3cN_c + 0.68rN_\gamma) \,.$$

The maximum tractive effort as assumed in equation (222) is

$$H = W \tan\phi + n_o \Delta c \,,$$

where $\Delta = \pi r^2/2$. Hence,

$$\mu_a = \frac{H}{W} = \tan\phi + \frac{c}{2(1.3cN_c + 0.68rN_\gamma)} \,. \tag{381}$$

Equations (379) and (381) lead to a comparative evaluation of the tractive properties of a track and the assumed wheel at low sinkage, when the external motion resistance may be neglected.

The correctness of the basic structure of these equations may be confirmed by the interpretation of numerous experimental data hitherto unexplainable in terms of the physics and geometry of vehicles and the soil. Consider, for instance, the attempt made by Franke to assign an order of merit to various types of vehicles, assessing them on the basis of the drawbar pull-weight ratio.[214] The experimental results of such an assessment are shown in Figure 137. Since the tests were performed

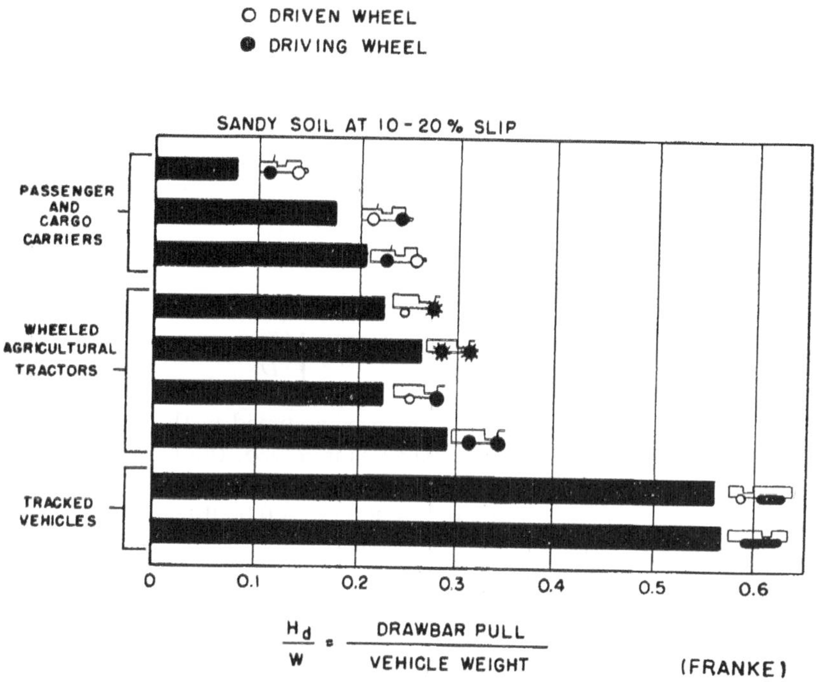

Fig. 137

in sand, the tractive effort-weight relation for tracked vehicles will be represented by equation (379), and for a wheel by equation (382), derived from equation (381) when assuming $c = 0$:

$$\mu_a = \frac{H}{W} = \tan \phi \,. \tag{382}$$

However, the vehicles reported were moving at an unspecified speed with 10–20% slip. Therefore, the motion resistance cannot be neglected and equations (379) and (382) have to be modified accordingly. What Franke was measuring was not the tractive effort H, but the drawbar pull H_d equal to H minus the resistance to motion R. The unit resistance $f^o = R/W$ was previously determined for both wheels and tracks. For Bernstein's case of $n = \frac{1}{2}$, according to equations (303) and (309),

$$f^o{}_{\text{track}} = \tfrac{2}{3}\,\frac{z_o}{s}$$

$$f^o{}_{\text{wheel}} = \tfrac{4}{5}\sqrt{\frac{z_o}{D}}\,.$$

Hence, the drawbar pull H_d per unit of vehicle weight is

$$(\mu_a)_{\text{track}} = \left\{1 + 0.64\left[\frac{h}{2l}\cot^{-1}\left(\frac{h}{2l}\right)\right]\right\}\tan\phi - \tfrac{2}{3}\,\frac{z_o}{s} \tag{383}$$

$$(\mu_a)_{\text{wheel}} = \tan\phi - \tfrac{4}{5}\sqrt{\frac{z_o}{D}}\,. \tag{384}$$

For the average conditions which may be assumed in the assessment of the order of magnitude of the quantities involved, the following typical figures, based on observation in sand and on accepted design trends, may be proposed: $N_\gamma = 20$ (for $\phi = 30°$); $\gamma = 0.06$ lb/cu in.; $l \cong 10$ in.; $h/2l \cong 0.1$; $z_o/s \cong 0.05$; $z_o/D = 0.2$. Accordingly,

$$(\mu_a)_{\text{track}} = [1 + 0.64\,(0.1 \times 1.47)]\,0.58 - 0.033 \cong 0.60$$

and

$$(\mu_a)_{\text{wheel}} = 0.58 - \tfrac{4}{5}\sqrt{0.2} \cong 0.22\,.$$

It will be seen that there is good agreement between the magnitude of the obtained ratios and the experimental results plotted in Figure 137. The computed $(H_d/W) = \mu_a$ value refers to propulsion by any number of wheels. In the case of the front- or rear-wheel drive, H values have to be corrected in accordance with the load resting upon the driven axle. In this way, the differences between the drawbar pull-weight ratio of vehicles with all-wheel and part-wheel drive may be explained.

Plane Motion of Wheeled Vehicles; Stability, Steering

The plane motion of a wheeled vehicle enhances problems which, like stability and steering, become critical, particularly at high speeds when the geometry and dynamics of the vehicle decide its roadworthiness.

Although the speed of cars was remarkably increased in the past 50 years (Figure 44), an analysis which would attempt an exhaustive and possibly rigorous study of the plane motion of vehicles appears to be lacking. Innumerable works which deal with particular aspects of vehicle stability present a broad but rather incomplete picture. Further work in this field is undoubtedly justified, particularly because the stability of vehicle motion is a primary safety factor in traffic.[215] Many questions related to racing cars also belong to the category of stability problems and offer a wide field for investigation.[216] Since, in the present work, the exploration of slow-running, off-the-road vehicles is the main target, the stability of fast-moving road vehicles will be only briefly reviewed.

The studies performed with highway vehicles usually have been based on the assumption of a certain picture of forces acting between the wheels and the pavement, and on the establishment of equations of equilibrium, which produced answers in regard to vehicle stability and general behavior. In this way, Heywang determined the conditions of a skid.[217] Hösl, in a similar way, investigated the distribution of forces acting on vehicles with front-, rear-, and all-wheel drives, and determined the effort required to steer it.[218] Since a study of this type does not help very much to elucidate the kinematics and geometry of the translation of the vehicle in unstable conditions, and becomes more complex, particularly when multi-axle vehicles are considered, the geometrical and kinematic problems had to be tackled separately. Thus, Schuboth investigated the geometry of steering as well as the forces involved, and produced an exhaustive list of references on the subject.[219] An experimental study of the geometry of turning, braking, and propelling multiwheel vehicles will be found in References 204, 220, and 215. The control of the plane motion and its kinematics for single vehicles and tractor-trailer units will be found in References 221 and 222. The mechanism of the steering of a motorcycle was investigated by Mühlfeld.[223]

In all these studies, the most uncertain assumption has been that regarding the nature of forces which generate between the hard surface and vehicle wheels. The deformation of pneumatics and the subsequent load change open a separate research field not only for experimental studies but for theoretical ones as well.[224–226] The unevenness of the ground, sidewind, etc. further complicate the problem.[227] In particular, airflow is the strongest factor which may destroy the stability and steerability of a fast-moving car. The difficulties encountered in a rigorous solution of the problem have led to the complex experimental procedures in a wind tunnel which are described in Reference 228. The methods used in the study of aerodynamic stability, whose principles were established by Kamm, will be found in Reference 229.

A general study of the nonuniform vehicle motion due to acceleration or deceleration forces was performed by Sartoris [230] and Zimielev.[231] The approaches made by both investigators are different. The first set the equations of motion of the vehicle center of gravity by using polar coordinates. The general form of this equation, written in terms of radial force F_r and tangential force F_φ, is as follows:

$$F_r = m\left[\frac{d^2r}{dt^2} - \varrho_g\left(\frac{d\varphi}{dt}\right)^2\right]$$

$$F_\varphi = I_g\frac{d^2\varphi}{dt^2} + \frac{2}{r}(I_o + mh^2)\frac{d\varphi}{dt}\frac{dr}{dt},$$

where m is vehicle mass, ϱ_g radius of gyration, and h the height of the center of gravity. I_g and I_o are the moments of inertia about the center of gravity and the instantaneous steering center respectively; r is the distance between the CG and the steering center; φ is the angular displacement of the car about the vertical axis which passes through the center of gravity. General conclusions in regard to steering, braking, and stability of motion were deduced from these equations.

The study of a four-wheel vehicle by Zimielev is based on the determination of instantaneous centers of acceleration, i.e., on the determination of points of the system moving with the vehicle which have zero acceleration. The location of the center was determined and the acceleration (a) of any point of the moving system was defined by an equation having the following general form:

$$(a) = r\sqrt{\omega^4 + \left(\frac{d\omega}{dt}\right)^2},$$

where r is the distance between the given point and the center of acceleration; ω is the angular velocity of the system. When the acceleration (a) is known, the dynamic forces acting upon the center of gravity of the vehicle may be investigated and the stability of motion studied.

However, as Lozano [232] truly asked, what is the criterion of vehicle stability? The differences in considering the motion stable and unstable appear to be large indeed. Julien discussed the problem [233] by analyzing the concepts of stability by Sèze [234] and Gratzmüller,[235] which represent the European point of view, and compared them with Olley's concept of "oversteering" and "understeering."[236] He arrived at the conclusion that apparently because of the excellent highway systems and long travel distances, the American idea of stability means a hyperstable, "understeered" vehicle which is being produced at the cost of a certain lag in response to steering; an American driver is apt to relax during a long journey and the car has to be very stable. European conditions, i.e., winding roads and short distances, apparently call for a quick steering response, and a less stable car. It will be seen from the foregoing that the problem offers an ample opportunity for studying not only the mechanics but also the psychological factors involved.

Stability of Tracked Vehicles in a Plane Motion

The stability of a tracked vehicle which translates in a nonuniform motion depends on the propelling forces F, resistance forces R, and on the dynamic structure of the vehicle as defined by the moment of inertia I, mass m, and relevant dimensions (Figure 138).

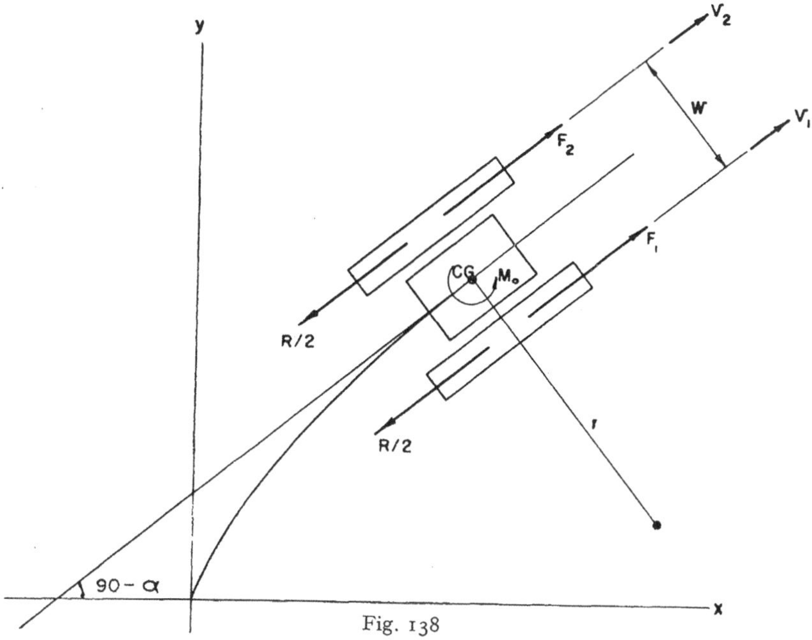

Fig. 138

A general equation for the movement of a track-laying vehicle which is subjected to all the forces involved has not been established. The complex form of such an equation would make its discussion difficult and may be of little practical use.

The following attempt to find a simple form of the discussed equation, however, gives some clues for determining the movement of the center of gravity of a vehicle, so that a general qualitative analysis of motion under assumed conditions may be offered.

Reference is made to the uniformly loaded vehicle shown in Figure 138. Forces F_1 and F_2 are propelling forces; $R/2$ is the movement resistance of each track; M_o is the moment of resistance to the rotation around the center of gravity. The resulting motion of the vehicle may be considered to be a combination of the rotation around the center of gravity and the straight-line displacement in the plane parallel to those lines formed by the circumferences of the rotating tracks.

The equation of movement of the center of gravity, as referred to the tangent to the path of this movement, is

$$\left.\begin{aligned} m\frac{d^2s}{dt^2} &= F_1 + F_2 - R \\ I\frac{d^2\alpha}{dt^2} &= \frac{w}{2}(F_2 - F_1) - M_o\,. \end{aligned}\right\} \quad (385)$$

If at the time $t = 0$, the linear velocity of the center of gravity is v_o and the angular velocity ω is zero, then

$$\frac{ds}{dt} = \frac{F_1 + F_2 - R}{m}\,t + v_o$$

$$\frac{d\alpha}{dt} = \frac{0.5\,w\,(F_2 - F_1) - M_o}{I}\,t\,.$$

Assume that

$$\left.\begin{aligned} \frac{F_1 + F_2 - R}{m} &= (a) \\ \frac{0.5\,w\,(F_2 - F_1) - M_o}{I} &= (b)\,. \end{aligned}\right\} \quad (386)$$

Then,

$$\left.\begin{aligned} \frac{ds}{dt} &= (a)t + v_o \\ \frac{d\alpha}{dt} &= (b)t \end{aligned}\right\} \quad (387)$$

Since the radius of curvature of the considered path is $r = ds/d\alpha$, equation (387) gives

$$r = \frac{(a)}{(b)} + \frac{v_o}{(b)t} \,. \tag{388}$$

The length of the path element is

$$ds^2 = dx^2 + dy^2 \tag{389}$$

and the tangent

$$\frac{dy}{dx} = \tan\,(90 - \alpha) \,. \tag{390}$$

By integrating the last of equations (387), it will be found that (if at $t = 0$, $\alpha = 0$)

$$\alpha = \tfrac{1}{2}(b)t^2 \,.$$

When combining this equation with equations (387), (389), and (370), the following functions will be obtained:

$$\frac{dx}{\sin \frac{1}{2}(b)t^2} = [(a)t + v_o]\,dt$$

$$\frac{dy}{\cos \frac{1}{2}(b)t^2} = [(a)t + v_o]\,dt \,.$$

The integration of these equations gives

$$\left.\begin{aligned} x &= -\frac{(a)}{(b)}\cos \tfrac{1}{2}(b)t^2 + v_o \int \sin \tfrac{1}{2}(b)t^2\,dt + \frac{(a)}{(b)} \\ y &= \frac{(a)}{(b)}\sin \tfrac{1}{2}(b)t^2 + v_o \int \cos \tfrac{1}{2}(b)t^2\,dt \,. \end{aligned}\right\} \tag{391}$$

Consider that

$$\int \sin \tfrac{1}{2}(b)t^2\,dt = \int \frac{\left[\sin \frac{\pi}{2}\left(\sqrt{\frac{(b)}{\pi}}\,t\right)^2\right]\sqrt{\frac{(b)}{\pi}}}{\sqrt{\frac{(b)}{\pi}}}\,dt$$

$$\int \cos \tfrac{1}{2}(b)t^2\,dt = \int \frac{\left[\cos \frac{\pi}{2}\left(\sqrt{\frac{(b)}{\pi}}\,t\right)^2\right]\sqrt{\frac{(b)}{\pi}}}{\sqrt{\frac{(b)}{\pi}}}\,dt \,.$$

If $t\sqrt{(b)/\pi} = u$, then $du = \sqrt{(b)/\pi}\,dt$ and

$$\left.\begin{aligned} x &= -\frac{(a)}{(b)}\cos\tfrac{1}{2}(b)t^2 + v_o\sqrt{\frac{\pi}{(b)}}\int\sin\frac{\pi}{2}u^2\,du + \frac{(a)}{(b)} \\ y &= \frac{(a)}{(b)}\sin\tfrac{1}{2}(b)t^2 + v_o\sqrt{\frac{\pi}{(b)}}\int\cos\frac{\pi}{2}u^2\,du\,. \end{aligned}\right\} \quad (392)$$

The integrals $\int\sin\frac{\pi}{2}u^2\,du$ and $\int\cos\frac{\pi}{2}u^2\,du$ are known as Fresnel integrals and have been determined. Equations (392) may be written in the following form:

$$\left[x - \left(\frac{(a)}{(b)} + v_o\sqrt{\frac{\pi}{(b)}}\int\sin\frac{\pi}{2}u^2\,du\right)\right]^2 = \left[-\frac{(a)}{(b)}\cos\tfrac{1}{2}(b)t^2\right]^2$$

$$\left[y - v_o\sqrt{\frac{\pi}{(b)}}\int\cos\frac{\pi}{2}u^2\,du\right]^2 = \left[\frac{(a)}{(b)}\sin\tfrac{1}{2}(b)t^2\right]^2.$$

By adding these formulas, the following may be obtained:

$$\left[x - \left(\frac{(a)}{(b)} + v_o\sqrt{\frac{\pi}{(b)}}\int\sin\frac{\pi}{2}u^2\,du\right)\right]^2 + \left[y - v_o\sqrt{\frac{\pi}{(b)}}\int\cos\frac{\pi}{2}u^2\,du\right]^2 = \left[\frac{(a)}{(b)}\right]^2. \quad (393)$$

The above is the equation of a family of circles, all having a constant radius $(a)/(b)$. The centers of the circles are located on a curve determined by Fresnel's integrals, which has the following parametric equation:

$$\left.\begin{aligned} x' &= v_o\sqrt{\frac{\pi}{(b)}}\int\sin\frac{\pi}{2}u^2\,du + \frac{(a)}{(b)} \\ y' &= v_o\sqrt{\frac{\pi}{(b)}}\int\cos\frac{\pi}{2}u^2\,du\,, \end{aligned}\right\} \quad (394)$$

and which is known as a Cornu spiral, as shown in Figure 139. It follows, therefore, that equation (393), which determines the path of the vehicle center of gravity, may be considered as the equation of a curve that may be obtained graphically by plotting circles which have a radius $(a)/(b)$ and whose centers are located on the Cornu spiral given by equations (394) (Figure 139). After a lengthy period, the path of a vehicle would be the circle

$$\left[x-\left(\frac{(a)}{(b)}+\frac{v_o}{2}\sqrt{\frac{\pi}{(b)}}\right)\right]^2+\left[y-\frac{v_o}{2}\sqrt{\frac{\pi}{(b)}}\right]^2=\left[\frac{(a)}{(b)}\right]^2.$$

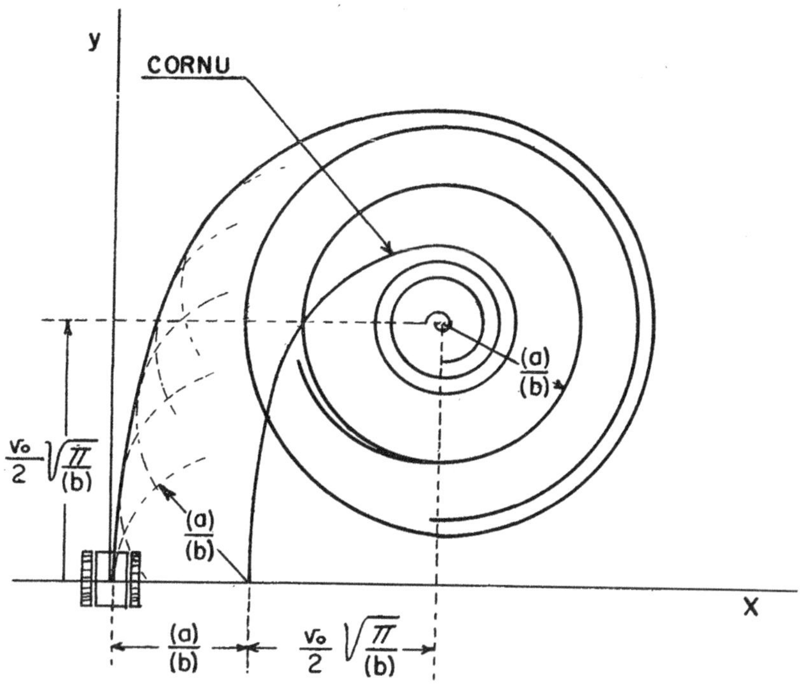

Fig. 139

This case, however, has little practical meaning; the initial part of the curve expressed by equation (392) or (393) is of particular interest.

When $(a) = (b) = 0$, equations (391) may be written in the following form: $x = 0$, $y = v_o t$. The above determines a uniform straight-line motion along the yy coordinate. If, for $t = 0$, $da/dt = \omega_o$, then when $(a) = (b) = 0$, $da/dt = \omega_o$ and $ds/dt = v_o$, which determines the radius of curvature: $r = v_o/\omega_o$. In these conditions, the vehicle moves on the circumference of a circle whose radius is v_o/ω_o and which is tangent to the yy coordinate at the zero point. The F_1 and F_2 forces may be determined in this case by means of equation (386):

$$\left.\begin{aligned} F_1 &= \frac{R}{2} - \frac{M_o}{w} \\ F_2 &= \frac{R}{2} + \frac{M_o}{w}. \end{aligned}\right\} \quad (395)$$

The above are the well-known equations first published by Zaslavski [42] and Beck [237] and later on by Kristi [33, 40] and Koessler and Glaubitz.[238] It can be seen that they are a particular case of a general phenomenon considered in this discussion and that they describe the behavior of a vehicle at a particular stage of equilibrium.

When $(a) > 0$ and $(b) > 0$, the vehicle path is as shown in Figure 139. If $(a) \lesseqgtr 0$ and $(b) = 0$, the vehicle moves along the yy coordinate with an acceleration or deceleration (a). When $(b) > 0$ and $(a) = 0$, equations (392) may be written in the following form:

$$x = v_o \sqrt{\frac{\pi}{(b)}} \int \sin \frac{\pi}{2} u^2 \, du$$

$$y = v_o \sqrt{\frac{\pi}{(b)}} \int \cos \frac{\pi}{2} u^2 \, du$$

and the discussed curve becomes identical with the Cornu spiral.

In the case when $(a) < 0$ and $(b) > 0$, the radius of the circle in equation (393) is negative. This means that the curve in question will be obtained by tracing circles inside the Cornu spiral.

If $v_o = 0$ but $(a) > 0$ and $(b) > 0$, then equation (393) takes the following form:

$$\left[x - \frac{(a)}{(b)}\right]^2 + y^2 = \left[\frac{(a)}{(b)}\right]^2,$$

which means that the path curve is transformed into a circle, the center of which is located on the xx coordinate at a distance $(a)/(b)$ from the zero point.

It might be observed that the change in sign by the F_1 force when braking involves the same change in sign of the corresponding $R/2$ force, which produces new values for (*a*) and (*b*). In consequence, the movement continues under conditions different from the initial ones. In such a case, the considered vehicle completes the motion on a circular path by turning around the center of the stopped track F_1.

The above discussion refers to simplified assumptions. In reality, the forces F_1 and F_2 are not constant and depend on engine torque and transmission. The effect of centrifugal force also cannot be neglected, especially at higher speeds. These facts are to be taken into consideration if more accurate results are to be obtained. In this case, however, although the method remains the same, the rigorous solution of the problem becomes so complex that it loses its general meaning.

The foregoing considerations indicate, however, that the steering executed through the change of forces F_1 and F_2 is unstable. Any change in movement resistance R also may cause perturbations in the stability of locomotion. For this reason, the force steering was abandoned in the early thirties and was replaced by various speed-steering schemes in which the geometry of the vehicle path is predetermined by rigidly controlled speeds v_1 and v_2 of both tracks. Only slow-running tracked vehicles are built today with brakes and clutches which provide the force-steering mechanism.

Steering of Tracked Vehicles

The steerability of tracked vehicles and the response to the steering mechanism depend largely on the geometric structure of these vehicles as well as on the load distribution along the ground contact area.

Consider equations (395), assuming that $R/2 = f^o W/2$:

$$F_1 = f^o \frac{W}{2} - \frac{M_o}{w}$$

$$F_2 = f^o \frac{W}{2} + \frac{M_o}{w}.$$

As was mentioned before, these equations represent a particular case of a uniform turn when no acceleration of motion occurs. This case is analyzed closely because it offers many clues to the understanding of the problem of steering tracked vehicles.

In order to use equation (395), the value of M_o should be determined first. If it is assumed that a vehicle has to turn by a certain angle while being resisted by the sum of uniformly distributed elementary forces

dR_x acting upon an element dx of the track located at a distance x from the center of rotation, then (Figure 140a)

$$M_o' = 4 \int_0^{s/2} x \, dR_x \, .$$

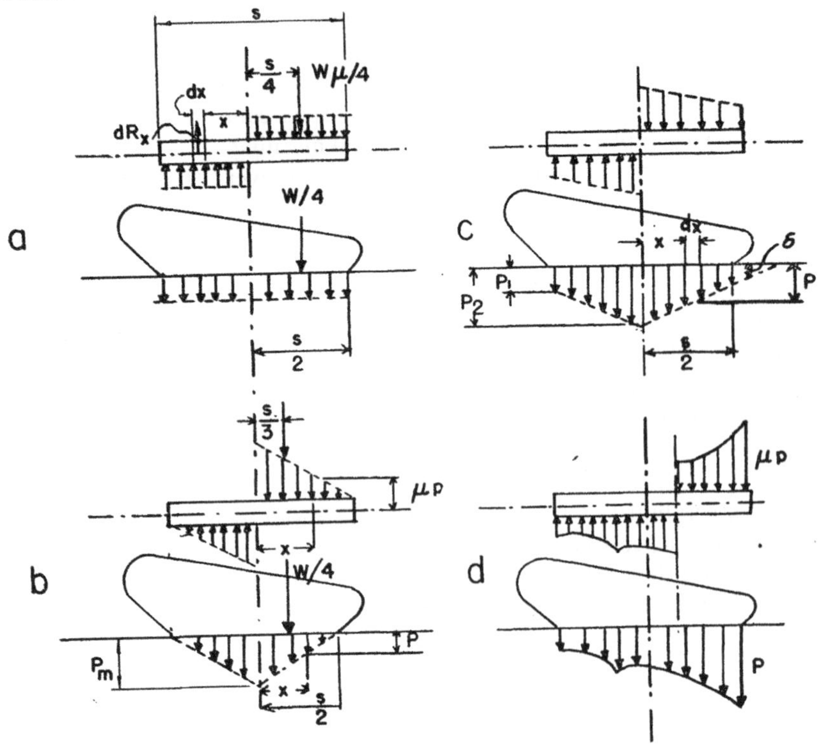

Fig. 140

The force dR_x resisting the turn may be assumed to be equal to that portion of the vehicle weight W which rests upon the track element dx times the coefficient of lateral friction μ of the track and the ground:

$$dR_x = \mu \frac{W}{2s} dx \, .$$

Combining the above equations gives

$$M_o' = 4\frac{W\mu}{2s}\int_0^{s/2} x\,dx = \frac{W\mu s}{4}, \tag{396}$$

and, accordingly, equations (395) may be written in the following form:

$$\left.\begin{aligned} F_1 &= f^o\frac{W}{2} - \mu\frac{Ws}{4w} \\ F_2 &= f^o\frac{W}{2} + \mu\frac{Ws}{4w}. \end{aligned}\right\} \tag{397}$$

It should be stressed that the value of M_o' as determined by formula (396) refers to the uniform distribution of load underneath the track (Figure 140a). If the distribution of ground pressure is triangular, as shown in Figure 140b, then the peak pressure p_m may be expressed by the following equations:

$$\tfrac{1}{2}\,p_m\frac{s}{2} = \frac{W}{4}$$

and

$$p_m = \frac{W}{s}.$$

Also,

$$\frac{p}{p_m} = \frac{\frac{s}{2} - x}{\frac{s}{2}} = \left(1 - \frac{2x}{s}\right).$$

Thus, the pressure at an arbitrary point

$$p = \frac{W}{s}\left(1 - \frac{2x}{s}\right).$$

Accordingly, the moment of resistance against turning is

$$M_o'' = 4\int_0^{s/2}\mu\,px\,dx = \frac{4\mu W}{s}\int_0^{s/2}\left(1 - \frac{2x}{s}\right)x\,dx = \frac{\mu Ws}{6},$$

and equation (395) may be written as follows:

$$\left.\begin{aligned} F_1 &= f^o \frac{W}{2} - \mu \frac{Ws}{6w} \\ F_2 &= f^o \frac{W}{2} + \mu \frac{Ws}{6w} \,. \end{aligned}\right\} \quad (398)$$

If the load distribution follows the scheme shown in Figure 140c, then again the moment of turn resistance is

$$M_o''' = 4 \int_0^{s/2} \mu\, px\, dx \,,$$

where, assuming that $\tan \delta = \dfrac{p_2 - p_1}{s/2}$, the pressure p equals

$$p = p_1 + \left(\frac{s}{2} - x\right) \tan \delta = p_1 + \left(\frac{s}{2} - x\right) \frac{p_2 - p_1}{s/2} \,.$$

If it is assumed that $p_2/p_1 = \lambda$ or that $p_1 = p_2/\lambda$, then

$$p = \frac{p_2}{\lambda} + 2\left(\frac{s}{2} - x\right) \frac{p_2 - p_2/\lambda}{s}$$

or

$$p = \frac{p_2}{\lambda} \left[1 + \frac{(s - 2x)\,(\lambda - 1)}{s}\right] .$$

Since, however,

$$\frac{W}{4} = \frac{p_1 + p_2}{2} \frac{s}{2}$$

and

$$p_2 = \frac{W\lambda}{s(1 + \lambda)} \,,$$

then, finally,

$$p = \frac{W}{s(1 + \lambda)} \left[1 + \frac{(s - 2x)\,(\lambda - 1)}{s}\right]$$

and

$$M_o''' = \frac{4\mu W}{s(1 + \lambda)} \int_0^{s/2} \left[1 + \frac{(s - 2x)\,(\lambda - 1)}{s}\right] x\, dx$$

$$= \frac{\mu}{6} \frac{Ws(\lambda + 2)}{1 + \lambda} \,.$$

Assume that $\lambda = 2$. Then, after substituting M_o''' into formulas (395),

$$\left.\begin{aligned} F_1 &= f^o \frac{W}{2} - \mu \frac{Ws}{4.5w} \\ F_2 &= f^o \frac{W}{2} + \mu \frac{Ws}{4.5w} . \end{aligned}\right\} \qquad (399)$$

In general, it will be seen that the moment of turning resistance of an arbitrary load distribution as shown in Figure 140d may be determined in the same way as previously if the location of the axis of rotation is known. This axis, for the case when no forces other than those considered are acting, passes through the "center of ground pressure" as previously discussed. If the load distribution is irregular, the integration leading to the determination of M_o in such cases may be performed graphically.

It should be noted that quite a sizable reduction in the amount of turning resistance may be achieved by proper shaping of the distribution of the ground pressure. It is evident that the triangular-type pressure (Figure 140b) is much more convenient than the uniform (Figure 140a) or trapezoidal pressure (Figure 140c).

Formulas (395) lead to important conclusions in regard to the steerability of a vehicle with reference to its over-all dimensions. Assume, in accordance with equation (267), that the maximum propelling force F_2 of the outside track of a turning vehicle equals or is smaller than

$$F_2 \leqslant c\Delta + \frac{W}{2} \tan \phi ,$$

where Δ is the track area which remains in contact with the soil, W is the vehicle weight, and ϕ is the angle of internal frictions. When combining the above equation with the second formula of equations (397), for instance, then

$$f^o \frac{W}{2} + \mu \frac{Ws}{4w} \leqslant c\Delta + \frac{W}{2} \tan \phi$$

or

$$\frac{s}{w} \leqslant \left(c\Delta + \frac{W}{2} \tan \phi - \frac{W}{2} f^o \right) \frac{4}{\mu W} ,$$

and, finally,

$$\frac{s}{w} \leqslant 4 \frac{c}{\mu} \frac{\Delta}{W} + 2 \frac{\tan \phi}{\mu} - 2 \frac{f^o}{\mu} .$$

If it is assumed that $\Delta/W = 1/p$, where p is the "mean ground pressure," the critical ratio s/w may be determined as follows:

$$\frac{s}{w} \leqslant \frac{2}{\mu}\left(\frac{2c}{p} + \tan\phi - f^o\right). \tag{400}$$

It defines the largest s/w at which the vehicle will steer without spinning the propelling track located outside the turning center.

In sand, for instance, where $c = 0$,

$$\frac{s}{w} \leqslant \frac{2(\tan\phi - f^o)}{\mu},$$

and for an average vehicle: $f^o = 0.1$; $\tan\phi = 0.7$; $\mu = 0.7$,

$$\frac{s}{w} \leqslant 1.7 .$$

Such is the condition of steerability of an average tracked vehicle in sandy soil. If the length-width ratio of the vehicle is larger than 1.7, the machine will not steer.

The corresponding limitation for a hard ground is slightly different. Under this condition, the following values may be assumed: $\tan\phi = 0.7$, $f^o = 0.05$, and $\mu = 0.6$. Then

$$\frac{s}{w} \leqslant 2.16 .$$

These examples illustrate the relationship between the terrain characteristics and the s/w ratio. The above further clarifies the previously discussed problem of the form of tracked vehicles (Chapter IV).

The lower limit of s/w is determined by the stability of steering. The first of equations (395) indicates that, in the case of an average vehicle, the force F_1 is always negative. This means that the track, adjacent to the center of turning, has to be braked in order to steer. However, a mere declutching of this track will steer the vehicle without brakes being applied if, in the case of a uniform load distribution,

$$F_1 = f^o\frac{W}{2} - \mu\frac{Ws}{4w} \leqslant 0$$

or if

$$\frac{s}{w} \geqslant \frac{2f^o}{\mu}. \tag{401}$$

For a soft ground ($f^o = 0.1$, $\mu = 0.7$),

$$\frac{s}{w} \leqslant 0.3 .$$

For a hard ground ($f^o = 0.05$, $\mu = 0.6$),

$$\frac{s}{w} \leqslant 1.66 .$$

It will be seen then that a vehicle may become unstable in soft ground if $s \cong 0.3w$, and in hard ground if $s \simeq 1.6w$.

Equations (400) and (401) as well as figures based on these equations indicate the order of magnitude of the s/w ratio which is desirable from the point of view of steerability and stability.

In reality, however, the mechanism of steering is more complex than the above discussion would indicate because the "center of pressure" and thus the moment of turning resistance M_o vary when external forces other than those considered in Figure 138 are applied. The most important of these external forces are the lateral and centrifugal forces and the longitudinal pulls which act on a drawbar.

Consider the lateral forces which will have to be taken into account when a vehicle turns on a slope. In the case of an uphill turn (Figure 141), the vertical ground reactions are

$$\left.\begin{aligned} Q_1 &= \frac{W}{2}\left(1 - \frac{2h}{w}\tan\alpha\right) \\ Q_2 &= \frac{W}{2}\left(1 + \frac{2h}{w}\tan\alpha\right). \end{aligned}\right\} \quad (402)$$

Accordingly, the respective movement resistances R_1 and R_2 will be obtained by multiplying equations (402) by the coefficient of resistance f^o.

However, when a uniform weight distribution is assumed, the moment of turning resistance will not be $\mu W s/4$, as was previously determined by formula (396). Now, because of the presence of a lateral force $W \sin\alpha$, which tends to move the vehicle down the slope, the distribution of frictional forces between the soil and the ground will be such that, at the moment of equilibrium, it also will balance $W \sin\alpha$. If it is assumed that the vehicle moves in the direction shown by the arrow "a" in Figure 141, the location of the new center of rotation will have to be moved a distance u toward the front of the vehicle. This distance is measured from the "center of pressure" which would exist on level ground in static conditions if a symmetrical uniform load were considered. Equation (396)

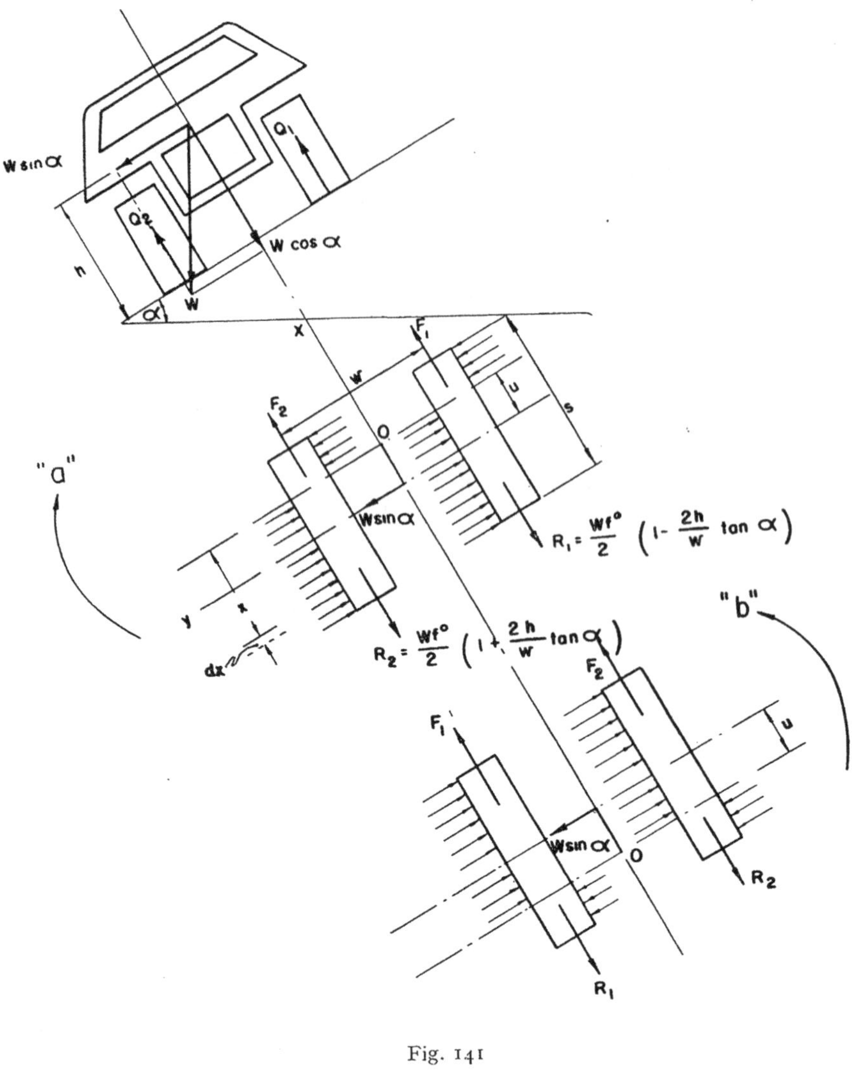

Fig. 141

remains basically valid, and only the limits of integration are to be adopted in accordance with the changed conditions:

$$-\left(\frac{s}{2}-u\right);\ \left(\frac{s}{2}+u\right).$$

The distance x is measured from the new center of rotation 0. Hence,

$$M_o = \frac{W\mu}{s}\left[\int_0^{s/2+u} x\,dx + \int_0^{-(s/2-u)} x\,dx\right] - Wu \sin \alpha$$

or

$$M_o = \frac{W\mu}{2s}\left(\frac{s^2}{2} + 2u^2\right) - Wu \sin \alpha .$$

The displacement u of the center of rotation will be only as large as it is required to keep the moment M_o at its minimum value. Hence,

$$\frac{dM_o}{du} = \frac{2W\mu}{s} u - W \sin \alpha = 0$$

and

$$u = \frac{s \sin \alpha}{2\mu},$$

and the moment of resistance M_o equals

$$M_o = \frac{W\mu}{2s}\left(\frac{s^2}{2} + \frac{s^2 \sin^2 \alpha}{2\mu^2}\right) - \frac{Ws \sin^2 \alpha}{2\mu}$$

or

$$M_o = \frac{W\mu s}{4}\left[1 - \left(\frac{\sin \alpha}{\mu}\right)^2\right].$$

It will thus be seen that the slope decreases the moment of resistance when the vehicle turns uphill. According to the denotations made on Figure 141, the steering forces F_1 and F_2 are

$$\left.\begin{aligned} F_1 &= \frac{Wf^o}{2}\left(1 - \frac{2h}{w} \tan \alpha\right) - \frac{W\mu s}{4}\left[1 - \left(\frac{\sin \alpha}{\mu}\right)^2\right] \\ F_2 &= \frac{Wf^o}{2}\left(1 + \frac{2h}{w} \tan \alpha\right) + \frac{W\mu s}{4}\left[1 - \left(\frac{\sin \alpha}{\mu}\right)^2\right]. \end{aligned}\right\} \quad (403)$$

In the case of a vehicle turning downhill in the direction "b" (Figure 141), the displacement u of the center of rotation will extend toward the rear of the vehicle. It may be shown in the same way as before that the moment of resistance in this case increases to the same extent as it decreased in the case of a vehicle turning downhill in the "a" direction. However, the values of the track resistance R_1 and R_2 will be reversed and

$$\left.\begin{aligned} F_1 &= \frac{Wf^o}{2}\left(1+\frac{2h}{w}\tan\alpha\right)-\frac{W\mu s}{4}\left[1+\left(\frac{\sin\alpha}{\mu}\right)^2\right] \\ F_2 &= \frac{Wf^o}{2}\left(1-\frac{2h}{w}\tan\alpha\right)+\frac{W\mu s}{4}\left[1+\left(\frac{\sin\alpha}{\mu}\right)^2\right]. \end{aligned}\right\} \quad (404)$$

It is interesting to consider formulas (397), (403), and (404) from the point of view of stability and steerability previously discussed in connection with driving on a plane. It will be seen that a vehicle that is stable on level ground may be entirely unstable in the hills. Equations (403) and (404) illustrate the effect of the lateral forces which may disturb the stability of locomotion. It is obvious that, in any case, the distance $u = (s \sin\alpha)/2\mu$ cannot be larger than $s/2$ because otherwise the turning point would be moved beyond the track contact area. Thus, when driving along a slope, steering depends on the condition that

$$\frac{s\sin\alpha}{2\mu} \leqslant \frac{s}{2}$$

or

$$\sin\alpha \leqslant \mu\,. \quad (405)$$

This is the same condition which enables the vehicle to have a straight-line motion along a slope without the danger of slipping sidewise down the hill.

The effect of the centrifugal force may be investigated in a similar way. Assume that the radius of curvature r_o of the path which is followed by the center of gravity of a uniformly loaded symmetrical vehicle is large enough compared to the length s of the ground contact area (Figure 142). Under these circumstances, β would be very small and r_o may be considered as being independent of the displacement u of the center of rotation 0. Accordingly, $\cos\beta \cong 1$ and $r_o \cong$ constant. Since

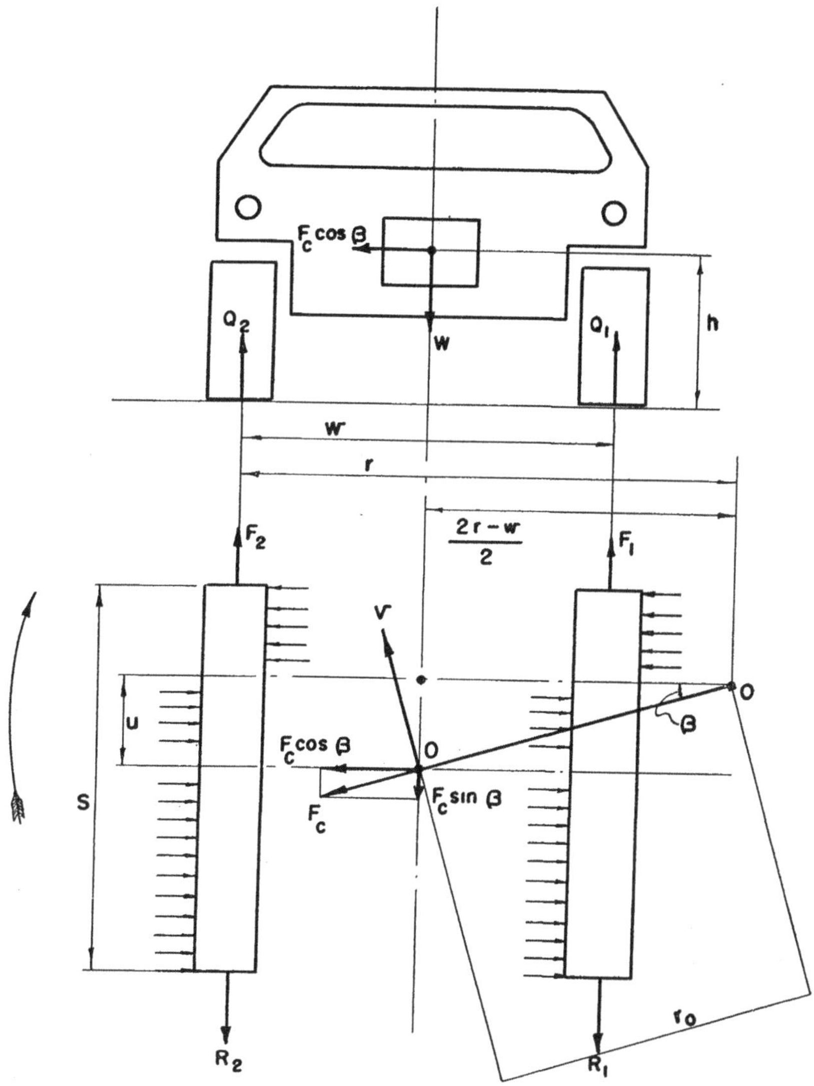

$F_c \cos \beta$
W
Q_2
Q_1
h
w
r
$\frac{2r-w}{2}$
F_2
F_1
V
O
β
u
$F_c \cos \beta$
O
S
F_c
$F_c \sin \beta$
r_0
R_2
R_1

Fig. 142

$F_c = Wv^2/gr_o$, where v is the linear speed of the center of gravity, the equation of equilibrium of moments is

$$M_o = \frac{W\mu}{s}\left[\int_0^{s/2+u} x\,dx + \int_0^{-(s/2-u)} x\,dx\right] - F_c u \cos\beta$$

$$= \frac{W\mu}{s}\left[\int_0^{s/2+u} x\,dx + \int_0^{-(s/2-u)} x\,dx\right] - \frac{Wv^2u}{gr_o}$$

or

$$M_o = \frac{W\mu}{2s}\left(\frac{s^2}{2} + 2u^2\right) - \frac{Wv^2u}{gr_o}.$$

The value u may be determined as previously:

$$\frac{dM_o}{du} = \frac{W\mu}{2s}4u - \frac{Wv^2}{gr_o} = 0$$

and

$$u = \frac{sv^2}{2gr_o\mu}.$$

Thus, the moment of turning resistance is

$$M_o = \frac{W\mu s}{4}\left(1 - \frac{v^4}{g^2r_o^2\mu^2}\right). \tag{406}$$

The movement resistance of tracks will be as follows:

$$R_1 = \frac{W}{2}f^o - \frac{h}{w}F_c \cos\beta$$

$$R_2 = \frac{W}{2}f^o + \frac{h}{w}F_c \cos\beta .$$

When $F_c = Wv^2/gr_o$ is substituted into the above equations, the driving forces $F_1 = R_1 - M_o/w$ and $F_2 = R_2 + M_o/w$ will be expressed by the equations

$$\left.\begin{aligned} F_1 &= \frac{W}{2}f^o - \frac{h}{w}\frac{Wv^2}{gr_o} - \frac{W\mu s}{4w}\left[1 - \left(\frac{v^2}{gr_o\mu}\right)^2\right] \\ F_2 &= \frac{W}{2}f^o + \frac{h}{w}\frac{Wv^2}{gr_o} + \frac{W\mu s}{4w}\left[1 - \left(\frac{v^2}{gr_o\mu}\right)^2\right]. \end{aligned}\right\} \tag{407}$$

Equation (407) does not contain the longitudinal component $F_c \sin \beta$ of the centrifugal force F_c, since this component was assumed to be negligible ($\sin \beta \cong 0$). Thus, it was assumed that it would not increase the forces F_1 and F_2. The existence of a sizable component of this nature,

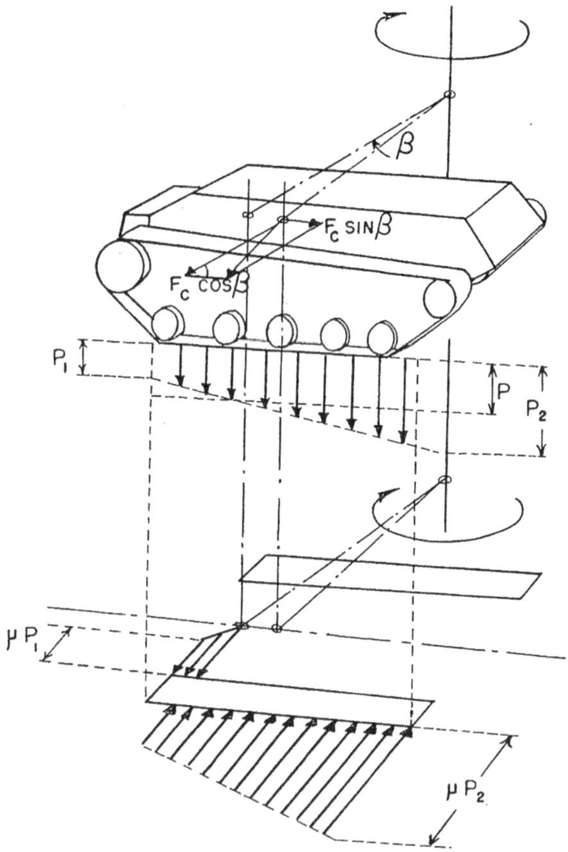

Fig. 143

however, would disturb the uniform pressure distribution p which, in this case, would have a trapezoidal pattern, starting with a smaller unit load p_1 in the front and a larger one p_2 in the rear (Figure 143). Accordingly, the moment of turning resistance M_o would not be the same as that expressed by formula (406). M_o could be evaluated by following the method in which a trapezoidal pressure distribution is assumed (Figure 140). It may be assumed, however, that the effect of $F_c \sin \beta$ is not considerable and equation (407) gives a fair approximation.

The relationship between the driving forces F_1 and F_2 and the radius of curvature r_0 may be determined according to equation (407). Kristi included in his collective work respective calculations and published

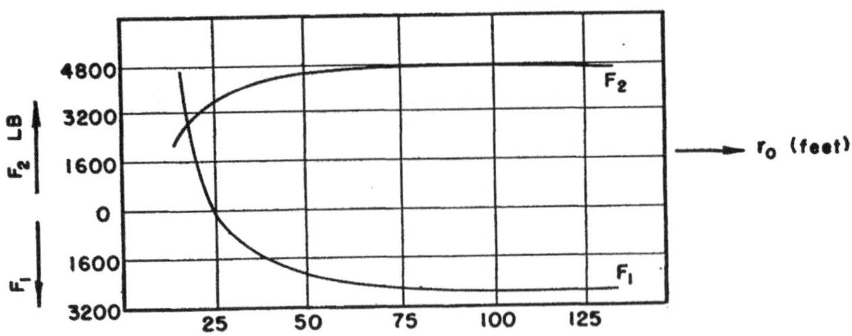

VEHICLE WEIGHT	W = 20,000 lb.
SPEED	v = 11 m.p.h.
GROUND CONTACT LENGTH	s ≅ 117"
TRACK TREAD	w ≅ 78"
LATERAL FRICTION	μ = 0.5
MOVEMENT RESISTANCE	f* = 0.1
HEIGHT OF THE LOCATION OF CG	h ≅ 31"

(KRISTI)

Fig. 144

nomograms which help to evaluate the variation of the driving forces as a function of the radius of turning.[40] The general form of this relationship is shown in Figure 144. Although experimental measurements are not quite in agreement with the above graph,[238] they do show the same over-all character. It is to be realized that the relationship between small r_0 values and F forces is not very accurately expressed by the discussed equations, mainly because of the assumption that β is very small.

A vehicle will skid when, as previously, $u \geqslant s/2$. In this case,

$$\frac{sv^2}{2gr_o\mu} \leqslant \frac{s}{2}$$

and

$$\frac{v^2}{r_o} \leqslant g\mu .$$

If speed v is the limiting factor, then

$$v \leqslant \sqrt{gr_o\mu} \tag{408}$$

and the critical radius of curvature

$$r_o \geqslant \frac{v^2}{g\mu} . \tag{409}$$

It will be seen that the friction μ between the ground and a track plays a decisive role. Equations (408) and (409) may be presented in a different form which reflects the physical constants of soil c and ϕ.

According to Coulomb's equation (123), the shearing stress $\tau = c + \sigma \tan \phi$. If it is assumed that σ is equal to the uniform pressure distribution p, and that μ may be substituted for by τ, then $\mu = c + p \tan \phi$. Accordingly, equations (408) and (409) may be written as follows:

$$v \leqslant \sqrt{gr_o(c + p \tan \phi)} .$$

and

$$r_o \geqslant \frac{v^2}{g(c + p \tan \phi)} . \tag{410}$$

Thus, the critical speed at a curvilinear motion, as well as the critical radius of turning at which a vehicle does not skid, may be qualitatively related to the same soil properties that were previously used for the determination of the tractive effort and "flotation" of a vehicle on the surface of soil.

Another problem of the steerability of a vehicle may be encountered when a tractor hauls a trailer (Figure 145). The moment of resistance against turning may be determined as in the previous cases. If it is

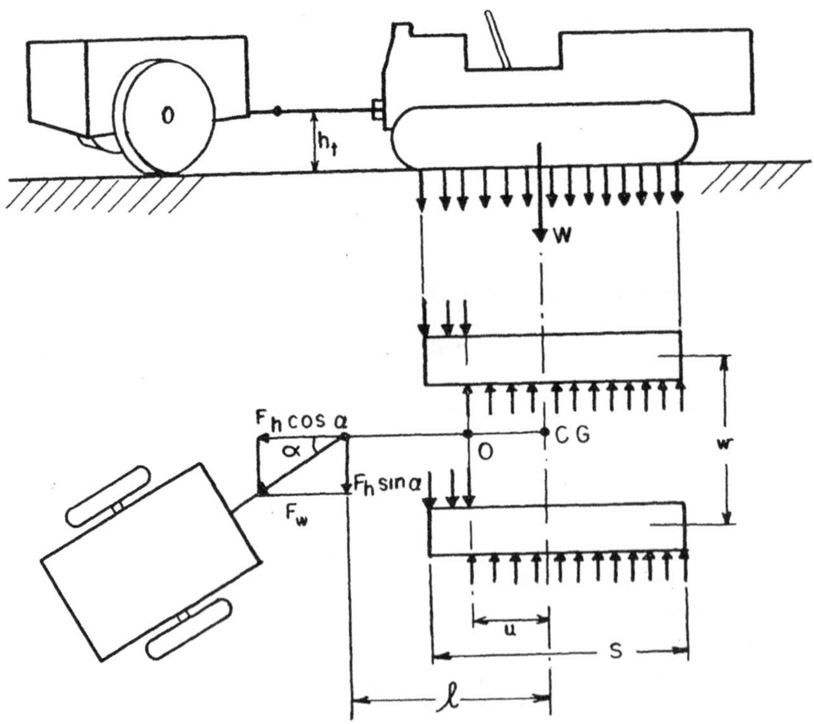

Fig. 145

assumed that the uniform pressure distribution is not disturbed by the location of the hook at height h_t above the ground, then

$$M_o = \frac{W\mu}{s}\left[\int_0^{s/2+u} x\,dx + \int_0^{-(s/2-u)} x\,dx\right] - F_h(l-u)\sin\alpha$$

or

$$M_o = \frac{W\mu}{2s}\left(\frac{s^2}{2} + 2u^2\right) + F_h(l-u)\sin\alpha$$

and

$$\frac{dM_o}{du} = \frac{2W\mu}{s}\,u - F_h\sin\alpha = 0\,.$$

Hence,

$$u = \frac{sF_h \sin \alpha}{2W\mu}$$

and

$$M_o = \frac{W\mu s}{4} - \frac{sF_h^2 \sin^2 \alpha}{4W\mu} + F_h l \sin \alpha . \tag{411}$$

It may be assumed that the maximum pulling force $F_h \cos \alpha$ cannot exceed the tractive effort of the tractor H minus its own movement resistance Wf^o. If the tractive effort is expressed, for example, by equation (267),

$$H = 2\Delta c + W \tan \phi ,$$

where 2Δ is the ground contact area, W the vehicle weight, and c and ϕ the soil constants. Then

$$F_h \cos \alpha = 2\Delta c + W \tan \phi - Wf^o$$

and

$$F_h = \frac{2\Delta c + W \tan \phi - Wf^o}{\cos \alpha} . \tag{412}$$

Let it be assumed that the negotiated soil has a hard surface or is of a sandy type, so that c may be assumed to be equal to zero. In such a case, after substituting F_h from equation (412) into equation (411), the moment of turning resistance may be expressed as follows:

$$M_o = \frac{W\mu s}{4} \left\{ 1 + \frac{(\tan \phi - f^o) \tan \alpha}{\mu} \left[\frac{4l}{s} - \frac{(\tan \phi - f^o) \tan \alpha}{\mu} \right] \right\} .$$

It will be seen that the trailer increases the turning moment as long as the second member of the last term of the equation is larger than zero. Beyond this point, the turning moment decreases. This takes place when

$$\frac{4l}{s} - \frac{(\tan \phi - f^o) \tan \alpha}{\mu} < 0$$

or when

$$\tan \alpha > \frac{4l\mu}{s(\tan \phi - f^o)} . \tag{413}$$

The critical value from the skidding point of view also has to be taken into consideration. This value may be determined as previously from the equation: $u \leqslant s/2$. In this case,

$$u = \frac{sF_h \sin \alpha}{2W\mu} \leqslant \frac{s}{2}$$

or, since it was assumed that $F_h = W(\tan \phi - f^o)/\cos \alpha$,

$$\frac{s(\tan \phi - f^o)\tan \alpha}{2\mu} \leqslant \frac{s}{2}.$$

Hence,

$$\tan \alpha \leqslant \frac{\mu}{\tan \phi - f^o}. \tag{414}$$

For an average tractor-trailer unit, it may be assumed that $l/s = 0.5$. Accordingly, equation (413), which determines the condition of reduced turning moment, may be written as follows:

$$\tan \alpha > \frac{2\mu}{\tan \phi - f^o}. \tag{415}$$

It will thus be seen that the occurrence of the reduced moment of turning resistance due to the trailer action, as expressed in equations (413) and (415), will not take place because the tractor would first start to skid at an angle α, which is one-half the angle that was determined by equation (414).

It may be demonstrated that in the general case of soil displaying both friction ϕ and cohesion c, the situation will be the same and equations (414) and (415) would comprise only the cohesive factor $2\Delta c/W$ in the denominator. This may be deduced from equation (411), which may be written in the following form:

$$M_o = \frac{W\mu s}{4} + F_h \sin \alpha \left(l - \frac{sF_h \sin \alpha}{4W\mu}\right).$$

By repeating the same reasoning, it may be shown that the skid will occur before M_o starts to decrease, and a tractor-trailer unit will always operate at an increased turning moment. Thus, the loads which have to be sustained by the steering mechanisms of tractors are higher than those of self-propelled single vehicles, and the steering conditions of a tractor-trailer aggregate are more rugged than those of a single vehicle. The driving forces F_1 and F_2 of the tractor are as follows:

$$F_1 = \frac{W}{2} f^o - \frac{W\mu s}{4w} - \frac{F_h \sin \alpha}{w}\left(l - \frac{sF_h \sin \alpha}{4W\mu}\right) + \frac{F_h \cos \alpha}{2}$$

$$F_2 = \frac{W}{2} f^o + \frac{W\mu s}{4w} + \frac{F_h \sin \alpha}{w}\left(l - \frac{sF_h \sin \alpha}{4W\mu}\right) + \frac{F_h \cos \alpha}{2}.$$

The increased moment of turning resistance may cause the outside track to spin if the angle α surpasses a certain value. The critical angle α at which the tractor will no longer steer may be determined from the equation expressing the force F_2.

The condition of steerability is $F_2 \leqslant (W/2)(H/W)$, where (H/W) for sand, for instance, is expressed by equation (379). Accordingly,

$$\frac{W}{2} f^o + \frac{W\mu s}{4w} + \frac{F_h \sin \alpha}{w}\left(l - \frac{sF_h \sin \alpha}{4W\mu}\right) + \frac{F_h \cos \alpha}{2}$$
$$< \frac{W}{2}\left\{1 + 0.64\left[\frac{h}{2l}\cot^{-1}\left(\frac{h}{2l}\right)\right]\right\} \tan \phi \qquad (416)$$

and the critical angle α of turn may be determined in terms of soil properties and other constants related to tractor weight and its geometry.

A more accurate study of the steerability of tracked vehicles may be performed only at the cost of laborious computations.[40] Original work done on this subject by Merritt [239] and Steeds [240] shows this point clearly.

In the previous analysis of the turning of a tracked vehicle, a uniform ground pressure was assumed and the vehicle was to rotate around the center of this pressure while moving forward. In the case of the existence of lateral forces, the center in question had to be displaced by a distance u toward the front or rear of the vehicle, depending on the direction of these forces. By determining the distance u, the propelling forces F_1 and F_2 and radii r_o of turning could be computed for the critical conditions in which no longitudinal slip between tracks was assumed.

The study of turning performed by Steeds refers to a similar but more general case in which both tracks slip longitudinally. This slip, which is an inseparable phenomenon of vehicle motion and which may prevent the vehicle from following the curve prescribed by the velocities of its sprockets, was defined as the rotation around instantaneous centers 0_o

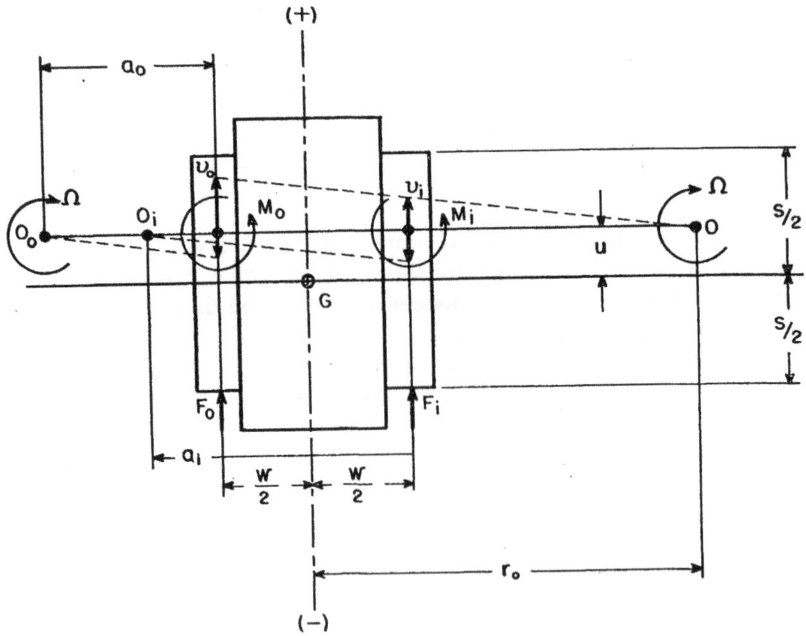

$$\upsilon_o = (r_o + \frac{W}{2})\Omega$$
$$\upsilon_i = (r_o - \frac{W}{2})\Omega$$
Speeds of the hull

$$\upsilon'_o = (r_o + \frac{W}{2})\Omega + a_o\Omega$$
$$\upsilon'_i = (r_o - \frac{W}{2})\Omega + a_i\Omega$$
Speeds of the track relative to the hull

$a_o\Omega$, $a_i\Omega$ — Slip velocities

a_o, a_i — Slip radii

(STEEDS)

Fig. 146

and 0_i located on a perpendicular to the track axis. These centers usually are positioned on the track side opposite the center of rotation 0 of a vehicle (Figure 146) because the slip speeds $a_o\Omega$ and $a_i\Omega$ and the velocities of the hull $[r_o + (w/2)]\Omega$ and $[r_o - (w/2)]\Omega$ are of opposite sign. However, there is no limitation to the location of the instantaneous centers of rotation 0_o and 0_i since slips may be oriented in the same direction as speeds. Radii of slip a_o and a_i will be called positive, or negative, depending on whether the centers 0_o and 0_i are located on the opposite or on the same side of the track with reference to the center 0. In the case of Figure 146, both a_o and a_i are positive.

Upon making such assumptions, it will be seen that the radius r_o of the curvature and the path followed by the vehicle center are not defined by the geometry which results from the angular velocities ω_o and ω_i of the sprockets,

$$r_o = \frac{w(\omega_i + \omega_o)}{2(\omega_o - \omega_i)}, \tag{417}$$

but depends on the slip of both tracks as defined by slip radii a_o and a_i and the angular velocity of turning Ω. In order to determine r_o, the angular velocities of the sprockets ω_o and ω_i are introduced into equations defining the track velocities relative to the hull (Figure 146):

$$\left(r_o + \frac{w}{2}\right)\Omega + a_o\Omega = \frac{d}{2}\,\omega_o$$

$$\left(r_o - \frac{w}{2}\right)\Omega + a_i\Omega = \frac{d}{2}\,\omega_i\,,$$

where $d/2$ is the sprocket radius. When dividing these equations, it will be found that

$$r_o = \frac{\dfrac{w}{2}\left(\dfrac{\omega_o}{\omega_i} + 1\right) + a_o - a_i\dfrac{\omega_o}{\omega_i}}{\dfrac{\omega_o}{\omega_i} - 1}. \tag{418}$$

The forces acting upon a single track due to slip may be determined by integrating the forces acting upon the track elements, in a way similar to that used before. Figure 147 shows an element of a track whose unit load is $W/2s$. If the adhesion μ is assumed to be isotropic, it is the same

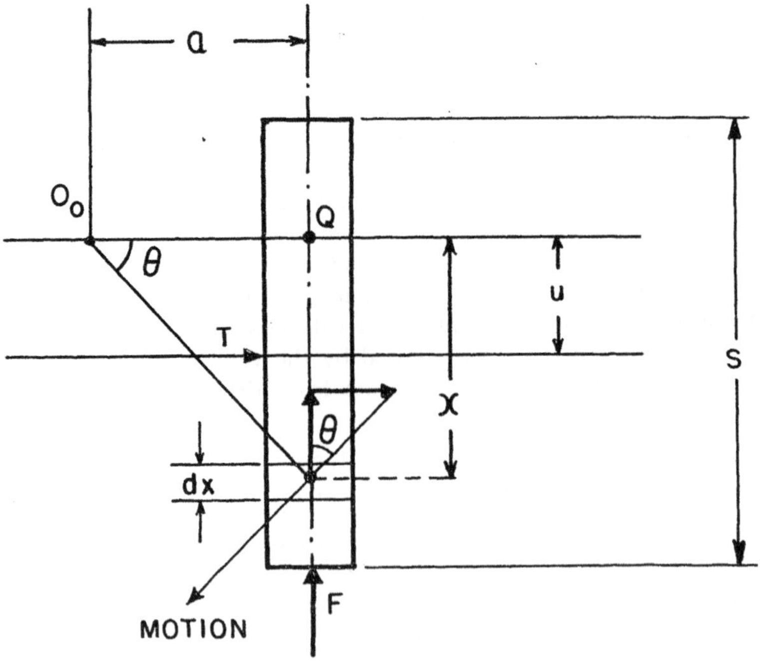

Fig. 147

in all directions of the track skid; then the elementary frictional force is $\mu W\, dx/2s$ and the total longitudinal force is

$$F = \frac{\mu W}{2s} \int_{-s/2+u}^{s/2+u} \cos \theta \, dx \, . \tag{419}$$

The lateral force can be found in a similar way:

$$T = \frac{\mu W}{2s} \int_{-s/2+u}^{s/2+u} \sin \theta \, dx \, . \tag{420}$$

On the other hand, $\cos \theta = a/\sqrt{a^2 + x^2}$ and $\sin \theta = x/\sqrt{a^2 + x^2}$. By substituting these values into equations (419) and (420) and integrating,

$$F = \frac{\mu W a}{2s} \log_e \frac{\psi_2 + \sqrt{\psi_2^2 + (2a/s)^2}}{\psi_1 + \sqrt{\psi_1^2 + (2a/s)^2}} \tag{421}$$

and

$$T = \frac{\mu W}{4}\left[\sqrt{\psi_2^2 + (2a/s)^2} - \sqrt{\psi_1^2 + (2a/s)^2}\right], \qquad (422)$$

where $\psi_1 = (2u/s) - 1$ and $\psi_2 = (2u/s) + 1$. The moments of the forces referred to point Q yield

$$M_o = \frac{\mu W}{2s}\int_{-s/2+u}^{s/2+u} \frac{x^2\,dx}{\sqrt{a^2 + x^2}},$$

or, after integrating,

$$M_o = \frac{\mu W s}{16}\left[\psi_2\sqrt{\psi_2^2 + (2a/s)^2} - \psi_1\sqrt{\psi_1^2 + (2a/s)^2}\right] - \frac{a}{2}F, \qquad (423)$$

where F is a value represented by formula (421).

The functions (421), (422), and (423) were computed by Steeds and his staff for various values of $2a/s$ and $2u/s$ when taking into consideration actual experience with tracked vehicles (see Figures 148, 149, and 150). The solution of any problem may be obtained by trial and error if an arbitrary value a which satisfies these equations can be represented by the following graphs.

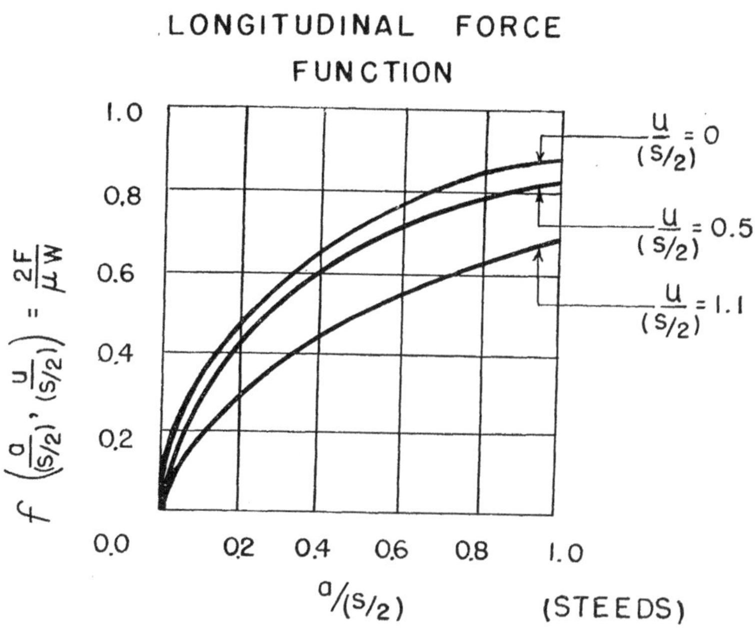

Fig. 148

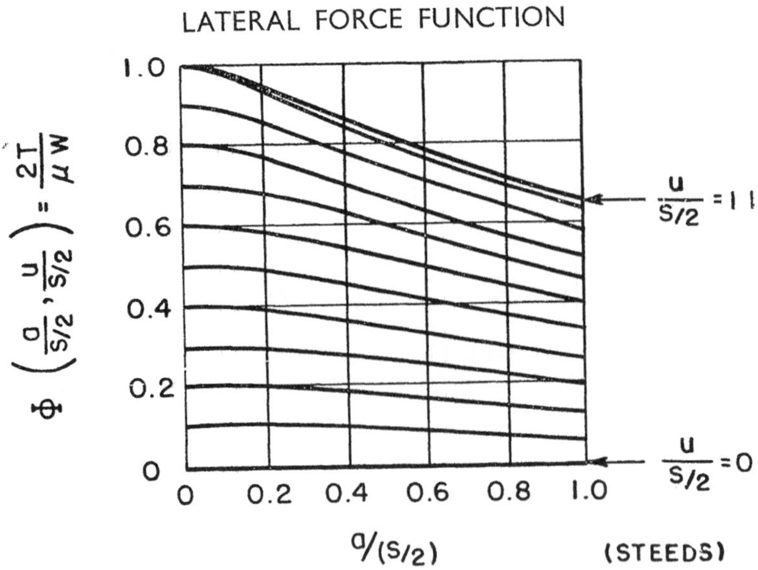

Fig. 149

Steeds' method is particularly adaptable to the assumption of anisotropic coefficients of adhesion. When considering equation (379), which encompasses the "grouser action," it becomes evident that the longitudinal adhesion of a track $\mu_L = (H/W)$ is larger than that measured in the transverse direction μ_T. Following the suggestion by Micklethwait,[88] the British school proposed the definition of track adhesion which is based on the assumption that skidding occurs in an arbitrary direction. It can be seen on Figure 151 that if skidding takes place in the longitudinal direction, then $\mu_L = \overline{OA}$; if, in the lateral one, $\mu_T = \overline{OB}$. Usually it is assumed that $\mu_T < \mu_L$, although this is not necessarily true. If the track shoe skids in an arbitrary way, then the resultant frictional force follows the direction $\overline{OG}$, which may be obtained as an intersection of lines $\overline{CE}$ and $\overline{DE}$. The locus of the G points, as the direction of the relative motion varies from OB to OA, is an ellipse. Thus it will be seen that the resultant adhesion force $W\mu$ is oriented in a different direction than the direction of the skid, a factor which is essential in the whole analysis. Under these conditions, equation (421) will contain μ_L, whereas equations (422) and (423) will use the value of μ_T.

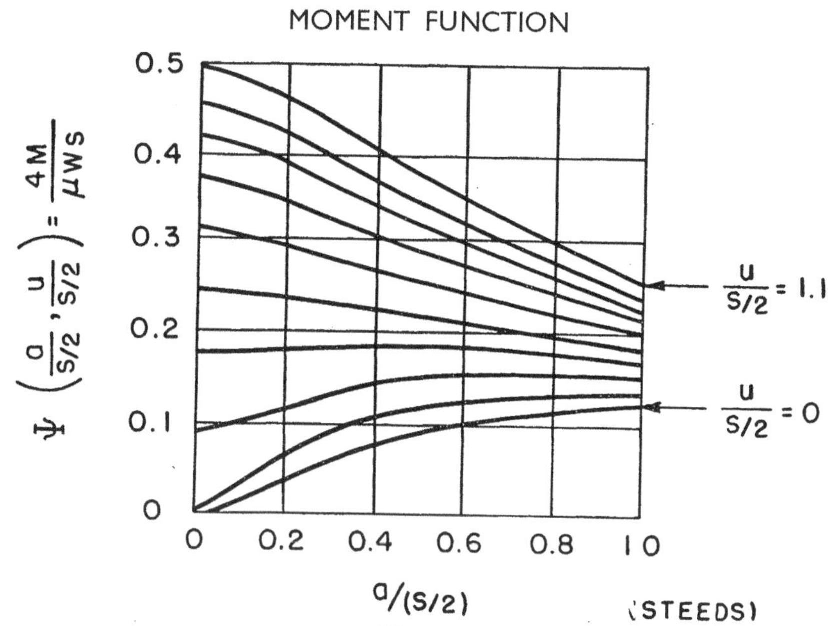

Fig. 150

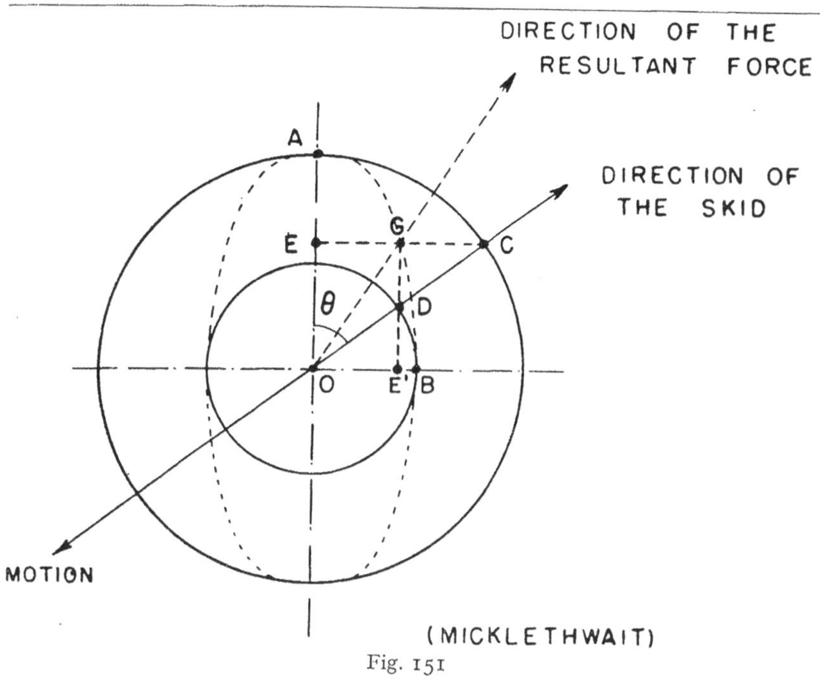

Fig. 151

Speed of Locomotion and Vehicle Vibrations

Should cross-country vehicles be improved beyond the present limits of the "go and no-go" criterion, then the speed of locomotion would become a general measure of vehicle mobility, and the attempts to increase the mobility would be synonymous with the attempts to increase the average speed over given terrain conditions.

It was learned on many occasions that the average speed of modern vehicles cannot be increased by a mere increase in engine power. As a matter of fact, a great majority of contemporary cross-country vehicles are overpowered. They are equipped with larger engines than those required for developing maximum speeds, of the accepted order of 35 mph, and yet, as experience shows, they are unable to ride in a typical terrain faster than from 5 to 10 mph.

The explanation of this paradox lies in the fact that the speed is governed not by the power train alone, but also by the geometry and dynamic properties of the vehicle structure. These factors, in conjunction with the unevenness of the terrain surface, produce vehicle vibrations which may be prohibitive for human endurance, and fatigue stress of metals if the riding speeds reach certain limits. Recent experience gained in cross-country locomotion through the conquest of vast spaces on a major scale by land vehicles stresses the ride discomfort situation very emphatically.

The disadvantages which stem from such a situation are serious. So far, only vehicles which can develop a maximum average cross-country speed of approximately 10 mph and less are available, and there is no experience at all with average speeds reaching a maximum limit of, say, 20 mph without causing critical pitching and/or bouncing. The losses caused by such a forcibly imposed speed limit need no comment, and yet the problem appears to have been practically neglected. This may be seen when the riding qualities of some existing vehicles are compared on a common basis. Figure 152 shows, for instance, the angle of pitch of two tracked vehicles built in large quantities in recent years. It illustrates that, if, for example, the pitch amplitude enclosed between 2.5° and 3.5° is considered for certain reasons to be the critical one, then vehicle *A* cannot ride faster than 5 mph, whereas vehicle *B* may exceed speeds of 20 mph. The inconsistency in the design of the vehicles is apparent. The need for more uniform and better riding qualities is also stressed by the fact that they would result in an improved economy of fast-moving

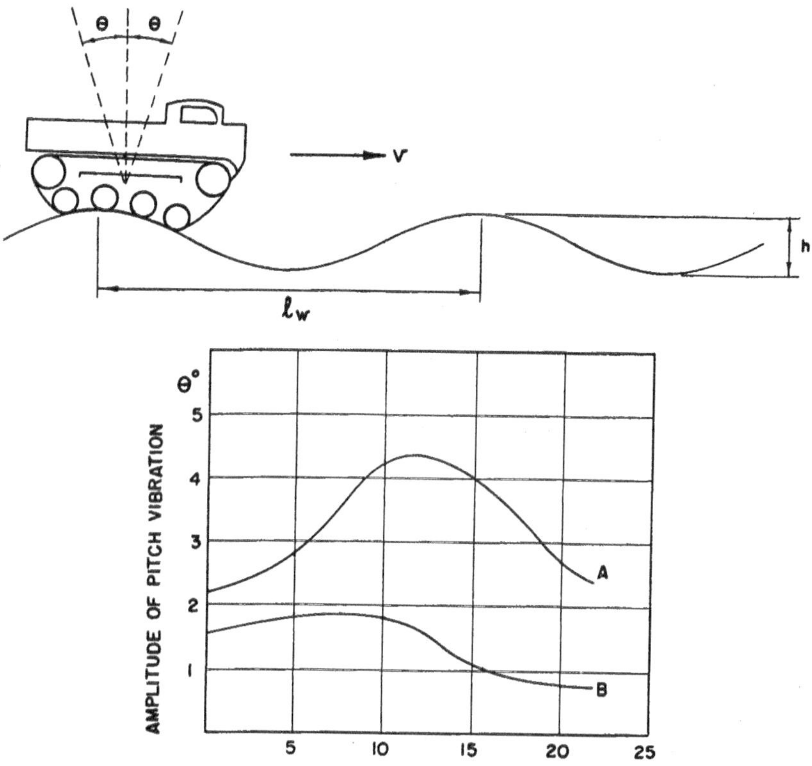

Fig. 152

vehicles since the problem of fuel consumption is closely linked with the speed of locomotion: a vehicle whose engine power is throttled in order to decrease the speed to 5 mph and less consumes several per cent more fuel than a comparable vehicle driving at a speed of 20 mph over the same terrain. The study of the problem is twofold: (a) a study of the principles of the mechanics of vehicle vibration and (b) an analysis of what may be considered a typical unevenness of the terrain in which a given vehicle is supposed to operate. By combining the results of the research in these two directions, an optimum solution may be reached, and a yardstick for the assessment of the economic values of a given operation may be obtained.

Methods applicable to the analysis of vibration problems have been well defined. Textbooks on vibrations by Den Hartog,[213] Timoshenko,[206] and particularly by Lehr [207] present a broad outline of the analytical approach. However, a direct application of this information to a vehicle appears to be impossible without a considerable amount of additional study. This may be seen in the work by Kamm [204] or in the treatise by Serruys.[241] The problem is further complicated by the psychological and physiological factors which come into the picture when the effect of vibrations on human beings is evaluated. The work done by Rowell,[242] Guest,[243] Jacklin,[244] Moss,[245] and Brown and Dickinson [246] rather indicates the complexity of the whole issue than clarifies the basic points, as was shown in an excellent review of the situation by Janeway.[247] The work by Janeway, Bourton-Douglas, and Goldman stressed in a recent publication by the American Society of Automotive Engineers,[248] which describes a general picture of the present status of the field of vehicle vibration, deals with the experience gained with railway cars and aircraft and highway vehicles and does not refer to cross-country operations. The vibrations and "ride comfort" of cross-country vehicles, particularly those equipped with tracks, present somewhat different questions and necessitate a different approach than those mentioned above. The idea of "ride comfort" also is to be modified in order to include the requirements which exist in cross-country locomotion and which are not encountered on a highway.

A rational analytical approach to the mechanics of vehicle vibration was not made until recently. Although the work by Lehr [249–251] and others [206, 213] contributed to a rigorous understanding of the vibrations of motor vehicles during the past 20 years, not until recently has the work by Lehr and Bertschinger been published in a series of articles which gives a complete outline of the problem.[252]

In accordance with Lehr, the pitch angle attained at a given speed is considered to be the measure of critical conditions which may restrict the ride over given terrain. Pitching amplitude appears to have a broader application than acceleration in the assessment of critical vibrations because, besides being easily interpreted in terms of acceleration (which gives the measure of the "ride comfort," as defined in automotive engineering), it also provides means for the assessment, for instance, of the possibility of road observation through optical instruments, or the performance of other special functions.

Since the pitch angle depends not only on the characteristics of suspension but also on the type of ground wave, a study of the character of this wave must be considered carefully. Unfortunately, there is practically no information available as to what types of terrain ruggedness are to be considered typical, or what the most severe test case is for the designer of a suspension.

The sinusoidal shape of the ground wave considered in one of the previous sections dealing with dynamic loads due to vehicle action is favored in all the investigations performed so far. For the time being, a ground wave approximately 10 ft long and 5 in. deep may be suggested as a typical contour of the surface of average areas in which the modern motor transport may be contemplated.

Although more research work is needed in order to establish statistical data of ground waves in various regions, it may be surprising to note that there is a definite pattern of waviness both on highways and in virgin terrain, particularly of specific geophysical origin. Work by Bangold [253] shows this point in regard to sand. Experience with snow indicates the existence of the same regularity, although it may not be visible at first glance.[254] A collection of a large number of data comprising length and amplitudes of ground waves and a statistical evaluation of these data may undoubtedly determine the geometric properties of terrain surface more rigorously than is now possible. However, a more or less arbitrary adaptation of a "typical wave" based on the available limited experience may serve the purpose of a comparative evaluation of vehicle vibrations, although it will not determine the absolute values that may be encountered in actual practice.

The general equation of the road wave was expressed in formula (364):

$$x_1 = \frac{h}{2}(1 - \cos \omega_g t) ,$$

where $\omega_g = 2\pi v / l_w$. The symbol l_w is the "wave" length, h its height, and v the vehicle speed.

The determination of vehicle pitch vibrations over a given terrain wave, as expressed in equation (364), may be performed in the following way. Assume an oscillating system, as shown in Figure 153. The equation of motion of a vehicle body having mass m is

$$m \frac{d^2 z}{dt^2} + \varrho \frac{d}{dt}(z - h \cos \omega_g t) + c_s (z - h \cos \omega_g t) = 0$$

or

$$m \frac{d^2 z}{dt^2} + \varrho \frac{dz}{dt} + c_s z = h(c_s \cos \omega_g t - \varrho \omega_g \sin \omega_g t) , \tag{424}$$

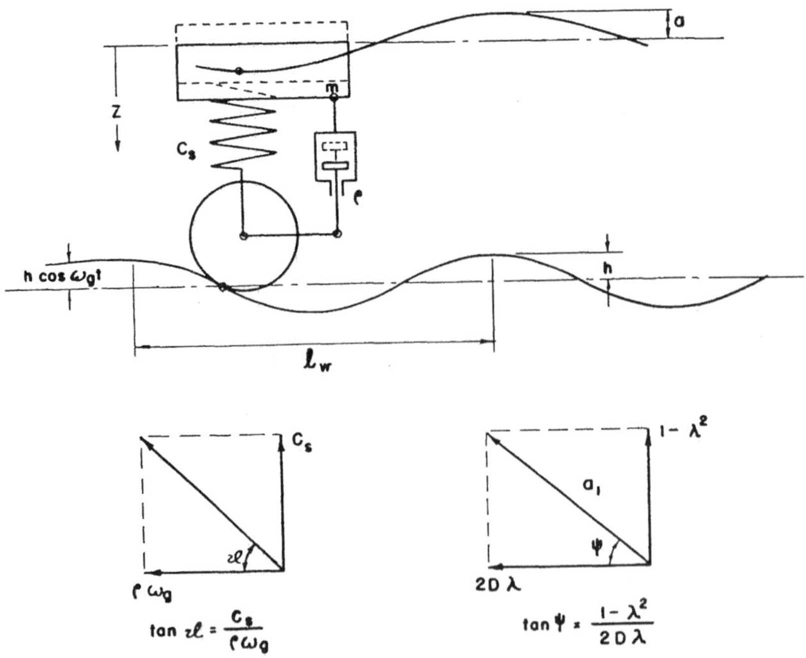

Fig. 153

where ϱ is the damping factor, c_s the spring constant, and ω_g is the frequency of the ground wave as previously defined.

The particular solution of equation (424) is

$$z = a \cos (\omega_g t - \psi) \ ,$$

where a is the amplitude of the forced vibration of mass m and ψ is the phase angle.

Differentiate the last equation twice. Then

$$\left.\begin{aligned} \frac{dz}{dt} &= - a\omega_g \sin (\omega_g t - \psi) \\ \frac{d^2z}{dt^2} &= - a\omega_g{}^2 \cos (\omega_g t - \psi) \ . \end{aligned}\right\} \quad (425)$$

The difference $c_s \cos \omega_g t - \varrho\omega_g \sin \omega_g t$ in equation (424) may also be presented in the following form, if the vectors c_s and $\varrho\omega_g$ are added (Figure 153):

$$c_s \cos \omega_g t - \varrho\omega_g \sin \omega_g t = -\sqrt{c_s^2 + \varrho^2\omega_g^2} \sin (\omega_g t - \vartheta) . \qquad (426)$$

Combine equations (424), (425), and (426):

$$-\omega_g^2 \cos (\omega_g t - \psi) - \frac{\varrho}{m} \omega_g \sin (\omega_g t - \psi) + \frac{c_s}{m} \cos (\omega_g t - \psi) +$$
$$+ \frac{h}{ma} \sqrt{c_s^2 + \varrho^2\omega_g^2} \sin (\omega_g t - \vartheta) = 0 .$$

Since, according to the established definition, $\sqrt{c_s/m}$ is the natural frequency ω_n of the system,

$$(\omega_n^2 - \omega_g^2) \cos (\omega_g t - \psi) - \frac{\varrho}{m} \omega_g \sin (\omega_g t - \psi) +$$
$$+ \frac{h}{ma} \sqrt{c_s^2 + \varrho^2\omega_g^2} \sin (\omega_g t - \vartheta) = 0 .$$

Divide this equation by ω_n^2 and denote the ratio of ω_g/ω_n by λ.

If it is assumed further, in accordance with the theory of vibrations, that the damping factor is defined by

$$D = \frac{\varrho}{2m\omega_n} ,$$

the considered equation may be written in the following form:

$$(1 - \lambda^2) \cos (\omega_g t - \psi) - 2D\lambda \sin (\omega_g t - \psi) +$$
$$+ \frac{h}{ma\omega_n^2} \sqrt{c_s^2 + \varrho^2\omega_g^2} \sin (\omega_g t - \vartheta) = 0 .$$

However, $\sqrt{c_s^2 + \varrho^2\omega_g^2} = m\omega_n^2 \sqrt{1 + 4D^2\lambda^2}$, which may be shown by the direct substitution of $D = \varrho/2m\omega_n$. Accordingly, the equation of vibration may be written in a simplified form:

$$\left[(1 - \lambda^2) \cos (\omega_g t - \psi) - 2D\lambda \sin (\omega_g t - \psi)\right] +$$
$$+ \left[\frac{h}{a} \sqrt{1 + 4D^2\lambda^2} \sin (\omega_g t - \vartheta)\right] = 0 . \qquad (427)$$

The first part of equation (427) enclosed in the bracket is the equation of free damped vibrations of the system, whereas the second part defines the forced damped vibrations.

According to the theory, the amplitude a_1 of free vibrations (Figure 153) is [206]

$$a_1^2 = (1 - \lambda^2)^2 + (2D\lambda)^2$$

and

$$-\sqrt{(1 - \lambda^2)^2 + (2D\lambda)^2} \sin(\omega_g t - 2\psi) = -\frac{h}{a}\sqrt{1 + 4D^2\lambda^2} \sin(\omega_g t - \vartheta) .$$

The response factor of forced vibrations will thus be

$$\frac{a}{h} = \frac{\sqrt{1 + 4D^2\lambda^2}}{\sqrt{(1 - \lambda^2)^2 + 4D^2\lambda^2}} . \tag{428}$$

By the use of similar methods, an analogical relationship may be deduced for the angular amplitudes θ_p of pitching and θ_l of the exciter wave, if D_1 denotes the torsional damping factor and λ_1 is the ω_g/ω'_n ratio, where ω'_n is the natural frequency of pitch vibration:

$$\frac{\theta_p}{\theta_l} = \frac{\sqrt{1 + 4D_1^2\lambda_1^2}}{\sqrt{(1 - \lambda_1^2)^2 + 4D_1^2\lambda_1^2}} . \tag{429}$$

The value of θ_l may be determined graphically for a given ground wave and length s of the effective ground contact by measuring the maximum angle θ at which the vehicle may tilt when crossing the terrain with a locked suspension. Figure 154 shows that for an assumed sinusoidal ground surface, the exciter angle θ decreases with an increase in vehicle length, which is self-explanatory.

The damping factor D_1 may be determined or assumed when considering the location and damping characteristics of the shock absorbers. The natural frequency of the angular vibration ω'_n may be found by following the procedure proposed by Lehr.[212]

To this end, the radius of gyration r_g may be assumed to be equal to

$$r_g = \zeta \sqrt{\frac{l + h}{12}} ,$$

where l and h are the maximum length and width of the vehicle, respectively (see Figures 33, 34, and 35), and ζ is an experimental factor which takes care of the nonuniformity of the mass distribution neglected

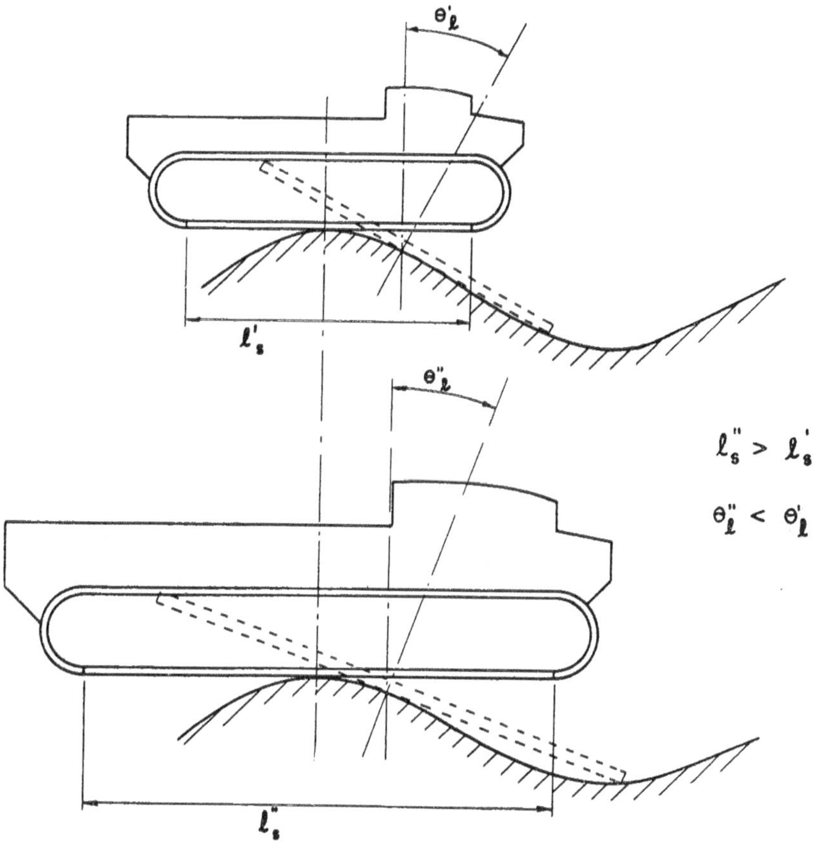

Fig. 154

in the above equation. The symbol ζ is assumed to be equal to approximately 1.15 for heavy tracked vehicles.

The center of pitch vibration 0 is located underneath the center of gravity, approximately at the level of the wheel centers (Figure 155). If it is assumed that the distance between 0 and CG is h_v, the moment of inertia I of vehicle mass becomes

$$I = mr_g^2 + mh_v^2 .$$

The torsional spring rate which produces pitching may be determined by assuming an imaginary suspension in which each wheel, or bogie, is

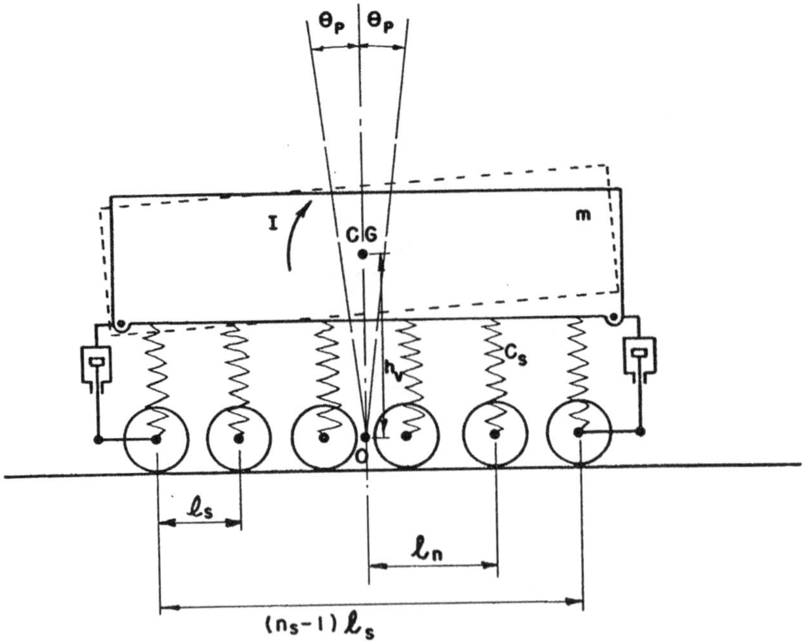

Fig. 155

replaced by springs having a c_s constant equivalent to the original spring system. The unit moment per radian M_c of the spring constants c_s will be obtained by taking the summation of the products $c_s l_n \times l_n$, where l_n is the distance between a given imaginary spring and the axis of pitch vibration (Figure 155):

$$M_c = \Sigma\, c_s l_n^2 = \frac{n_s(n_s + 1)\,(n_s - 1)}{12}\, c_s l_s^2\,,$$

where l_s is the distance between two adjacent wheels assumed to be constant throughout the whole vehicle.

The unit moment (per radian) $M_{CG} = mgh_v$ has to be subtracted from M_c in order to obtain the net torsional spring rate of vibration M:

$$M = M_c - M_{CG} = \frac{n_s(n_s + 1)\,(n_s - 1)}{12}\, c_s l_s^2 - mgh_v\,.$$

Now, the natural frequency of the pitch vibration may be determined from the equation

$$\omega'_n = \sqrt{\frac{M}{I}}.$$

Since, as was previously deduced, $\omega_g = 2\pi v/l_g$, $\lambda_1 = \omega_g/\omega'_n$ may be determined and substituted into equation (429). Thus, the pitch amplitude θ_p may be determined as a function of vehicle speed v, in terms of which ω_g and, in consequence, λ_1 were expressed.

The effect of vehicle length upon a smooth ride may be clearly seen when θ_p is computed by means of equation (429). Figure 156 shows the θ_p values calculated for three tracked vehicles having different lengths of 5.0, 7.56, and 11.1 ft and a damping $D_1 = 0.3$ (see Figure 154). The vehicles cruise over a ground wave approximately 18.5 ft long and 0.3 ft deep. If a critical pitch angle of approximately 4° is imposed as a limit, then vehicles *A* and *B* must travel with a speed not exceeding 5 mph, whereas vehicle *C* may practically develop any speed up to 20 mph if the low peak vibration close to the resonance speed of approximately 7 mph is quickly passed.

Figure 156 relates quantitatively to the speed and form of the vehicle and leads to conclusions that are similar to those reached when the morphological characteristics of motor vehicles were discussed in Chapter IV: an elongated vehicle form is always more preferable than a stubby shape.

The effect of stronger damping ($D_1 = 0.5$) is shown by the dotted lines and indicates the importance of this factor as far as vehicle performance is concerned.

To summarize, it may be concluded that the problem of springing a vehicle not only entails the question of ride comfort and other psychological or physiological factors, but also decides the average cruising speed of a vehicle over an uneven terrain. In the problem thus formulated, springing and damping alone are not the main core of the whole question. It may be seen that the paramount problem is the form of the vehicle. Suspension design comes next.

Obstacle Performance

The "waviness" of the ground obviously may degenerate into such irregular and deep-cut forms that it may stall the vehicle motion entirely. If this type of ground obstacle could be overcome, in most cases there

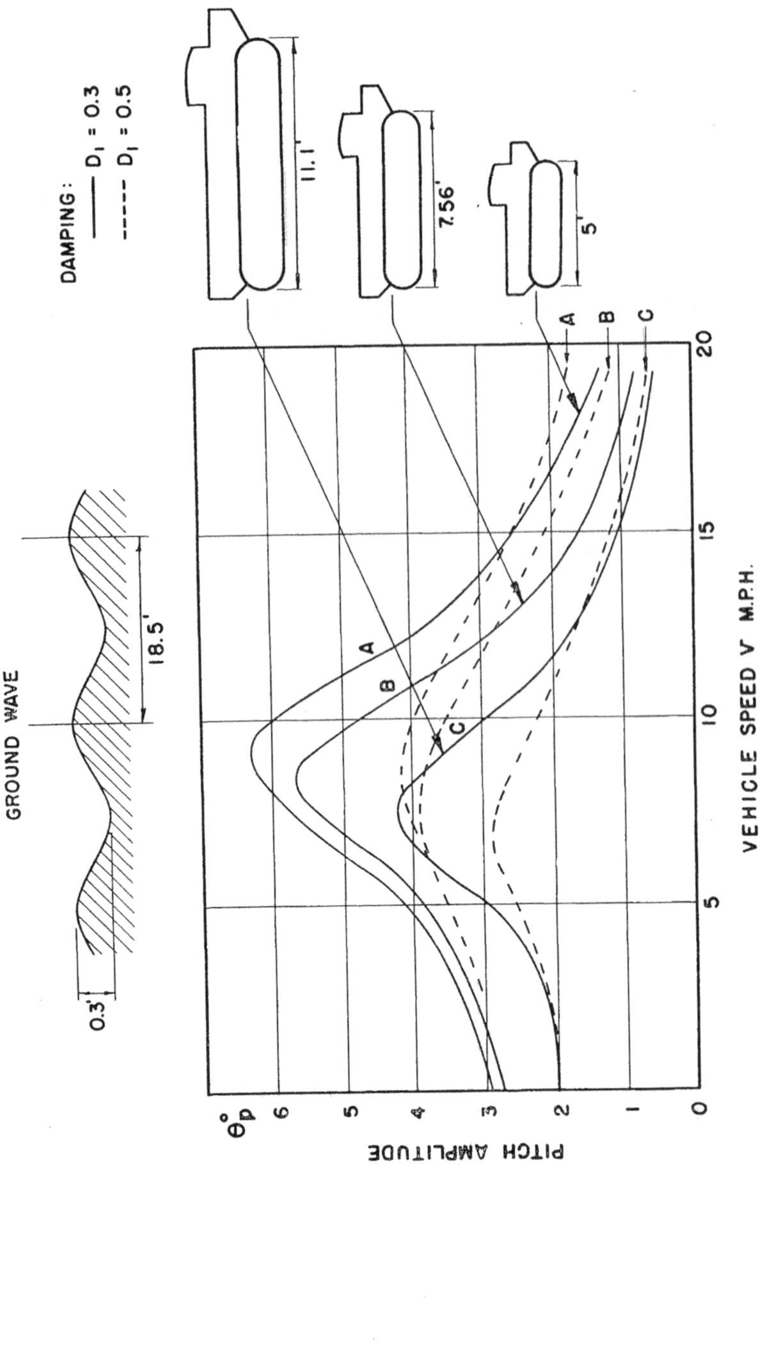
GROUND WAVE
0.3'
18.5'
DAMPING:
D1 = 0.3
D1 = 0.5
11.1'
7.56'
5'
A
B
C
PITCH AMPLITUDE
Θp°
0
1
2
3
4
5
6
0
5
10
15
20
VEHICLE SPEED V M.P.H.

Fig. 156

would be no question of speed of locomotion or vibrations—only the problem of "go or no go" would have to be considered.

Broadly speaking, there are two elementary types of obstacles related to the geometry of terrain surface: horizontal cavities and vertical walls. Usually, these two types of terrain hindrances are accepted in their ideal form as the measure of obstacle performance. This form, defined by plane surfaces intersecting at right angles, automatically postulates undeformable hard ground (Figure 157a).

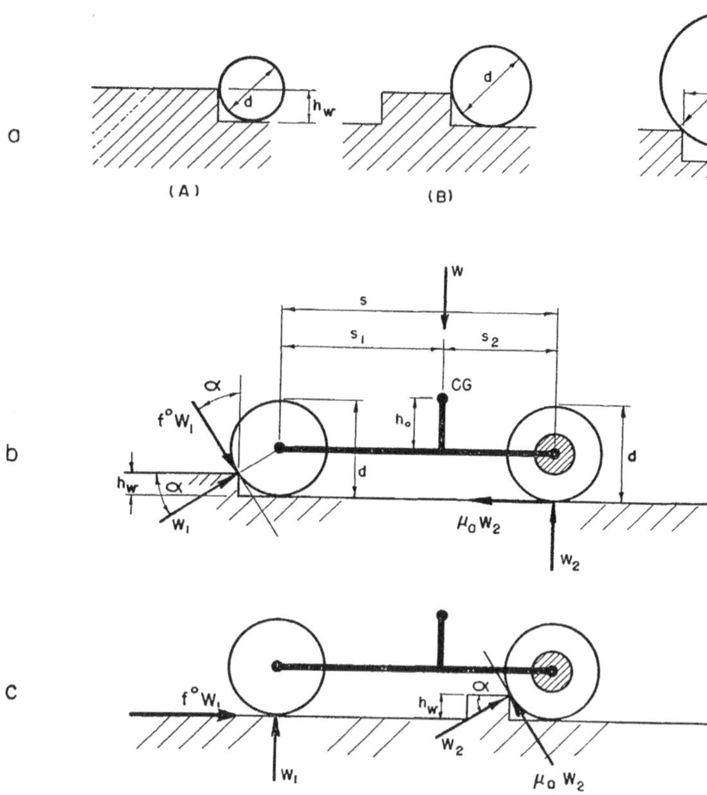

Fig. 157

It is known from daily experience that the obstacle performance of various types of vehicles is different as far as wall and ditch crossing is concerned. The fact that quantitative evaluation of such performance appears to be often neglected leads to imperfect designs. Kühner theoretically investigated the problem of obstacle crossing of two-axle vehicles with front-, rear-, and all-wheel drive as a function of the empirical "coefficient of adhesion" μ_a, which has been strictly defined in conjunction with equation (384). The investigated performance of the vehicle was plotted in dimensionless terms of obstacle height or width divided by wheel diameter.[255] Similar studies were reported by Kristi,[33] Chudakov,[256] and others.

It seems, however, that these studies do not close the case completely since there are other factors of vehicle geometry which affect the negotiable height h_w of the wall or ditch width l_d (Figure 157a) besides the wheel diameter d. In particular, the location of the center of gravity and the wheel base could be considered separately, instead of in the form of axle loads as was done by Kühner.

If a rigid suspension is assumed, and a free body of a rear-driven vehicle is considered, as shown in Figure 157b, then the equilibrium of forces and moments involved will yield the following equations:

$$\left.\begin{aligned} &W_1 \cos\alpha + f^o W_1 \sin\alpha - \mu_a W_2 = 0 \\ &W_1 \sin\alpha + W_2 - f^o W_1 \cos\alpha - W = 0 \\ &f^o W_1 \frac{d}{2} + W_2 s - W s_1 - \mu_a W_2 \frac{d}{2} = 0\,. \end{aligned}\right\} \tag{430}$$

The solution of equations (430) gives the following dimensionless relation:

$$\begin{aligned} &\left(\frac{\mu_a + f^o}{\mu_a}\frac{s_1}{s} - \frac{f^o}{\mu_a} + \frac{1}{2} f^o \frac{d}{s}\right)\sin\alpha - \\ &- \left(\frac{1}{\mu_a} - \frac{1 - f^o\mu_a}{\mu_a}\frac{s_1}{s} - \frac{1}{2}\frac{d}{s}\right)\cos\alpha = \frac{1}{2} f^o \frac{d}{s}\,, \end{aligned} \tag{431}$$

where $\sin\alpha = 1 - 2h_w/d$. Equation (431) gives the relationship between the vertical obstacle performance h_w/d and the geometrical parameters of vehicle structure: s_1/s and d/s. The explicit solution of this equation may be obtained if it is assumed that f^o is very small, which would be correct for hard ground. Then,

$$\left(\frac{h_w}{d}\right)_{\substack{\text{front}\\ \text{wheel}}} = \frac{1}{2}\left\{1 - \frac{1}{\sqrt{1 + \mu_a^2\left[\dfrac{s_1/s}{1 - (s_1/s) - (\mu_a d/2s)}\right]^2}}\right\}. \tag{432}$$

For $\mu_a = 2s_2/d$, $h/d = \frac{1}{2}$, which is identical to the case considered by Kühner [255] and which agrees with the experiment. The front wheels will surmount higher obstacles the smaller the s/d and the larger the s_1/s.

It may be shown that when the obstacle is crossed by the rear wheel, the critical case will have to be that of obstacle B (Figure 157a). Following the denotations of Figure 157c, the equations of equilibrium may be written in such a case as follows:

$$\left.\begin{aligned} & f^o W_1 + W_2 \cos\alpha - \mu_a W_2 \sin\alpha = 0 \\ & W_1 + W_2 \sin\alpha + \mu_a W_2 \cos\alpha - W = 0 \\ & \mu_a W_2 \frac{d}{2} + W_1 s - W s_2 - f^o W_1 \frac{d}{2} = 0\,. \end{aligned}\right\} \quad (433)$$

For $\sin\alpha = 1 - 2h_w/d$, the solution of equations (433) and the assumption of $f^o \cong 0$ gives

$$\left(\frac{h_w}{d}\right)_{\substack{\text{rear}\\\text{wheel}}} = \frac{1}{2}\left(1 - \frac{1}{\sqrt{1+\mu_a^2}}\right). \quad (434)$$

Equation (434) indicates that the performance of the rear wheel is independent of the structural geometry of the vehicle. By plotting equations (432) and (434), it may be shown that for all μ_a values, the rear wheel limits vehicle performance.

When an identical method is applied to the problem of ditch crossing, it will be found that the values of l_d/d differ from the above-determined values of h_w/d [equations (432) and (434)] by a conversion factor only, which results from the relation,

$$\frac{l_d}{d} = 2\sqrt{\frac{h_w}{d} - \left(\frac{h_w}{d}\right)^2}\,.$$

Thus, ditch performance may be presented on the same graph on which wall performance was plotted if the "ditch" scale is adjusted accordingly. The graphical way of conversion is shown in Figure 158.

In the case of both axles being driven, the plan of forces may be assumed as shown in Figure 159a. Accordingly, the equation analogous to equation (431) will be as follows:

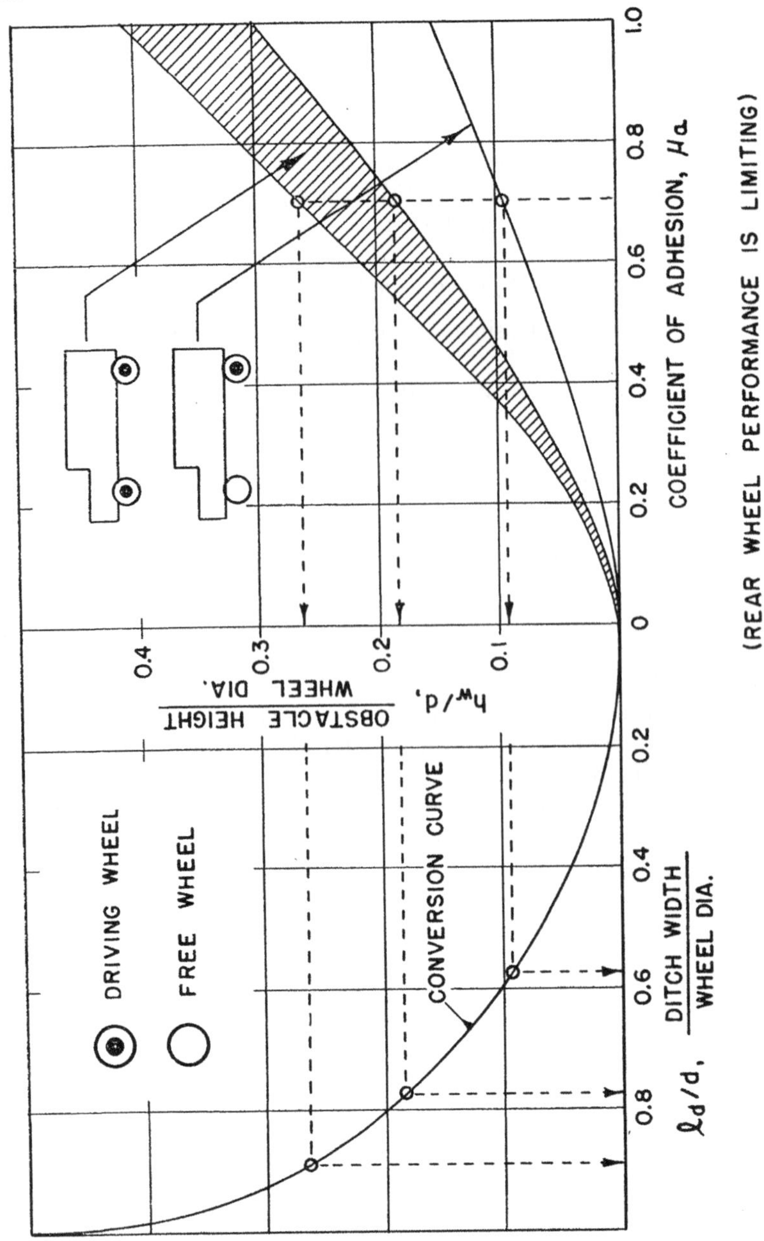
DRIVING WHEEL
FREE WHEEL
CONVERSION CURVE
OBSTACLE HEIGHT / WHEEL DIA., h_w/d,
0.1
0.2
0.3
0.4
0
0.2
0.4
0.6
0.8
1.0
COEFFICIENT OF ADHESION, μ_a
l_d/d, DITCH WIDTH / WHEEL DIA.
(REAR WHEEL PERFORMANCE IS LIMITING)

Fig. 158

$$\left(\frac{1}{\mu_a} - \frac{1 + \mu_a^2}{\mu_a}\frac{s_1}{s} - \frac{1}{2}\frac{d}{s}\right)\cos\alpha -$$

$$-\left(1 - \frac{1}{2}\mu_a\frac{d}{s}\right)\sin\alpha - \frac{1}{2}\mu_a\frac{d}{s} = 0\,, \qquad (435)$$

where, again, $\sin\alpha = 1 - 2h_w/d$. When this equation is plotted in terms of h_w/d, it will be seen that h_w/d decreases with an increase in s/d, and that the increase in s_1/s considerably improves the obstacle performance of the front wheel; this may even lead to the case when the vehicle will climb an obstacle higher than the radius of the wheel.

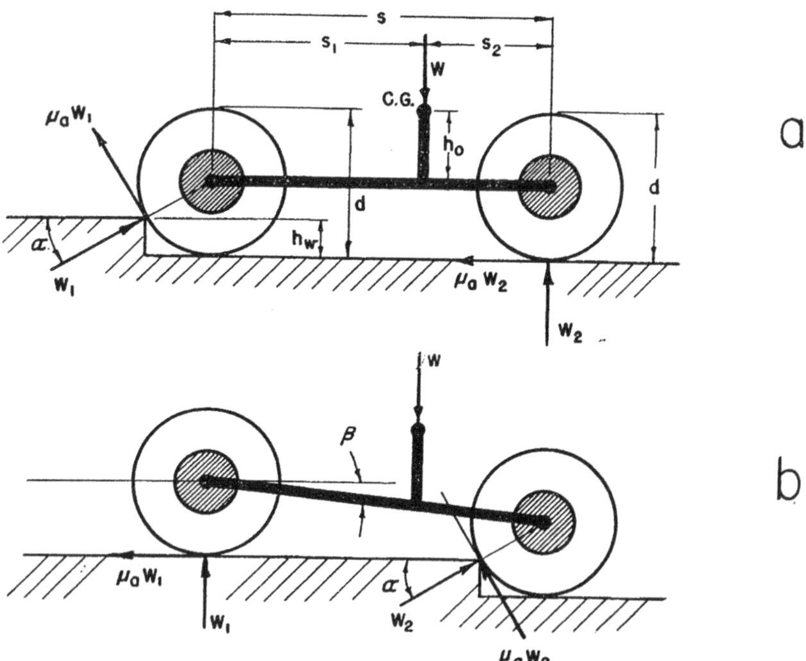

Fig. 159

The performance of the rear wheel will reach critical conditions when driving over the "plateau" type of obstacle (obstacle A, Figure 157a) and not over the "bump" type (B) as was in the case of a rear-axle drive.

By following the denotations of Figure 159b, the dimensionless equation which relates the geometry of the vehicle with the geometry of the considered ground surface may be deduced in the same way as equations (431) and (435), and yields

$$\left[(\cos\beta - \mu_a \sin\beta) + \frac{1}{2}\mu_a \frac{d}{s}\right] \sin\alpha - \left[\left(\frac{1+\mu_a{}^2}{\mu_a}\frac{s_1}{s} - \mu_a\right)\cos\beta - \right.$$

$$- \left(\frac{1+\mu_a{}^2}{\mu_a}\frac{h_o}{s} + 1\right)\sin\beta - \left.\frac{1}{2}\frac{d}{s}\right]\cos\alpha -$$

$$- \frac{1}{2}\mu_a\frac{d}{s}\left[(s - s_1)\cos\beta + h_o \sin\beta\right] = 0\,, \qquad (436)$$

where again, $\sin\alpha = 1 - 2h_w/d$.

The analysis of this equation indicates that the effect of s_1/s is opposite to that obtained when the front wheel of the four-wheel-driven vehicle negotiates the obstacle. A nose-heavy vehicle with a long wheel basis will perform better when crossing an obstacle with the rear wheels than when crossing it with the front wheels. Large s/d values improve the performance of the rear, regardless of the weight distribution. Since the front wheels and rear wheels respond in the opposite directions of vehicle performance with reference to changes in s/d and s_1/s, it may be concluded that there is an optimum of design condition where both wheels have the same performance and one does not limit the exploits of the other.

Such optima may be found when the intersections of the $(h_w/d) = f(\mu_a)$ lines are plotted for various s_1/s values. A graph of this type may serve the purpose of checking the consistency of various designs. A preliminary analysis of this problem indicates a certain amount of nonscatter of the existing values, which possibly could be improved by small changes in the s_1/s ratios.

Figure 158 shows the obstacle performance of available one- and two-axle-driven, four-wheel vehicles. It will be noted that the rear-wheel-driven vehicles are far below the all-wheel drive. The performance of the latter depends on the s_1/s ratio, as previously discussed, and embraces values which are enclosed in the hatched area, whose lower and upper limits are determined by the geometry of the investigated vehicles.

It may be concluded that at the coefficient of adhesion $\mu_a = 0.7$, the worst vehicle produces approximately $h_w/d = 0.18$ and $l_a/d = 0.77$, whereas the best one gives $h_w/d = 0.26$ and $l_a/d = 0.9$. A similar vehicle with a rear-axle drive only would climb walls and cross ditches determined by $h_w/d = 0.09$ and $l_a/d = 0.55$, respectively, i.e., its performance would be divided roughly in half, which is in agreement with the experiment.

It should be noted that the obstacle performance of wheeled vehicles is significant for high "adhesion" values of μ_a; however, the advantages of the all-wheel drive over the rear-wheel drive are the same throughout the whole range of road conditions.

The above considerations refer to negligible rolling resistance f^o. If this factor is included, considerable changes in conclusions may be expected. Their nature, however, and their practical value cannot be formulated until more research on wheels applied to soft ground is performed, in accordance with the problems discussed in Chapter VI. So far, very little is known about obstacle performance in soft soils. In particular, multiaxle vehicles, such as six or eight wheelers, would need special consideration. Kühner's work clearly illustrates this point.[255]

It should also be borne in mind that the study so far performed refers to rigidly suspended vehicles. A mathematical presentation of the effect of sprung wheels would undoubtedly be prohibitive, and the only way to analyze quantitatively the obstacle performance would be through research work on small-scale models, as will be discussed in Chapter XI.

The obstacle performance of tracked vehicles was analyzed by Kühner,[255] who suggested that if the center of gravity is located half way on the distance $s + 0.7\ (r_f + r_r)$, then the vehicle can cross a ditch $l_d = (4/9)\ [s + 0.7\ (r_f + r_r)]$. The denotations used in these formulas are shown in Figure 160a.

Half-tracked vehicles may have two types of suspensions: the track member may be either pivoted freely around one point (Figure 160b) or supported at two or more points (Figure 160c). In the first case, the ditch performance is limited to $0.7d$ or $0.8d$. In the second case, the ditch width may equal the smaller of two values, $0.35d + l_a$ and $l_b + 0.7r_r$, if the *CG* is located between the supporting points of the track unit.

The vertical obstacle crossing of tracked vehicles may be analyzed by establishing the equations of equilibrium of the forces involved. This method was adopted by Kühner and is described in Reference 255. As a matter of interest, there is a Russian method which gives a graphical solution of the maximum value of h_w.[33]

A tracked vehicle crossing a wall will level off when its center of gravity is right above the edge of the obstacle (Figure 161a). If, inversely, the vector of vehicle weight is rotated and perpendiculars to that

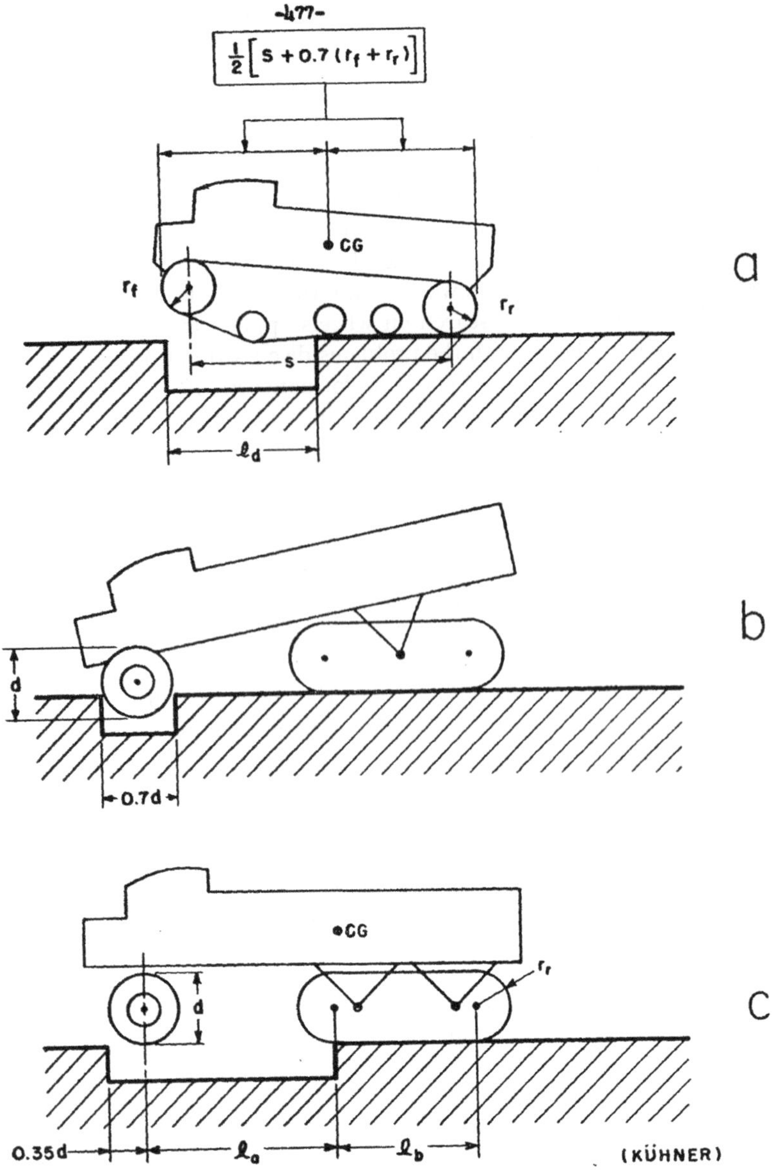
-477-
$\frac{1}{2}[s + 0.7(r_f + r_r)]$
CG
r_f
r_r
s
ℓ_d
a
d
0.7d
b
CG
r_r
d
0.35d
ℓ_a
ℓ_b
c
(KÜHNER)

Fig. 160

vector are tangent to the rear envelope of the track (Figure 161b), then the distances 1—1′, 2—2′, 3—3′, 4—4′ will be the respective obstacle height h_w which will cause the given vehicle tilt α.

From Figure 161b, the maximum $(h_w)_{max}$ may be determined and the maximum tilt α defined. Figure 161b directly gives the angle of departure

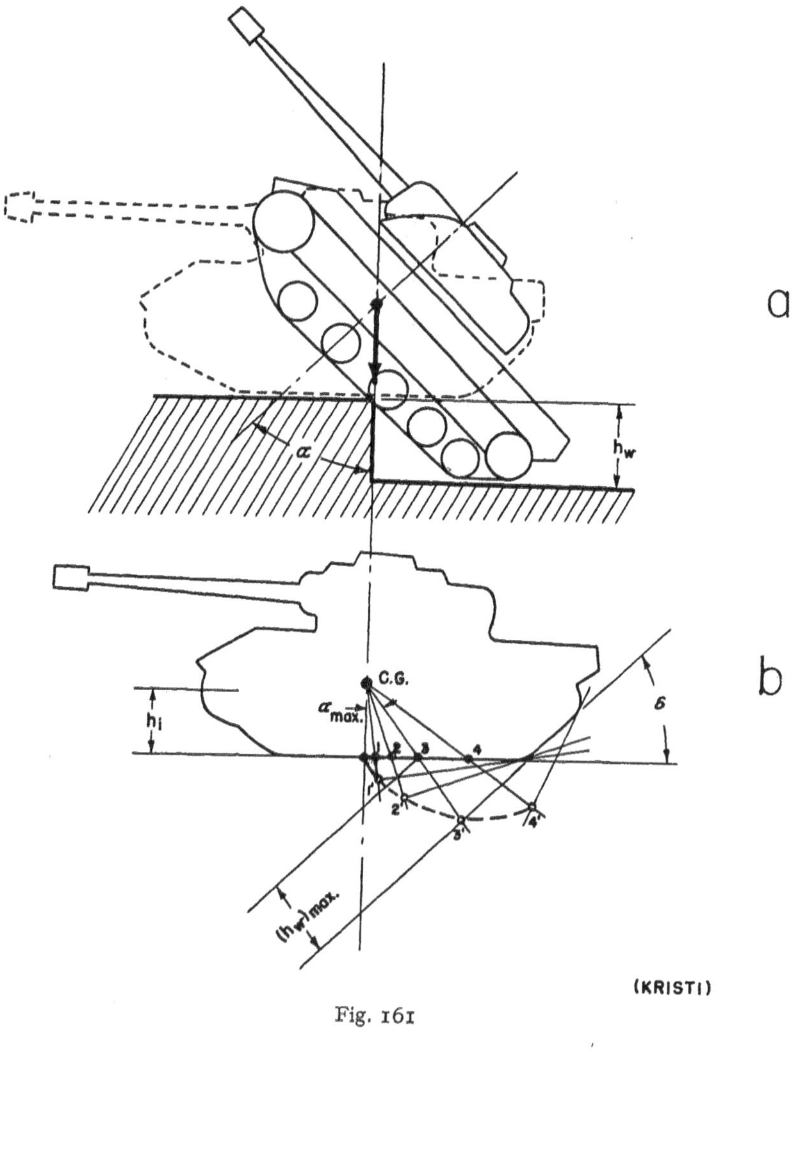

Fig. 161

δ of the track which should be selected for the given $(h_w)_{max}$, and also suggests the height h_i of the front wheel: it is generally accepted that $h_i \cong (h_w)_{max}$. Any obstacle higher than $(h_w)_{max}$ will overturn the vehicle.

Some Problems of Propeller-Driven Sleds

The mechanical weakness of snow structure, particularly in undisturbed conditions, led to the development of air-screw-driven sleds. The motion resistance of such a sled is overcome by the reaction forces of the air which, in most cases, supersede the inadequate, horizontal reaction of snow. Such propulsion is practical mainly because of the low motion resistance produced by the self-lubricating properties of ice.

The development of this type of transport mainly in Russia, Norway, and, during the last war, in Germany offers some points of interest which may be briefly analyzed as follows.[174, 257, 258]

There are basically three problems which should be looked into: the load distribution upon the skis, the stability of sleds, and the shock forces which occur when travelling over high-friction spots.

The problem of load distribution on front and rear skis basically does not represent anything new in comparison to the same problem related to wheeled and tracked vehicles, as considered at the beginning of this chapter. The nature of propulsion, however, creates a few distinct situations which may be described separately.

Consider a pusher-type, one-propeller sled, shown in Figure 162. The equations of equilibrium of moments in the longitudinal plane give the following loads acting upon front and rear skis:

$$W_1 = \frac{Wa_2 - H(a_1 \sin \alpha - h_p \cos \alpha)}{a_2 + a_3} \tag{437}$$

$$W_2 = \frac{Wa_3 + H\left[(a_1 + a_2 + a_3) \sin \alpha - h_p \cos \alpha\right]}{a_2 + a_3}, \tag{438}$$

and the distribution of the loads upon the left- and right-hand rear ski which are affected by propeller torque M and by centrifugal force when the sled travels on a turn having a radius r is

$$\left.\begin{aligned} G_1 &= \frac{W_2}{2} + \frac{W}{g}\frac{v^2 h_o}{ra_4} - \frac{M}{a_4} \\ G_2 &= \frac{W_2}{2} - \frac{W}{g}\frac{v^2 h_o}{ra_4} + \frac{M}{a_4}. \end{aligned}\right\} \tag{439}$$

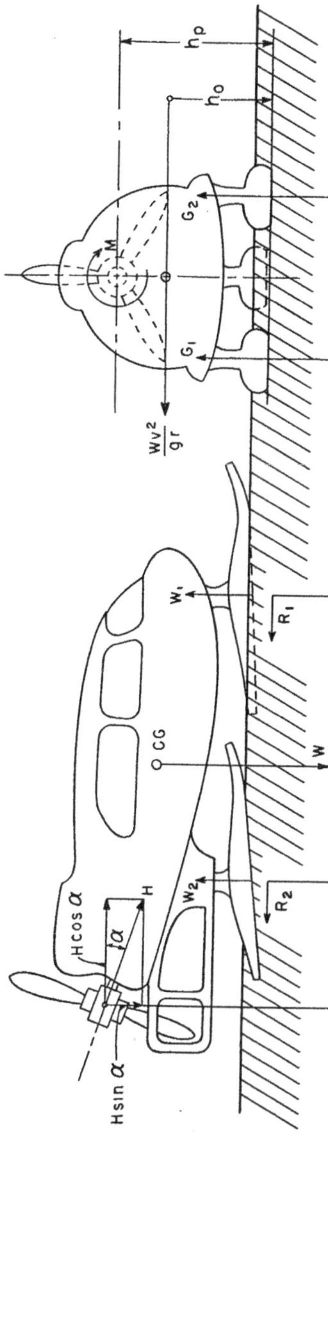

Fig. 162

Equations (439) indicate that the driving torque may increase the load on one side and reduce it on the other to a considerable extent. The consequences of such an occurrence in types of ground other than snow are perhaps not so serious. In snow, however, the sliding properties change very much with the ski load, and the stability of the whole vehicle may be disturbed.

Not enough is known about the effect of load distribution upon sliding conditions, as was discussed in Chapter VIII, and the ground reactions are rather unpredictable. Therefore, a theoretical analysis of the stability of propeller sleds, performed, for instance, by the same method that was applied to the study of the stability of wheeled vehicles,[189, 231] would be more uncertain. Experimental data regarding the cornering forces of a ski are scarcer than those related to a pneumatic tire.

Only very general indications in regard to the stable form of the propeller sled are available. These are based on Russian experience,[174] where it has been found that for a propeller sled to be stable, it should be designed in such a way that the base-to-tread ratio of the skis (Figure 162) equals approximately

$$\frac{a_4}{a_2 + a_3} \cong 0.58\,. \tag{440}$$

The uniformity of snow cover very often is interrupted by drift and terrain unevenness. Frequently, bare spots of hard ground may be encountered when the snow is blown off by winds. The suddenness and unpredictability of the occurrence of such spots cause sleds driving at speeds of the order of approximately 20 mph to be suddenly slowed down by high friction, in spite of the full propeller pull H, which involves shock forces practically nonexistent is any other type of locomotion.[174]

Assume that the length of the bare ground encountered in time t is s. The instantaneous vehicle speed which slows down from the original speed v_o is v.

$$s = \int_0^t v\,dt\,.$$

The value of speed v due to the deceleration a_d is

$$v = v_o - a_d t\,.$$

Therefore,

$$s = \int_0^t (v_o - a_d t)dt$$

$$s = v_o t - \frac{a_d t^2}{2}.$$

If it is assumed that, in critical cases, the sled will be brought to a stop, i.e., $v = 0$, then

$$v_o = a_d t$$

and

$$s = \frac{a_d t^2}{2} = \frac{v_o^2}{2a_d}.$$

Hence,

$$a_d = \frac{v_o^2}{2s}. \tag{441}$$

The deceleration a_d due to the shocks expressed in terms of gravity acceleration g is plotted in Figure 163 as a function of speed v_o and length s of the high-friction strip. This graph indicates that speeds above 20 mph lead to shocks having a deceleration of $2g$ which are caused by bare

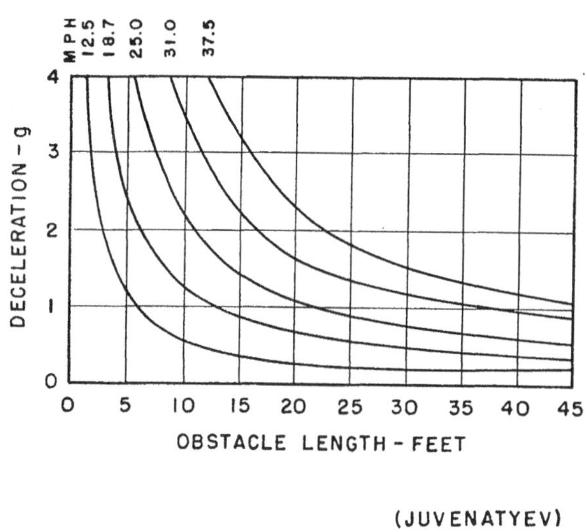

Fig. 163

strips 6 ft long. Such strips may make it impossible to drive propeller sleds above 20 mph. Higher speeds require an absolutely uniform snow or ice cover that is available, for example, when travelling over a frozen lake. Incidentally, frozen lakes offer the most economic routes of travel for propeller-driven sleds.

General Problems of Amphibian Vehicles

The successful development and use of amphibian vehicles, principally by the U.S. Armed Forces, were accomplished on a tremendous scale during World War II. Although this was the final proof under fire for this vehicle, several other successful crafts of this type were built between 1920 and 1939.[32, 259] However, the LVT's and the DUKW's, the latter being converted from the $2\frac{1}{2}$ T 6×6 U.S. Army workhorse, were undoubtedly the most successful. In a smaller way, the little tracked amphibious Weasel, built for light cargo and reconnaissance work, was also a success, and it demonstrated that a normal land track, even with the return portion of the track submerged, could usefully propel a tracked amphibian in water. This concept was widely applied to several flotation devices, which permitted regular medium tanks to land from off-shore vessels as temporary amphibians.

Although it is evident that many of the problems of amphibian design and construction are in the realm of naval architecture, and still others in that of vehicle design, there are several which are peculiarly amphibian.

The primary problem in amphibian design, one which is beyond the scope of the present discussion, is that of balancing the conflicting requirements for water and land operation to suit best the operational needs. Because of its dual function, an amphibian carries extra gear at all times, which makes its basic performance on land or water poor as compared to that of corresponding vehicles (or craft) designed for use in either situation alone. In effecting the necessary design compromises, the resulting amphibian may lie almost anywhere between a water craft which will barely crawl a few yards onto a good beach (and hence may, with proper design, be a relatively good boat) and a land vehicle which will scarcely float and propel itself in mill-pond conditions. Both of these extremes have been approached in actual designs, but generally as a result of the basic viewpoints of the design agencies, rather than as a result of a rational analysis of the total operational situation in which the vehicle is intended to be useful.

The more detailed technical problems related to naval architecture are concerned with the displacement, trim, stability, resistance, propulsion, and steering of the vehicle in water. Detailed methods for calculating displacement, trim, and static stability may be found in many good textbooks on Naval Architecture.[46, 260, 261] The simple geometric and static-mechanics principles governing ship calculations are directly applicable to amphibians, although the labor of calculation is frequently increased by the nonfair shapes and large appendages usually found in amphibians. Because of the desire to keep the land profile of an amphibian as low as possible, the freeboards of these vehicles when fully loaded are usually less than those of comparable boats. This fact appears to cause no hardship on the vehicle's seaworthiness if the vehicle is completely decked over. Tangential to these static naval-architecture problems are such mechanical problems as maintaining water tightness through the many hull openings necessary for land drive.

The motion resistance of amphibians in water operation can be accurately obtained in the design stage only through scale-model testing in a towing tank. The resistance will range from five to over ten times that of a normal boat designed to carry the same loads at about the same speeds. Before examining the reasons in detail, it will be necessary to outline the origins of motion resistance of a floating body.[46, 261, 262] There are generally considered to be four components of total resistance: surface wavemaking, skin friction, hull eddymaking, and appendage resistance. Surface wavemaking is a function of the Froude number at which the vehicle operates ($v/\sqrt{gl}$, where v = vehicle speed, g = gravity acceleration, and l = vehicle length at the waterline) and the form and loading of the vehicle. In general, at Froude numbers between about 0.25 and 0.35, the wavemaking resistance of vessels which do not have dynamic lift begins to increase rapidly with speed, increasing by as much as the fifth power of the speed, over a small range. Because of the limited size of amphibians (limited by consideration of land use), this point corresponds to speeds of from 5 to 8 mph, with the result that the amphibians must operate at speeds where wavemaking becomes a major factor, in order to be militarily acceptable. Skin friction, hull eddymaking, and appendage resistance, on the other hand, are all functions of the Reynolds number at which the vehicle operates ($Vl\bar{\gamma}/\eta$, where γ is the mass density of the water and η is its viscosity). Skin friction, in addition, is a function of the wetted area of the vehicle, and increases approximately as the 1.8 power of speed. Hull eddymaking is largely a function of hull form, and is reduced by careful design of a vessel's stem lines.

Appendage resistance, although a function of the vehicle's basic Reynolds number, is more particularly a function of the Reynolds numbers of individual axles, wheels, shafts, etc. (calculated on the basis of a characteristic length on the particular unit, such as the diameter of an axle, rather than on the basis of a selected characteristic of the vehicle as a whole). If the Reynolds numbers are high enough, and they normally are, appendage resistance will vary as the square of the speed.[178]

As already mentioned, amphibians, because of their small size, generally must operate at high Froude numbers in order to be useful. This fact, together with the generally full form of amphibians dictated by both size limitations and the necessity to support relatively large total weights buoyantly within a limited rectangular envelope, results in vessels with abnormally high wavemaking resistance. Coupled with this, and growing out of the same fullness of form, is a high hull-eddymaking resistance. Discontinuities in the hull for wheel cutouts and the like add greatly to the total resistance. Skin friction is high because of large surface areas and the low Reynolds number at which the vehicle operates in water; however, compared to the above components, it is only a minor contributor to total resistance. Advantage is taken of this, and of the small linear-scale ratios involved (6 : 1 to 10 : 1 as against more than 100 : 1 on some ship models), to simplify the expansion of model data to full-size figures. It has become customary to expand amphibian-model data simply by the linear-scale ratio cubed, without the skin-friction correction normally applied in ship prediction. This tends to overestimate the prototype resistance by a small per cent, but is found to be within the limits on accuracy imposed by errors of other sorts in designing or analyzing the results for a particular vehicle.

From model tests run to date, appendage resistance appears to be the largest component, unless special care is taken to reduce it. Thus, it has been found that a good boat form, with wheels and axles (or tracks) added as largely unhoused appendages, has a higher total resistance than a relatively boxy form which largely encompasses the wheels and axles within its envelope. It is evident that, in the latter case, the reduced appendage resistance more than makes up for the increased wavemaking, hull eddymaking, and, perhaps, skin friction due to the otherwise less favorable form.[259] This conclusion is in line with data on the air resistance of Grand Prix racing cars, which show that for the normal type of car with wheels fully exposed, the air drag of the body only is but 5% of the total air resistance of the complete car.[216]

Historically, propulsion of amphibians has been by means of screw propellers, the principles and design of which follow normal naval-architecture practice. Problems arising from their application derive from their vulnerability in land operation unless adequately protected, and in several ways they derive from the poor hull form with which they must work. Propeller protection first demands that propeller diameters be limited. This in itself imposes basic losses in efficiency when propelling a vehicle with high motion resistance. Additional protection may be obtained either by placing the propellers in a protected position in deep tunnels in the hull or close behind a boxier hull, or by making the propellers retractable. The first solution is mechanically simpler, but involves high propulsive losses due to unfavorable water flow to the propellers. The second solution sacrifices mechanical reliability for somewhat better propulsion. Regardless of propeller arrangements, no successful, useful load-carrying amphibians have yet been built in which the propulsive coefficient (resistance horsepower/gross engine horsepower) has exceeded about 40%. Figures for wartime amphibians are generally less than 30%, while in good boat practice they will run between 60% and 70%. Of course, relatively high mechanical losses in power transmission to the propeller, and auxiliary loads such as bilge pumps, contribute to the poor showing of amphibians.

Successful propulsion of tracked vehicles by running the tracks in water appears to have been achieved first by Roebling in the Alligator.[259] In this design, and in the LVT's which grew from it, the forward moving portion of the moving track, the return track, was brought clear of the water, and the track shoes had deep turbine-like blades attached. This vehicle developed the relatively good propulsive coefficient of about 25%. Subsequent track modifications to improve land-track life and remove side thrust from the propelling tracks lowered this value to approximately 20%, due to the fact that the altered track developed its thrust upon a smaller area of water.

During the war, track-propelled vehicles were developed on which the entire track was submerged while operating in the water. It was discovered that by limiting the water flow in the return track through the use of shrouding, ordinary land tracks could develop propulsive coefficients of about 10%, and special tracks could raise this value to about 15%. More recent work has increased this figure still more.

The advantages of track propulsion are a reduction in mechanical complexity and vehicle weight and increased reliability. Use of a normal track configuration, which implies a fully submerged track, further decreases weight and complexity. Track propulsion by means of fully submerged tracks made it readily possible to convert various standard tracked vehicles to temporary amphibians during World War II. The power available in tanks was generally so large that even a 10% propulsive coefficient could be borne for temporary service and short distances. These conversions approached the limiting Froude number speed, so that even an increase in propulsive coefficient to 20% would not have increased their speed by much more than 1 mph, although such an increase could have been used to advantage to decrease their fuel consumption while operating in the water.

Some experimental amphibians have been built with water-jet propulsion, the energy of the jet being supplied by an internally mounted pump. In proposing any alternate means of propulsion, the basic momentum principle of propellers, which adequately explains fundamental propeller action, cannot be grossly violated without loss in efficiency. This principle makes it clear that maximum efficiency is achieved by developing thrust upon a maximum area of water. A recent jet-propelled amphibian replaced two normal propellers having a total disc area of about 400 sq in. by jets with areas totaling about half this amount. The result, from fundamental principles, was foregone. The "experimental" vehicle did not perform as well as the original. Thus, it cannot yet be said that water-jet propulsion has been fairly tried in these applications.

Steering of amphibians presents the same sort of problems as propulsion. If rudders are used, they must be either very large or placed in the propeller (in a general sense) slipstream, and must be somehow protected during land operation. Because of their poor form, amphibians in general are not directionally stable underway. When this is combined with poor driver visibility, operation over even moderate stretches of open water becomes distinctly hazardous.

Performancewise, the principal problem is one of mobility in unfavorable soil conditions. The amphibian, in order to function in any but the most contrived circumstances, cannot avoid operation in soft sand, marshy, and otherwise difficult terrain, found predominantly on river banks, lakes, and seacoasts. Few roads run right down into the water. It is safe to say that no amphibian yet built has enough mobility to cope with more than a small fraction of the world's coastal lands. Present amphibians are almost entirely limited by their lack of mobility for landings on beaches. The most needed improvements in amphibians lie in this area.

It is evident from the foregoing that many problems properly in the sphere of vehicle design or naval architecture are so extreme as to be almost peculiar to amphibians. There are still others which must be definitely classified as such. The principal one is that of making a landing or departure from shore, particularly through surf. Experience has shown that amphibians, particularly the wheeled vehicles, are more able in this operation than landing boats of any sort. During wartime tests, skilled operators repeatedly took DUKW's through surf as great as 30 ft from trough to crest. In actual operations, these vehicles proved to be the only ones capable of supplying troops ashore in several amphibious landings when the weather became unfavorable and landing beaches were subjected to high surf.

Some of this ability was accidental, but accident or no accident, it deserves comment. Emphasis on low land profile has resulted in relatively low freeboards, even for the heaviest vehicles, with the consequence that the reserve buoyancy of most vehicles is only a fraction of their weight. Thus, when high surf overtakes or meets the vehicle in the surf zone, the vehicle responds vertically in a very sluggish fashion, so that during the time of the encounter, its vertical movement is quite small. The same low freeboard provides little area for surf impact. The net result is a sort of inertia damping, i.e., the period of the disturbance is so much less than the natural period of movement of the vehicle that its response is very low. Moreover, the medium and larger vehicles usually have their "feet on the ground" for a large part of the surf zone, and rapidly move on out of it (particularly in a landing), where a landing boat must wait and be buffeted for some time while it is being loaded or unloaded.

Perhaps the greatest danger in surf landings is broaching in the surf zone, which often leads to capsizing. The same inertia-damped response due to low reserve buoyancy helps in this case. If broaching occurs at the moment of making a landing, there is an added tendency to overturn the vehicle by a "tripping" action. As the vehicle starts to broach, the effect is that of a rear-wheel skid on ice, and the rudder-correcting attitude requires the same reaction from the driver, i.e., the "wheels are turned into the skid." When rudder and steerable front wheels are coupled, this automatically places the wheels in position to roll up on the beach should contact be made, and the danger of "tripping" is materially reduced. The problems of amphibian design for this type of operation, which have so far been rather well solved on some vehicles, are basically those of not losing ground in compromises with other requirements through a lack of understanding of the reasons, however accidental, for the successes of earlier vehicles. This is not to say that research on the relationships of detailed hull refinements and surf action is not or should not be carried out. Such studies, at the very least, serve to give information necessary in planning amphibian operations.

There is an additional problem in making landings and departures on steep banks. The partial support of the vehicle on the bank will tend to submerge the outboard end and, in extreme cases, may swamp the vehicle if cooling air or cargo openings are submerged by this action. The problem is similar to a ship launching, but may require repeated solution in the course of a design or design analysis.

The problem is essentially one of simple statics. Referring to Figure 164, it is evident that for static equilibrium, at each instant as the vehicle

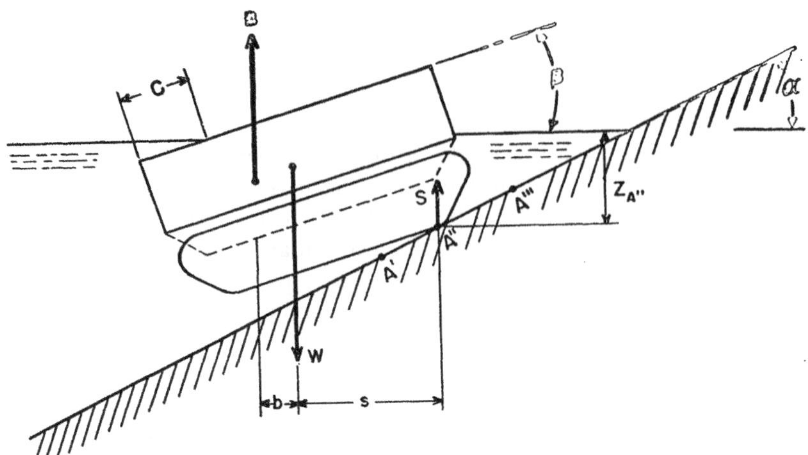

Fig. 164

progresses up the bank from A' to A'' etc., two conditions must be satisfied:

$$\left.\begin{aligned} Bb &= Ss \\ \text{and} \quad B + S &= W\,. \end{aligned}\right\} \quad (442)$$

The difficulty arises from the fact the buoyancy B and its horizontal separation from the center of gravity b are both functions of the depth of the point of contact $Z_{A''}$ and the trim angle β, while the horizontal separation of the point of contact A'' and the center of gravity s is also a function of β. The first two of these functions may generally be presented only graphically and, hence, the entire solution may be best obtained graphically.

For each contact depth ($Z_{A''}$), a likely point of contact on the vehicle is selected, and B and b are calculated for two or three values of β covering the likely range of the equilibrium β ($\leqslant \alpha$). From a knowledge of W, the gross vehicle weight S is calculated from the second of equations (442) corresponding to each value of B (and hence of β). Bb and Ss (readily calculated for each value of β) are each plotted as functions of β, and their intersection gives the value of the equilibrium β. A scaled vehicle profile may then be rotated about the assumed point of contact (A'') until this angle is reached, and any measurement of interest, such as c, taken. This procedure may be repeated for a number of contact points and a number of slopes α, and continuous curves of the desired information may be plotted as the vehicle moves up (or down) various slopes during a landing or departure.

Vehicle Forms and Types as Related to the Mechanics and Geometry of the Terrain

Soil properties, as discussed before, may lead to the establishment of certain general criteria in an assessment of the rationality of vehicle form and type. Consider two extreme cases of purely frictional ($c = 0$) and purely cohesive ($\phi = 0$) soils. The tractive effort H in the first case is expressed by equation (272) and depends solely on vehicle weight W. In the second case, H is defined by equation (273) and depends exclusively on the size of the ground contact area.

In Figure 165, the soil types are plotted on the horizontal axis; these are defined by the preponderance of c and ϕ, starting with a purely frictional soil such as dry sand ($c = 0$) and ending with a purely cohesive soil such as supersaturated clay ($\phi = 0$). Between these two extremes are represented all intermediate types of soils comprising both friction and cohesion. On the vertical axis, the tractive effort or drawbar pull is plotted. Since, according to equation (272), only weight counts as far as traction is concerned in frictional soils, the vehicle designed for operation in such a soil should be heavy enough to develop the required pull. The size of its contact area is irrelevant and should be reduced to the minimum

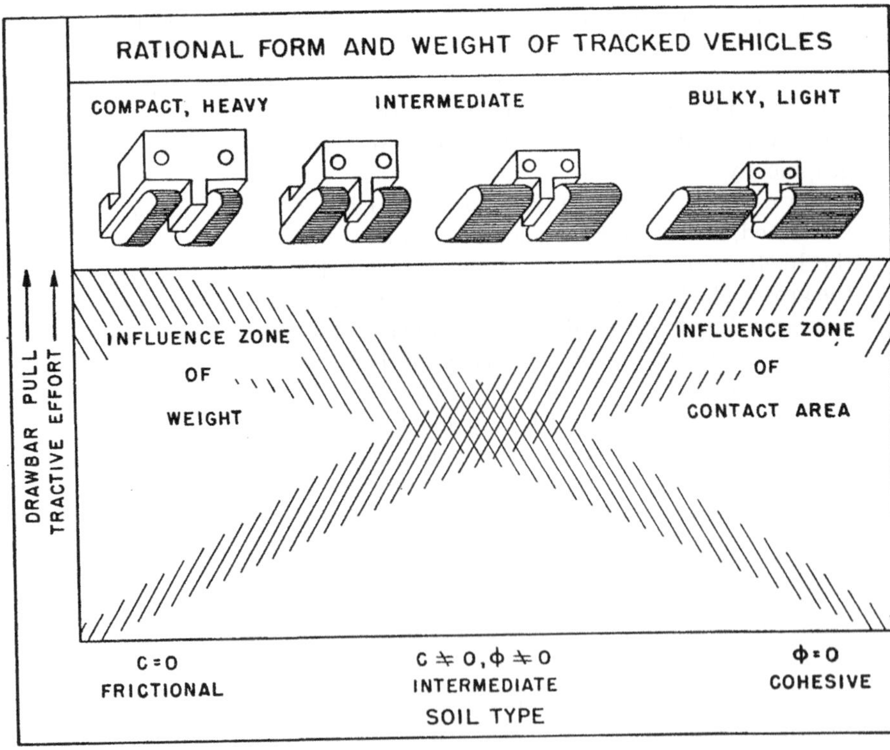

Fig. 165

defined by equation (263) in order to provide only the adequate "flotation." Such a requirement would lead to a rather small-size track. The rational type of a vehicle is thus characterized by the maximum of weight and the minimum of track size.

In the other extreme, i.e., in a plastic clay ($\phi = 0$), weight is irrelevant and only track dimensions count, as may be seen from equation (273). Thus, the vehicle type desired is determined by the minimum of weight and the maximum of track size. Between these two limits, the combination of size and weight, as expressed by equation (271), should affect the form and type of the vehicle.

Figure 165 indicates qualitatively to what extent the form and weight of tractors designed for operations in desert sands ought to differ from tractors destined to operate, for instance, in rice paddies, should such a differentiation be required for any reason. A fairly accurate quantitative evaluation of this subject may be made by applying previously discussed equations (271), (272), and (273).

It was shown when analyzing the problem of slippage that the definition of soil type in terms of friction ϕ and cohesion c alone is not sufficient to determine the amount of traction which may be developed by a given vehicle between zero and maximum pull potentially available in soil. Two other soil constants K_1 and K_2 were introduced, accordingly, when determining the relationship between the tractive effort and the slip of a track, or generally speaking, of any ground contact area [equation (288)]. The constant K_1 determines the state of looseness of soil, whereas K_2 indicates the extent to which a given soil preserves its original strength when disturbed (remolded) by the shearing action of a moving vehicle.

Figure 112, which was constructed on the basis of equation (288), shows that the tractive effort H per unit of width of the contact area may be obtained from the integral $H = 4l \int_0^s \tau_x \, dx$, where τ_x is the shearing strength of soil and s is the length of the contact area. In other words, H is represented for a given slip by the area enclosed between the horizontal axis and the τ_x line related to the considered slip. The area enclosed between a tangent to the peak of the given τ_x curve and the curve itself represents the force lost due to slip.

The integral $H = 4l \int_0^s \tau_x \, dx$ was plotted for silt soil in Figure 113 as a function of $i_o s$. Its ordinates show the unit tractive effort as a function of maximum soil distortion $j_m = i_o s$. The difference between the peak of the curve, $(H/4ls)_{\max}$, and a given ordinate also represents the loss due to slip i_o at a given $i_o s$. If various lengths s of the ground contact area are assumed, then $H/4ls$ may be plotted for various slips i_o. A graph constructed in such a way for the considered silt soil is shown in Figure 166a. On the basis of this graph, a curve showing the relation between slip i_o and the length of the ground contact area s may be plotted for the maximum values of $H/4ls$, i.e., for the maximum unit traction.

Such a plot was made in Figure 166b, which, in turn, indicates how long the ground contact areas of the driving gear, wheel, or track should be in order to produce the maximum available tractive effort $H/4ls$ at a given slip (i_o) and loads (σ) in a given soil (c, ϕ, K_1, K_2) [see curve (A)].

In the case of the silt soil previously described, and for a unit load of

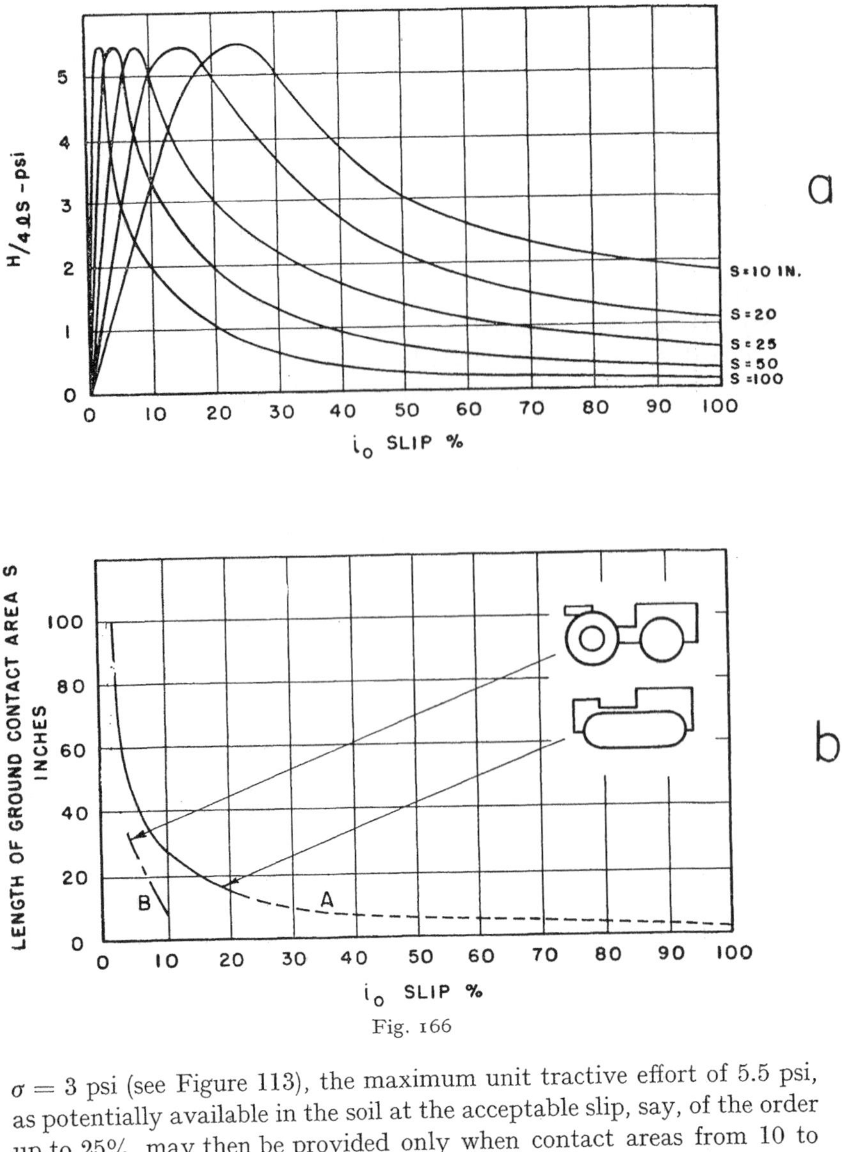

Fig. 166

$\sigma = 3$ psi (see Figure 113), the maximum unit tractive effort of 5.5 psi, as potentially available in the soil at the acceptable slip, say, of the order up to 25%, may then be provided only when contact areas from 10 to 100 in. long are used: in other words, when tracks are applied. The same tractive effort in the case of a wheel (contact area shorter than 10 in.)

would be produced at the cost of slip between 25% and 100%. Assuming that such slips are unacceptable in view of the previously discussed losses, a wheeled tractor may be rejected in the considered case as being uneconomical.

If, however, not the maximum traction of 5.5 psi available in soil strength is required but, say, 3 psi is considered to be adequate, then the slip–contact-area curve will be represented by the dotted line (B) in Figure 166b. It may be seen from the curve B that such tractive effort can be developed in the silty soils with a slip smaller than 10% by a contact area even shorter than 10 in., i.e., a large-diameter wheel may be safely used. In the final analysis, it may then be concluded that the performance curve (A) necessitates the application of a tracked tractor, whereas performance (B) allows the use of a wheeled tractor, which is shown diagrammatically in Figure 166b. Thus, the rational type of a vehicle may be correlated with the given type of soil, and the economy of operation achieved.

The choice between wheeled and tracked vehicles does not depend on only the mechanical strength of soil, as defined by c and ϕ, and on soil structure, as determined by K_1 and K_2. Figures 41 and 42 suggest indirectly that the geometry of the ground surface and the form of a vehicle, whether tracked or wheeled, decide "go and no-go" performance. Details of this performance were discussed in conjunction with Figures 158, 160, and 161. A study of the relationship between vehicle vibrations and the velocity of locomotion made in this chapter also shows how the ground-surface geometry affects the speed of the vehicle, depending on the length of the vehicle (Figure 156).

A tracked vehicle is basically a wheeled one and, as was mentioned before, it bridges the ground unevenness with the rail it carries. It is logical to expect, therefore, that the more irregular the ground surface, the more tracks needed. Figure 156 also indicates that higher average cross-country speeds require longer vehicles and more wheels in order to reduce the pitch amplitude of the exciter and thus reduce vehicle vibrations. The question as to whether or not these multiple wheels are to be wrapped with tracks is to be decided on the basis of the tractive effort H and slip i_o, as previously discussed, in addition to the assumption of the type of obstacles to be negotiated.

Ditch and wall crossing put tracked vehicles above the wheeled ones. The methods of determining speeds at critical vibrations for various types of vehicles and the methods of estimating obstacle performance, as

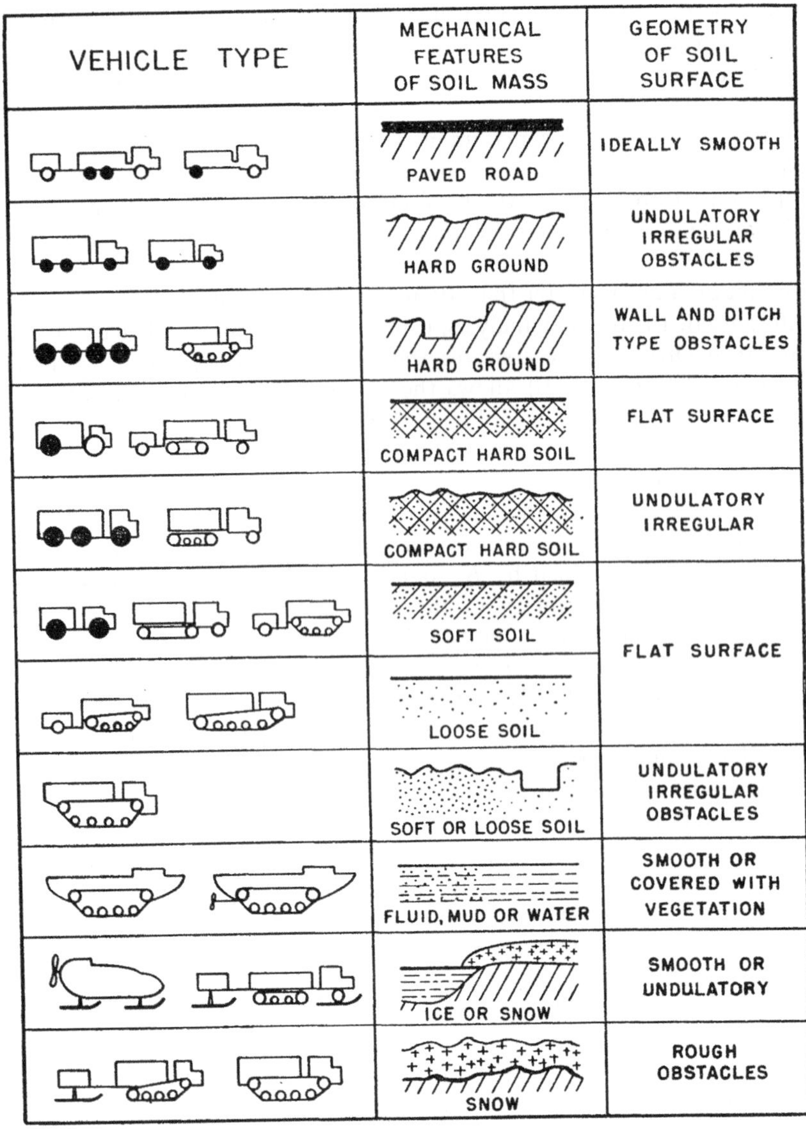
VEHICLE TYPE
MECHANICAL FEATURES OF SOIL MASS
GEOMETRY OF SOIL SURFACE
PAVED ROAD
IDEALLY SMOOTH
HARD GROUND
UNDULATORY IRREGULAR OBSTACLES
HARD GROUND
WALL AND DITCH TYPE OBSTACLES
COMPACT HARD SOIL
FLAT SURFACE
COMPACT HARD SOIL
UNDULATORY IRREGULAR
SOFT SOIL
LOOSE SOIL
FLAT SURFACE
SOFT OR LOOSE SOIL
UNDULATORY IRREGULAR OBSTACLES
FLUID, MUD OR WATER
SMOOTH OR COVERED WITH VEGETATION
ICE OR SNOW
SMOOTH OR UNDULATORY
SNOW
ROUGH OBSTACLES

Fig. 167

discussed in this chapter, offer clues for the assessment of the types of vehicles which would be most rational under the assumed conditions. The general conclusions on this subject, based on the material reviewed, may be summarized in a pictorial way, as shown in Figure 167. The black circles in Figure 167 mean driving wheels.

On this figure, the problem of a cargo carrier and a tractor-trailer unit is qualitatively formulated, assuming that the presently used, one- or two-axle trailers limit the speed performance of tractors very considerably because of the vibrations and bouncing of the trailer, which, on an uneven ground surface, impose low speed limits. Consequently, trailers cannot be recommended for use in any undulating or obstacle-infested terrain, and a cargo carrier would be in a position of developing comparatively higher speeds in such terrains.

The analysis of the wheel-vs-track problem performed in Chapter VII indicates that tracked trailers appear to be unjustified. Accordingly, in Figure 167, only wheeled trailers were considered and the one-axle units were sketched for the sake of simplicity. The number of wheels and axles needed in a trailer may be evaluated, however, on the basis of the required flotation by means of equations (211) or (212), if soil conditions are specified in terms of c and ϕ.

X. THE TRAFFICABILITY OF SOILS, PERFORMANCE OF VEHICLES, AND THE ECONOMY OF LOCOMOTION

In the previous chapters, elements of the relationship between soil and a vehicle were considered. These elements may be helpful now in the elucidation of fundamental problems of cross-country locomotion, in which the ability of soil to sustain the motion of motor vehicles and the performance of the vehicle jointly contribute to the economy of transport.

The Meaning of the Idea of Soil Trafficability

The literal meaning of the words "soil trafficability" signifies the ability of the ground to sustain the traffic of vehicles; however, the physical nature of such an ability remains to be defined if a trafficable soil is to be distinguished from nontrafficable ground.

The lack of this definition has been felt for a long time and repeated efforts have been made to determine trafficability in terms of various *ad hoc* conceived criteria. The most popular definition of the trafficability of soil has been in terms of the concept of its hardness. Since, however, a rigorous physical definition of this concept is lacking even in metallurgy,[70] the field in which this concept was established many years ago and in which it is still being used widely, the hardness of soil has never become a successful element in defining trafficability.

The main difficulty encountered in this respect is how to define the hardness and how to measure it. During the early attempts at solving the problem, when civil engineers were helping automotive engineers in the matters of soil mechanics, a method which had found only limited application in civil-engineering work with soils was suggested as a ready solution of trafficability problems. Since that time, the so-called Proctor needle [263] has been considered as a device which determines the trafficability of soil in terms of a single and simple index.

The Proctor needle is a rod with an enlarged tip which is forced into the soil at a rate of about ½ in./sec. The force needed to drive the rod into the ground is assumed to be correlative with the strength of the ground responsible for sustaining vehicle motion. The readings of the penetrative force are calibrated in pounds per square inch, depending on the tip area.

The study of soil mechanics made in Chapter V reveals that if there is any correlation between the indications of the needle and soil trafficability, it is so remote that no physical analysis of this question is possible at this stage. Furthermore, the use of psi units proved misleading because the behavior of the needle and of the vehicle depends on the form and size of the loaded areas and not on assumed pressures alone. Despite extensive modification of Proctor's needle by the introduction of circular plates fastened to the rod,[113] or the use of cones,[264, 265] the definition of soil trafficability in terms of the pressure needed to force a cone into the soil, evaluated in nonspecified physical units of the "cone index," did not enlarge the understanding of the mechanics of soil, which is instrumental in vehicle motion. Both the cone shape and the scale of the spring have been adopted more or less arbitrarily and the "cone index" has been correlated only empirically with the "go and no-go" soil conditions in respect to particular existing vehicles in a limited number of ground types.

Although such a method may be expedient, it does not exhaust all possible traffic conditions, it is not accurate, and does not permit the evaluation of the performance of nonexisting, unorthodox vehicles which are, for instance, in the stage of planning. Consequently, the designer is unable to predict on the basis of the "cone index" how a vehicle *in statu nascendi* will behave in a soil whose trafficability has been determined by means of a rather arbitrary empiricism.

It should be noted that the first known use of the "needle" in connection with vehicle problems was made by Bernstein in 1913.[124] This investigation, however, was more concerned with the determination of the uniformity and stability of given soil conditions than with the correlation of the needle "index" and wheel performance, as attempted by McKibben and Hull.[266]

The application of the penetrometer to the study of uniformity appears to be very useful and was made recently with great success. Similar application of the impact penetrometer built by the Swiss school also proved successful in the determination of snow stratification.[267]

The variability of soil properties and the inadequacy of the penetrometer to express this variability by means of a homologous scale led to the study of the shearing strength of soil and of the effect upon this strength of various factors such as moisture content, grain-size analysis, etc.[268] These factors, however, also did not provide the sought-for definition of trafficability because the theoretical knowledge of the relationship between shearing strength of soil and vehicle locomotion was not considered.

If it is assumed that this relationship, described in detail in Chapters VI, VII, VIII, and IX, is accepted, which may be agreed upon on the basis of the promising correlations so far made between theory and practice, then the new definition of soil trafficability may be proposed in terms of values which have a more real physical meaning: cohesion c, friction ϕ, specific gravity γ, and structural soil "constants" K_1, K_2, k, and n. The geometrical picture of soil surface should be specified in previously discussed terms, if the ground is not flat.

Before going into a more detailed discussion of the proposed scheme, it will be realized that soil trafficability cannot be defined in a single value of pounds per square inch, "cone index," moisture content, or even in a certain concept of a more complex value formed by an aggregate of component values. Just as the weather forecast, with which the navigability of the air is judged, cannot be expressed in terms of temperature or humidity alone, the trafficability of soils cannot be defined by a single unit.

Following equations (262) and (263), it will be seen that the ability of soil to support loaded surfaces increases with an increase in density γ, friction ϕ (N_γ, N_c, N_q), and cohesion c. Equations (271) and (222) show a similar relation between ϕ, c, and traction. Thus, the soil has a better chance to keep a vehicle "afloat" and let it overcome the encountered motion resistances, the higher the γ, ϕ, and c values, or, in other words, the higher the shearing strength τ as defined by equation (123). In plastic soils, yielding stress σ_{pl} in pure compression divided by $\sqrt{3}$ may be such a criterion if the condition of plasticity may be expressed by equation (111). However, the shearing stress alone does not define unequivocally the trafficable conditions, because the latter also depend on the form of the loading area as expressed with reference to flotation by equations (262), (263), and (265).

The form effect, as previously discussed, must be considered if shear strength τ is investigated, because the same size but different forms of contact areas when acting upon soils having the same strength τ, but different ϕ and c values, will give different results. Thus, only c and ϕ and not the shearing stress τ may be considered to be the real invariants which define the trafficability of soil in the broadest sense.

If the soil is not loaded to such an extent that its maximum shear strength $\tau = c + \sigma \tan \phi$ is exploited [see equation (123)], or if vehicle motion produces slip which disturbs the original soil structure as demonstrated by equation (284), the tractive effort available usually will be smaller than that determined by means of c and ϕ alone [equation (271)]. In order to assess the tractive effort, not only friction ϕ and cohesion c have to be considered, but also the two new soil coefficients K_1 and K_2 proposed in equation (288).

Thus, when all the values of γ, c, ϕ, K_1, and K_2 are known, then the ability of soil to support the vehicle weight W and to develop the tractive effort H may be considered as being theoretically defined. Motion resistance may be determined if values k and n are known.

The combined effect of γ, c, ϕ, k, n, K_1, and K_2 may be large enough to sustain locomotion on a flat area. If, however, the configuration of the ground surface is such that larger forces are needed to support the locomotion, the vehicle will stall. Accordingly, as far as trafficability is concerned, the geometry of the soil surface other than that of a level plane is to be listed as a negative factor whose effect should be subtracted from the positive action of friction, cohesion, and the K values. Even the ground "waviness" as defined by the x_1 values of equation (364) negatively affects soil trafficability since it reduces the speed of the traffic if critical vibrations are considered.

The indices of terrain roughness which reflect terrain trafficability are either encompassed in equation (364) of the ground wave or defined separately or jointly by the wall height h_w, ditch width l_d, and ground wave dimensions l_w and h.

Hence, soil trafficability (ST) may be considered as being defined if all the independent soil variables are given:

$$(\mathrm{ST}) = f(\gamma, c, \phi, k, n, K_1, K_2, h_w, l_d, l_w, h) . \tag{443}$$

It will be seen that in the case of homogeneous soil masses, eleven quantities are required in order to assess the degree of the ability of soil to support traffic. Since the quantities in question are not constant throughout a given area or time span, but are in most cases susceptible to large variation, only a statistical value of (ST) may be determined, in a general case; a weather forecast also is of a more or less statistical character. Apparently, there is not much difference in predicting the trafficability of soil and the state of atmospheric conditions.

Measurement of the Indices of Soil Trafficability

It results from the preceding section that soil trafficability may be considered as being defined if the following values are known: γ, c, ϕ, k, n, K_1, K_2, h_w, l_d, l_w, h. In such a case, by using the equations discussed in this work, flotation, tractive effort, speed limit due to pitching, and obstacle performance of any imaginable vehicle may be evaluated. The eleven soil "constants" quoted above are thus the indices of soil trafficability, and the definition of the latter is reduced to the measurement of the values of these indices.

Measurements of c and ϕ in the laboratory may be performed in accordance with standard methods developed in civil engineering.[86] Such tests, however, apply only in exceptional cases to field experience, because of the nonhomogeneity of actual soils and the scale effect. The methods developed by the writer are based on a shear test performed with a loading area large enough to produce conditions as close as possible to those created by a vehicle. The scale of such a test is rather flexible and depends on prevailing loads and soil characteristics. An apparatus of the discussed type to be used in the field is described in Reference 5 and is shown schematically in Figure 168a: a grouser plate 1 loaded with load W is moved by means of cable 2 by shearing force H. The record which may thus be obtained makes it possible to calculate the c and ϕ values in accordance with the denotations shown in Figure 65.

A portable apparatus of similar type, with the horizontal H and vertical W loadings acting simultaneously at a pre-set, constant H/W ratio, also was proposed for preliminary exploration of soil trafficability, and is described in Reference 269. The Naval Civil Engineering Research and Evaluation Laboratories at Port Hueneme, California, have adopted this principle for further exploration of the problem.[270] Procedures related to soil shear tests as performed by means of the basic grouser method were subsequently elaborated in a standardized form to the effect that comparative c and ϕ values could be obtained in varying conditions.[271] Experimental measurements of ϕ and c performed in this way permitted the evaluation of the tractive efforts of vehicles in given soils with an accuracy of a few per cent (Figure 168b).

In order to find the tractive effort at any rate of vehicle slip when the maximum of soil strength is not developed, determination of the K_1 and K_2 coefficients is required. This may be done by plotting experimentally the shear stress-strain curve for a given soil, as shown in Figure

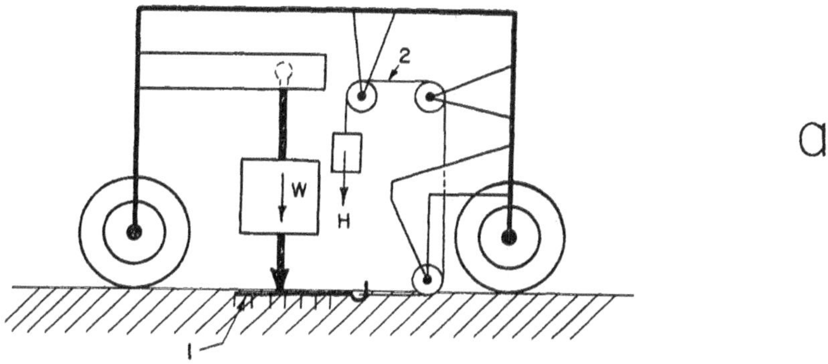

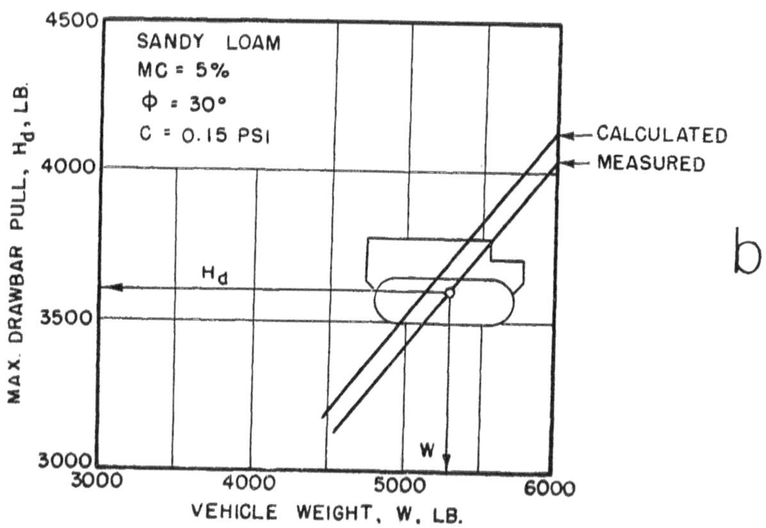

Fig. 168

111. By trial and error, the K values may be so selected that the experimental curve may be expressed by means of equation (279). The equation thus established would contain a complete description of soil from a trafficability point of view and, according to the present knowledge of the subject, may be considered as a definition of that part of the traf-

ficability concept which depends on the mechanical properties of soil. Values of k and n may be similarly obtained from measurements of load-sinkage curve.

The determination of the geometrical properties of the ground, which play an important role in the second part of the trafficability concept, necessitates the measuring of the heights h_w of the walls and breadth l_d of ditches, as well as of the lengths l_w and amplitudes h of the ground waves or terrain roughness. Such measurements were never seriously performed with reference to cross-country transport, although the determination of the unevenness of paved road surfaces has been pursued intensely during the last 20 years. This situation is somewhat paradoxical since cross-country locomotion is infinitely more affected by the roughness of the terrain than is the highway system of transport.

In order to show the nature of the problem, a brief review of relevant highway research would be useful. It would be impossible, however, to enumerate at this time all the apparatus designed since the early 1930's [272, 273] in both the United States and Europe for the purpose of measuring the waviness and roughness of pavements. They may be approximately divided into three groups: equipment which traces the exact profile of the surface, apparatus for recording certain dynamic effects of surface unevenness, and instruments which measure selected dimensions and integrate the measurement so that an assumed index of roughness is automatically deduced. The various types of equipment used are described in References 274 and 275, and an analysis of the methods involved is made in Reference 211. A recorder developed in recent days in the United States [276] appears to offer the most promising results and has been adopted for tests in Germany [211].

It is doubtful whether the methods employed in highway research would be proper for measuring the terrain surface from the viewpoint of cross-country locomotion. The various indices conceived in respect to the evaluation of the roughness of pavements [211, 274, 277, 278] deal mainly with the problem of ride comfort. In a cross-country drive, undoubtedly other problems besides comfort enter into the picture, as was mentioned in Chapter IX.

Exploratory measurements of terrain waves and roughness performed by the writer by means of a profilometer suggested by Croce [109] indicate serious difficulties which stem from the fact that highway roughness is measured in fragments of an inch, whereas terrain unevenness may be of the order of magnitude measurable in feet. Thus, not only new methods of measuring waves and bounces of the terrain are needed, but also new concepts in their evaluations are required so that they can fit the theoretical evaluation of the relationship between a vehicle and the geometry of the ground surface, as presented in this work.

Early attempts to measure ground waviness were based on the measurement of the amplitude of the ground wave h, arbitrarily defined as half the height from the lowest point to the highest occuring in a specified distance l_w (which might actually contain several wave lengths) (Figure 169b). Ground roughness was similarly defined in terms of the amplitude h' determined over a shorter selected distance. Analysis of only the frequencies of occurrence of given dimensions of obstacles h_w and l_d and of the amplitudes h and h' poorly defined as above proved far from satisfactory.

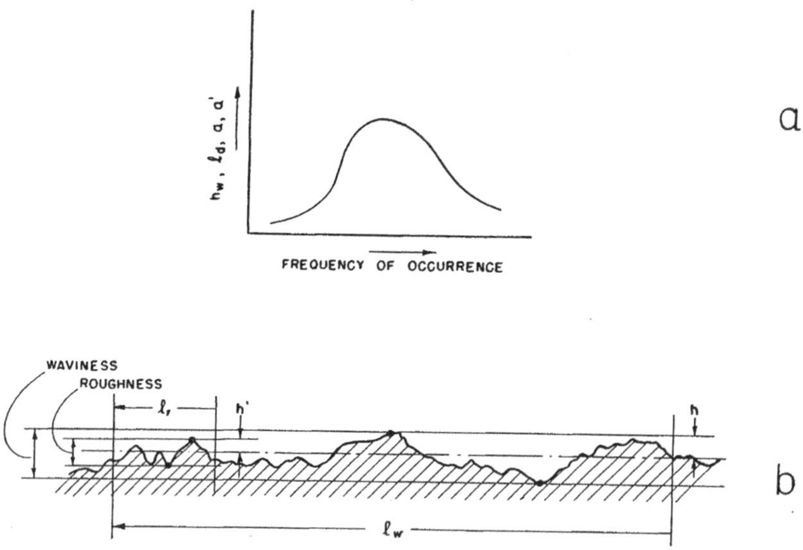

Fig. 169

The main difficulty is the question of what physical quantity should be adopted as being characteristic for the geometry of a given terrain surface. The ground wave by Lehr [212] fits the first approximation only. Even in highway research, this problem has not been entirely solved. The derivation of mean values of amplitudes is not satisfactory, because a small number of sizable bounces may produce the same "mean value" as a large number of smaller unevennesses; the first case, however, would produce strong vehicle vibrations, whereas the second may be entirely ineffective.[211]

The wide field open for the investigation of this problem, and the fact that a rational design of vehicle form and of its suspension depends on a quantitative knowledge of terrain properties indicate that a high priority should be given to a rational study aimed at defining the significant parameters of ground unevenness, possibly in statistical terms.

Classification of Soils and Snow

In the preceding section, various features of terrain were discussed with the purpose of defining what makes a given ground trafficable or not. A number of quantitatively defined factors were proposed, which, when combined, would yield a probable answer as to whether or not or to what extent the given ground is trafficable. The way in which these factors may be interlocked in the definition of a particular soil type, and the changing character of the factors themselves indicate that an "astronomical" number of combinations may be established when listing soil types from a trafficability point of view. Without some sort of a classification of this multitude of combinations, the over-all picture would be lost entirely and general conclusions would be impossible. Thus, the need for classification of soil and snow from a trafficability angle appears essential. It may be regarded as a synthesis of the analysis of the relationship between soil and a vehicle.

Such a need has been recognized clearly in the last 10 years.[268] The search for its fulfillment, however, has never been pursued in a systematic and continuous way. Attempts at fitting an "index" of trafficability to a soil classification established for civil-engineering purposes [279, 280] failed before they were seriously begun. It was suggested that pedological soil classification established to serve agriculture [281, 282] be used as a basis in a general evaluation of the trafficability of soils. This also became discouraging when the difference between the purpose of raising crops and moving machines across country was realized. For similar reasons, the existing international system of snow classification [283] seems to offer little opportunity for a solution of the present problem, particularly because the purpose of this classification appears to be rather undefined.

In the final analysis, then, the conclusion that there is an urgent need for a new type of soil classification which reflects ability to support vehicle traffic appears inescapable. Although it is impossible to foresee at the present stage of knowledge just how such a classification would look, it may be interesting to speculate on the subject within the limits of the material discussed in this work.

It results from the previous section that there are two groups of terrain features which affect soil trafficability: the first reflects the soil strength or consistency, γ, c, ϕ, k, n, K_1, and K_2, and the second relates locomotion to the geometry of the ground surface, h_w, l_d, l_w, and h.

If such is the case, it is logical to assume that any system of classification should include these values, directly or indirectly. For a start, take the first group of values. It may be recalled that in Chapter VII it was shown that a dimensionless value may be formed out of all the factors but one listed in the first group. This value may be derived from equation (283):

$$\frac{\tau}{c + \sigma \tan \phi} = \frac{e^{(-K_2 + \sqrt{K_2^2 - 1})K_1 j} - e^{(-K_2 - \sqrt{K_2^2 - 1})K_1 j}}{y_{\max}}.$$

Since this equation comprises the primary factors of soil trafficability, c, ϕ, K_1, and K_2, it may be helpful, as limited experience indicates, in the exploration of the effects of secondary factors, like moisture content, remolding of soils, etc., which affect the motion of vehicles. The dimensionless character of the equation may be considered to be helpful also in the establishment of certain indices which would distinguish particular types of trafficable or nontrafficable soils. The achievement of such a goal, however, is as far away as it ever was. The suggestion of using equation (283) as a tool in establishing soil classification is only intended to emphasize that such a classification can materialize only when a physical knowledge of the soil strength necessary to support locomotion is properly defined.

However, even if this condition were satisfied, the classification would be incomplete because, besides the soil strength, the geometry of its surface and k and n values must also be considered, as was previously stressed at several occasions.

The evaluation of terrain geometry in terms of "consolidated" units is nonexistent. Criteria of terrain roughness and waviness are to be established and methods of their measurement devised. The significant progress made in this respect in highway research may be helpful. It was shown that the problem is reducible to the determination of geometrical values in which use of a statistical approach for evaluation appears to be inevitable.

Although the case of snow should be treated in the same way as that of soil, it may bring into the relevant inquiry some factors which are not encountered in soils. The structure of snow, however, although different than that of soil, also necessitates the adoption of defined physical characteristics instead of descriptive qualitative features. As Bader and Niggli truly pointed out, strictly defined physical values may well lead to "an accurate and neat classification which is of importance to the further development of practical and scientific research work on snow." [80]

Bader and Niggli's classification, which may be a good start for future work, was established in terms of air permeability K_o and void ratio ε_o (see Figure 85b) or specific gravity γ of snow. A graph showing the approximate relationship between these quantities and the usual qualitative snow designations is reproduced in Figure 170. It reflects the changes which take place in snow cover due to metamorphosis and indicates the broad meaning of factors which so far have been little if at all considered in the trafficability of snow.

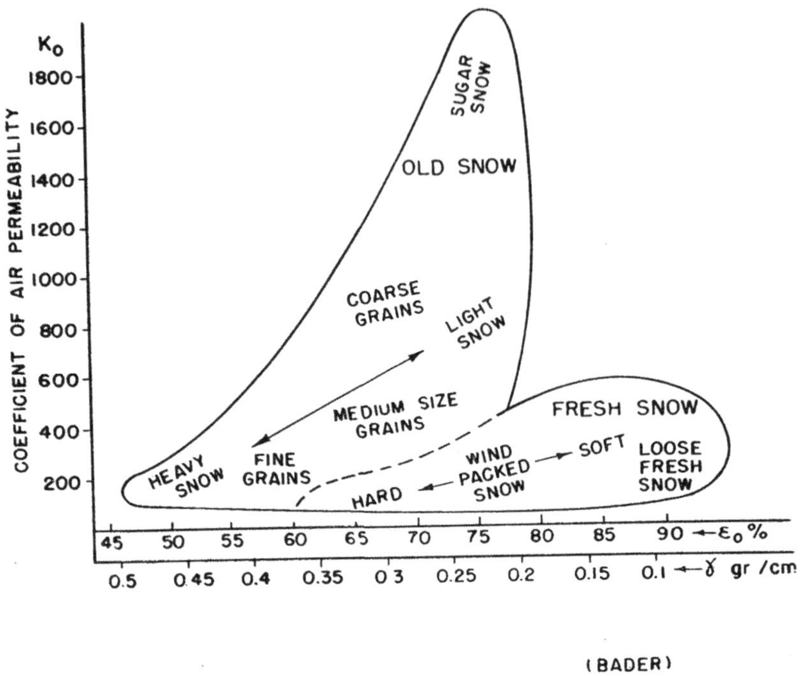

Fig. 170

A Measure of Vehicle Performance

Cross-country locomotion imposes such a variety of relevant obstacles that the measurement of vehicle performance obviously becomes a complex problem. A universal vehicle, which still appears to be the aim of research and development of automotive engineers, is supposed to perform satisfactorily on the highway and in mud, snow, and rough terrain. Thus, its performance must be assessed in all these conditions, which embrace soft-ground performance in an infinite variety of forms, and obstacle performance, as determined by the dimensions of walls, ditches, and ground unevenness. As a result, vehicle performance cannot be expressed in a single or a few values, but necessitates the evaluation of achievement under many assumed conditions.

Since these conditions are great in number and practically never have been "standardized," the performance indices determined in an arbitrarily chosen set of conditions are more or less haphazard and can seldom be compared with each other. This situation is mainly responsible for the difficulties in establishing goals and methods for future vehicle development; for how can the future be assessed if the past and present are not evaluated in accordance with a uniform rigid scale?

An unnecessary though historically unavoidable complication in the measurement of vehicle performance was introduced in the late 1920's when the mobility aspect of cross-country locomotion was interwoven with the purely functional performance of the vehicle. This situation was particularly distinct in the field of combat vehicles, where it was necessary for the vehicles to protect the crew from enemy fire and lend themselves as weapon platforms, as well as have soft-ground- and obstacle-crossing ability. A search for universal formulas which would express such a complex performance of the vehicle became widespread in 1925–30.

At that time, new terminology was introduced in order to define what was generally termed the "operational mobility" (*mobilité stratégique*) or, vaguely, the "tactical efficiency" of the vehicle. For instance, as a measure of this efficiency, Crompton proposed the product of squared speed by vehicle height, whereas Heigl insisted that a more appropriate measure would be the ratio of the squared speed of the tank over its frontal area.[32] In modern times, Jarret proposed an elaborate system of the numerical evaluation of particular tank operations which could be integrated in the summary term of the over-all performance.[284]

Countries in which motor vehicles were introduced in quantity much later than in the United States, England, Germany, France, and Italy established an elaborate system of credits which were assigned for the accomplishment of specific phases of performance in order to establish a scale of merit for vehicles which were to be either purchased from abroad or produced at home. Thus Russia, for instance, undertook in the early 1920's competitive rides in which foreign vehicle makers participated. Vehicle performance was assessed in numerical indices which reflected not only the mobility aspects, but also economic factors related to cost of maintenance, production, investment, etc.[34] A similar practice was followed in the analysis of racing-car performance.[216] From the foregoing discussion, it appears that all the yardsticks of vehicle performance so far assumed depend on a local and more or less arbitrary definition of what is to be included in the test operation.

In order to simplify the problem and yet to formulate it more rigorously, let it be assumed that the performance of a vehicle may be considered from two angles: (1) as a functional operation and (2) as a purely locomotive operation. The first type of performance refers to special tasks for which the vehicle was designed: fighting, cable laying, transport, etc. The second type is related to what may be called vehicle mobility in a physical sense; it embraces all the phenomena which relate to motion: acceleration, obstacle crossing, slip, tractive effort, flotation, etc.

The scope of this work does not permit dwelling on the first group of operations, which is large enough to become a separate study. Following the previous discussion on the relationship between soil and a vehicle, only the second angle of vehicle performance will be considered.

The mobility of a vehicle, which results from a combination of driving forces and the forces which oppose the motion, in an environment of a variable nature will be variable itself. If, therefore, this type of vehicle performance is to be measured, two assumptions have to be made: (1) the conditions in which mobility is to be determined and (2) the type of scale to be applied.

In modern specifications, these two assumptions are seldom clearly stated: operating conditions of the environment are usually defined in a descriptive way, and the yardstick of performance is, in most cases, the maximum speed or drawbar pull, without much justification of this choice. Experience has shown that such an approach to the problem is inadequate and that some standardization of the methods leading to the evaluation of vehicle mobility is needed.

From the previous discussions in Chapters IV and IX, it follows that a more tangible definition of the environment may be obtained by specifying the values of γ, c, ϕ, k, n, K_1, K_2, h_w, l_a, l_w, and h prevailing under given circumstances. These values will be called terrain parameters. However, the measurement of particular vehicle speeds in conditions characterized by any possible combination of these factors presents not only an insurmountable but also a useless work.

It appears more advisable to measure the average speed over a distance covering a certain terrain that has been defined in specific parameters and which is considered as being typical for the given operation, or which is merely a basis of reference serving to obtain values in comparable conditions.

The average speed considered as a measure of vehicle performance, and defined within the given area of locomotion, is expressed in number of miles per hour and is a function of terrain variables which are also defined in units of length, time, and mass.

The scale of vehicle performance thus conceived is identical with the measure by which the adaptation between vehicle form and environment may be evaluated, as was discussed in Chapter IV. The average vehicle speed determined during a given operation, which will now be called the operational speed, serves a broader purpose than the evaluation of the power train and transmission.

Evaluation of Vehicle Performance; Typical Operational Conditions

Experimental evaluation of operational speed does not present difficulties. As a matter of fact, it has been often determined, but, in most cases, the operation conditions were not kept constant or were not properly defined, if determined at all.

It is difficult indeed to determine what terrain conditions are to be considered as being typical: what types of soils should be included, what obstacles assumed. These difficulties, however, cannot prevent the approaching of the problem on a more rational and imaginative basis. Particularly, the evaluation of the mobility of vehicles which are in the stage of planning requires the establishment of an imaginary terrain cross section which could serve as the basis of comparison with existing vehicles. If a theoretical "proving ground" is assumed and assessed in terms of terrain parameters, then the theoretical determination of vehicle performance by means of the methods discussed in this work may be considered to be feasible, to a large extent. Although the accuracy and refinement of these methods offer a great deal for further improvement, even a rough approximation of such an evaluation appears to be somewhat closer to reality than most modern guesswork, based on undefined values.

As an example, consider two vehicles having unequal lengths of ground contact areas: $s_a > s_b$. Both travel over a route whose characteristics are shown in Figure 171. Assume that both vehicles are powered with the same number of hp per unit of weight and have a uniform ground pressure which is equal in both cases. Their route is composed of paved road I, hard, wavy ground II, loose soil III, and mud IV. The instantaneous speed of vehicles A and B will be the same on the paved road; it will obviously equal $v = \eta$ (hp)$/f^o$, where unit motion resistance f^o is expressed by equation (302), (303), or (304), and η is the efficiency of the power train. In hard, wavy ground II, the speed will be limited by the pitching amplitude. When starting with equation (429) and the method previously described in regard to the establishment of graphs (Figure 154), it will be found that the speed of vehicle A is, say, 25% higher than that of vehicle B. Such a difference is common in average terrain conditions for vehicles whose lengths of track differ by approximately 50%. In soft ground III, a further drop in speed of both vehicles must be expected, due to the inevitable slip. Since the length s_a of the ground contact area of vehicle A is longer than s_b of vehicle B, the slip i_o of vehicle A will be proportionately smaller than that of vehicle B. This follows from the study of the slip problem made in Chapter VII. Thus, in this zone, v_a is also larger than v_b. Finally, in zone IV of a very loose, soft mud, the motion resistance per unit of weight will be greater at vehicle B because its belly clearance will be lower than that of vehicle A, and the bulldozing effect [equation (305)] as well as the lateral effect [equation (306)] may be expected to be proportionally larger. This will cause a further drop in the speed of vehicle B.

Thus, the average speed of both vehicles along the assumed course will not be the same and $v_a > v_b$, as shown approximately by the dashed lines in Figure 171. It should be stressed that this difference in the performance of both vehicles would have remained undetected if only their maximum speeds had been specified, because, according to the original assumptions, these speeds are equal for both vehicles.

The degree of variation in the operational speeds of vehicles A and B depends on the type of the "typical" terrain adopted as the basis for evaluation. A single computation of the discussed type cannot cover all possible cases and several terrain types should be considered. Various

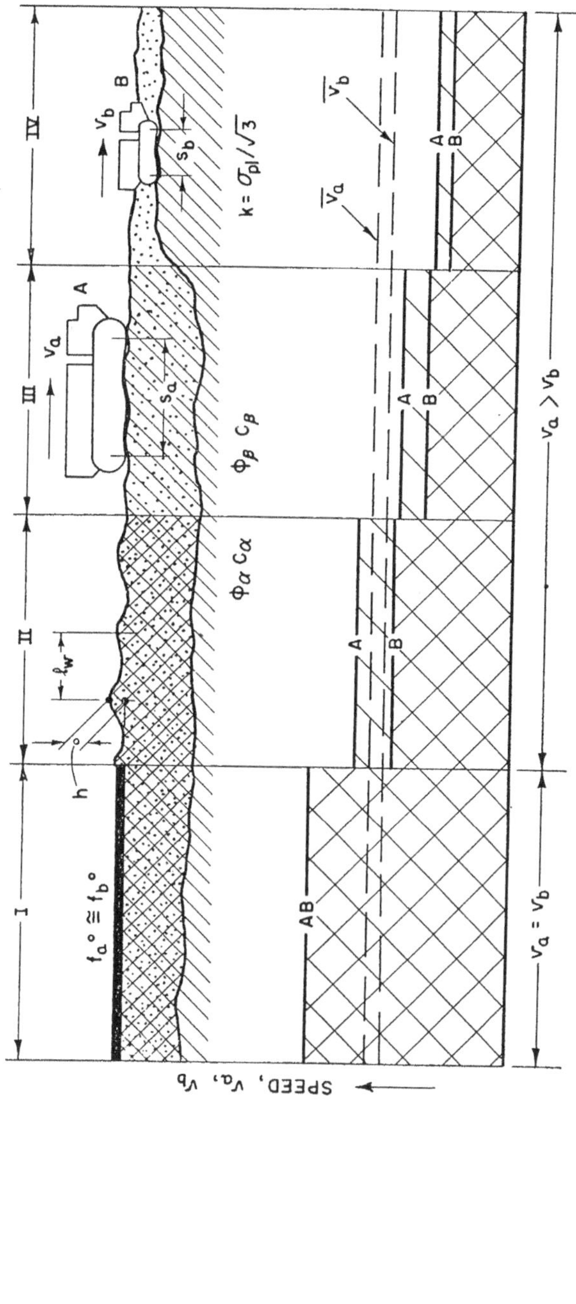
I
II
III
IV
A
B
v_a
v_b
s_a
s_b
$f_a^\circ \cong f_b^\circ$
h
ℓ_w
$\Phi_\alpha\ c_\alpha$
$\Phi_\beta\ c_\beta$
$k = \sigma_{pl}/\sqrt{3}$
$\overline{v_a}$
$\overline{v_b}$
AB
$v_a = v_b$
$v_a > v_b$
SPEED, v_a, v_b

Fig. 171

methods of integrating the results thus obtained have been proposed. An extension of these methods might yield a final index of mobility expressed in terms of the "integrated operational speed." However, for a complete picture, it would probably be necessary to state the lowest and highest speed limits in extreme conditions as well. Although attempts at providing the "typical terrain" conditions for the testing of vehicle mobility are being made at all the proving grounds, much work remains to be done in order to systematize and standardize the methods involved.

Speed and Engine Power

The example analyzed in the previous section illustrates that two vehicles of the same maximum speed and the same unit power develop different speeds in the same terrain. It may then be concluded that power and maximum speed do not directly affect operational speed. This conclusion has been confirmed by experience, and it would not be difficult to conceive terrain conditions in which even a more powerful and "fast" vehicle would travel more slowly over the given terrain than an "underpowered" vehicle.

In most cases, the engine power is not fully developed for most of the time because of the geometric properties of the ground surface, which create either discomfort for the riders [248] or dynamic loads (see Figure 154) beyond the endurance of men and material.

A survey of power increase and increase in the operational speed of cross-country vehicles indicates a wide disproportion between the two trends (Figure 172). Undoubtedly, the powerful engines of modern vehicles enable the driver to accelerate faster or to climb hills at a higher speed. This gain, however, cannot be appreciated in the same terms off the road as an overpowered car is preferred on a highway in heavy traffic. Although the significance of the power surplus in a cross-country vehicle should not be underestimated, its true meaning should be carefully analyzed in order not to widen further the gap between operational speed and power, as shown in Figure 172. Overpowered vehicles are uneconomical, as will be seen later on, and, in most cases, the economy of cross-country transport appears to be a more valuable asset than rapid acceleration.

It was shown in Chapter IV that the morphology of modern cross-country vehicles is static and that their dimensions are frozen within the framework of what is called the light-, medium-, or heavy-vehicle class. Accordingly, it appears logical to admit that the limits imposed

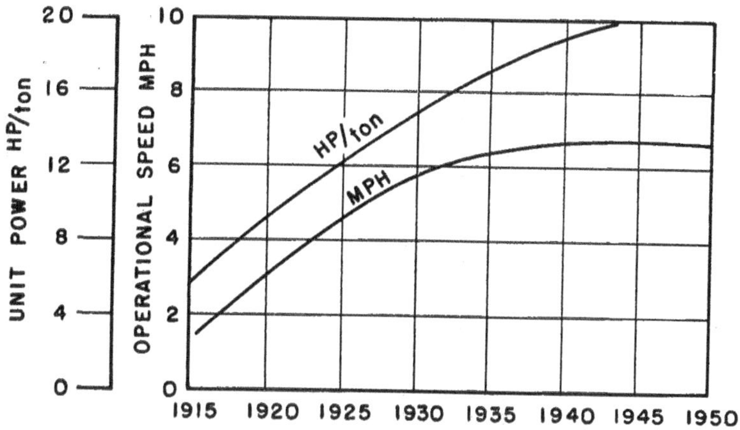

Fig. 172

upon speed by the geometry of the ground surface are also fixed to a large extent, since they mostly depend on the form of the vehicle, and that the present operational speeds, which, in the best cases, according to Lehr, do not exceed 10 mph,[212] have reached a maximum which will not be easily surpassed unless all the efforts concentrate on the improvement of the suspension and the form of the vehicle.

It may then be concluded that an increase in operational speed will not be achieved by a further increase in engine power, and a revolution in vehicle design will not be achieved by the invention of new, more powerful engines.

A modern land vehicle, unlike the airplane, is unable to take advantage of a powerful thrust, because the thrust cannot be larger than the strength of the ground. The speed of the land vehicle is limited by ground waves just as the speed of a ship is often limited by rough seas rather than by the lack of power. If this conclusion is true, then the volume of efforts made by automotive engineers in the study of engines, as shown in Figure 27, is highly disproportionate and perhaps futile.

Vehicle Performance and Engine Torque

Loads imposed upon the engine by the changing conditions of a cross-country drive are severe and require a great deal of flexibility of the torque transfer in order to adjust the driving forces to the suddenly mounting or disappearing resistance.[278] Such a situation exists only in a small degree on the highway and in a still smaller degree on the railroads.

Consider a vehicle which moves with speed v on soil having R_1 motion resistance (Figure 173a). In such a case, from the available torque $a + b$, only part b is used, whereas torque a remains in reserve. However, if the motion resistance increases up to the R_2 value, then the performance of the torque curve of the engine becomes of great importance. It will be seen that of curves 1, 2, and 3, only torque 3 permits further drive without gear change. Flatter curves 1 and 2 would necessitate shifting to a lower gear, although the torque may be produced by an engine having the same maximum power as the engine which gives torque 3. In conclusion, then, flat or falling torque curves are not convenient if the regular gear box is used. The shocks, braking, and acceleration which are associated with gear change do not let the vehicle operate on optimum slip and, in extreme cases, may lead to stalling. Engines which produce flat torque curves are called inelastic and should be avoided in cross-country vehicles.

The problem of engine "elasticity" was first defined by Flössel.[285] In an exhaustive study which embraces not only motor vehicles but also steam-, electric-, Diesel-, and turbine-driven locomotives with mechanical and hydraulic transmissions, Flössel made a survey of the definitions of engine elasticity by Becker, Kluge, Richter, Zeman, Wien, and others, and arrived at the following conclusions.

The elasticity of the torque e_t is defined by the ratio of the maximum torque M_2 to torque M_1 which prevails at the maximum power (Figure 173b):

$$e_t = \frac{M_2}{M_1}.$$

The elasticity of revolutions of the engine e_r is defined by the ratio of rpm's corresponding to the considered torques:

$$e_r = \frac{n_1}{n_2}.$$

The elasticity of the engine is the product of e_r and e_t:

$$E_e = \frac{M_2}{M_1}\frac{n_1}{n_2}. \qquad (444)$$

The index of elasticity defined in such a way varies for various engines.

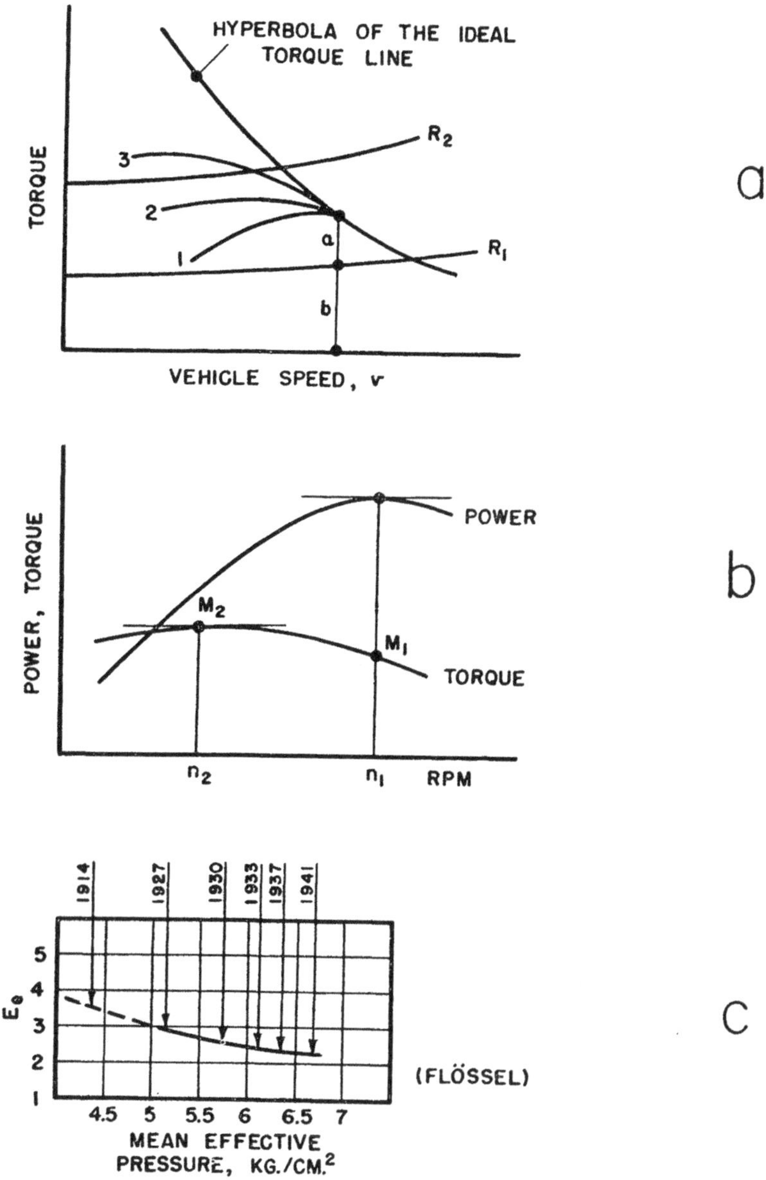
HYPERBOLA OF THE IDEAL TORQUE LINE
TORQUE
VEHICLE SPEED, v
R2
R1
3
2
1
a
b
a
POWER, TORQUE
POWER
TORQUE
M2
M1
n2
n1
RPM
b
1914
1927
1930
1933
1937
1941
Ee
5
4
3
2
1
4.5
5
5.5
6
6.5
7
MEAN EFFECTIVE PRESSURE, KG./CM.2
(FLÖSSEL)
c

Fig. 173

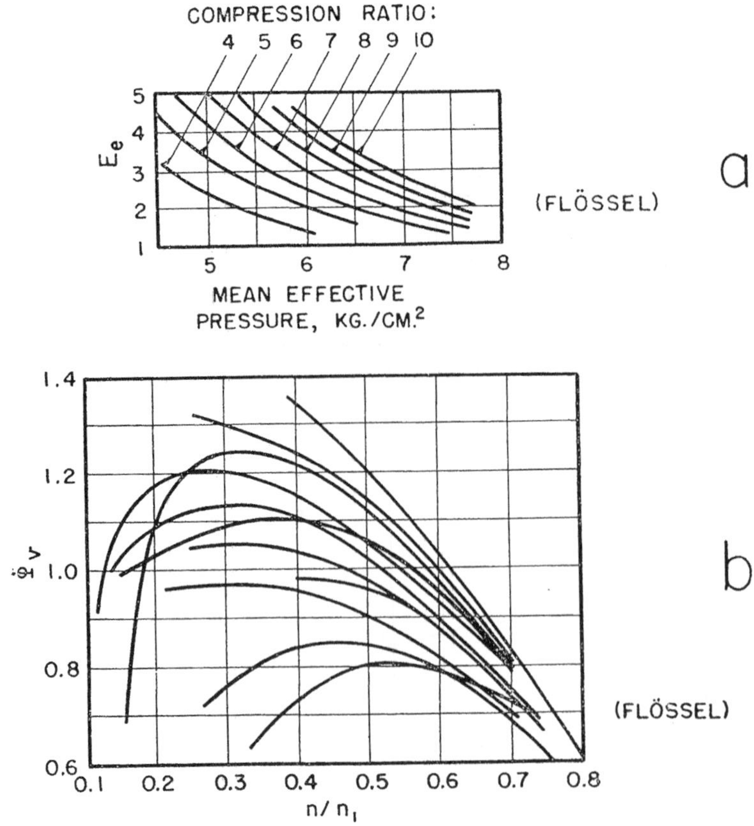

Fig. 174

According to Flössel, the trend in the development of American engines may be described in terms of E_e and the mean effective pressure, as shown in Figure 173c. It indicates that the elasticity steadily decreases, which appears plausible since the power of automobile engines steadily increases. The increase in power compensates for the reduced elasticity, because, as Kluge pointed out, the easiest way of effectively getting an elastic torque curve is to use an oversized engine.[285]

Whether this is an economical way, particularly as far as cross-country vehicles are concerned, appears to be doubtful. It is interesting to note that the present trend toward high-compression engines tends to increase the elasticity E_e of the engine (Figure 174a). In general, however, the form of the torque curve of an engine depends on design factors like rpm, valve type, bore-stroke ratio, mean effective pressure, carburetion, combustion-chamber type, cooling, mufflers, etc.[285]

The concept of engine elasticity by Flössel may be applied directly to the evaluation of vehicle performance such as hill climbing or acceleration. From this evaluation, the concept of the elasticity function Φ_v of a whole vehicle was derived. The definition of Φ_v is expressed by the following equation:

$$\Phi_v = e_t - \frac{1}{e_r^2}. \tag{445}$$

The larger the Φ_v, the higher the acceleration, climbing power, etc.

Figure 174b shows values of Φ_v plotted as a function of engine rpm, expressed as a fraction of revolutions developed by the engine at the maximum power. The wide span between the elasticities of the various cars shown in this figure raises the question as to how a corresponding picture would look in the case of cross-country vehicles. The question is open. It appears that this problem has not been sufficiently stressed in vehicle design. Engines have been selected mostly because of the available power rather than because of suitable elasticity. In these cases, the final goal was undoubtedly achieved, but only at the cost of higher fuel consumption or at the expense of using high-grade fuels.

A study of the engine torque by Kühner [286] analyzes the elasticity of various types of engines and discusses their applicability to cross-country vehicles. In Figure 175a, the torque curves are plotted in the dimensionless form of $M/M_{n(\max)}$, where $M_{n(\max)}$ is the moment developed at the maximum rpm. Torque 1 is typical for a nonelastic, highly stressed sport car or airplane engine. Curve 2 characterizes an engine which is basically of the same construction as 1, but which has slightly reduced maximum power. Curve 3 is the typical torque line for a European passenger-car engine. Curve 4 shows the torque of an average truck engine. The most suitable for cross-country operation would be an engine with characteristics 5, which surpasses the ideal torque curve defined by the constant power.

The advantages of an elastic torque curve in a cross-country vehicle equipped with a regular gear box are shown in Figure 175b. The left-hand graph indicates a close fit between the ideal torque M_i and the torque line obtained by the gear box. The engine is suitably elastic The right-hand graph shows large differences between both torques,

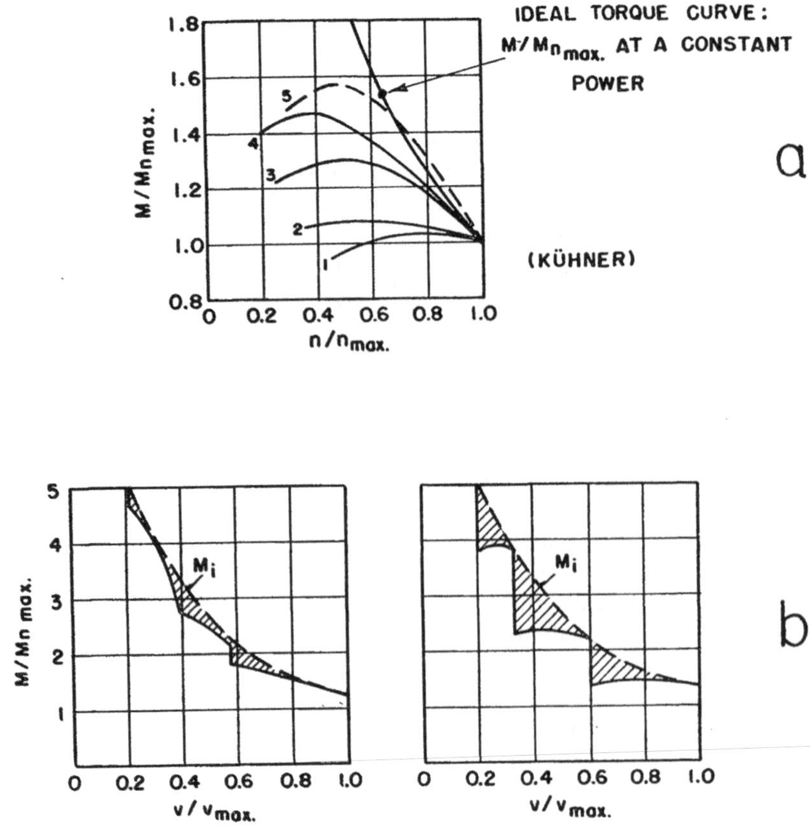

Fig. 175

which leads to the losses due to an incomplete exploitation of the inelastic engine. To avoid these losses, a larger number of gears would be needed.

The elasticity of the engine has practically no meaning when a hydraulic transmission is used. The theoretical principles of the operation of such a transmission applied to motor vehicles have been published in References 287, 288, and 289 and a more popular presentation is given in Reference 290. Hydrodynamic principles may be found in Reference 291.

When a hydraulic torque converter is used, the engine rpm is practically independent of the output load of the converter, and the introduction of a mechanical transmission between the hydraulic transmission and the sprocket or driving wheel may bring the output torque curve very close to the ideal torque line M_i (Figure 176a). Thus, the losses of the fluid gear, shown by the hatched area, may be reduced to a minimum with a small number of gears (two gears, I and II, shown in Figure 176a).

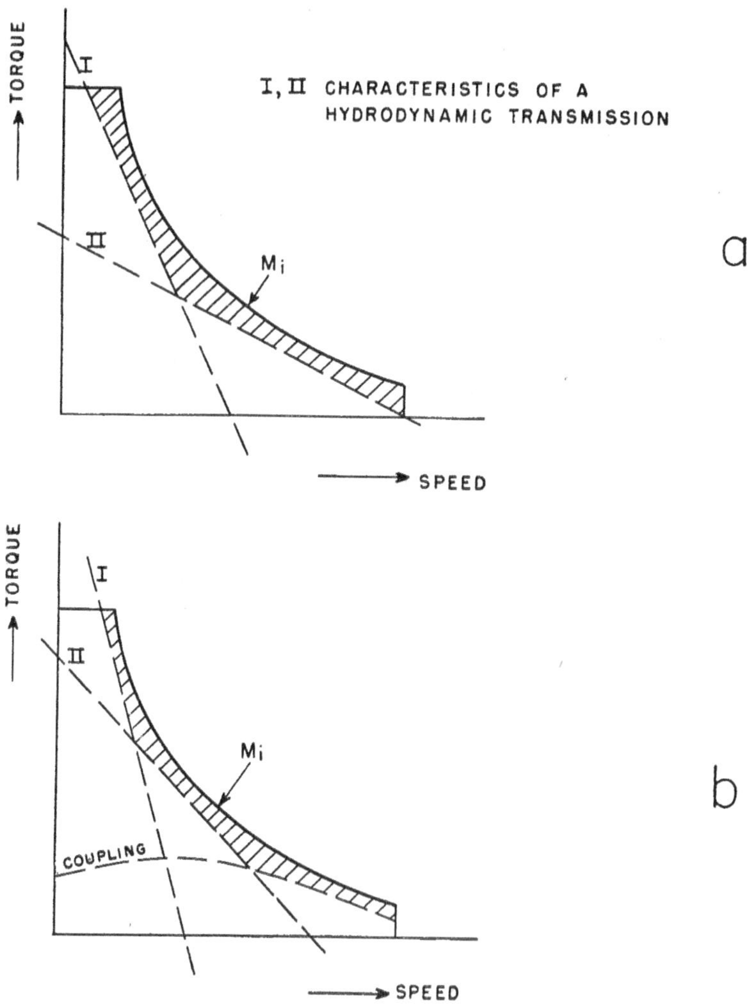

Fig. 176

The introduction of a direct coupling at high speed together with two indirect gear ratios, for instance, will have a still better effect (Figure 176b). The adoption of a many-cycle torque converter, instead of the single-stage torque converter used in the most popular automotive pump-stator-turbine arrangement, obviously leads to mechanical complication.

An improvement in torque transmission can be made by dividing the engine torque into parallel branches, only one of which is transformed by a converter.[292] Some authors have hinted that further development of a hydrostatic transmission also may lead to better performance.[293] The reason is that the hydrostatic transmission has the same hyperbolic characteristics as the ideal torque M_i, whereas the output torque of a hydrodynamic torque converter follows an approximate straight line I or II, as shown in Figure 177. Thus, a hydrostatic transmission may be adjusted to M_i directly; this cannot be done with a hydrodynamic converter.

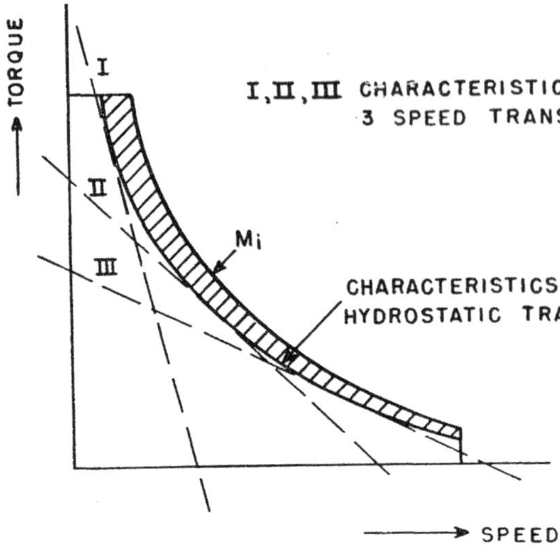

Fig. 177

The necessity of introducing automatic controls on torque in cross-country vehicles is indicated not only by everyday experience, but also by theoretical evidence.[294, 295] A review of the analyses available in the field of vehicle transmissions [296–298] indicates the necessity of an over-all theoretical study of the problem of torque conversion in order to systematize the present multitude of attempts in this direction,[299] particularly for off-the-road operations.

Fuel Consumption and Power Trains

It is known from daily experience that two vehicles of the same weight, but different power trains, will consume different amounts of fuel in the same conditions.[300] If the vehicle weights are not equal, then some increase in fuel consumption is to be assigned on account of the heavier weight. In this case, a comparative unit of fuel consumption will be the number of miles per gallon times the weight, or, as it is usually termed, the number of ton miles per gallon.

The number of miles per gallon or ton miles per gallon in given terrain conditions depends on the fuel characteristics of the engine, the ratio of engine torque to driving torque, efficiency of the transmission, internal resistance of the running gear, air resistance, and speed of locomotion.

Consider the fuel characteristics of a gasoline engine shown in Figure 178. This chart gives the consumption w in lb/hp-hr with reference to the maximum power and fractional engine powers ψ for 75%, 50%, and 25% of throttle opening as a function of rpm. The optimum of fuel consumption is $w = 0.52$ lb/hp-hr at n rpm and full engine power. It is evident that when the engine runs at part throttle, its economy decreases rapidly. Thus, at n rpm and 50% power, fuel consumption will amount approximately to $w = 0.6$ lb/hp-hr and at n_1 rpm, $w = 0.67$ lb/hp-hr.

If the power plan of the vehicle is traced for various speeds developed at a given transmission gear (Figure 179), then the power required for propulsion with speed v' and given motion resistance R' may be graphically determined as distance b. Assume that the dashed curves denote the power which is available after subtracting power consumed by air resistance, engine accessories, transmission, and internal losses; then the power surplus available for acceleration, hill climbing, etc., will be expressed by distance a. Thus, the percentage of engine power at which the vehicle runs is

$$\psi = \frac{b}{a + b} 100\% \, .$$

Since speed v' is determined by a corresponding scale n' rpm of the engine, fuel consumption w lb/hp-hr may be determined for given ψ% and n' from Figure 178, as was previously discussed. The total consumption

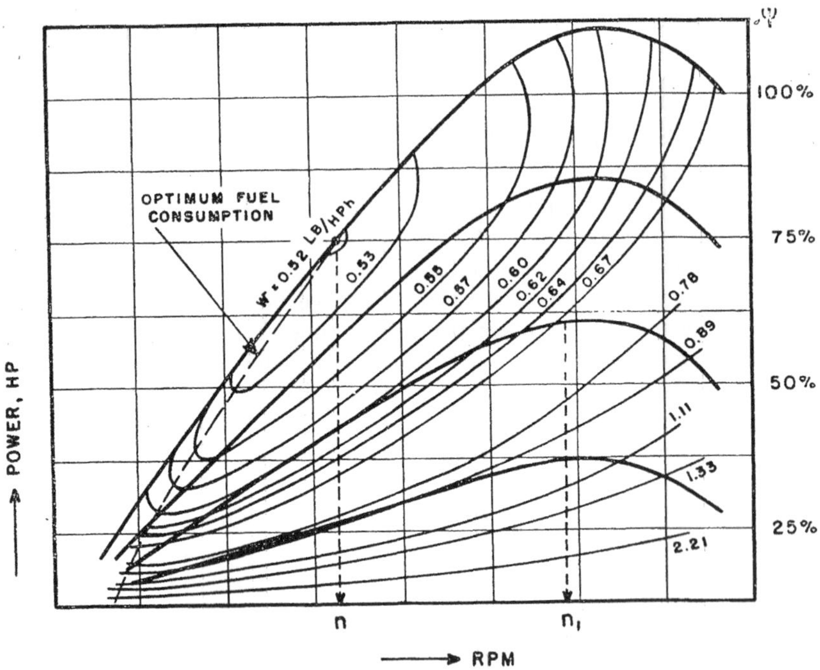

Fig. 178

per hour will be obtained when multiplying w by the number b of hp:

$$w_t = bw \,.$$

Since, at this consumption, the vehicle travels at v' mph, the miles per gallon may be obtained as follows:

$$\text{mpg} = \frac{6.2\, v'}{bw} \,,$$

where 6.2 is the conversion factor from pounds to gallons. Thus, for a given motion resistance R' which reflects the type of soil negotiated, and for a given speed v', the economy of the vehicle may be expressed in mpg. By assuming various R', R'', R''', etc. values, the economy in various types of terrains may be evaluated.

An analysis of cross-country vehicles made by means of the discussed methods indicates that the gear ratios of these vehicles often do not give the optimum economy, and that a slight change in the number of gears, or in gear ratio in the final drive, which changes the shape of power curves I, II, and III, may improve the situation very considerably. This conclusion ties up with the necessity of adjusting the over-all ratios of vehicles to the types of terrain in which they operate. Such an adjustment has to be made very carefully, however, in order not to affect adversely the dynamic properties of vehicle motion.[301–304] It may be helped a great deal by the theoretical methods discussed in References 305 and 306. The problem of fuel economy of a motor vehicle was theoretically considered, among others, by the authors whose works are quoted in References 256, 307, 308, and 309. A study of these works discloses general

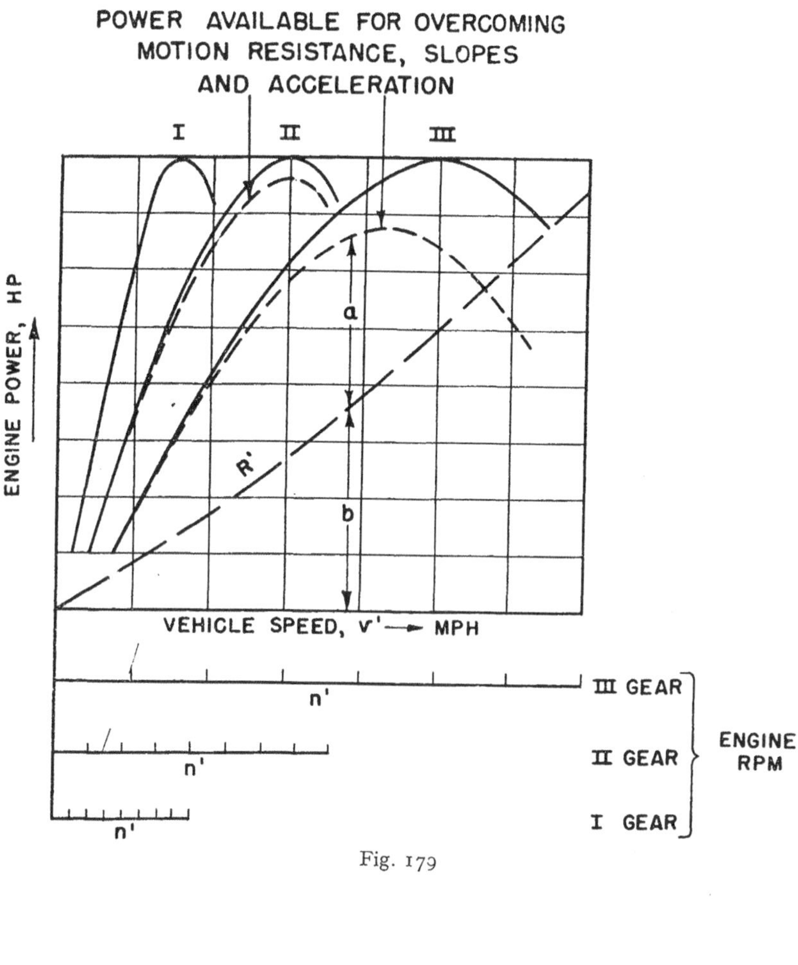

Fig. 179

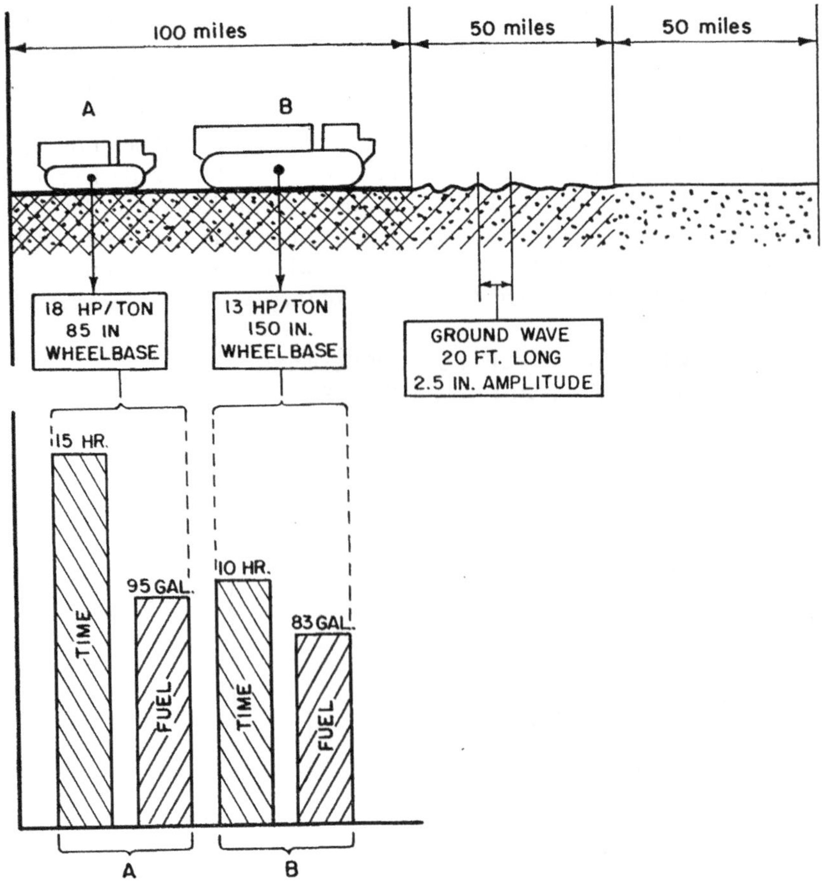

Fig. 180

methods in the approach to one of the very complex problems in which the technical performance and economy of a machine are complicated by the psychological factors of driving.

The increase in fuel consumption of a vehicle which is forced to slow down because of the prohibitive vibrations due to riding over a rough terrain may be determined theoretically when using the method previously described.

A decrease of the economy in such conditions may be illustrated in an example of two vehicles of differing forms, one of which was overpowered in comparison to the other (by virtue of being lighter, but using the same engine) and thus had to throttle the engine to a larger extent. Vehicle *A* (Figure 180) was powered with approximately 18 hp/ton, vehicle *B* with 13 hp/ton. In a cruising distance of 100 miles of hard road and 100 miles of a terrain as shown in Figure 180, vehicle *A* would get 2.1 mpg, whereas vehicle *B*, 2.4 mpg. The travel time and the total fuel consumption of the overpowered vehicle would be approximately 15 hours and 95 gallons. The larger, "underpowered" vehicle would be faster, covering the stated distance in approximately 10 hours with 83 gallons of fuel. These findings do not exaggerate the differences between the performances of both vehicles because both the highway and terrain were considered in equal proportions. It may be easily calculated that in the case of travel in the terrain alone, or with a smaller proportion of highways, the deterioration of fuel economy would be much more distinct. Figure 180 shows that fuel economy also may be entirely in consonance with time saving because the underpowered vehicle is faster by 5 hours than the overpowered one. This difference in the time of travel reflects the effect of the form of both vehicles: transporter *A* has a shorter wheel base than *B*. The form obviously affects the fuel consumption, because the longer vehicle may travel faster, prior to reaching critical vibrations, and its engine does not need to be throttled too much. This example illustrates the nature of the relationship between vehicle form, power, and transmission on one side, and the economy of fuel and time on the other.

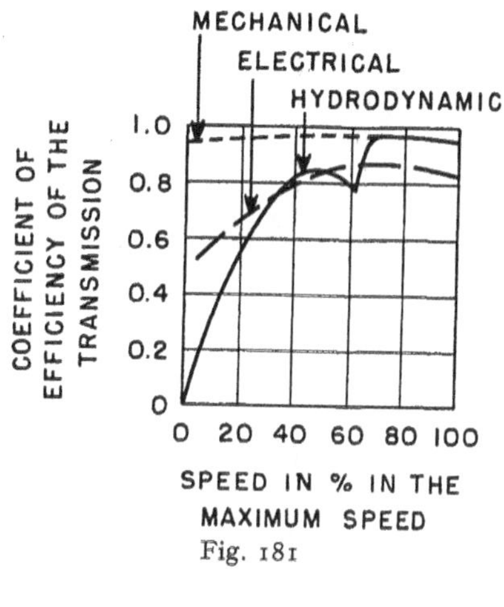

Fig. 181

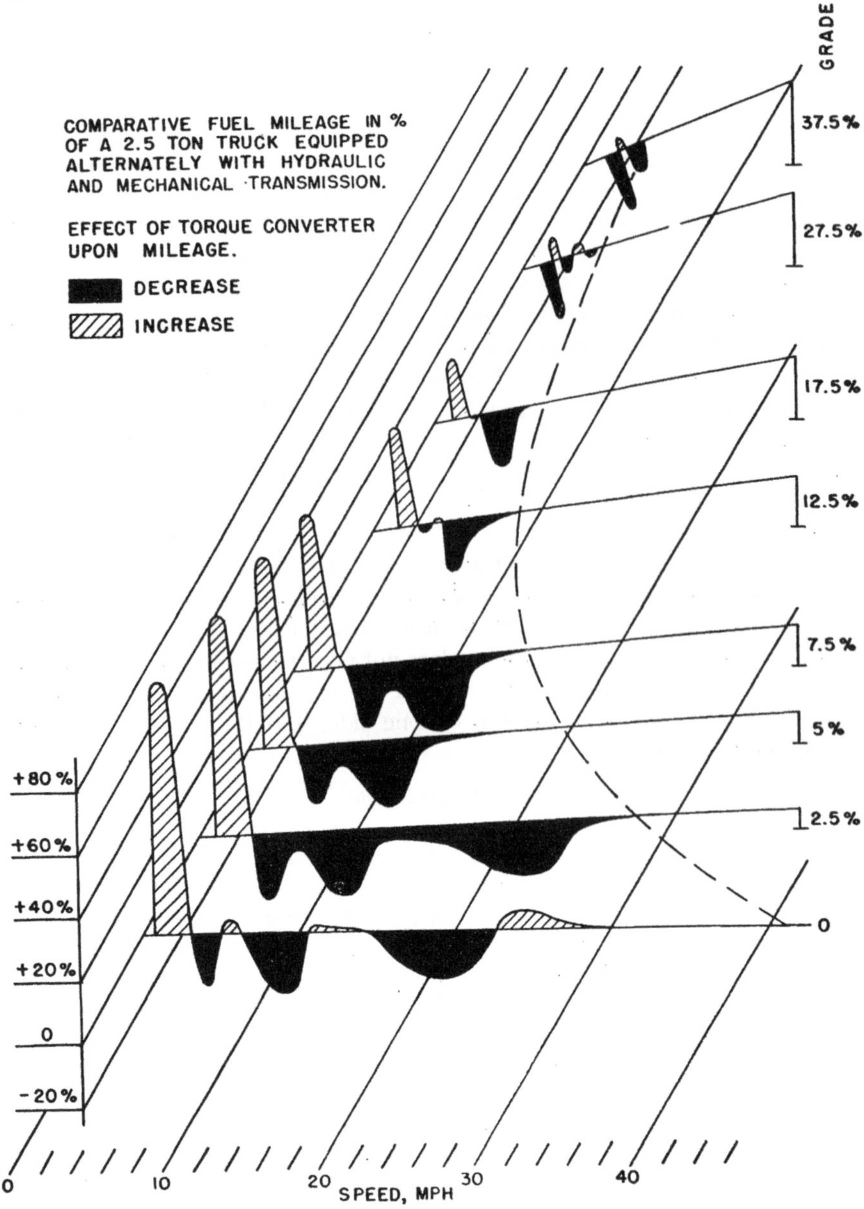

COMPARATIVE FUEL MILEAGE IN % OF A 2.5 TON TRUCK EQUIPPED ALTERNATELY WITH HYDRAULIC AND MECHANICAL TRANSMISSION.
EFFECT OF TORQUE CONVERTER UPON MILEAGE.
DECREASE
INCREASE
GRADE
37.5%
27.5%
17.5%
12.5%
7.5%
5%
2.5%
0
+80%
+60%
+40%
+20%
0
-20%
0
10
20
30
40
SPEED, MPH

Fig. 182

Similar principles govern the economy of vehicles in the case of hydraulic transmissions. However, the evaluation of fuel consumption and speeds is more cumbersome. The methods of such an evaluation may be directly deduced from fundamental works on automobile torque converters.[287] They have been particularly elaborated by Prokofiev,[289] who assembled and analyzed much of the information from all the available sources.

As was mentioned before, the torque characteristics of the transmission may be matched very closely with the ideal torque line if a proper speed change is selected in a supplementary mechanical transmission located between the torque converter and final drive. Under the assumption of an optimum gear shifting in the supplementary transmission, a very convenient fuel characteristic of the vehicle may be obtained. Since it is impossible, however, to eliminate entirely the operation of the converter at low efficiency points, certain areas of operation will be less efficient in the case of the converter than in the case when a purely mechanical or electrical transmission is used (Figure 181). These areas depend on speed and loads since the above factors affect vehicle economy at any transmission. The comparative fuel consumption of two identical 2.5-ton trucks, one equipped with a regular 5-speed gear box and the other with a standard type of torque converter with a 3-speed supplementary gear box, was computed as a function of the grade which may be negotiated on a hard-surface road and the speed which may be developed in given conditions. The results are shown in Figure 182, which indicates clearly the zones of speeds and grades where the hydraulic transmission will be less economical than a regular gear box. It is interesting to note that with an increase in grade, i.e., motion resistance, the advantages of the converter diminish, which is confirmed by experience. In average conditions, when speed and load vary within normal limits, the increase in fuel consumption by the torque converter is of the order of magnitude of 10%.

Effective and Nominal Fuel Consumption in Cross-Country Operations

In a highway operation, fuel is available at any point required. In most cases, it is transported to these points by all available means of transportation: ship, railroad, pipeline, or truck. The cost of transportation is reflected by the fuel price and is not directly measurable in terms of the amount of fuel burned in transporting the fuel to the point of purchase.

Assume that a certain amount of payload is to be moved across country through an uninhabited area. A caravan of special vehicles which must be used for such a purpose has to carry a definite amount of fuel because there are no refilling stations. Thus, the total consumption of fuel per payload carried across country will be higher than the nominal fuel consumption of the cargo carriers alone: it has to be increased by the consumption of the fuel carriers themselves. In the final analysis, the total or effective fuel consumption increases very rapidly in adverse terrain conditions and may be evaluated as follows.

Consider that the distance of transport is l miles long, the number of cargo carriers is N_c, number of fuel carriers N_f, fuel consumption of the cargo carrier u_c lb/mile, fuel consumption of the fuel carrier u_f lb/mile, total fuel required by cargo carrier W_c lb, and that required by fuel carriers W_f lb; the fuel capacity of a tank of the cargo carrier is w_c lb; the payload and tank capacity of a fuel carrier is w_f.

Assume that the entire convoy arrives at the end of the l mile trip with empty tanks. Also assume that the fuel consumption of cistern vehicles does not change substantially although they gradually empty along the route. The total fuel required is

$$W_t = W_c + W_f \,.$$

The fuel required by cargo carriers is

$$W_c = lN_c u_c \,,$$

and the fuel required by fuel carriers is

$$W_f = lN_f u_f \,.$$

Hence,

$$W_t = l(N_c u_c + N_f u_f) \,.$$

The number of fuel carriers

$$N_f = \frac{W_t - N_c w_c}{w_f} \,.$$

Then,

$$W_t = l\left(N_c u_c + \frac{W_t - N_c w_c}{w_f} u_f\right)$$

or

$$W_t = \frac{lN_c u_f \left(\frac{u_c}{u_f} - \frac{w_c}{w_f} \right)}{1 - l\frac{u_f}{w_f}}.$$

Thus the effective fuel consumption of cargo carriers, which is composed of the nominal fuel consumption of all vehicles, is

$$u_e = \frac{W_t}{lN_c} = \frac{u_f \left(\frac{u_c}{u_f} - \frac{w_c}{w_f} \right)}{1 - l\frac{u_f}{w_f}}.$$

Since the nominal consumption of the cargo carrier is assumed to be u_c, the ratio of this consumption to the effective consumption is

$$\eta = \frac{u_e}{u_c} = \frac{1 - \left(\frac{u_f}{w_f} \Big/ \frac{u_c}{w_c} \right)}{1 - l\frac{u_f}{w_f}}. \tag{446}$$

Equation (446) may be rewritten in the following form, assuming that the distance which may be covered by one cargo carrier on one tankful is $l_c = w_c/u_c$. The distance covered by one fuel carrier, on the fuel it contains, is $l_f = w_f/u_f$; then,

$$\eta = \frac{1 - \frac{l_c}{l_f}}{1 - \frac{l}{l_f}}.$$

In this equation, l_c is very small as compared to l_f; hence,

$$\eta \cong \frac{1}{1 - \frac{lu_f}{w}}$$

and the effective fuel consumption u_e in lb/mile may be determined in terms of the nominal fuel consumption u_c as follows:

$$u_e \cong u_c \frac{1}{1 - \frac{lu_f}{w_f}}. \tag{447}$$

Equation (447) indicates that u_e becomes infinite when $1 - (lu_f/w_f) = 0$ or when the critical distance

$$l_{crit} = \frac{w_f}{u_f}$$

is reached. This means that the movement of the convoy beyond that distance is impossible because of lack of fuel. If $w_f = 3000$ lb and $u_f = 5$ lb/mile, then $l_{crit} = 3000/5 = 600$ miles. At any distance shorter than that, say 500 miles, the effective fuel consumption will be

$$u_e = u_c \frac{1}{1 - \dfrac{500 \times 5}{3000}} = 6u_c \,.$$

In other words, if the nominal fuel consumption u_c of a cargo carrier is the same as that of a fuel carrier (identical vehicles used in a convoy) i.e., when, for instance, $u_f = u_c = 5$ lb/mile, then the effective consumption will be 30 lb/mile.

This example illustrates the importance of the economy of fuel of cross-country vehicles. The smallest saving in mpg will be multiplied by a large factor and, hence, the actual saving in fuel will be much greater than the nominal saving. In this connection, a detailed study by Andreau, on the utilization of wind power for propulsion of motor vehicles, may be mentioned.[310]

The Efficiency of Cross-Country Transport

The great truck lines and bus companies carefully select the vehicle type which will achieve the best economy and efficiency of operation. Their problem, however, is relatively simple, for they have to make the choice between a number of truck or bus models, all of which are subject to only minor design differences. If an error in judgment is made, then the losses are not damaging and may be corrected easily.

A similar problem in an operation across country poses graver questions. The present knowledge in this field is so inadequate that some will propound a tracked vehicle as being the most suitable means of locomotion, whereas others may recommend a wheeled or half-tracked vehicle. The differences in cost and performance of the various types of cross-country vehicles which may be considered prohibit making such a mistake, because the losses involved may result in the complete failure of the enterprise. Thus, the adoption of adequate measures in a rational evaluation of vehicle problems in off-the-road operation are paramount.

As an illustration of what may be involved, consider the idea of off-the-road transport efficiency. Such efficiency is assumed to be expressed by the ratio x, whose value is determined as follows:

$$x = \frac{\text{delivery rate of cargo (lb/hr)}}{\text{fuel consumption (lb/hr)}}.$$

In order to estimate x values for various types of vehicles in various terrain conditions, assume that the time in which payload W of a vehicle is delivered over a distance of 1 mile is $1/v$, where v = vehicle speed. If the delivery is continuous by using a column of trucks, each spaced 1 mile apart, a load W will be delivered every $1/v$ hour. Thus the hourly delivery rate is

$$\frac{W}{1/v} = Wv \text{ (lb/hr)}.$$

If the hourly fuel consumption is u_h, then index x, as previously defined, will be

$$x = \frac{Wv}{u_h}.$$

Index x may be made independent of the type of fuel used if u_h is expressed in terms of the caloric energy it represents.

It may be seen that the value of x for a single vehicle depends on the length of the route. If the latter is l miles long, then the rate of delivery is

$$\frac{W}{l/v} = \frac{Wv}{l}.$$

Hence,

$$x' = \frac{Wv}{u_h l} \tag{448}$$

or

$$x' = \frac{x}{l}. \tag{449}$$

Assume that the terrain is not uniform but composed of various types distinguished by various motion resistances which, of course, produce various fuel consumptions u_h', u_h'', u_h''' ... u_h^n at corresponding speeds of travel $v_1, v_2, v_3 \ldots v_n$. The above values of u_h and v reflect all sorts of motion resistance as considered in this chapter and in Chapters VI and VII, including the slowdown due to the vibrations analyzed in Chapter IX. Accordingly, the efficiency of transport x at particular distances of the whole route will be

$$x' = \frac{Wv_1}{u_h' l_1}, \quad x'' = \frac{Wv_2}{u_h'' l_2}, \quad x''' = \frac{Wv_3}{u_h''' l_3}, \quad \text{etc.,}$$

where l_1, l_2, l_3, etc., are the lengths of particular types of road. The average speed v over the whole route $l = l_1 + l_2 + l_3 + \dots l_n$ is

$$v_{\text{average}} = \frac{l_1 + l_2 + l_3 + \dots l_n}{\dfrac{l_1}{v_1} + \dfrac{l_2}{v_2} + \dfrac{l_3}{v_3} + \dots \dfrac{l_n}{v_n}}.$$

The average fuel consumption per distance l is

$$(u_h)_{\text{average}} = \frac{u_h' \dfrac{l_1}{v_1} + u_h'' \dfrac{l_2}{v_2} + u_h''' \dfrac{l_3}{v_3} + \dots u_h^n \dfrac{l_n}{v_n}}{\dfrac{l_1}{v_1} + \dfrac{l_2}{v_2} + \dfrac{l_3}{v_3} + \dots \dfrac{l_n}{v_n}}.$$

Hence,

$$x_{\text{average}} = \left(\frac{Wv}{ul}\right)_{\text{average}}$$

$$= \frac{W}{u_h' \dfrac{l_1}{v_1} + u_h'' \dfrac{l_2}{v_2} + u_h''' \dfrac{l_3}{v_3} + \dots u_h^n \dfrac{l_n}{v_n}}. \tag{450}$$

As an example, consider the terrain cross section shown in Figure 183. Two types, a tracked and a wheeled vehicle of the same cargo capacity, are considered. Preliminary computations performed in accordance with the methods discussed in this work indicate that the speeds and fuel consumptions of both vehicles are as shown in Figure 183. Accordingly, the efficiency of transportation, in accordance with equation (450), will be as follows:

$$x_A = \frac{W}{30 \times \frac{10}{50} + 6.5 \times \frac{10}{5} + 25 \times \frac{20}{10}} = 0.0145W$$

$$x_B = \frac{W}{35 \times \frac{10}{35} + 12 \times \frac{10}{10} + 22.5 \times \frac{20}{15}} = 0.0192W\,.$$

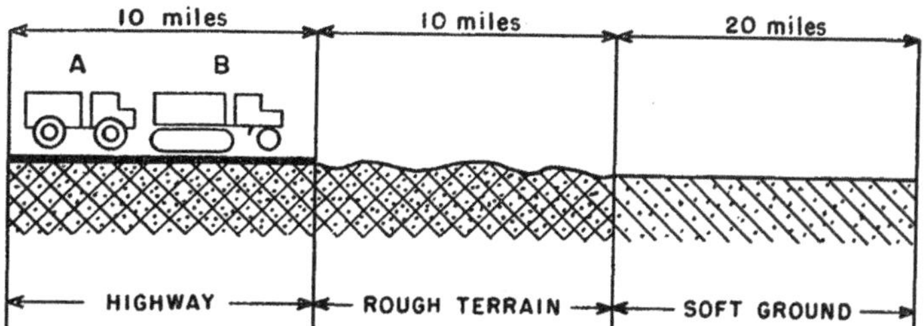

		HIGHWAY	ROUGH TERRAIN	SOFT GROUND
V mph	A	50	5	10
	B	35	10	15
u lb/m	A	0.6	1.3	2.5
	B	1.0	1.2	1.5

Fig. 183

Hence, the payload of vehicles *B* will be delivered at an efficiency approximately 33–1/3% higher than that of vehicles *A*. Accordingly, *B* should be preferred for the given operation.

For the sake of further elucidating the problem of transport efficiency, amphibians are compared with boats, using another measure. In Figure 184, the (ton miles per hour/gross horsepower) $\times$ 5.34 at the reported top speeds of a number of amphibians has been plotted as a function of Froude number. The (ton miles per hour/gross horsepower) $\times$ 5.34 (using short tons) is equivalent to the over-all propulsive efficiency or propulsive coefficient (n) times the lift-drag ratio (Δ/R) of the vehicle, where lift is interpreted as the total load-carrying ability (the total weight) of the vehicle. It is thus a dimensionless index of the transport efficiency of the vehicle in water. It will be higher with higher propulsion efficiencies and with better lift-drag ratios. Plotted in the same figure are the corresponding data for a normal round-bottom boat of about the same general size. It will be noted that, despite a good deal of scatter, some introduced no doubt by unrealistic maximum-speed claims rather than by reliable data, the indices of all the amphibians obviously leave much to be desired.

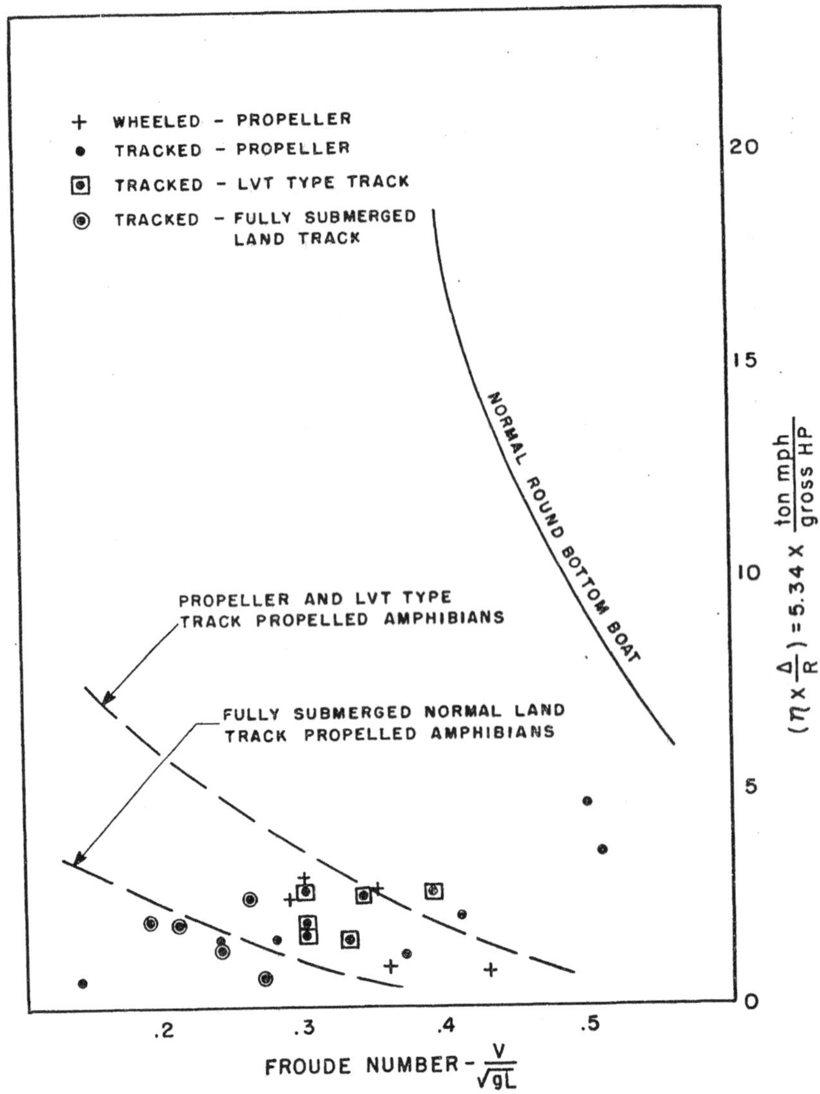

+ WHEELED - PROPELLER
• TRACKED - PROPELLER
TRACKED - LVT TYPE TRACK
TRACKED - FULLY SUBMERGED LAND TRACK
NORMAL ROUND BOTTOM BOAT
PROPELLER AND LVT TYPE TRACK PROPELLED AMPHIBIANS
FULLY SUBMERGED NORMAL LAND TRACK PROPELLED AMPHIBIANS
20
15
10
5
0
.2
.3
.4
.5
FROUDE NUMBER - $\frac{V}{\sqrt{gL}}$
$(\eta \times \frac{\Delta}{R}) = 5.34 \times \frac{\text{ton mph}}{\text{gross HP}}$

Fig. 184

Trafficability Maps

Conditions of off-the-road locomotion change not only with the type of ground but also with climatic conditions. The effect of climate is perhaps the strongest in the Arctic, where unconquerable marshes and lakes of the summer turn into rough frozen terrain in winter.[311] Even in moderate climates, the change from one season of the year to another may create conditions in which even the most advanced transport engines fail.[312] For the same reason, the Allies encountered tremendous difficulties in the operations spread across the globe when the same vehicle was supposed to operate the year round in all climatic conditions. Approximately 22 large organizations were engaged at that time and millions of dollars were spent for projects aiming to predict and improve the trafficability of motor vehicles. When progress in this field proved

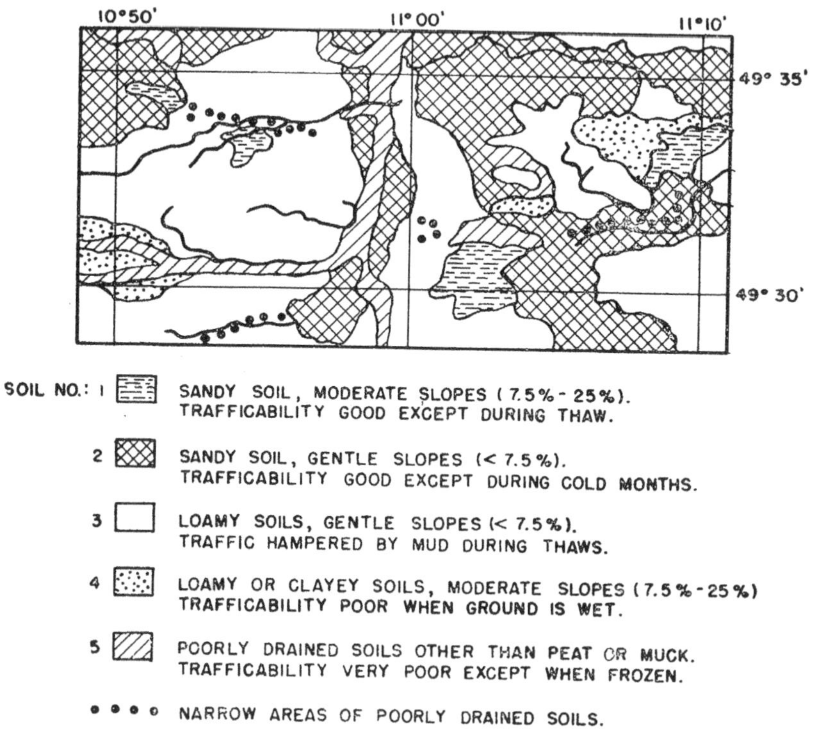

Fig. 185

slow and tedious, substantial efforts were concentrated almost exclusively on the prediction of soil trafficability so that existing vehicles could avoid innumerable terrain and meteorological traps. A survey of this work is made in Reference 313. The attempts at introducing climatological factors in trafficability forecast, which were made extensively during the last war, are described in Reference 314.

A striking feature of the discussed work is the haste with which it was performed, which resulted from the emergency at that time. Apart from the penetrometer, previously described, and the vague interpretation of soil-vehicle relationship, there was no scale in which a trafficable soil could be distinguished from a nontrafficable one. From this point of view, the tracing of maps to help an operation move along passable areas and avoid those which cannot be negotiated has been of debatable value.

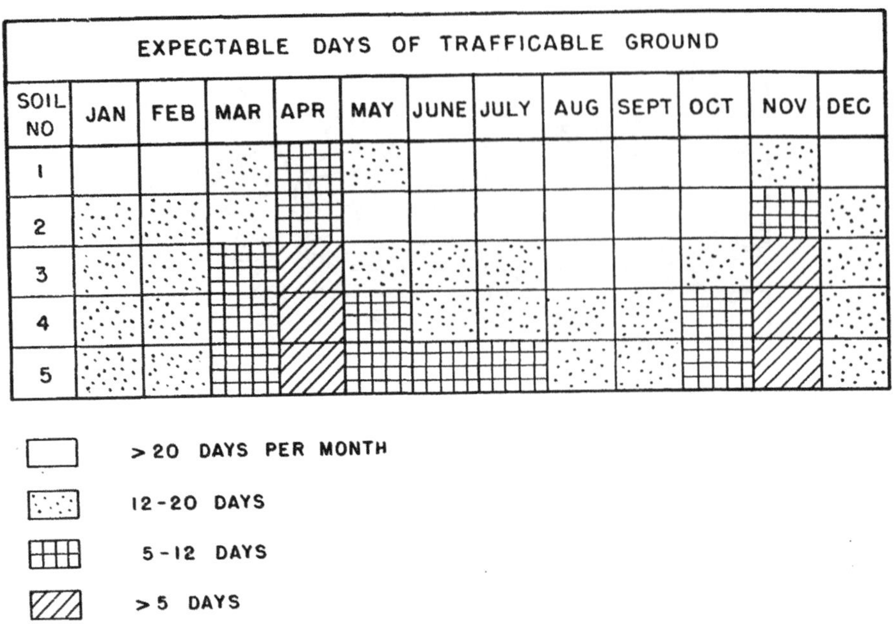

Fig. 186

Figures 185 and 186 represent a sample of such a map with the climatic "key" for the interpretation of trafficability. On the background of the outline of soil mechanics and soil-vehicle relationship discussed in this work, it appears clear that the presentation of trafficability in the form shown in Figure 185 constitutes a real step forward, toward the systematization of the concepts involved. However, it cannot solve the problem on the level of the vehicle driver, although it is undoubtedly useful when used at higher echelons of locomotion planning.

The main difficulty in establishing a "road" map of a terrain for use similar to that of the road map of an automobile association is the lack of an accepted method of measuring trafficability. Until this step is agreed upon, the mapping of soil surface from the locomotion point of view will always be handicapped. The other question is whether the measurement of trafficability is possible in a form that can be readily used. A discussion on the subject presented in this chapter suggests that techniques in the prediction of trafficability will be as involved as those of meteorological forecasts. In this case, the organization of soil mapping and trafficability forecasts would have to equal the organization of meteorological service, and the cost of such an enterprise would have to be confronted with its usefulness.

The only escape from these difficulties seems to be to improve the performance of vehicles so that they will be less sensitive to climatic and textural soil variations. If this is done, it may not be necessary to record and analyze many data, since they would not affect vehicle performance adversely. Thus, a radical improvement in cross-country vehicles may make it easier to produce trafficability maps and forecasts. If the present vehicles are not to be improved, the cost of the climatological approach to trafficability may be prohibitive, and the vicious circle thus closed.

Radical Improvement in Economy and Performance of Motor Vehicles

To summarize the discussion on the improvement in vehicle performance and economy, it may be concluded that the present limitations in this respect are rooted in the extremely complex mechanism of the phenomena involved, particularly in the realm of soil trafficability, and in the lack of a systematic study of the relevant fields.

Although it would be premature to make definite conclusions at this time, it may be agreed, with reasonable accuracy, that the performance and economy of cross-country vehicles will not be increased substantially by a mere increase in engine power or by a possible improvement in engine thermodynamic efficiency. A better adaptation of the vehicle form to the environment of operation is inevitable if radical progress is contemplated. However, the engine and power train remain the crux of the problem.

An attempt was made in Chapter IV to show how vehicle form was frozen and how the vehicle concept became standardized in terms of a few combinations of standard elements (Figure 46). Should there be any substantial change in the morphology of motor vehicles, then it would be unthinkable that the form of the component elements would remain the same. This refers first of all to the engine and transmission which affect vehicle form to the largest extent.

The internal-combustion engine, despite the variety of shapes, may

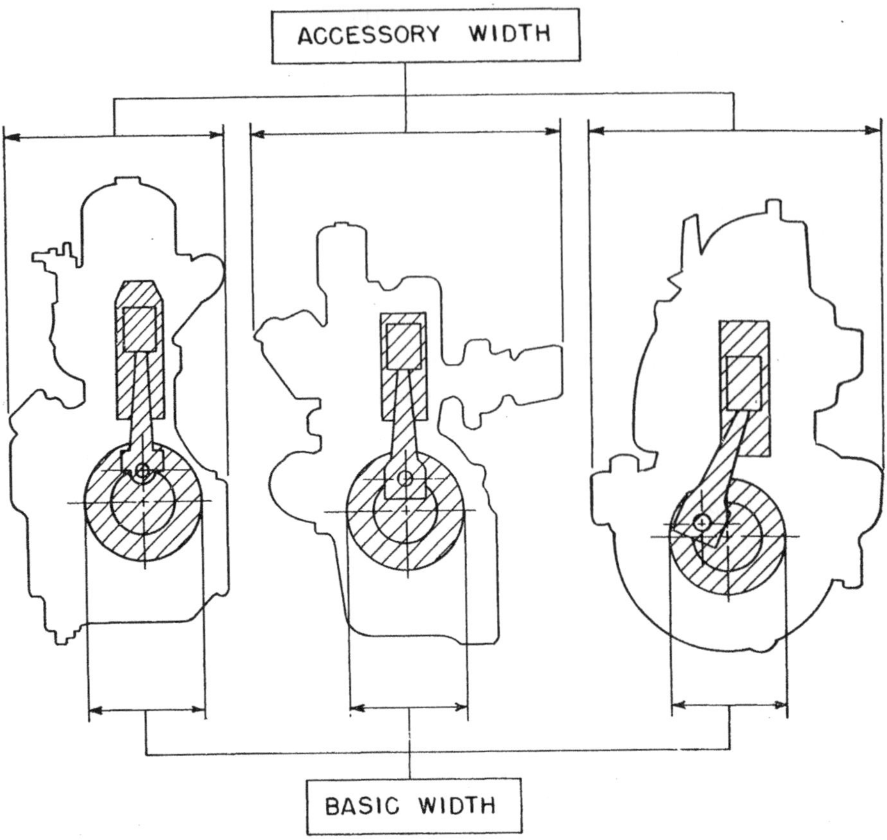

Fig. 187

always be seen in the fundamental form of the cylinder-crankcase unit, whose shape and dimensions, defined basically by the bore and stroke, have been widened and heightened by the accessories to such an extent that it forms an inflexible bulky block much larger than the fundamental unit (Figure 187). The installation of such a block may be possible only in the front, in the rear, or in the middle of the vehicle. It is unthinkable to spread it thinly, for instance, on both, or one side of the vehicle as was proposed in one of the Ledvinka patents (Figure 188). To this end, a flat engine of basic morphology determined by the bore-stroke characteristics should be used, i.e., an engine whose form has been augmented by the accessories in a functional way, different from present standards.

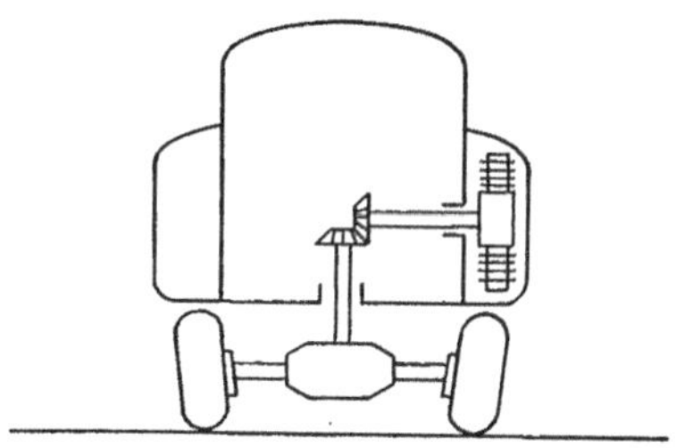

(LEDVINKA)
Fig. 188

The discussed example indicates that any radical change in the basic outline of a vehicle ties up very closely with a change in engine and power-train form, whose present standardized rigidity and nonfunctional bulkiness may be made responsible, to a large extent, for the stagnation in any substantial change of performance and economy of off-the-road motor vehicles.

It appears that, at present, only one possibility exists: breaking down the single power units into smaller elements and thus reshaping the engine-transmission aggregates in such a way that they fit the desired form of the vehicle. The multi-engine tank units built during World War II as emergency solutions [31] cannot be considered as an argument against this concept because they did not recognize the morphological requirements. An attempt to design a multi-engine aggregate of relatively small motors, with the full recognition of these requirements, although still incomplete, has been nonetheless encouraging.[315]

The concept of such an engine is in line with the rational trend in the development of internal-combustion motors. It is known that the power obtainable per unit of cylinder displacement increases with decreasing cylinder size.[204] Accordingly, the breaking down of 6, 8, or 12 cylinders into a compound unit of a total of, say, 24 or more cylinders would help to produce more power from one cubic inch of the displacement. A similar conclusion may be reached by assuming that the limit cylinder output is determined by the combined thermal and mechanical stresses which may be expressed in terms of the "specific speed" of the engine [316] equal to rpm $\sqrt{\text{power}}$, which is comparable to the notion of the "specific speed" of turbo machinery.[291]

A multi-engine aggregate would be, in addition, undoubtedly less expensive in production than a single unit of the same power. A partial use of aggregate elements may lead to fuel economy whenever full power is not required. This development therefore appears to be worthy of further investigation.

There may be some doubt as to whether any other radical engine development, e.g., the gas turbine, would bring substantial changes in vehicle performance. There are several doubts as to whether or not it will be economical to power the land vehicle with a gas turbine,[319] particularly in off-the-road operation. The question is of paramount importance if the effective fuel consumption, as defined in equation (447), is considered. The problem certainly is not a simple one and much remains to be done before a satisfactory solution, if any, is reached.[317, 318]

As far as regular reciprocating engines are concerned, no other radical change can be expected. These engines are the subject of development laws which were formulated as early as 1905.[320] The power, weight, and dimensions have been enclosed within inflexible limits.[316, 321] The volume of effort allocated to their study does not seem to be entirely warranted although some combinations of thermocycles may appear interesting.[322]

The engine and transmission, however, do not comprise the whole vehicle. A radical improvement in its economy and performance in an off-the-road operation necessitates a further study of the relationship between soil and a vehicle in order to find out whether or not the full potential strength of the ground is used in developing traction and "flotation." The present study of the problem discussed in Chapter VII indicates that this question is not answered at all and that little or no progress can be achieved in this respect with conventional vehicles. This conclusion has led to a search for more fundamental knowledge of the soil-vehicle relationship. The preliminary results indicate that there are good reasons to expect that some radical departure from conventional trends may substantially improve the methods and means of cross-country locomotion.

In any case, it appears evident that the more "universal" the vehicle is to be, the worse its efficiency and economy. The trend toward the differentiation of vehicle types in accordance with the requirements of the environment in which they have to operate is inevitable. A morphological analysis, whose principles were outlined in Reference 323, might be of some help in this subject. The trend toward standardizing vehicles as a whole should be limited to the standardization of the component elements of a vehicle only, while preserving the diversity of forms of the vehicles themselves. This is the way in which nature has built locomotion in the animal world.

XI. SCALE-MODEL TESTING AND DIMENSIONAL THEORY

The preceding chapters have stressed not only what is known and has been formalized in the field of vehicle mechanics, but also the large areas which are unsettled and often not yet explored. These areas lie largely, but not entirely, in the realm of off-the-road performance. Various analytical methods, for the most part borrowed from other more advanced fields, have been outlined, in the hope that if they are further developed and properly exploited, they should make significant advances possible at a relatively lower cost than a purely empirical approach.

In this chapter, scale-model testing will be considered as a tool for vehicle research and development. Scale-model testing, which has proven invaluable in aerodynamic and hydrodynamic studies for many years, is, of course, but one of many types of exploration which should prove useful in formulating and checking analytical work and in providing semi-empirical or fully empirical solutions to particular problems and classes of problems which resist straightforward analytical treatment.

Full-Size, Analogue, and Scale-Model Testing; Advantages and Limitations

Testing of full-size prototypes and their components has been the backbone of vehicle dynamics studies to date and is likely to remain of considerable importance primarily because of their relatively small size and ease in testing compared to ships and aircraft. In the many cases where an existing production or pilot model may be tested with a minimum of adaptation and special instrumentation, the desired tests may be run on the prototypes, provided the desired test conditions may be found or established. The difficulties, other than those already indicated, are for the most part obvious. Extensive changes are likely to be costly and time consuming. Control of test conditions is apt to be either poor or expensive. Perhaps the most important drawback from an analytical viewpoint is the mechanical and general complexity of any available full-size vehicles, which, on the one hand, will usually obscure the behavior under study, and, on the other, preclude the testing of simplified concepts so essential in building a sound analysis.

The alternative to full-size testing is some sort of analogue testing. The most obvious analogues are scale models, although experience in other fields indicates that distorted models may be useful for some problems, such as structures, while completely dissimilar models of the electrical analogy or even the high-speed computer type may be helpful in still others.

In general, all analogue testing has certain advantages, provided the development of basic apparatus, techniques, and interpretive understanding does not have to be done at the expense of a single job. The time and cost of obtaining a given amount of data, exclusive of initial model overhead, will be less with the analogue than with the prototype. These savings will accumulate rapidly in a long or extensive program involving many changes of the vehicle and the test conditions. The corollary to this is that in practice, more and hence better data will be obtained from the analogue.

Exclusion of "model overhead" in this consideration is of particular importance in vehicle testing. The cost of a "dissimilar" model might be relatively little, consisting perhaps of only a circuit diagram which may be readily executed in the existing apparatus. The cost of what is more usually considered a model, whether to scale or distorted, is likely to be high, and in certain instances, prohibitive. This is obviously a function of complexity and, less obviously, of scale. All but the simplest objects of interest in motor-vehicle mechanics are almost uniformly mechanically complex. Reduced scale requires increased precision, and precision costs time and money. It is for this reason that some differential in other matters may be accepted in testing a prototype when it is readily available, rather than constructing and testing a scale or distorted model of it. Of course, the first cost of a scale model is still but a fraction of that of a special pilot prototype. Therefore, where performance tests of the sort determined only by the configuration and running-gear details of the vehicle (as distinct from wear, ruggedness, engine cooling, etc.) are desired, at least prior to commitment of larger sums for other purposes, scale-model tests may still show great savings in both time and money.

In a laboratory devoted to scale-model testing, the "initial model overhead" problem may be largely solved by the construction of a few precise, highly flexible components from which an endless variety of configurations may be built up by a "meccano"-like operation. Such models then become apparatus and truly overhead. One field where it is thought that even this approach will be inadequate is in the important one of detailed testing of pneumatic-tire designs. Certain approximations may be made for the purpose of testing a complete vehicle configuration with tires; but it is likely that the cost of accurate scale-model tires with scaled flexibility and deflection characteristics under scaled loads would be so little different from the cost of full-size experimental tires (and, of course, many, many times greater than that of any tire in production) that their use would not be justified. This becomes particularly apparent when it is recalled that full-size experimental tires can be used to test the design from every conceivable viewpoint, wear, rupture strength, etc., in addition to obtaining its performance from a vehicle-mechanics viewpoint. Thus, apparatus for determining the detailed performance of tires should be constructed to test full-size tires.

In summary, it may be said that better control of all test variables, particularly the test conditions, usually will be achieved in analogue testing. Simplified concepts may be readily and economically evaluated, and extraneous effects arising from any cause whatsoever may be eliminated or at least minimized.

Aside from the model-cost problem already discussed at length, there are two other difficulties in analogue testing, neither of which is insuperable. First, a certain degree of understanding of the phenomenon under study is prerequisite to fruitful analogue experiments. Second, the precision, particularly with scale models, must carry over into the test apparatus and instrumentation and into techniques for controlling the conduct of the tests. Scale or distorted models are often sensitive to minor variations in intrinsic test conditions which would not measurably change the behavior of a full-size object. The reasons for this will become clearer as the discussion proceeds.

Before beginning a study of the theory of scale-model testing and the dimensional theory underlying it, a review of the types of problems in which some sort of testing in general, and scale-model testing in particular, might be used is in order. In the absence of a full analysis, scale models may be used in research to help formulate the basic hypotheses upon which an analysis may be based and to check any conclusions reached by analysis at any stage through to its completion. If, because of its complexity, the phenomenon does not lend itself to full analysis, scale-model testing, dimensional theory, and analysis may team together to provide semi-empirical prediction equations. Scale models may also be used to solve immediate, particular problems, such as the prediction of prototype performance under stated conditions, or of the performance changes to be expected as a result of contemplated design changes. Even when a full analysis is available, if it is extremely complex, model tests may yield particular design solutions more quickly and economically than calculation.

Principles of Dimensional Analysis

> Dimensional analysis is a method by means of which we deduce information about a phenomenon from the single premise that the phenomenon can be described by a dimensionally correct equation among certain variables. The generality of the method is both its strength and its weakness. With little effort, a partial solution to nearly any problem is obtained. On the other hand, a complete solution is not obtained, nor is the inner mechanism of a phenomenon revealed, by dimensional reasoning alone.
>
> The result of a dimensional analysis of a problem is a reduction of the number of variables in the problem... [324]

Dimensional analysis has been developed principally by British and American scientists, and its first extensive use is credited to Lord Rayleigh at about the turn of the century. He is also credited with first stating, largely intuitively, some of its first principles, although, according to Murphy,[325] M. Dupré preceded him in this respect by approximately twenty-five years. Buckingham in 1914 enunciated but did not entirely prove his basic theorem which embraces all of dimensional analysis, thus beginning the process of replacing intuition by rigor, mysticism by mathematics. This process was nearly completed in 1922 by Bridgman.[326] In the time since, numerous publications have dealt with the subject, often from the viewpoint of a particular field in which the author is applying general principles.[132, 327] In the past few years, two new texts have appeared;[324, 325] the one by Langhaar, already quoted, represents a still further step in the direction of bringing rigor to the theory of dimensional analysis.

The quotation at the beginning of this section contains the essence of dimensional analysis, which may be restated as two axioms:

(a) absolute numerical equality may exist only when quantities are similar qualitatively; and

(b) the ratio of the magnitudes of two like quantities is independent of the units used in their measurement, provided that the same units are used for evaluating each.

The second statement may be considered either as axiomatic to dimensional analysis or as a rule of operation in setting up the systems of units used throughout the physical sciences, for it is certainly a quality present in all physical measurements, except those which are in themselves normally expressed as ratios, as, for example, angles, temperature, etc.

Quantities measured in the physical sciences may be loosely divided into two classes: primary quantities, which are those that may with more or less ease be measured by comparing the quantity to be measured with a standard unit and forming the ratio; and secondary quantities, which are most conveniently obtained by combining by some rule of operation two or more measurements of primary quantities. As a consequence of the axioms, it may be shown that the rule of operation by which a secondary quantity must be formed from the primary quantities involved, or, more generally, by which any dependent measurement of a phenomenon is evaluated in terms of the independent factors causing it, is as follows:

$$\alpha = C_\alpha \, . \, a_1^{C_1} \, . \, a_2^{C_2} \, . \, a_3^{C_3} \, \ldots \, a_n^{C_n} \, , \tag{451}$$

where α is the dependent factor, C_α is dimensionless, and $a_i^{C_i}$ is a power of the independent variable. It should be noted that the primary or independent quantities are not necessarily the fundamental dimensional units which will be dealt with later; for example, mass M, length L, and time T.

Equations describing a physical system may be classed for dimensional purposes as either homogeneous or nonhomogeneous. An equation is said to be dimensionally homogeneous if the form of the equation does not depend on the fundamental units of measurement. From this definition, it follows that an equation of the form

$$x = a + b + c + \ldots \tag{452}$$

is dimensionally homogeneous if and only if the variables x, a, b, c, ... all have the same dimensions. All other equations, however useful or valid, and there are numberless ones, are classed as dimensionally nonhomogeneous. When it is further stipulated that, in general, the numerical coefficients in an equation may not be considered to have dimensions, the class of valid nonhomogeneous equations is more readily seen to be large.

If, in the definition of dimensional analysis on page 457, the term "dimensionally homogeneous" is substituted for "dimensionally correct," then the basic theorem of Buckingham, known as the Pi Theorem (no relation of this pi to that of a circle), may be approached. This theorem states that if an equation is dimensionally homogeneous, it can be reduced to a relationship among a complete set of dimensionless products. It may be shown further that a complete set of such products will consist of $(n - r)$ independent functions, where n is the number of variables in the original equation and r, as Langhaar has shown, is the rank of the dimensional matrix of the n variables. The dimensional matrix is a rectangular array of numbers formed as follows:

$$\begin{array}{c|ccccc} & a_1 & a_2 & a_3 & \cdots & a_n \\ \hline b_1 & c_{1,1} & c_{1,2} & c_{1,3} & \cdots & c_{1,n} \\ b_2 & c_{2,1} & c_{2,2} & \cdot & \cdots & \cdot \\ \vdots & \vdots & \vdots & \vdots & \cdots & \vdots \\ b_s & c_{s,1} & c_{s,2} & \cdot & \cdots & c_{s,n} \end{array} \quad (s < n) \qquad (453)$$

where the a's are the variables in the original equation, the b's are the fundamental dimensional units of which they are composed (mass M, length L, and time T, for example, in the usual mechanical system), and c_s expresses the exponent of the fundamental units in the product, which defines the ultimate dimensions of the respective original variables. The rank of such a matrix is equal to the highest order of any nonzero minor within any determinant which may be formed from any selection of s columns. Usually, in dimensional analyses, the rank r will equal the number of fundamental dimensional units s.

In performing a dimensional analysis, the procedure, in brief for the moment, is first to list the independent variables which are thought to affect the phenomenon under study, and the dependent variables by means of which the behavior is to be described. From these, the $(n - r)$ independent dimensionless coefficients are formed, usually without great difficulty, and subsequently rearranged and transformed into what appears to the practitioner to be the most illuminating form. Finally, the functional equations are written by equating each dependent coefficient to an undefined function of all the independent coefficients.

Evidently, the crux of the operation is the selection of the independent variables "thought to affect the phenomenon under study." There is no rule by which these may be selected infallibly. They are exactly the same factors which would appear in a dimensionally homogeneous, analytical solution of the same problem to the same degree of refinement. Since an analytical solution is purely the result of mathematical operations upon certain premises stated before the analysis began, these variables are precisely the same factors with which such an analysis would begin. The same necessity to formulate hypotheses concerning the mechanism of the behavior studied exists for a dimensional analysis as for a more formal analysis, exept that, for the former, the hypotheses need not be formulated mathematically. Except in degree, there is the same basic concern to eliminate trivial factors from the analysis, in order to avoid unnecessary complication and, at the same time, not to overlook important variables, whose omission would lead to false conclusions.

After both the independent and dependent variables to be considered are selected (it is convenient to keep these separated), the dimensional matrix already referred to can be formed. To do this, a selection of fundamental units must be made. In mechanical systems, mass, length, and time are usually used, although force may readily and often profitably be used in place of mass. In certain instances in which no part of the phenomenon depends on Newton's relation between mass and force, as in certain cases of very slow motion, both force and mass may be considered as being fundamental, with a resultant increase in the rank of the dimensional matrix and, hence, a decrease in the number of dimensionless variables controlling the system behavior. In order to form the matrix, the so-called dimensional formulas for the variables selected must be known. For convenience, these are given below for the several variables which might enter the vehicle-mechanics picture.

		MLT *System*	*FLT* *System*
Independent Variables:			
Geometric Variables			
l	length	L	L
b	width	L	L
h (etc.)	grouser height	L	L
a_1, a_2 (etc.)	center-of-gravity locations	L	L
r_g	radius of gyration	L	L
θ	angular measurements	—	—
n_o	number of wheels, etc.	—	—

		MLT *System*	*FLT* *System*
Other Object Properties			
p	parameter of wheel flexibility*	$ML^{-1}T^{-2}$	FL^{-2}
q	parameter of wheel flexibility*	MT^{-2}	FL^{-1}
c_s	suspension spring rate	MT^{-2}	FL^{-1}
μ	coefficient of friction soil to wheel material	—	—
Soil Properties			
r	soil particle size	L	L
c	cohesion	$ML^{-1}T^{-2}$	FL^{-2}
ϕ	angle of internal friction	—	—
ϱ	mass density ($=\gamma\, g$)	ML^{-3}	$FL^{-4}T^{2}$
E	bulk modulus of elasticity	$ML^{-1}T^{-2}$	FL^{-2}
η	viscous property	$ML^{-1}T^{-1}$	$FL^{-2}T$
System Properties			
v	velocity	LT^{-1}	LT^{-1}
F	applied force	MLT^{-2}	F
M	applied torque	$ML^{2}T^{-2}$	FL
g	acceleration of gravity	LT^{-2}	LT^{-2}
Dependent Variables:			
R	measured force	MLT^{-2}	F
z_1, z_2 (etc.)	measured displacements	L	L
i_o	slip ratio	—	—

* The curve of axle vertical displacement (δ) as a function of vertical load when a flexible wheel (such as a tire) is loaded on a hard surface may be approximated by an equation of the form

$$P = p\delta^2 + qf,$$

where P = vertical load, f = axle "deflection," and p and q are flexibility parameters. In order for this equation to be dimensionally homogeneous, each of the terms on the right-hand side must have the dimensions of force, i.e., the dimensions of P on the left-hand side. Thus the dimensions of p are FL^{-2} or $^{1}ML^{-}T^{-2}$, and those of q are FL^{-1} or MT^{-2}.

After the matrix is formed, its rank is checked. An obvious example of a case where the rank is less than the number of fundamental units is in the study of a purely static situation, but the MLT system is used in the matrix. Since time is not really involved, despite its constant recurrence in the dimensional formulas, it should be expected *a priori* that the number of fundamental dimensions really involved, and hence the number by which the variables are reduced when the situation is described dimensionlessly, should be two rather than three. Of course, the question does not arise if the *FLT* system is used for this type of problem.

Formation of the dimensionless products may be accomplished in various more or less formal ways. No method is guaranteed to yield the results in the most useful form and, hence, none is "better" than the others. Usually, they may be formed by inspection, with only the following preparation. Select a number of the independent variables equal to the rank of the dimensional matrix and containing among them all the fundamental dimensions used. It is wise to make this selection from among those variables which are most basic to the system and, hence, are least likely to be later judged negligible. If possible, variables which are going to receive special study should not be included in this group. Next, combine each of the remaining independent, then dependent, variables singly with the base group in such a way as to yield a dimensionless product. This will automatically produce $(n - r)$ dimensionless products which must be independent because each of the $(n - r)$ nonbase variables appears separately in only one of the products.

If the choice of base variables has been a fortunate one, the resulting dimensionless products will be helpful for either analytical or experimental purposes. However, if not, they may be readily transformed, often to more attractive forms. Inverting a coefficient or raising it to any power does not alter its dimensionless qualities or its independence, or add to or subtract from the total number of such coefficients $(n - r)$. These operations may therefore be performed on individual products at will. Again, multiplying or dividing any two products does not alter the fact that they, and their result, are dimensionless. It does create three products where only two are wanted, and moreover, only two of the three, but any two, are independent. From this, it may be gathered that any two products may be multiplied together and their product taken in place of either one, but not both, of the two original coefficients. These operations in any combination may be carried out to any length. The number, independence, and dimensionlessness of the resulting system of products will not differ from those resulting from the original formation of the products, and the new system is precisely as good theoretically as the old. As an aid to analysis or experiment, of course, the new system may be more or less useful depending on the skill with which the transformation is done.

Finally, the functional equations relating the dependent dimensionless products to the independent products are written. The dimensional analysis is completed. Usually, the difficulties are only beginning.

Theory of Testing of Models

The understanding which makes the testing of models a valid procedure for obtaining data on a prototype system stems directly from dimensional theory. Buckingham's Pi Theorem states that if

$$\alpha_1 = f(\alpha_2, \alpha_3, \alpha_4 \dots \alpha_n) \tag{454}$$

expresses the relationship between a dependent variable (α_1) and a number of independent variables (α_2, α_3, $\alpha_4 \dots \alpha_n$), and is a dimensionally homogeneous equation, it may be replaced by the equation

$$\Pi_1 = F(\Pi_2, \Pi_3 \dots \Pi_{n-r}) , \tag{455}$$

where the Π's are independent dimensionless products. Equation (455) is equally valid for predicting the behavior of the system.

The problem to be solved might be most simply stated as follows: to find α_1, given α_2, α_3, $\alpha_4 \dots \alpha_n$. If the functional relationship is unknown, or exceedingly complex, α_1 may be found by means of measuring it in a test in which α_2, $\alpha_3 \dots \alpha_n$ have all their specified values. This, of course, is a full-size test. The Pi Theorem states that equation (455) expresses the system performance as truly as equation (454). In equation (455), the unknown is combined with certain independent variables whose values are known (Π_1), and is a function of the $(n - r - 1)$ other independent dimensionless products of the $(n - 1)$ known variables. Thus, α_1 may be found by running a test to find Π_1.

This test will be described by Π_2, Π_3, ... Π_{n-1}, each of which is a product of known values of α_2, $\alpha_3 \dots \alpha_n$ and, hence, have known numerical values. These are of the type of the well-known Reynolds number of hydrodynamics:

$$(\Pi_j =)\ Re = \frac{vl\varrho}{\eta} , \tag{456}$$

where v, l, and ϱ have the same or similar meanings as used in soil-vehicle studies (see tabulation on pages 460–61), and η is the viscosity of the test fluid. The test conditions α_1, α_2, $\alpha_3 \dots \alpha_n$, or in this case, ... v_1, l_1, ϱ_1, $\mu_1 \dots$, are such that the numerical value of $\Pi_j = Re$ is specified. There are no restrictions in setting the individual test conditions ... v_1, l_1, ϱ_1, μ_1, ..., provided only that they are arranged so that the values of each of the Π's are preserved. If it is possible to arrange it so that during the test, each Π of the test equals the corresponding Π of the given conditions, then the measured Π_1 of the test must equal the Π_1 under the desired conditions, even though the tests, to the lay eye, bear little apparent relationship to the prototype and prototype conditions.

Where the dimensional analysis is the most complete analytical solution available for the problem, scale models are generally necessary for any such tests, including the scale 1 : 1. By this means, the Π's formed from the geometric properties of the prototype are all made equal in the model and prototype systems. Thus,

$$\left(\frac{b}{l}\right)_m = \left(\frac{b}{l}\right)_p, \quad \left(\frac{h}{l}\right)_m = \left(\frac{h}{l}\right)_p, \quad \left(\frac{r_g}{l}\right)_m = \left(\frac{r_g}{l}\right)_p, \quad \left(\frac{r}{l}\right)_m = \left(\frac{r}{l}\right)_p, \text{ etc.} \tag{457}$$

Distorted models, i.e., models in which geometric similarity is not adhered to, are feasible only where analysis is adequate to reconvert the results obtained on the distorted model to the prototype conditions, or where there is a great deal of experimental background which, in effect, serves the same purpose. Distorted models probably will not find wide application in vehicle-mechanics studies in the near future.

Conditions for Scale-Model Vehicle Testing

Just prior to World War II, scale-model vehicle testing to determine such factors as the high-speed stability of road trains, and the effects of various design changes on the cornering performance of road racing cars, was successfully started in Germany.[328]

The relationships between a prototype and a geometrically scaled vehicle model were derived by Lutz by the more cumbersome and less general method of considering the physical laws involved, which is popular in Continental Europe.[329] It is instructive to follow this derivation and to compare the results with those obtained by dimensional analysis. Let

M =	prototype	mass	m =	model	mass
F =	,,	force	f =	,,	force
T =	,,	time	t =	,,	time
L =	,,	length	l =	,,	length
A =	,,	area	a =	,,	area
P =	,,	contact pressure	p =	,,	contact pressure
I =	,,	moment of inertia	i =	,,	moment of inertia
HP =	,,	power	hp =	,,	power
ϱ_p =	,,	mass density	ϱ_m =	,,	mass density

V	= prototype speed		v	= model speed	
N	= ,,	rpm	n	= ,,	rpm
μ_p	= ,,	coefficient of adhesion, wheels to roadway	μ_m	= ,,	coefficient of adhesion, wheels to roadway
g	= ,,	acceleration of gravity	g	= ,,	acceleration of gravity.

By Newton's Law,

$$F = \frac{ML}{T^2} \text{ and } f = \frac{ml}{t^2}$$

or

$$\frac{F}{f} = \frac{MLt^2}{mlT^2}.$$

Now,

$$M = \varrho_p L^3 \text{ and } m = \varrho_m l^3 ;$$

if

$$\frac{\varrho_p}{\varrho_m} = K ,$$

then

$$\frac{M}{m} = K \frac{L^3}{l^3}.$$

Also, since g is, for practical purposes, invariant between the two systems, and since the acceleration of gravity obviously would influence the behavior of a vehicle passing over even a slight bump, all corresponding accelerations in the two systems must be adjusted in such a way that they too will be equal. Thus,

$$\frac{L}{T^2} = \frac{l}{t^2} \text{ or } \frac{t^2}{T^2} = \frac{l}{L}$$

and

$$\frac{F}{f} = K \frac{L^3}{l^3}. \tag{458}$$

If now the condition is imposed that contact pressures between the vehicle and roadway are to be equal in model and prototype systems, i.e.,

$$P = p, \text{ or } \frac{F}{A} = \frac{f}{a},$$

and noting that, by geometric similarity,

$$\frac{A}{a} = \frac{L^2}{l^2},$$

then

$$\frac{F}{f} = \frac{L^2}{l^2}.$$

From the last equation and (458),

$$K = \frac{\varrho_p}{\varrho_m} = \frac{l}{L} \text{ or } \frac{M}{m} = \frac{L^2}{l^2}. \tag{459}$$

Similarly, for speeds,

$$\frac{V}{v} = \frac{L/T}{l/t} = \frac{L}{l}\frac{t}{T} = \frac{L}{l}\sqrt{\frac{l}{L}} = \sqrt{\frac{L}{l}}; \tag{460}$$

for rpm's,

$$\frac{N}{n} = \frac{t}{T} = \sqrt{\frac{l}{L}};$$

for moments of inertia,

$$\frac{I}{i} = \frac{ML^2}{ml^2} = \frac{L^2}{l^2} \cdot \frac{L^2}{l^2} = \frac{L^4}{l^4}; \tag{461}$$

and for powers,

$$\frac{HP}{hp} = \frac{FLt}{flT} = \frac{L^2}{l^2}\frac{L}{l}\sqrt{\frac{l}{L}} = \left(\frac{L}{l}\right)^{5/2}. \tag{462}$$

The forces arising from adhesion of the vehicle wheels to the road (F_a and f_a) are given by

$$F_a = Mg\mu_p \text{ and } f_a = mg\mu_m.$$

It is necessary that these forces bear the same relationship one to the other as the forces arising from the action of Newton's Law, i.e., that

$$\frac{F_a}{f_a} = \frac{L^2}{l^2}, \text{ but } \frac{F_a}{f_a} = \frac{Mg\mu_p}{mg\mu_m} = \frac{L^2}{l^2}\frac{\mu_p}{\mu_m},$$

from which it may be concluded that

$$\frac{\mu_p}{\mu_m} = 1\,. \tag{463}$$

In summary, by letting $L/l = \lambda$ and remembering that the model and prototype are presumed to be geometrically similar,

$$\left.\begin{aligned}
&\frac{F}{f} = \lambda^2 && \text{(a)}\\
&\frac{\varrho_p}{\varrho_m} = \frac{1}{\lambda} && \text{(b)}\\
&\frac{V}{v} = \lambda^{\frac{1}{2}} && \text{(c)}\\
&\frac{N}{n} = \frac{1}{\lambda^{\frac{1}{2}}} && \text{(d)}\\
&\frac{I}{i} = \lambda^4 && \text{(e)}\\
&\frac{HP}{hp} = \lambda^{5/2} && \text{(f)}\\
&\frac{\mu_p}{\mu_m} = 1\,. && \text{(g)}
\end{aligned}\right\} \tag{464}$$

These relationships, of course, govern both the model setup and the interpretation of measured model behavior in terms of prototype behavior. Note that condition (464d) is actually completely determined by (464c), and that (464e) is more usually thought of as part of geometric similarity through the mechanism of the radius of gyration—the mass associated with that radius being determined by the relationship (464a).

By the more general dimensional-analysis approach, the procedure is as follows, assuming geometric similarity at the outset:

	l	F	ϱ	V	I	HP	P	N	g	μ
M	0	1	1	0	1	1	1	0	0	0
L	1	1	—3	1	2	2	—1	0	1	0
T	0	—2	0	—1	0	—3	—2	—1	—2	0

(465)

Note: $n - r = 7$.

If ϱ, l, and g are selected as the base variables, then seven terms may be formed. If the term containing, say, v is selected as the dependent term, the following equation may be written:

$$\frac{v^2}{gl} = \Phi\left(\frac{F}{\varrho g l^3}, \frac{I}{\varrho l^5}, \frac{P}{\varrho g l}, \frac{HP^2}{\varrho^2 g^3 l^7}, \frac{N^2 l}{g}, \mu\right). \tag{466}$$

Now, in order to run a theoretically valid scale-model test, it is necessary that

$$\left.\begin{aligned}
&\left(\frac{F}{\varrho g l^3}\right)_m = \left(\frac{F}{\varrho g l^3}\right)_p && \text{(a)}\\
&\left(\frac{I}{\varrho l^5}\right)_m = \left(\frac{I}{\varrho l^5}\right)_p && \text{(b)}\\
&\left(\frac{P}{\varrho g l}\right)_m = \left(\frac{P}{\varrho g l}\right)_p && \text{(c)}\\
&\left(\frac{HP^2}{\varrho^2 g^3 l^7}\right)_m = \left(\frac{HP^2}{\varrho^2 g^3 l^7}\right)_p && \text{(d)}\\
&\left(\frac{N^2 l}{g}\right)_m = \left(\frac{N^2 l}{g}\right)_p && \text{(e)}\\
&\mu_m = \mu_p\,, && \text{(f)}
\end{aligned}\right\} \tag{467}$$

where the subscripts "m" refer to the model system and the subscripts "p" refer to the prototype system. If these conditions are fulfilled, then

$$\left(\frac{v^2}{gl}\right)_m = \left(\frac{v^2}{gl}\right)_p. \tag{468}$$

By imposing the same condition as used previously, i.e., $P_m = P_p$, letting $l_p/l_m = \lambda$ as before, and recalling that g is for practical purposes a constant, (467c) becomes

$$\frac{\varrho_p}{\varrho_m} = \frac{1}{\lambda} . \tag{a}$$

From (469a) and (467a),

$$\frac{F_p}{F_m} = \lambda^2 ; \tag{b}$$

(469a) and (467b) give

$$\frac{I_p}{I_m} = \lambda^4 ; \tag{c}$$

(469a) and (467d) give

$$\frac{HP_p}{HP_m} = \lambda^{5/2} ; \tag{d}$$

(467e) gives

$$\frac{N_p}{N_m} = \frac{1}{\lambda^{\frac{1}{2}}} ; \tag{e}$$

and (467f) gives

$$\frac{\mu_p}{\mu_m} = 1 . \tag{f}$$

If these conditions are met, then, from (467e),

$$\frac{v_p}{v_m} = \lambda^{\frac{1}{2}} . \tag{g}$$

(469)

The equivalence between the conditions expressed by equations (464) and those expressed by (469) is exact, as, of course, it must be if the two methods are both correct. In general, the dimensional-analysis approach will prove to be shorter and less subject to careless errors.

Note that this "on-road" test setup neglects any consideration of air lift and drag forces, even though relatively high speeds are not ruled out by any of the discussion. It is convenient to consider these as being entirely distinguishable components of the total behavior which may be separately studied in a wind tunnel, with either a model or prototype. This is facilitated by running the road-behavior model tests on a moving roadway, with the model stationary relative to the ambient atmosphere.[330]

Wind-tunnel model tests have their own problems, which will not be dealt with, except to say that vehicle problems have proved to be largely solvable as a result of the vast experience with similar problems in aircraft testing.[48] It should be noted that if road-behavior and aerodynamic tests are to be run concurrently, aerodynamic lift forces cannot bear the proper relationship to inertia and gravity forces under the condition $P = p$ unless the density of the air follows the relationship (469a) (derived for the density of the model material, but valid nonetheless). In any event, barring both impractical tests in atmospheres of special viscosities and trivial cases (scale = 1 : 1), the air drag forces on the model will not be in the proper prototype relationship to the other forces but this introduces an error which is generally either correctable or negligible.

Consider next the problem of testing model vehicles in soil. The on-road tests just discussed might be loosely classified as "acceleration" tests, i.e., tests in which the dynamic behavior was determined largely by forces arising out of inertia, gravity, and friction. When the vehicle goes into soil, the "acceleration" test is compounded with a structural test, moreover, one in which the material is worked well beyond its elastic limit. The material so abused is the soil. Such a compound test invariably gives birth to difficulties due to the theoretical necessity of scaling the mechanical properties of the "structural" material, which is usually impractical if not impossible. Examination of the conditions imposed for model tests on-road shows that if air forces are neglected, there are no conflicting scaling requirements. Condition (469a) is not particularly practical, but further examination of the problem reveals that it can be fully handled by proper ballasting of moving parts and the vehicle as a whole, to achieve the correct masses and mass distributions. It will be shown that off-road model tests require many conflicting conditions for theoretical precision.

The preceding chapters have given a certain insight into many of the factors which might be expected to play a role in the soft-ground performance of vehicles. These have already been summarized. It should be noted that further work may show that additional soil factors will be required to describe adequately the soil and its condition for vehicle purposes, such as factors proposed in the discussion of track and wheel slip in Chapter IX. For the moment, however, it is possible to proceed without them, because they have, so far, not been evaluated experimentally in any way. With this possible exception, it is thought the previously quoted list of factors in the preceding table is complete. For illustrative and discussion purposes, it includes several factors which have to date proved to be of negligible importance, so that this list errs on the side of being too complete, bringing with it to the analysis all the attendant difficulties of a great many variables.

Considering that all the independent geometric properties are taken care of by the statement of geometric similarity, the dimensional matrix of the remaining variables may be written as follows:

	Object Properties				*Soil Properties*					
	l	q	ψ	μ	r	c	ϕ	ϱ	E	η
M	0	1	1	0	0	1	0	1	1	1
L	1	0	0	0	1	—1	0	—3	—1	—1
T	0	—2	—2	0	0	—2	0	0	—2	—1

	System Properties				*Dependent Variables*		
	v	F	M	g	R	z	i_o
M	0	1	1	0	1	0	0
L	1	1	2	1	1	1	0
T	—1	—2	—2	—2	—2	0	0

$n - r = 14$ (470)

By selecting F, ϱ, and l as the base variables (three variables which will always be present in any experiment), the Π terms will be as follows:

$$\text{Object Properties: } \frac{pl^2}{F}, \ \frac{ql}{F}, \ \frac{\psi l}{F}, \text{ and } \mu \tag{471)–(474}$$

$$\text{Soil Properties: } \frac{r}{l}, \ \phi, \ \frac{cl^2}{F}, \ \frac{El^2}{F}, \text{ and } \frac{\eta^2}{\varrho F} \tag{475)–(479}$$

$$\text{System Properties: } \frac{v^2 \varrho l^2}{F}, \ \frac{M}{Fl}, \text{ and } \frac{\varrho g l^3}{F} \tag{480)–(482}$$

$$\text{Dependent Variables: } \frac{R}{F}, \ \frac{z}{l}, \text{ and } i_o\,. \tag{483)–(485}$$

It should be observed that ϱ appears in three terms only. Two of these are associated with dynamic forces $v^2\varrho l^2/F$ and $\eta^2/\varrho F$, and in the third, ϱ appears in combination with g as ϱg, which is, of course, the weight density or specific weight (γ) of the soil. Thus, if the two truly dynamic factors are neglected, ϱ, as a controlling variable, may also be neglected and γ substituted. In this form, it becomes clear that the effect of neglecting g is that of neglecting forces arising from the weight of the soil (as distinct from its mass). Incidentally, transforming the three products under discussion gives the following forms familiar in hydro- and aerodynamic research:

$$\frac{v^2\varrho l^2}{F} \overset{D}{=} \frac{F}{\varrho v^2 l^2} \text{ the pressure coefficient} \tag{486}$$

$$\frac{v^2\varrho l^2}{F} \cdot \frac{F}{\varrho g l^3} = \frac{v^2}{gl} \text{ the Froude number, and} \tag{487}$$

$$\frac{\varrho F}{\eta^2} \cdot \frac{v^2\varrho l^2}{F} \overset{D}{=} \frac{vl\varrho}{\eta} \text{ the Reynolds number,} \tag{488}$$

where the symbol $\overset{D}{=}$ signifies dimensional equality.

Before proceeding to the implications of these products on setting up and interpreting an actual model test, their meaning will be considered. In a system of the sort which is dealt with here, it may be visualized that the forces, considered as separable components arising from various sources, are as follows:

forces generated by soil cohesion $\alpha\ cl^2$,
forces arising from acceleration of the soil particles $\alpha\ \varrho^2 l^2 v^2$,
forces arising from viscous shearing of the soil $\alpha\ \eta v l$,
forces generated by moving soil with or against gravity $\alpha\ \varrho g l^3$,
forces from elastic deformation of the soil $\alpha\ El^2$,
forces resulting from applied loads such as model weight $\alpha\ F$,
forces resulting from applied torque $\alpha\ M/l$,
forces generated by soil friction α tan ϕ, and
forces generated by soil-to-wheel friction α tan μ.

Thus, in a theoretically correct model test, in which all the products for the model system are made equal to the corresponding products from the prototype system, this equality may be interpreted as expressing the fact that ratios of components of net forces to each other, or to the total, acting at corresponding points in the two systems and, hence, in the entire system, are the same in the two systems. The importance of this should be evident. In a scale-model test, only net forces are measurable. If the ratio of each component to the net is the same, model and prototype, the expansion factor to be applied to the measured net force to predict prototype net force may be determined from a knowledge of the ratio which must be applied to any one force in the system, such as the weight. When the components must be separated and each expanded by a different ratio, due to unavoidable departures from equality of corresponding and important dimensionless products, either a rather well-advanced analysis must be used, or some sort of empirical division made, based on extensive experiment and correlation.

To illustrate this simply, consider equation (141), which is concerned with the safe strip load on a soil per running unit of strip length:

$$V = (2l)cN_c + (2l)qN_q + \frac{\gamma}{2}(2l)^2N_\gamma ,$$

where $(2l)$ = the width of the strip in, say, inches and q = the surcharge in, say, pounds per inch (per inch). If it is considered that q arises only from sinkage z, then

$$q = \gamma z . \tag{489}$$

Also, in order to make the equation apply to a three-dimensional case, let $b \geqslant 15 \times 2l$ be the length of a strip footing and, by neglecting end effects,

$$(bV) = b(2l)cN_c + \gamma\,[bz(2l)N_q + \tfrac{1}{2}b(2l)^2N_\gamma] . \tag{490}$$

Assume now that a scale-model test of some prototype footing is to be made in a given soil. The relationship between the measured model load and the prototype load obviously would be given by

$$\frac{(bV)_p}{(bV)_m} = \frac{\{b(2l)cN_c + \gamma\,[bz(2l)N_q + \frac{1}{2}b(2l)^2N_\gamma]\}_p}{\{b(2l)cN_c + \gamma\,[bz(2l)N_q + \frac{1}{2}b(2l)^2N_\gamma]\}_m} . \tag{491}$$

Recall that N_c, N_q, and N_γ are dimensionless functions of ϕ alone. If the values of these separate coefficients are known, there is no need to run the test; if not, the evaluation of the expansion factor, as it stands, is impossible. By introducing geometric similarity, (491) can be rewritten as follows:

$$\frac{(bV)_p}{(bV)_m} = \frac{k_1 l_p^2 c_p (N_c)_p + \gamma_p \, [k_2 l_p^3 (N_q)_p + k_3 l_p^3 (N_\gamma)_p]}{k_1 l_m^2 c_m (N_c)_m + \gamma_m \, [k_2 l_m^3 (N_q)_m + k_3 l_m^3 (N_\gamma)_m]}$$

$$= \frac{l_p^2 \, \{k_1 c_p (N_c)_p + \gamma_p l_p \, [k_2 (N_q)_p + k_3 (N_\gamma)_p]\}}{l_m^2 \, \{k_1 c_m (N_c)_m + \gamma_m l_m \, [k_2 (N_q)_m + k_3 (N_\gamma)_m]\}} \,. \qquad (492)$$

If $\phi_m = \phi_p$, then $(N_c)_m = (N_c)_p$, etc. Then the subscripts may be dropped, and the equation reduces to the form

$$\frac{(bV)_p}{(bV)_m} = \left(\frac{l_p}{l_m}\right)^2 \cdot \frac{K_1 c_p + \gamma_p l_p K_2}{K_1 c_m + \gamma_m l_m K_2} \,. \qquad (493)$$

This reasoning, of course, does not provide any insight into the values of K_1 and K_2, but the right-hand factor, which contains K_1 and K_2, can now be reduced to a value independent of them. Thus, if $\gamma_p = \gamma_m$, and $c_p/l_p = c_m/l_m$, then

$$\frac{(bV)_p}{(bV)_m} = \left(\frac{l_p}{l_m}\right)^3 , \qquad (494)$$

or by letting $\gamma_p l_p = \gamma_m l_m$ and $c_p = c_m$,

$$\frac{(bV)_p}{(bV)_m} = \left(\frac{l_p}{l_m}\right)^2 . \qquad (495)$$

In this way, by modifying the model test conditions in the proper way, the expansion coefficient may be reduced to a form whose value can be computed without any knowledge of the values of N_c, N_q, and N_γ, or without factoring the net model result into components, which is the same thing. Equation (494) or (495) and the corresponding conditions on c, ϕ, and γ could have been derived with no knowledge of equation (141). Of the dimensionless products already derived for a more general situation, the following would be applicable to this case:

$$\phi \,, \quad \frac{cl^2}{F} \,, \quad \frac{\gamma l^3}{F} \,, \quad \frac{z}{l} \,. \qquad (496)\text{–}(499)$$

In order to run the model test with geometrically similar models, the sinkage z may be used as a criterion of safe load. Then

$$\left(\frac{z}{l}\right)_m = \left(\frac{z}{l}\right)_p$$

if

$$\left.\begin{aligned} \phi_m &= \phi_p \\ \left(\frac{cl^2}{F}\right)_m &= \left(\frac{cl^2}{F}\right)_p \\ \left(\frac{\gamma l^3}{F}\right)_m &= \left(\frac{\gamma l^3}{F}\right)_p . \end{aligned}\right\} \tag{500}$$

and

From this, it is evident that if $c_m = c_p$, $\gamma_m l_m$ must equal $\gamma_p l_p$, and then

$$\frac{F_p}{F_m} = \left(\frac{l_p}{l_m}\right)^2 ; \tag{501}$$

or, if $\gamma_m = \gamma_p$, then c_m/l_m must equal c_p/l_p and

$$\frac{F_p}{F_m} = \left(\frac{l_p}{l_m}\right)^3 , \tag{502}$$

which are entirely the same conditions previously derived.

The general idea of proportionality of component forces during scale-model tests is embodied in the concept of dynamic similarity, which appears to have been first stated by Stokes in 1856.[324] Dynamic similarity exists between two systems if corresponding parts of the systems experience similar net forces. Two systems in which homologous particles lie at corresponding points at corresponding times, i.e., systems which are "kinematically similar," will be dynamically similar if their mass distributions are similar. By these definitions, scale-model testing of dynamic models becomes possible when a given model system is arranged in such a way that it is dynamically similar to the given prototype system. As has already been demonstrated, the net result of the application of Buckingham's Pi Theorem to model testing leads to this same conclusion.

Reduction to Practice

After the dimensional analysis has been performed and dimensionless products have been obtained which theoretically must govern the phenomena under study and, more particularly, the model testing of these phenomena, it is desirable to check the results experimentally. Where the complete set of products introduces conditions which cannot be entirely fulfilled, tests must be run with a view to determining, if possible, which of the conflicting conditions appears to have the largest influence on the results. It will sometimes be found that, to achieve the desired degree of accuracy in a test, some of the products may be ignored completely. The experimental procedure for such investigations is simple in concept, though frequently difficult in practice. A number of experiments is planned, each of which consists of several tests which, in terms of the dimensionless products, are identical, or identical except for certain conflicting factors. Within any one experiment, for example, the size of the object may be varied from one test to another, and the other factors adjusted by various means so as to preserve equality of some or all of the dimensionless products. The results of the several tests are put into the proper dimensionless form and compared. It is desirable that these comparisons be made on the basis of statistical methods for testing equality, which are briefly discussed in Reference 331. If the results of the several tests produce the same dimensionless results, the analysis, or that part of the analysis which has been applied, is checked for the type of application represented by the particular group of tests. By repetition of this procedure, the analysis may be explored for a variety of types of physical situations within its general scope.

This procedure obviously includes, as an end point, correlation between model and full-size performance, and indeed, it is this last check which will always be the most revealing. It is generally found that within a given laboratory, the range of sizes which can be tested at all is limited, and, moreover, unless several sizes of apparatus are constructed for the tests, which implies that all the tests utilize the same apparatus, tests of the smallest objects will be difficult because of apparatus insensitivity, while those of the largest will introduce problems due to equipment overloading. In the case of the early tests of soil-vehicle scale relationships, it was found possible to run experiments in which the over-all scale ratio within the group reached 8:1, through the use of two separate setups.[132] For the more complex types of check tests which followed, only 4:1 was practicable when using only one setup, and these tests were made difficult by the factors already enumerated. From this experience, it appears that tests within one setup are not desirable over a range of much more than 2:1.

It should be noted that in vehicle scale-model testing in general, an over-all scale ratio of much greater than 20:1 probably will never be necessary, and scale ratios of the order of 8:1 to 12:1 probably will serve most needs. This is fortunate from the viewpoint of precision in meeting the theoretically required equality of dimensionless products. For example, in a system of towed rigid wheels, rigidly mounted, operating in a nonviscous soil, i.e., pl^2/F, ql/F, $\psi l/F$, $\eta^2/\varrho F$, and M/Fl not influencing the behavior, the equation for one of the dependent variables may be written, assuming that the geometry is fixed:

$$\frac{R}{F} = \theta_1\left(\mu\,,\ \frac{r}{l}\,,\ \phi\,,\ \frac{cl^2}{F}\,,\ \frac{El^2}{F}\,,\ \frac{v^2\varrho l^2}{F}\,,\ \frac{\varrho g l^3}{F}\right). \tag{503}$$

Thus, in order to run proper scale-model tests of the system so as to determine R/F, it is necessary that

$$\mu_m = \mu_p$$

$$\left(\frac{r}{l}\right)_m = \left(\frac{r}{l}\right)_p, \quad \text{etc.}$$

The coefficients involve five soil properties and the acceleration of gravity. Of these, g and ϱ are for all practical purposes not alterable at the will of the experimenter through a sufficient range to be of interest. Thus, the only way in which

$$\left(\frac{\varrho g l^3}{F}\right)_m = \left(\frac{\varrho g l^3}{F}\right)_p \tag{504}$$

is by making

$$\frac{F_m}{F_p} = \left(\frac{l_m}{l_p}\right)^3 = \frac{1}{\lambda^3}\,. \tag{505}$$

This condition, taken with the remaining products, dictates that

$$\frac{c_m}{c_p} = \frac{1}{\lambda}\,, \tag{506}$$

$$\frac{E_m}{E_p} = \frac{1}{\lambda}\,, \tag{507}$$

and

$$\frac{v_m}{v_p} = \frac{1}{\lambda^{\frac{1}{2}}}\,. \tag{508}$$

In addition, of course, it is necessary that

$$\mu_m = \mu_p \,, \tag{509}$$

$$\frac{r_m}{r_p} = \frac{1}{\lambda} \,, \tag{510}$$

and

$$\phi_m = \phi_p \,. \tag{511}$$

Thus a rather strictly defined soil is required in which the model test must be run in order to obtain valid predictions of prototype performance. Even though accurate prototype conditions are not specified, these conditions must still be respected, lest the model test predict performance in unlikely or noncritical soil conditions. It is probable that many small "demonstration" models have "shown" performance superior to that which would be achieved by the full-size vehicle because this fact was not understood.

The soil for the model tests must have the same angle of internal friction as that of the prototype system, but reduced grain size, reduced cohesion, and reduced elasticity. It is probable, and indeed experience of experiments to date indicates, that the grain-size condition, properly a part of the geometric similarity already stated, but kept separate for discussion purposes, may be neglected provided the largest grains are small in relation to the smallest detail of interest on the model. In fact, grain size is so intimately connected with mechanical properties in fine-grain soils that if it could not be neglected with the stated proviso, scale-model vehicle tests in soil would probably have to be abandoned. The relationship whereby soil cohesion and elasticity must be reduced at constant friction is nearly as difficult. Again, experience to date has indicated that the elasticity of soils in the nearly impassable soil conditions with which the vehicle research worker is concerned plays little part in the total behavior and may be neglected. It should be stressed at this time that the experience referred to previously, and in what will follow, is presently very limited, and that the tentative conclusions drawn from it, quoted here, must be the subject of further study before they can be completely accepted. It appears likely that the condition that the soil-to-wheel material coefficient of friction be the same in the two systems [equation (509)] could be dealt with by a proper selection of wheel material or surface. Some such artifice probably would be required, because the soil specified for the model tests under the other conditions would differ in several respects from the prototype soil, and could be expected to change this frictional property in some dependent fashion.

Even the remaining condition that soil cohesion be reduced at constant friction presents many difficulties, except at the end points (i.e., $\phi = 0$ or $c = 0$). To achieve this even approximately would require a more extensive cataloging of soils than now exists, as well as ready access to a number of these soils which (in their various states) provide the proper ranges of c and ϕ more or less independently.

One approach to this, which has been briefly and inconclusively explored, is the use of synthetic soils. One series of such soils might be made up of graded glass beads (available in graded sizes from 0.005 in. to 0.040 in. in diameter) mixed with varying quantities of silicone fluids of varying viscosities. Modeling clay with various admixtures of oils to reduce its strength might be used. Some manufacturers of earth moving equipment have used a mixture of foundry sand and motor oil with some success in investigating the behavior of self-loading, scraper-type, earthmoving machinery. It is conceivable that colloidal iron in a liquid binder and subject to a variable, controllable field might be developed wherein the mechanical properties could be controlled over some range by means of field control only. Use might be made of shot of varying materials, perhaps with binders, in order to vary the soil density for certain tests.

It should be noted that in all these proposals, a secondary but important advantage other than control of properties would be gained, i.e., stability of properties. Most natural soils have different cohesion and friction at different moisture contents and compactions, so accurate control of these two parameters is necessary in order to obtain reproducible test conditions and, hence, reproducible tests. Control of the mechanical properties is obviously necessary for successful experimentation and is one of the most difficult problems in scale-model vehicle testing in soils.

In tests run in accordance with the scheme just outlined, it is evident that small variations in, say, soil cohesion in the model test setup are equivalent to variations λ times as large in full size. Thus, a variation of 0.1 psi in soil cohesion might make a negligible difference in full-size performance, but the same variation in model conditions would produce a change in performance equivalent to a change in full-size cohesion of 1.0 psi (if $\lambda = 10$, say), which probably would not be at all negligible.

It is also evident from this discussion that the instrumentation used to measure soil mechanical properties in conjunction with model tests must be of high precision. Soil instruments and measurements in general were discussed in Chapters V and X. At present, it has been found that for model work, it is necessary to have some sort of direct shear device which may be used upon the soil bed *in situ*. For a given soil bed, accurate control of moisture content, measured by weighing samples, even drying and reweighing, and of compaction, measured by some sort of super-accurate penetrometer, have been found adequate for the actual controlling of the soil conditions in soils below the Atterberg plastic limit in moisture content. The Atterberg test is discussed in References 63 and 280. The problems of still wetter soils remain to be investigated.

It is not the intent in this discussion to indicate that any of the problems mentioned is insurmountable, but rather merely to call attention to their range and general difficulty. In fact, many of them disappear for model tests that can be run with the effects of soil weight neglected (as distinct from soil mass). This is equivalent to removing the product $\varrho g l^3/F$ from the functional relationship (503), reducing it to

$$\frac{R}{F} = \theta_2\left(\mu, \frac{r}{l}, \phi, \frac{cl^2}{F}, \frac{El^2}{F}, \frac{v^2\varrho g l^2}{F}\right).$$

From this, it is evident that if

$$\frac{F_m}{F_p} = \frac{1}{\lambda^2}, \tag{512}$$

the same soil (subject to the earlier proviso on grain size), in the same condition, and the same wheel material may be used for the model and prototype systems; then

$$v_m = v_p\,. \tag{513}$$

Equation (513) imposes no hardship in the test procedure provided that slow-speed behavior only is to be studied. For higher speeds, it means that long model soil b ns would be required.

This greatly simplified model test procedure has been found adequate for tests on wheeled vehicles with wheels of normal proportions, in level, homogeneous soils. Although tests in nonhomogeneous soils have not been run as yet, the distribution of the several types of soils, or soil properties, should follow geometric similarity. If the nonhomogeneity is disposed in other than horizontal strata, or if the surface is other than level, gravity effects certainly cannot be neglected, and the more exact and difficult model procedure outlined by equation (503) must be employed. In addition, tests of certain types of running gear which, compared to normal wheels, are superior in exploiting the weight of the soil in getting traction or bearing may require the more exact procedure. It is probable that even wheels, in very highly compacted soils in which bearing failure takes place by the classical general failure, will require this more thorough treatment. This last case, however, is one of interest perhaps to airplane landing-gear research, but not generally to vehicle-mobility research.

In a homogeneous, liquid mud, and possibly in one with some small plasticity as defined by Bingham,[332] underlain by a very hard surface, both c and ϕ would be zero, and the problem would be reduced to the hydrodynamic test procedure. In the event that there are surface waves created by the passage of the vehicle, which necessarily involve gravity action, the classic dilemma of hydrodynamic model testing is faced. The theory states that it is necessary that

$$\frac{\left(\dfrac{\eta}{\varrho}\right)_m}{\left(\dfrac{\eta}{\varrho}\right)_p} = \frac{1}{\lambda^{3/2}}. \tag{514}$$

Since, however, the viscosities of liquid muds are all higher than that of water by an amount which increases with decreasing water content,[332] it appears entirely possible to meet condition (514) for most tests of the type described by increasing the water content somewhat. Considerable work of interest on the viscosities of the muds would be required in order to obtain the proper scaling, but the procedure certainly seems feasible at this time.

For soils intermediate between liquid and nonplastic, for which an appreciable viscous property (but not true viscosity in the usual sense) is certainly extant, the problem is evidently of the type discussed earlier in connection with scaling c with ϕ being constant, but of another order of complexity. The very complexity of soil in general, i.e., its wide range of properties, actually makes solutions to these problems possible, whereas they are impossible in tests with simpler media. The hydrodynamic dilemma just mentioned has never been experimentally solved for ship tests, but has been overcome by means of empirical corrections. The lack of a thorough, fundamental research on all the soil properties of all types of soil, in all conditions of moisture content and compaction—in short, of the cataloging earlier mentioned—is sorely felt in work such as this.

Secondary problems, such as soil stickiness and slipperiness, which are on occasion of extreme importance in vehicle performance, probably cannot be studied by means of models, at least not simultaneously with the primary vehicle behavior. Problems of combined soil-vegetal support must await adequate techniques for defining and measuring the mechanical properties of the vegetation, or the soil-vegetal mass, before model tests may be contemplated. Finally, tests in which so-called "remolding" soils are important must involve special soil-bed preparation techniques if tests are desired in remolding soils which are not completely remolded. In such soils, in their unremolded condition, the mechanical properties change rapidly as the soil is worked by the vehicle running gear so that, in effect, the last part of the running gear continuously operates in a soil having mechanical properties distinctly different than those encountered by the leading members of the gear.

Although this section has dealt largely with the problems, inadequacies, and unexplored territories of vehicle-soil scale-model testing, it should be emphasized, before concluding, that even without the solution of these problems, much can now be done in nonplastic soils by the simplified procedure which is based on scaling forces as λ^2, and by the use of the more thorough procedure with cohesionless or frictionless soils. Further development of the existing techniques leading to a most useful tool for exploring the most complex vehicle-terrain relationship appears entirely feasible. Fundamental research on such soil properties as quasi-viscosity must form a part of any future development in this field. The practical problems of vehicle-soil scale-model tests center upon control of soil conditions, and all appear largely solvable.

Dimensional Theory in Analysis

As an example of the usefulness of dimensional reasoning in practical analysis, consider the general Bernstein equation for the rolling resistance (R) of a wheel in soil loaded with load W, as discussed in Chapter VI:

$$R = b\,\frac{k z_o^{n+1}}{n+1},$$

where

$$z_o = \left[\frac{3W}{(3-n)bkd^{\frac{1}{2}}}\right]^{\frac{2}{2n+1}}$$

or

$$R = \frac{bk}{n+1}\left[\frac{3W}{(3-n)bkd^{\frac{1}{2}}}\right]^{\frac{2n+2}{2n+1}}, \tag{515}$$

in which Bernstein suggested (Chapter VI) that

$$k = 2a' + ba'' . \tag{185}$$

If this equation is combined with Bernstein's principal assumption,

$$p = kz^n , \tag{163}$$

then

$$p = (2a' + ba'')z^n . \tag{516}$$

In examining the dimensions of this expression, it is evident that n is dimensionless, and that

$$2a' \overset{D}{=} FL^{-(n+2)} \tag{517}$$

$$a'' \overset{D}{=} FL^{-(n+3)} . \tag{518}$$

Thus, $2a'$ has the dimensions of c/d^n, and a'' has the dimensions of γ/d^n, where c = soil cohesion, as before, and γ = soil specific weight or weight density ($= \varrho g$). Thus,

$$2a' = \frac{\delta c}{d^n} \tag{519}$$

and

$$a'' = \frac{\sigma\gamma}{d^n}, \tag{520}$$

where δ and σ are dimensionless constants of proportionality. Equation (185) then becomes

$$k = \frac{1}{d^n}(\delta c + \sigma\gamma b) . \tag{521}$$

Substituting (521) into (515), and reducing, gives

$$\frac{R}{W} = \frac{\left(\frac{1}{\delta}\right)^{\frac{1}{2n+1}} \left(\frac{3}{3-n}\right)^{\frac{2n+2}{2n+1}}}{1+n} \left(\frac{1}{\frac{1}{{}_bC_c} + \frac{\sigma/\delta}{{}_bC_\gamma}}\right)^{\frac{1}{2n+1}}, \tag{522}$$

where

$${}_bC_c = \frac{W}{cbd} \tag{523}$$

and

$${}_bC_\gamma = \frac{W}{\gamma b^2 d}. \tag{524}$$

Note that (523) is dimensionally equivalent to condition (477) previously derived; (524) is equivalent to (482). By dropping the detailed coefficients, equation (522) may be rewritten in the following form:

$$\frac{R}{W} = A\left(\frac{1}{\frac{1}{{}_bC_c} + \frac{B}{{}_bC_\gamma}}\right)^m. \tag{525}$$

For a cohesionless soil, such as dry sand ($c = 0$), equation (525) immediately reduces to

$$\frac{R}{W} = K\,{}_bC_\gamma{}^m \tag{526}$$

since $1/{}_bC_c = 0$. Preliminary tests with rigid wheels indicate that in sand with $\phi = 34°$ and at slow speeds, this equation becomes

$$\frac{R}{W} = 0.5\, C_\gamma^{0.4}, \tag{527}$$

where

$$C_\gamma = {}_bC_\gamma \left(\frac{b}{d}\right)^{3/2} = \frac{W}{\gamma b^{\frac{1}{2}} d^{5/2}}. \tag{528}$$

Application of the discovery already mentioned, that in mildly cohesive soils the behavior of wheels depends primarily on the dimensionless product cl^2/F, leads to the conclusion that for such soils, equation (525) may be rewritten as follows:

$$\frac{R}{W} = A\ {}_bC_c{}^m\,. \tag{529}$$

Again, preliminary tests indicate that for slow-moving rigid wheels in moist, highly compacted loam, $\phi = 24°$ to $25°$, the rolling resistance of smooth plastic wheels is given by

$$\frac{R}{W} = 0.044\ C_c{}^{0.4}\,, \tag{530}$$

and for deeply grousered wheels by

$$\frac{R}{W} = 0.065\ C_c{}^{0.4}\,, \tag{531}$$

where

$$C_c = {}_bC_c\left(\frac{b}{d}\right)^{1/3} = \frac{W}{cb^{2/3}d^{4/3}}\,. \tag{532}$$

Thus, by dimensional reasoning and some intuition, Bernstein's rather cumbersome equation, which involves numerous constants of a nature foreign to the more rigorous soil-mechanics treatises, is reduced to a simpler form involving familiar dimensionless products. The application next of general soil knowledge and particular model experience further reduces it to forms which may be readily evaluated by experiment. It should be noted, of course, that the several constants determined by the preliminary tests cited are probably themselves the result of operations on several dimensionless products—in equation (526) for example, K and m may be functions of ϕ, n, b/d, etc.

From this example, it may be seen that over and above its use in model test work, dimensional analysis can be useful in analysis generally, through reducing the number of variables involved; in setting up test procedure (model or full size); in presenting and generalizing data; and in building up empirical or semi-empirical equations for complete systems.

Examples of Vehicle Scale-Model Research

The first published information on the use of a scale model of the vehicle for test purposes may be found in Reference 333. In this paper,

Bradley and Wood analyze the behavior of a vehicle with locked wheels. Although the dimensional analysis was not applied in a strict sense, the above-mentioned paper shows the advantages which may be gained in following the discussed method.

A real start in scale-model research was made in the early 1930's in Germany, after Lutz [328] and Kamm [204] published the principles of similarity between vehicles. Problems of steerability and braking of various combinations of wheel drive were investigated under Kamm's direction.[215] The most famous achievement of the scale-model approach was the improvement of the stability of trailers and trailer trains, which, in usual conditions, have a tendency toward "fishtailing." By using stationary scale models on a "mobile road," Dietz and Huber [334, 335] produced a complete solution of the problem. A new test technique on a "mobile runway" was established, with the model weight reaching 100 lb in special cases.[329]

An analytical approach by Marquard to the question of vehicle vibrations led to the proposal of testing the vibrational properties of vehicles by means of simplified skeleton models.[336] A method of producing any desired type of exciter vibrations, caused by the selected type of ground unevenness, was explored by the present writer with the purpose of reproducing a given type of vehicle vibration in a laboratory without the need of road tests.[337]

It appears evident that any tests of obstacle performance, particularly of complex spring and self-propelled vehicle models, may be successful if the principles of dimensional analysis discussed in this chapter are applied.

The Statistical Treatment of Experimental Data

It results from the foregoing discussion that the required precision of experiments is in conflict with the number of variables involved and with the great many accidental factors which inseparably follow the experiment. Nonsystematic evaluation of results may lead to mistaken conclusions, or conclusions entirely obscured by a multitude of unknown and/or uncontrolled factors.

This consideration is of prime importance in research phases of vehicle-soil studies where the validity of given hypotheses or formulations must be determined with some confidence despite the obscuring influence of lack of numerical precision. Thus, a theoretical prediction which agrees with field data within 10% might be entirely acceptable as a final result of certain studies, but in the research phase, such variation might lead to doubts as to the truth or completeness of the hypothesis on which the analysis was based. Systematic treatment of data by statistical methods will often resolve such questions and help to avoid improper conclusions.

Perhaps the best way to explain the statistical approach to experimentation is to illustrate the concepts involved by a simple situation. Imagine that a model wheel is pulled over a compact soil at such a speed and under such a load that theoretically it should take a force of 10 lb to pull it through the soil. The experiment is performed five times, and the forces recorded. The five observations, in pounds, are:

$$\begin{array}{r} 10.15 \\ 9.75 \\ 9.65 \\ 9.55 \\ \underline{10.15} \\ 49.25 \end{array}$$

$\bar{x} = 9.85$ (mean of the observations).

The question is: can it be assumed on the basis of these data that the theoretical force of 10 lb has been verified?

Suppose it is known from previous experience that the precision of the observations in experiments of this type is no worse than ± 0.4 lb. The standard engineering approach then might be to observe that 9.85 falls within the range 10 ± 0.4. The fact that the mean reading differed from 10 lb by 0.15 lb could be attributed to experimental error. The experimenter might very well conclude that the theoretical force of 10 lb had been verified.

It cannot be argued that this sort of an approach necessarily leads to incorrect conclusions. On the contrary, most of the time it probably leads to correct conclusions; but there are some objections to making decisions based on numerical data in the above-described manner. In the first place, no consideration was given to the number of observations, but certainly a mean of 9.85 lb for 100 observations would be more cause for concern than a mean of 9.85 lb for five observations. In the second place, there is no information on the likelihood of coming to a wrong decision. For example, it is quite possible that, just by chance, the mean of observations could be 10.5 lb even though the true theoretical force is 10 lb. Had this occurred, the theoretical force of 10 lb would have been denied even though it was true. It is possible to accept 10 lb even though the true theoretical force is 9.5 lb. The possibility of making such errors exists regardless of the approach, but what is more serious is that the customary engineering approach does not take into consideration at all the probabilities of making such errors. It is a surprising fact that an experiment can be carried out by the most rigorous scientific principles until the very end when a conclusion is reached on an uncertain and arbitrary basis. By bringing the probabilities of making incorrect decisions into the problem, the statistical method replaces the uncertainty with a degree of assurance, and it is this property which is perhaps the chief reason for its use in the treatment of experimental data.

In order to explain the statistical approach to the problem, several concepts must first be discussed, the first being the notion of a measure of experimental error.

There are actually many measures of experimental error. The simplest is the range, the largest observation minus the smallest observation. The difficulty with the range is its dependence on the extreme observations to the virtual exclusion of the middle observations. More effective measures are the variance and the standard deviation. The variance of a set of observations is the average of the squared deviations from the mean of the observations. It is often represented by the symbol s^2. In the illustrative example,

$$s^2 = \frac{(10.15 - 9.85)^2 + (9.75 - 9.85)^2 + (9.65 - 9.85)^2}{5} + \\ + \frac{(9.55 - 9.85)^2 + (10.15 - 9.85)^2}{5} .$$

The square root of the variance is called the standard deviation. In this example, $s = 0.25$. (For reasons not germane to the discussion, this is often multiplied by a correction factor. See Reference 338.) In the following discussion, the terms "standard deviation" and "experimental error" will be used synonymously. Referring to the illustrative example again, the deviation of the sample mean, 9.85, from the theoretical force, 10, is 0.15 lb. The answer to whether this is an appreciable difference certainly must lie in the experimental error. This fact is often overlooked, but the idea of measuring deviations in terms of a strict measure of experimental error is one of the significant, though simple, contributions of statistics to the interpretation of data.

The next objective is to establish a mathematical model to serve as a basis for the treatment of the data. Consider the elements which are involved in the experimental error. Each reading is affected by factors such as unevenness of the soil, atmospheric conditions which cause slight changes in soil mechanical properties, out-of-roundness of the wheel, error in reading the instrument, etc. It is assumed then that an observation, x, is composed of several components, i.e.,

$$x = \mu + \varepsilon_1 + \varepsilon_2 + \dots + \varepsilon_k,$$

where μ is the force required if the soil is perfectly smooth, the wheel is perfectly round, etc., and the ε's are the contributions of the above-mentioned miscellaneous factors.

It would be practically impossible to measure the contribution of any of these factors to a particular measurement, and they will contribute different amounts to different observations. The mathematical model hypothesizes that the contributions of the ε's enter the observations in a random manner, and that underlying each ε is a set of probabilities by means of which its long-run behavior could be predicted. This set of probabilities is called the distribution of the random factor.

It would be as difficult to obtain these probabilities as it would have been actually to measure the contribution of a random factor in the first place. However, it is a remarkable fact that no matter what the distribution of an individual ε, the distribution of the *sum* of the ε's can be assumed to have what is called a normal distribution. The probabilities involved with this distribution are well known and tabled in almost all books on statistics, including Reference 339. The mathematical theorem which underlies the above statement is called the central limit theorem. (See Reference 340 for a proof.)

The normal distribution is often described by a curve, the equation of which is

$$f(x) = \frac{1}{\sqrt{2\pi}\sigma} e^{-\left[\frac{1}{2}\frac{(x-\mu)^2}{\sigma^2}\right]}.$$

The symbol μ is called the population mean or true mean. The problem of the illustrative example is actually: does $\mu = 10$? It is understood, of course, that an observation, being equal to a constant plus a sum of the ε's, is also normally distributed. The symbol σ is called the population standard deviation or true standard deviation. In the illustrative example, the population could be considered as consisting of all the trials that could be made on the wheel; in other words, virtually an infinite population. The standard deviation obtained from the data, 0.25, is not the true standard deviation but merely an estimate. It would be extremely unlikely that a sample standard deviation would be equal to the true standard deviation. In most situations, the true standard deviation is not known, but sometimes, because of a wealth of data in which the same types of measurements have been made under the same conditions, it can be assumed to be known. It will be assumed that the true standard deviation for the wheel-timing experiment is known to be 0.21. This is only to make the explanations of the important points clearer; there is a well-known treatment of the problem for the case where σ is not known (Reference 331).

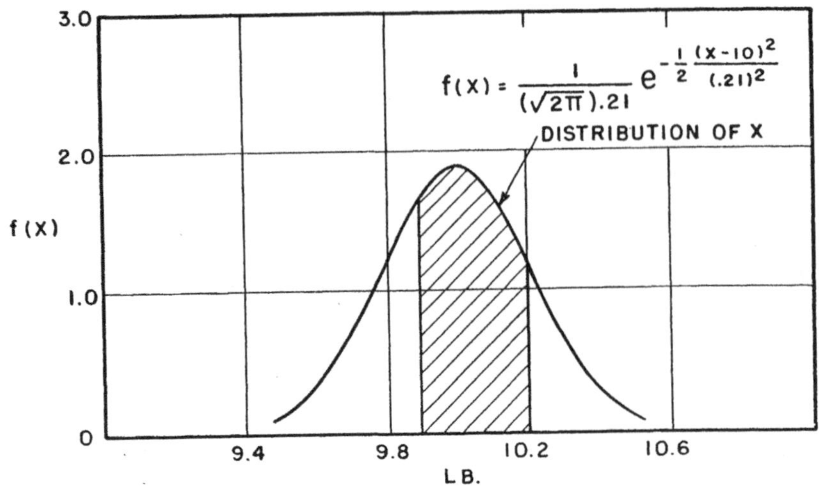

Fig. 189

A normal distribution with $\mu = 10$ and $\sigma = 0.21$ is illustrated by Figure 189. Areas under the normal curve correspond to probabilities, and the cross-hatched area in Figure 189 is the probability that an observation will fall between 9.90 and 10.20. Naturally, the total area under the curve, i.e., the total probability, is unity.

It is accepted, then, that an observation is to be treated as a random variate with mean μ and standard deviation σ. However, the number of interest obtained from the experiment is not a single observation but the sample mean, $\bar{x}$, of the observations. The mean, being a function

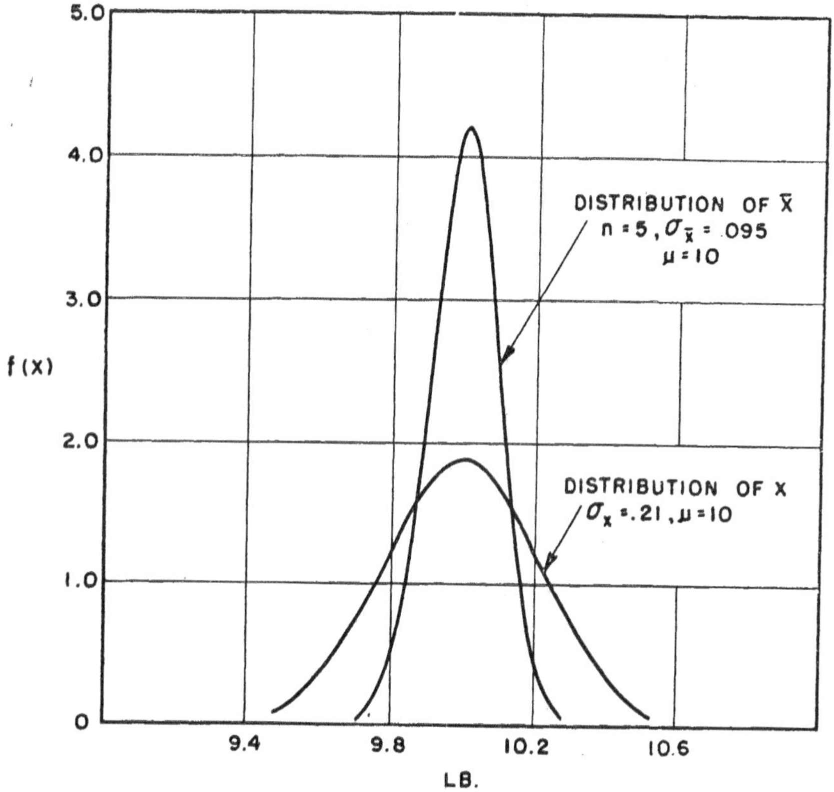

Fig. 190

of random variables, is itself a random variable. As might be expected, it is normally distributed, and the true mean of its distribution is μ. In fact, in most experimentation, $\bar{x}$ can be considered to be normally distributed, even though the x_i are not. It is natural to expect that the true standard deviation of $\bar{x}$ would be less than the true standard deviation of x. In other words, it would be expected that, for example, five experimentally obtained $\bar{x}$'s would fall closer to μ than five experimentally obtained x's. It can be shown that the true standard deviation of $\bar{x}$ is equal to the true standard deviation of x divided by the square root of the number of observations; symbolically, $\sigma_{\bar{x}} = \sigma/\sqrt{n}$ (Reference 341). Note that the standard deviation of $\bar{x}$ decreases as n increases. Figure 190 shows the distribution of $\bar{x}$ superimposed on x. In this problem,

$\sigma_{\bar{x}} = 0.21/\sqrt{5} = 0.095$, and it is against the number 0.095 that the difference $9.85 - 10 = 0.15$ must be measured. Thus $0.15/0.095 = 1.58$. In other words, a value of 9.85 lb for the sample mean is 1.58 standard-deviation units from the theoretical true mean of 10.

Formally, the procedure is to establish the hypothesis $\mu = 10$ and then to test this hypothesis. What are needed, however, are boundaries such that if a sample mean falls within these boundaries, the hypothesis that $\mu = 10$ could be accepted; if the sample mean falls beyond, the hypothesis would be rejected. The difficulty is that even if the true mean is 10, a sample mean can deviate from 10 considerably, due to a peculiar combination of the random factors. In fact, no matter what the boundaries, there is some chance of rejecting the hypothesis even though it is true. The statistical method, however, permits the experimenter to choose this probability, and it is called the level of significance of the test. The temptation is to choose a small level of significance, but this raises the probability of accepting the hypothesis when it is false. Levels of significance of 0.05 or 0.01 are often used, but the statistician may choose to raise or lower the level of significance if one type of error is regarded as being much more serious than the other for the problem under discussion. Suppose an 0.05 level of significance is chosen for this problem. From the tables of the normal distribution, it is found that the probability of a normally distributed variate falling more than 1.96 standard deviations from its true mean is 0.05.

It is required, then, to establish boundaries such that if the mean falls within these boundaries, the hypothesis that $\mu = 10$ will be accepted, and if μ is outside these boundaries, the hypothesis is rejected. Since the variate of interest is $\bar{x}$ rather than x, the boundaries will be $10 \pm 1.96\sigma_{\bar{x}}$. Recall that $\sigma_{\bar{x}} = 0.095$. Then

$$1.96\sigma_{\bar{x}} = (1.96)\,(0.095) = 0.186 \text{ lb}$$

$$10 + 1.96\sigma_{\bar{x}} = 10 + 0.186 = 10.186 \text{ lb}$$

$$10 - 1.96\sigma_{\bar{x}} = 10 - 0.186 = 9.814 \text{ lb}$$

If the test is set up so that the hypothesis $\mu = 10$ will be rejected if $\bar{x}$ is greater than 10.19 or less than 9.81, the probability is 0.05 that the hypothesis will be rejected when it is false. The probabilities of wrongfully accepting the hypothesis can also be calculated (Reference 331). The test is illustrated by Figure 191. The sample mean obtained from the observations in this experiment is 9.85. This number falls within the ac-

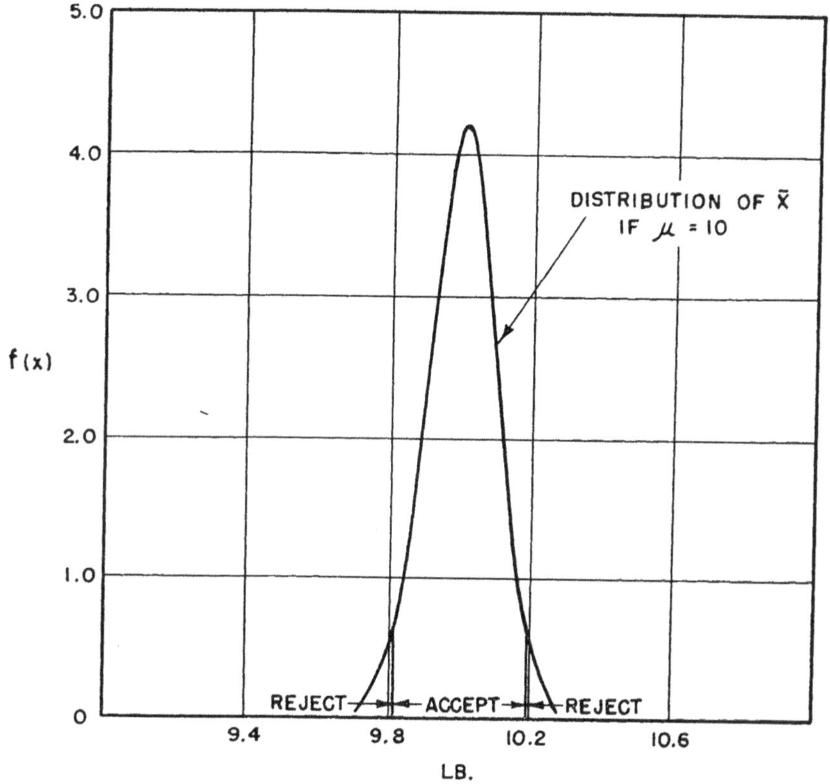

Fig. 191

ceptance region and therefore the hypothesis that the true mean is 10 is accepted. In other words, the investigation shows that the deviation $10 - 9.85 = 0.15$ could easily be attributed to the random factors.

Although the illustrative example so far used is of the very simplest type, it serves to illustrate the following:

(1) the use of the experimental error (standard deviation) for measuring deviations,

(2) the classification of difficult or impossible-to-measure variables as random variables, and

(3) the importance and usefulness of the central-limit theorem.

Linear Relationship

Most problems that arise in vehicle mechanics are more complex. Suppose the characteristic of interest is R/F, as in equation (503), and the results of eight observations for eight sets of values of the dimensionless variable are as shown in the following table:

Model R/F	*Full-Scale R/F*
0.120	0.132
0.121	0.120
0.115	0.128
0.135	0.140
0.122	0.107
0.125	0.112
0.147	0.128
0.140	0.130

Figure 192 illustrates this table.

If there were no experimental error, all the points in Figure 192 would lie on a line of a slope 1, passing through the origin. Of course, all the points do not lie on this line. Should this be attributed to experimental error or must it be said that the experimentation does not verify the theory? The answer lies in a procedure in which some of the basic principles are similar to those already discussed. A line is fitted to the data by the method of least squares (Reference 341). This method fits the line so that the sum of the squares of the deviations from the line is a minimum. The slope of this line is a function of the observations and, hence, is a random variable. Therefore, it possesses a distribution and this distribution has a standard deviation σ_m which is a function of the experimental error σ. The experimental error is estimated on the basis of the deviations of the points, not from their mean, but from the fitted line. Under assumptions (Reference 341) usually satisfied well enough to make the method effective, the distribution of the slope of the fitted line is known.

The slope obtained by calculation is subtracted from 1 and this deviation is compared to σ_m. Formally, the hypothesis that the true slope equals 1 is being tested at whatever level of significance the experimenter desires.

The above-described analysis is an example of what is called a regression problem. Such problems arise often. In fact, any situation in which a dependent variable is thought to be a function of one or several dependent variables can be approached by regression methods. Further information on problems of this type is given in References 339 and 342.

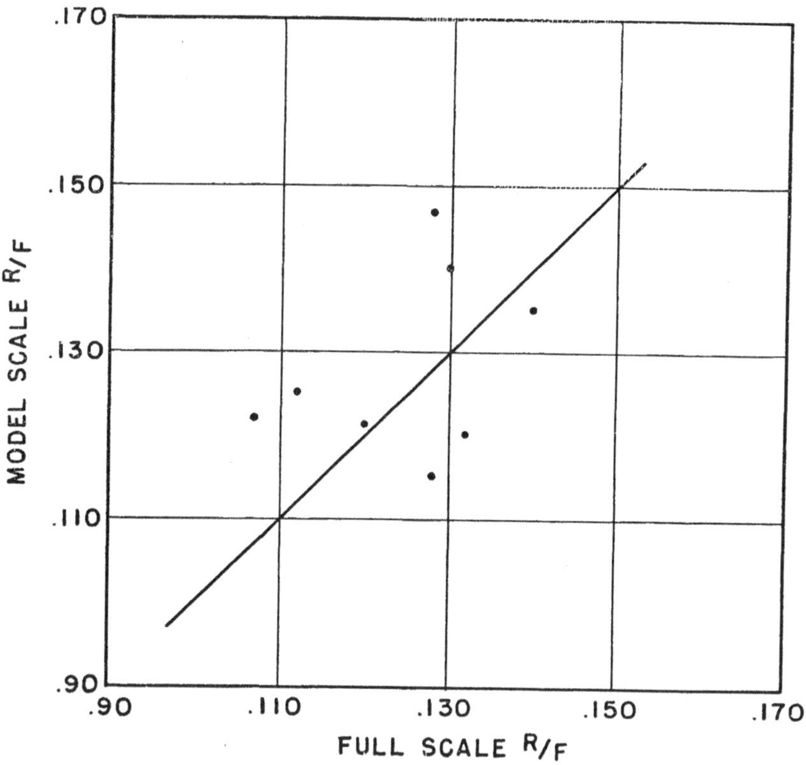

Fig. 192

Experimental Design

Another powerful statistical tool which can be applied to problems in vehicle mechanics is the theory of experimental design, first developed by Fisher (Reference 343). Suppose, for example, that it is required to determine whether tread, shape, and wheel material affect the rolling resistance of wheels in a given soil. A properly designed experiment can estimate the experimental error, render a decision on the effect of all these variables, and do it with surprisingly few observations. A design suitable for this problem is called the Latin Square. It is assumed that four shapes of wheels are available, and that these wheels can be covered with as many different materials as required and treaded in as many different ways as required. Four materials and four treads will be used. The shapes are represented by Roman numerals I, II, III, IV; the treads by Arabic numerals 1, 2, 3, 4; and the materials by letters A, B, C, D. Consider observations on tires having the tread, shape, and material properties described by the following square:

Tread	*Shape*			
	I	II	III	IV
1	C	B	D	A
2	D	A	C	B
3	B	D	A	C
4	A	C	B	D

This will require sixteen observations in all. It is noteworthy that the experimental error can be estimated from this design whether or not there are tread, shape, or material effects, and in spite of the fact that two observations are never made on the same tire. The method of estimating the experimental error in such a design is described in Reference 344. In order to determine whether tread has an effect on rolling resistance, the mean of each tread row is obtained. The deviations of these means from the mean of the means provide an estimate of experimental error if there is no tread effect. But, if there is tread effect, this estimate will be enlarged. Hence, the test of tread effect is to form the ratio

$$\frac{s_1^2}{s_o^2},$$

where s_1^2 = estimate of experimental error, valid only if there is no tread effect, and s_o^2 = estimate of experimental error, valid whether or not there are tread, shape, or material effects. If this ratio is considerably greater than 1, it is likely that tread has an effect on rolling resistance. The distribution of the ratio is known and a test of significance can be carried out. Similar procedures apply for shape and material effects.

The subject of experimental design has undergone considerable expansion in recent years. More complicated experimental situations require more complicated design, and many have been developed. References 344 and 345 contain most of the more widely used designs.

The foregoing is only a brief summary of statistical techniques which have been evolved to meet frequently occurring experimental problems. Not only do the techniques herein mentioned have wide ramifications, but many other powerful statistical approaches have been developed. Furthermore, the wide variety of research problems which require statistical planning and investigation has continually stimulated the expansion of the subject. Results already obtained on the performance of wheels in soil (Reference 346) indicate that statistical methods can be of continuing aid in the field of vehicle mechanics.

REFERENCES

(WITH A FREE TRANSLATION OF FOREIGN TITLES)

1. CLEVELAND, R. M., AND WILLIAMSON, S. T. *The Road Is Yours*, New York, 1951.
2. NEESEN, F. *Gestaltung und Wirtschaftlichkeit der Land–, Wasser– und Luftfahrzeuge* (Form and Economy of Land, Water, and Air Vehicles), Jena, 1940.
3. GABRIELLI, G., AND VON KÁRMÁN, TH. "Maximum Speed and Specific Power of Vehicles," *ATA*, Turin, January 15, 1948.
4. DAVIDSON, K. S. M. *Notes on the Power-Speed-Weight Relationship for Vehicles*, Stevens Inst. of Tech., ETT Tech. Memo. No. 97, Hoboken, N. J., 1951.
5. BEKKER, M. G. "Relationship Between Soil and a Vehicle," *SAE Quarterly Trans.*, Vol. 4, No. 3, 1950.
6. PACOTTE, J. *La Pensée Technique* (The Spirit of Technology), Paris, 1931.
7. KING, H. *Hydraulics*, New York, 1948.
8. RASHEVSKY, N. *Mathematical Biophysics*, Chicago, 1948.
9. MCFARLAND, E. *Textbook of Ordnance and Gunnery*, New York, 1929.
10. RASHEVSKY, N. "On the Locomotion of Mammals," *Bulletin of Mathematical Biophysics*, Vol. 10, 1948.
11. APPELL, P. *Traité de Mécanique Rationnelle* (Outline of Theoretical Mechanics), Vol. II, Paris, 1927 (quoted by Rashevsky).
12. MARKS, L. S. *Mechanical Engineers' Handbook*, New York, 1941.
13. MOSAUER, W. "The Locomotion of Snakes and Its Anatomical Basis," *Zool. Jahrb., Abt. f. allg. Zool. u. Physiol. d. Tiere*, Vol. 52, 1932, p. 191 (quoted by Rashevsky).
14. PRANDTL, L., AND TIETJENS, O. G. *Applied Hydro- and Aeromechanics*, New York, 1937.
15. LVOV, I. D. *Traktory* (Tractors), Moscow, 1933.
16. GAWN, R. W. L. "Fish Propulsion in Relation to Ship Design," *Shipbuilding & Shipping Record*, December 28, 1950.
17. REYNOLDS, O. "On Rolling Friction," *Phil. Trans. of the Royal Soc. of London*, Vol. 166, 1876, p. 1.
18. SARTIAUX, F. *La Civilisation* (Civilization), Paris, 1938.
19. GORDON, A. *A Treatise Upon Elemental Locomotion*, London, 1836.
20. LEFEBRE DE NOETTES. *L'Attelage de cheval de selle à travers les âges* (Harness of the Saddle Horse Throughout the Ages), Paris, 1931.
21. MAXIME-SCHUHL, P. *Machinisme et Philosophie* (Technology and Philosophy), Paris, 1938.

22. USHER, A. P. *A History of Mechanical Invention,* New York, 1929.
23. MANTOUX, P. *Industrial Revolution of the Eighteenth Century,* London, 1928.
24. ANONYMOUS. "The Evaluation of the Chain Tractor," *The Engineer,* Vol. CXXIV, 1917.
25. YOUNG, C. F. F. *The Economy of Steam Power on Common Roads,* London, 1860. (Quoted from *The Engineer,* Vol. CXXIV, 1917.)
26. COULOMB, C. A. *Essai sur une application des règles de maximis et de minimis à quelques problèmes de statique* (Essay on the Application of the Rules of Maxima and Minima to Some Problems of Statics), Paris, 1776.
27. BOUSSINESQ, J. *Application des potentials à l'étude de l'équilibre et du mouvement des solides élastiques* (Application of Potential Theory in a Study of Equilibrium and Motion of Elastic Solids), Paris, 1885.
28. TERZAGHI, K. "Old Earth Pressure Theories and New Test Results," *Engineering News Record,* Vol. 85, 1920, p. 632.
29. SWINTON, E. D. *Eye Witness and the Origin of Tanks,* New York, 1933.
30. SUTTER, M. F. *The Evolution of the Tank,* London, 1937.
31. ICKS, R. J. *Tanks and Armoured Vehicles,* New York, 1945.
32. HEIGL, F. *Taschenbuch der Tanks* (Tank Handbook), Munich, 1926, 1927, 1930, 1934, 1935, and 1938.
33. KRISTI, M. K. (Collective Work). *Avtotraktorny spravotchnikh* (Tractor Handbook), Moscow, 1938.
34. Periodical Publications and Magazines published between 1925–1940: *Agricultural Engineer* (USA), *Avtotraktornye dielo* (Russian), *Army Ordnance* (USA), *ATZ* (Automobiltechnische Zeitschrift) (German), *Cavalry Journal* (British), *Der Motorwagen* (German), *Die Technik in der Landwirtschaft* (German), *Die Kraftfahrkampftruppe* (German), *Farm Machinery Equipment* (USA), *Infantry Journal* (USA), *Militärwochenblatt* (Swiss), *Militärwissenschaftliche Mitteilungen* (German), *Mechanizacya i Motorizacya* (Russian), *Przeglad Wojskowo-Techniczny* (Polish), *Royal Tank Corps Journal* (English), *Revue de Cavalerie* (French), *ZVDI* (*Zeitschrift der Vereine der Deutschen Ingenieure*) (German).
35. KUHNE, G. *Handbuch der Landmaschinentechnik* (Handbook of Agricultural Machinery Technology), Berlin, 1930.
36. BECKER, G. *Motorschlepper* (Tractors), Berlin, 1926.
37. GORIATCHKIN, B. P. (Collective Work). *Teoria i proisvodstvo sielskohoziaynih mashin* (Theory and Development of Agricultural Machinery), Moscow, 1936.
38. MCKIBBEN, E. G. *Some Kinematic and Dynamic Aspects of Rigid Transport Wheels,* Ames, Iowa, 1938.
39. SMITH, C. W., AND HURLBUT, L. W. *A Comparative Study of Pneumatic Tires and Steel Wheels,* Ne. Agr. Exp. Sta., 1934.
40. KRISTI, M. K. *Tanki* (Tanks), War Academy, Moscow, 1937.
41. MEDVEDEV, M. I. *Kinematika gusienitchnyh tsepi* (Kinematics of Caterpillar Chains), Moscow, 1934.
42. ZASLAVSKI, B. I. *Kratki kurs rastchota tankov* (Short Course on Tank Design), Moscow, 1932.

43. ALLHANDS, J. L. "Tools of the Earth Mover," Sam Houston College Press, 1951.
44. SHEPPARD, E. W. *Tanks in the Next War*, London, 1938.
45. DUNPHIE, Maj. Gen. C. A. L. "Tanks of the Future," *Canadian Army Journal*, August, 1947.
46. ROSSEL, H. E., AND CHAPMAN, L. B. *Principles of Naval Architecture*, Vol. I, New York, 1939.
47. MEYER-BERKHOUT, B. *Typenbeschränkung und Rationalisierung in der deutschen Kraftwagenindustrie* (Limitation of the Number of Types and Rationalization of the German Motor Vehicle Industry), Verlag Dt. Motorzeitschrift, Dresden, 1942.
48. KOENIG-FACHSENFELD, R. *Aerodynamik des Kraftfahrzeuges* (Aerodynamics of a Motor Vehicle), Frankfurt, 1951.
49. RIECKERT, P., AND GUSS, F. "Die Strassenlage des Kraftwagens" ("Roadability of Motor Vehicles"), *ATZ*, No. 2, 1951.
50. KÜHNER, K. "Geländefahrzeug und Gelände" ("The Off-the-Road Vehicle and Terrain"), *VDI*, No. 7, 1937.
51. Personal communication by Mr. W. J. MAZE of Brewer Ltd. and Mr. A. THOMSON of Pacific Car and Foundry Co., August 11, 1950.
52. GOUGE, A. "Size in Transport," 36th Wilbur Wright Memorial Lecture, *J. of the Royal Aero. Soc.*, July, 1948.
53. DRIGGS, I. H. "Aircraft Design Analysis Methods as Employed by the Research Division of the Bureau of Aeronautics, U.S. Navy Department," BuAer NAVAER DR Report No. 1139, presented at the First Fall Meeting of the Royal Aero. Soc., London, October, 1949.
54. TERZAGHI, K. *Theoretical Soil Mechanics*, New York, 1944.
55. LOVE, A. E. H. *A Mathematical Treatise on the Theory of Elasticity*, Cambridge Univ. Press, 1927.
56. TIMOSHENKO, S. *Theory of Elasticity*, New York, 1937.
57. PRESCOTT, J. *Applied Elasticity*, New York, 1946.
58. SOUTHWELL, R. V. *Theory of Elasticity*, Oxford Univ. Press, 1941.
59. NADAI, A. *Theory of Flow and Fracture of Solids*, Vol. 1, New York, 1950.
60. COKER, E. G., AND FILON, L. N. G. *A Treatise on Photo Elasticity*, Cambridge, 1931.
61. FROCHT, M. M. *Photoelasticity*, New York, 1948.
62. PAASWELL, G. "Transmission of Pressure Through Soils...," *Trans. Am. Soc. of Civil Eng.*, Vol. LXXXV, 1922.
63. KRYNINE, D. *Soil Mechanics*, New York, 1941.
64. TIMOSHENKO, S., AND MACCULLOUGH, G. H. *Elements of Strength of Materials*, New York, 1940.
65. LEE, G. H. *An Introduction to Experimental Stress Analysis*, New York, 1950.
66. HAEFELI, R. *Erdbaumechanische Probleme im Lichte der Schneeforschung* (Problems of Soil Mechanics in the Light of Snow Research), Zurich, 1944.
67. RUTLEDGE, P. C. "Theories of Failure of Materials Applied to the Shearing Resistance of Soils," paper presented at SPEE Civ. Eng. Div. Conf. on Soil Mechanics, State College, Penn., June, 1939.

68. Cummings, A. E. "Distribution of Stresses under a Foundation" (discussion by W. S. Housel), *Proc. Am. Soc. of Civil Eng.*, Vol. 62, No. 8, 1936.

69. Burmeister, D. M. "Graphical Distribution of Vertical Pressure Beneath Foundations" (discussion by A. E. Cummings), *Trans. Am. Soc. of Civil Eng.*, Vol. 103, 1938.

70. Drucker, D. C., and Prager, W. "Soil Mechanics and Plastic Analysis or Limit Design," *Quarterly of Appl. Math.*, Vol. 10, 1952.

71. Love, A. E. H. "The Stress Produced in a Semi-Infinite Solid by Pressure on Part of the Boundary," *Phil. Trans. of the Royal Soc. of London,* Series A, Vol. 228, 1928–29.

72. Newmark, N. M. "Simplified Computations of Vertical Pressures in Elastic Foundations," Univ. of Ill. Eng. Exp. Sta., Circular No. 24, 1935.

73. Griffith, J. H. "Pressures under Substructures," *Engineering and Contracting*, March, 1929.

74. Fröhlich, O. K. *Druckverteilung im Baugrunde* (Distribution of Pressure in Soil), Berlin, 1934.

75. Krynine, D. P. "Pressures Beneath a Spread Foundation," *Trans. Am. Soc. of Civil Eng.*, Vol. 103, 1938.

76. Haefeli, R. *Schneemechanik* (Snow Mechanics), Zurich, 1939.

77. Kondratiev, A. C., Krahelskij, I. V., and Shakov, A. A. *Increase of Snow Density under the Action of a Compressive Load* (English translation), Acad. of Science of the USSR, Moscow-Leningrad, 1945.

78. Bucher, E. *Beitrag zu den theoretischen Grundlagen des Lavinenverbaus* (Contribution to the Theoretical Foundation of Avalanche Structure), Bern, 1948.

79. Dobrowolski, A. B. *Historia Naturalna Lodu* (Natural History of Ice), Warsaw, 1923.

80. Bader, H., and Niggli, P. *Mineralogische und strukturelle Charakterisierung des Schnees und Schneemetamorphose* (Mineralogical and Structural Characteristics of Snow and Snow Metamorphosis), Zurich, 1938.

81. Meyerhof, G. G. "An Investigation of the Bearing Capacity of Shallow Footings on Dry Sand," *Proc. Int. Conf. Soil Mechanics*, Vol. 1, Amsterdam, 1948.

82. Terzaghi, K., and Peck, R. B. *Soil Mechanics in Engineering Practice*, New York, 1948.

83. Huber, M. T. *Czasopismo Techniczne,* Lwow, 1904. See also: *Przeglad Techniczny* (Technical Review), Nos. 5–10, Warsaw, 1947.

84. Hencky, H. "Über das Wesen der plastischen Verformung" ("The Nature of a Plastic Deformation"), *ZVDI*, Vol. 69, 1925.

85. Jürgenson, L. "The Shearing Resistance of Soils," *J. Boston Soc. of Civil Eng.*, Vol. 21, 1934.

86. Am. Soc. for Testing Materials. "Symposium on Shear Testing of Soils," *Proc. Am. Soc. for Testing Materials*, Vol. 39, 1939.

87. Rutledge, P. C. "Cooperative Triaxial Shear Research," Soil Mechanics Fact-Finding Survey Progress Report – Waterways Exp. Sta., Vicksburg, 1947.

88. Micklethwait, E. W. E. *Soil Mechanics in Relation to Fighting Vehicles*, Military Coll. of Science, Chobham Lane, Chertsey, 1944.
89. Highway Res. Board. "Compendium on Soil Testing Apparatus," *Proc. 18th Annual Meeting, Highway Res. Board*, Washington, 1938.
90. Bekker, M. G. "Photographic Method of Determining the Soil Action beneath Footings," *Proc. Second Int. Conf. Soil Mech.*, Vol. 3, Rotterdam, 1948.
91. Maag, E. *Grenzbelastung des Baugrundes* (Limit Loads of the Ground), Zurich, 1938.
92. Seligman, G. *Snow Structure and Ski Fields*, London, 1936.
93. Tammann, G. "Die lineare Kristallisationgeschwindigkeit des Eises" (The Linear Speed of Ice Crystallization"), *Zeits. f. anorganische und allgemeine Chemie*, No. 2, 1935.
94. Tammann, G. "Die Rekristallisation" ("Recrystallization"), *Forschungen u. Fortschritte*, No. 24, 1931.
95. Nakaya, U., and others. "Physical Investigation of Snow. General Classification of Snow Crystals," *J. of Faculty of Science*, Hokkaido Univ., Tokyo, 1934 and 1936.
96. Weinberg. *Led* (Ice), Moscow, 1940.
97. Kragelskij, I. V. *Analysis of Equipment for Snow Compaction, Method of Determining Hardness and Compactness of Snow Cover, Physical and Mechanical Properties of Snow...* (English translation), Acad. of Science of the USSR, Moscow, 1945.
98. Kondratiev, A. S. "Teploprovodnost snegovogo pokrova..." ("Heat Conductivity of Snow Cover"), *Phys. and Mech. Prop. of Snow*, Acad. of Science of the USSR, Moscow, 1945.
99. Kragelskij, I. V., and Shakhov, A. A. "Change of Mechanical Properties of Snow Cover..." (English translation), *Phys. and Mech. Prop. of Snow*, Acad. of Science of the USSR, Moscow, 1945.
100. Afanasjev, A. D. "Mechanisatzya snegouborotchnyh rabot..." ("Mechanization of Snow Handling"), *Phys. and Mech. Prop. of Snow*, Acad. of Science of the USSR, Moscow, 1945.
101. Hawkins, L. A. *Adventure into the Unknown*, Chap. 24, "Artificial Precipitation," New York, 1950.
102. DeQuervain, M. "Schnee als kristallines Aggregat" ("Snow as an Aggregate of Crystals"), *Experientia*, Vol. 1/7, 1945.
103. Dorsey, N. E. *Properties of Ordinary Water Substance*, New York, 1940.
104. Rutledge, P. C. "Theories of Failure of Materials Applied to the Shearing Resistance of Soils," paper presented at SPEE Civ. Eng. Div. Conf. on Soil Mechanics, State College, Penn., June, 1939.
105. Freudenthal, A. M. *The Inelastic Behavior of Engineering Materials and Structures*, New York, 1950.
106. Houvink, R. *Elasticity, Plasticity, and Structure of Matter*, New York, 1938.
107. Dobrowolski, A. B. *Glaciers' Structure and Movement Theories*, Warsaw, 1948.
108. Croce, K. "Technisch wichtige Eigenschaften von Schnee" ("Technically Important Snow Properties"), *Forschungsarbeiten*, Strassenwesen, Berlin, 1941.

109. BEKKER, M. G. "Snow Studies in Germany," NRC of Canada, Ottawa, 1951.
110. GORUBUNOV, A. L. Publications of the Leningrad Air Force Academy, Acad. of Science of the USSR, quoted by Kragelskij, Moscow, 1945.
111. NORTON, A. E. *Lubrication*, New York, 1942.
112. BOWDEN, F. P., AND HUGHES, L. L. "The Mechanism of Sliding on Ice and Snow," *Proc. of the Royal Soc. of London*, Series A, 1939, pp. 172, 280.
113. KLEIN, G. J. "Aircraft Ski Research in Canada," NRC Report No. MM 225, Ottawa, 1950.
114. SCHLEICHER, F. "Zur Theorie des Baugrundes" ("Theory of Soil"), *Der Bauingenieur*, Vol. 7, 1926.
115. SOUTHWELL, R. V. *Relaxation Method...*, Oxford Univ. Press, 1940.
116. GLOVER, R. E., AND CORNWELL, F. E. "Stability of Granular Materials," *Proc. Am. Soc. of Civil Eng.*, Vol. 69, 1943.
117. KÖGLER, F. Discussion re "Soil Mechanics Research" by Gilboy, *Proc. Am. Soc. of Civil Eng.*, Vol. 59, 1933.
118. HOUSEL, W. S. *Applied Soil Mechanics*, Ann Arbor, 1940.
119. POLLITT, H. W. W., AND LEWIS, W. A. "A Laboratory Study of the Settlement of Loaded Rectangular Plates into Soft Soil," *Proc. Int. Conf. Soil Mechanics*, Vol. 3, Amsterdam, 1948.
120. TERZAGHI, K. "Principal Theories of Settlement," *Proc. Int. Conf. Soil Mechanics*, Vol. III, 1936.
121. MEYER-PETER, E. *Berechnung der Setzung von Bauwerken* (Estimation of the Settlement of Structures), Zurich, 1938.
122. THOMPSON D'ARCY, W. *On Growth and Form*, Cambridge Univ. Press, 1952.
123. VOELLMY, A. *Die Bruchsicherheit eingebetteter Rohre* (Strength of Embedded Pipes), Zurich, 1937.
124. BERNSTEIN, R. "Probleme zur experimentellen Motorpflugmechanik" ("Problems of the Experimental Mechanics of Motor Ploughs"), *Der Motorwagen*, Vol. 16, 1913.
125. GABRIELLI, G., AND VON KÁRMÁN, TH. "What Price Speed ?" *Mech. Eng.*, Vol. 72, No. 10, 1950.
126. FÖPPL, L. *Die strenge Lösung für die rollende Reibung* (Rigorous Solution of Rolling Friction), Munich, 1947.
127. HERTZ, H. *Gesammelte Werke* (Collected Works), Leipzig, 1895.
128. LETOSHNEV, M. N. "Kolesnye povozki..." ("Wheeled Carts and Their Performance on Country Roads"), *Teoria i. proisvod-stvo Sielskohozaynih Mashin*, Moscow, 1936.
129. MORIN, M. A. "Memoir sur le tirage des voitures..." ("Memorandum on Vehicle Hauling"), *Comptes Rendus* de l'Acad. des Sciences, Vols. 10 and 11, Paris, 1840 and 1841.
130. MEYER, H. "Untersuchungen der Fahrwiderstände von Ackergeräten" "Investigation of Movement Resistance of Farm Implements"), *Die Technik in der Landwirtschaft*, Vol. 13, 1932.
131. SWIEZAWSKI, T. "Toczenie Kola" ("The Rolling of a Wheel"), *Czasopismo Techniczne*, Nos. 2–6, Lwow, 1932.

132. NUTTALL, C. J. "Scale Model Vehicle Testing in Non-Plastic Soil," Experimental Towing Tank, Rep. 394, Hoboken, N. J., 1949.
133. MCKIBBEN, E. G., AND THOMPSON, H. J. "Comparative Performance of Steel Wheels and Pneumatic Tires," *Agr. Eng.*, Vol. 20, No. 11, 1939.
134. MCKIBBEN, E. G., AND DAVIDSON, J. BROWNLEE. "Rolling Resistance of Individual Wheels," *Agr. Eng.*, Vol. 20, No. 12, 1939.
135. MCKIBBEN, E. G., AND DAVIDSON, J. BROWNLEE. "Effect of Inflation Pressure on the Rolling Resistance of Pneumatic Implement Tires," *Agr. Eng.*, Vol. 21, No. 1, 1940.
136. MCKIBBEN, E. G., AND DAVIDSON, J. BROWNLEE. "Effect of Outside and Cross-Section Diameters on the Rolling Resistance of Pneumatic Implement Tires," *Agr. Eng.*, Vol. 21, No. 2, 1940.
137. SYBEL, H. "Standort, Wesenheiten und Gegenwartsaufgaben bäuerlicher Maschinenforschung" ("State Problems and Present Tasks of Development of Agricultural Implements"), *VDI*, No. 21, 1934.
138. RIES, L. W. "Das Förderproblem in der Landwirtschaft" ("The Problem of Transport in Agriculture"), *VDI*, No. 21, 1935.
139. BRUNNER, E. F. "Problems in the Design ... of Pressure Tires for Farm Tractors," *Agr. Eng.*, 1933, p. 45.
140. SHIELDS, J. W. "Pneumatic Tires for Agricultural Tractors," *Agr. Eng.*, 1933, p. 39.
141. MCCUEN, G. W. "Ohio Tests of Rubber Tractor Tires," *Agr. Eng.*, 1933, p. 41.
142. HURLBUT, L. W. "Comparative Test Results of Rubber Tires and Steel Wheels for Tractors," *Agr. Eng.*, 1933, p. 217.
143. SMITH, C. W., AND HURLBUT, L. W. "A Comparative Study of Pneumatic Tires and Steel Wheels on Farm Tractors," *Agr. Eng.*, 1934, p. 35.
144. WILEMAN, R. H. "Pneumatic Tires vs Steel Wheels for Tractors," *Agr. Eng.*, 1934, p. 62.
145. MOSES, B. D., AND FROST, K. R. "Tractive Performance of Pneumatic Tires and Steel Wheels," *Agr. Eng.* 1934, p. 55.
146. JONES, W. B. "A Summary, Pneumatic Rubber Tires on Farm Equipment," *Agr. Eng.*, 1934, p. 49.
147. WILEMAN, R. H. "Effect of Tire Size on Tractor Efficiency," *Agr. Eng.*, 1938, p. 27.
148. SAUVE, E. C. "Single Versus Dual Pneumatic Tires," *Agr. Eng.*, 1940, p. 105.
149. SMITH, C. W. "Field Studies with Dual Tractor Tires," *Agr. Eng.*, 1940, p. 277.
150. GARAVAGLIA, F. "Pneumatici per Trattori Agricoli" ("Pneumatic Tires for Agricultural Tractors"), Mem. ed Atti del Centro di Studi per l'Ingeneria Agraria, 1948, p. 39.
151. MCKIBBEN, E. G., AND DAVIDSON, J. BROWNLEE. "Effect of Wheel Arrangement on Rolling Resistance," *Agr. Eng.*, Vol. 21, No. 3, 1940.
152. ROTTA, J. "Zur Statik des Luftreifens" ("Statics of a Pneumatic Tire"), *Ingenieur-Archiv*, Vol. XVII, 1949.

153. Martin, H. "Druckverteilung in der Berührungsfläche zwischen Reifen und Fahrbahn" ("Load Distribution between a Tire and the Road").
154. Wedemeyer, E. A. "Reifenprofil Geheimnisse" ("Problems of a Tire Profile"), *ATZ*, 1940, p. 429.
155. Schmidt, C. "Der Kraftschluss zwischen Reifen und Fahrbahn" ("Grip between Tires and the Road"), *ATZ*, 1938, p. 392.
156. Dietz, O., and Huber, L. "Reifenverschleiss bei Zwei und Dreiachs Lastwagenanhängern" ("Tire Wear of Two or Three Axle Trailers"), *VDI–DKF*, Heft 22, 1939.
157. Gough, V. E., and Jones, P. W. B. "Tire Testing," *Automobile Eng.*, Vol. 42, No. 549, 1952.
158. Kluge, H., and Haas, E. "Rollwiderstand von Luftreifen" ("Rolling Resistance of Pneumatic Tires"), *VDI–DKF*, Heft 26, 1939.
159. Literature Reviews: *ATZ*, No. 6, 1948; *The Engineer*, June 2, 1948.
160. Bourcier de Carbon, C. "Aspects Nouveaux de la Méchanique du Pneumatique" ("New Aspects of the Mechanics of a Pneumatic Tire"), *SIA*, Dec., 1949.
161. Ford, W. L. "Type of Soil Factor in Selection of Tires," *SAE Journal*, Dec., 1951. See also "Selection of Tires for Earthmoving Equipment for Best Traction, Lowest Rolling Resistance and Good Ground Support," Spec. Publication Dept., SAE, 1951.
162. Nuttall, C. J., and Bekker, M. G. "Discussion on the Effect of Lug Height...by Reed and Shields," ETT, Technical Note, Hoboken, N. J., 1950.
163. Reed, J. F., and Shields, J. W. "The Effect of Lug Height and of Rim Width on the Performance of Farm Tractor Tires," SAE Nat. Tractor Meeting, Milwaukee, 1950.
164. Stewart, R. C., Weiss, S. J., and Magill, D. V. "Study of the Mobility of Vehicles as Related to Trafficability of Traction Media," Tech. Memo M–020 USN Civil Engineering & Research Lab., Port Hueneme, Calif., 1951.
165. Rathje, J. "Schnittvorgang im Sande" ("Process of Sand Cutting"), *VDI*, Forschungsheft, 1931.
166. Kanafojski, C. "Przyczynek do laboratoryjnych badań odkształceń i oporów gleby..." ("Contribution to the Laboratory Study of Strain and Soil Resistance"), *Czasopismo Techniczne*, Nos. 3–8, Lwow, 1934.
167. Halkinov, A. D. "Selskohoziaystvennye Tractory" ("Agricultural Tractors"), quoted by Letoshnev, Leningrad, 1931.
168. Wintergerst, E. *Die technische Physik des Kraftwagens* (Technical Physics of a Motor Vehicle), Berlin, 1940.
169. Boyd, J. E. *Strength of Materials*, New York, 1950.
170. Foppl, A. *Vorlesungen über technische Mechanik* (Lessons on Technical Mechanics), Leipzig, 1900.
171. Buchholz, W. "Erdwiderstand auf Ankerplatten" ("Soil Resistance of Anchor Plates"), *Jahrbuch der Hafenbautechnischen Ges.*, Vol. XII, 1930/31.
172. Myklestad, N. O. *Vibration Analysis*, New York, 1944.

173. LITTLE, L. F. "Some Notes on the Performance of Tracked Vehicles," Rep. Vickers Armstrong Exp. Works, Chertsey, England, 1944.
174. JUVENATIEV, J. N. *Aerosanye* (Propeller Sleds), Moscow and Leningrad, 1939.
175. "Lastannahmen und Konstruktionsrichtlinien für Schneekufen" ("Design Data and Load Capacity of Skis"), Rundschreiben des Reichsministers der Luftfahrt, Berlin, 1938.
176. "Rundschreiben No. 33/40" ("Circular No. 33/40"), Festigkeitsprüfstelle Alderschof, Berlin, 1940.
177. FERRIER, A. "Winter Landing Gear," *J. of the Royal Aero. Soc.*, June, 1934.
178. ROUSE, H. *Elementary Mechanics of Fluids*, New York, 1946.
179. BIKERMAN, J. J. *Surface Chemistry for Industrial Research*, New York, 1948.
180. ERIKSSON, R. "Medens friction mot snö och is" ("Friction of Snow and Ice"), Meddelande nr 34–35 (with an English summary), Stockholm, 1949.
181. ADAM, N. K. *The Physics and Chemistry of Surfaces*, London, 1940.
182. GEMANT, A. *Frictional Phenomena*, Brooklyn, 1950.
183. EPSTEIN, L. A. "Quelques nouvelles données experimentales sur le phenomène du glissement" ("Some New Data on the Phenomenon of Sliding"), *Comptes Rendus* (Doklady) de l'Acad. des Sciences de l'USSR, Vol. XXVI, No. 8, Moscow, 1940.
184. MURRAY, A. B. "The Hydrodynamics of Planing Hulls," paper presented at February, 1950, Meeting of the New England Section of NAME, Weymouth, Mass., 1950.
185. KLEIN, G. J. "The Snow Characteristics of Aircraft Skis," NRC of Canada, Aeronautical Rep. AR–2, Ottawa, 1947.
186. RICHTER, G. D. *Snow Cover, Its Formation and Properties* (English translation), Acad. of Science of the USSR, Moscow, 1945.
187. VON KÁRMÁN, TH. *The Impact on Seaplane Floats During Landings*, NACA Rep. No. 321, Washington, 1929.
188. BOUCHARD, M. J. "La Dynamique du Saut à Skis" ("Dynamics of a Ski Jump"), *Mécanique*, No. 278, 1938.
189. VIETORIS, L. "Der Schi im Licht der Festigkeitslehre" ("A Ski in the Light of Stress Analysis"), *Zeitsch. f. Math. u. Naturwiss., Unterricht aller Schulgattungen*, Heft 1 and 2, 1939.
190. YNTEMA, R. T., AND MILWITZKY, B. *An Impulse—Momentum Method for Calculating Landing-gear Contact Conditions in Eccentric Landings*, NACA T.N. 2596, Washington, 1952.
191. WEISS, S. J. "A Rational Approach to Heavy Duty Sled Runner Design," T.N. N–033, U.S. Naval Civil Eng. Res. & Evaluation Lab., Port Hueneme, Calif., 1951.
192. HILL, R. *The Mathematical Theory of Plasticity*, Oxford, 1950.
193. STROHHÄCKER, P. "Ausgleich-Übersetzungsgetriebe für mehrachsangetriebene Kraftfahrzeuge" ("Unequal Torque-Splitting Differentials for Multi-Axle Vehicles"), *ATZ*, No. 13, 1938.

194. STROHHÄCKER, P. "Hinterachsantrieb oder Allradantrieb für Schwerlaststrassenschlepper" ("Rear-Axle or All-Axle Propulsion for Heavy Road Tractors"), *ATZ*, No. 13, 1942.

195. MEYER, H. "Untersuchung der Achsdrucke im Fahrbetrieb in der Ebene, in der Steigung und in Gefälle von zwei-und dreiachsigen Lastkraftwagenzügen." ("Investigation of Axle Loads..."), *ATZ*, No. 21, 1934.

196. STROHHÄCKER, P. "ATZ Konstructionstafeln No. 37" ("Design Tables No. 37"), *ATZ*, No. 13, 1939.

197. STROHHÄCKER, P. "Steigfähigkeit bei Hinterachs und Allradantrieb" ("Climbing Performance of Rear- and All-Axle Propulsion"), *ATZ*, No. 21, 1941.

198. JANTE, A. "Einfluss horizontaler Kräfte auf Achsdruck, Federbelastung und Fahrzeugneigung" ("The Effect of Horizontal Forces upon Axle Load, Spring Load, and Vehicle Tilt"), *ATZ*, No. 7/8, 1944.

199. STEEDS, W. "Rigid Six Wheeled Vehicles,—A Study of the Effect of Torque Reaction on Weight Distribution," *The Automobile Engineer*, March, 1931, p. 172.

200. BARGER, E. L., CARLETON, WM., MCKIBBEN, E. G., and BAINER, R. *Tractors and Their Power Units*, New York, 1952.

201. GOLDBECK, G. *Die Fahrtmechanik des Kraftfahrzeuges* (The Mechanics of the Ride of a Motor Vehicle), Stuttgart, 1949.

202. CASAGRANDE, A., AND SHANNON, W. L. "Strength of Soils Under Dynamic Loads," *Proc. Am. Soc. of Civil Eng.*, Vol. 74, 1948, p. 591.

203. KLÖPPEL, K., AND MOPPERT, H. "Zur Grösse des Dynamischen Raddruckes luftbereifter Kraftfahrzeuge beim Überfahren von Bodenunebenheiten" ("The Magnitude of Dynamic Axle Loads of Pneumatic–Tired Vehicles while Traveling over Road Unevennesses"), *VDI*, No. 28, 1950.

204. KAMM, W. *Das Kraftfahrzeug* (The Motor Vehicle), Berlin, 1936.

205. POPP, C. *Über die Stosswirkung unrunder Eisenbahnräder* (The Shock Performance of Non-round Wheels), Munich, 1944 (quoted from Reference 203).

206. TIMOSHENKO, S. *Vibration Problems in Engineering*, New York, 1929.

207. LEHR, E. *Die Schwingungstechnik* (Vibrations in Engineering), Berlin, 1934.

208. WEDEMEYER, E. A. "Fahrzeugfederung" ("Vehicle Suspension"), *ATZ*, 1935, p. 272.

209. WEDEMEYER, E. A. "Der Schwingende Reifen als Ursache der Strassenstörung" ("Wheel Bouncing as the Cause of Road Damage"), *ATZ*, 1935, p. 558.

210. MORGAN, A. D. "Correlation of Roughmeter and Skid Tests with Pavement Type ..." Highway Res. Bd. Bull. No. 37, Washington, 1937.

211. NEUMANN, E. "Die Begriffsbestimmung von Welligkeit und Rauhigkeit und die Verfahren zur Messung dieser Werte" ("The Idea of Waviness and Roughness and the Methods of Measuring Relevant Values"), *Strasse und Autobahn*, Nos. 10 and 11, 1951.

212. LEHR, E. "Springing and Damping of the Suspension of Armored Vehicles" (translated from the German), School of Tank Technology, Chobham Lane, Chertsey, 1945.

213. DEN HARTOG, J. P. *Mechanical Vibrations*, New York, 1937.

214. FRANKE, R. "Heereszugwagen" ("Military Tractors"), *VDI*, No. 22, 1937.

215. KAMM, W. "Die Entwicklungsrichtungen im Kraftfahrwesen" ("Development Trends of Automotive Engineering"), *ATZ*, No. 33, 1938.

216. POMEROY, L. *The Grand Prix Car*, London, 1949.

217. HEYWANG, F. "Beitrag zur Stabilität von Kraftfahrzeugen" ("Contribution to the Study of the Stability of Motor Vehicles"), *ATZ*, No. 18, 1939.

218. HÖSL, O. "Die Kurvenfahrt des Kraftwagens" ("Curvilinear Travel of a Motor Vehicle"), *ATZ*, No. 14, 1939.

219. SCHUBOTH, H. "Beitrag zur Systematik der Kraftfahrzeuglenkung. Kinematic, Dynamik und Arbeitsaufwand bei Einachs- und Mehrachslenkung" ("Contribution to the Systematization of Automobile Steering. Kinematics, Dynamics, and Energy Requirement of One- and Multi-Axle Steering"), *ATZ*, Nos. 21 and 22, 1939.

220. WEDEMEYER, E. A. "Der Wendekreis" ("The Radius of Turning"), *ATZ*, No. 2, 1940.

221. HATZ, M. D., MOSER, E., PFYL, F., AND RUBI, H. "Géométrie de la Direction" ("Geometry of Steering"), *SIA*, 1950, p. 131.

222. DEAN-AVERNS, R. "Steering Design," *The Automobile Engineer*, 1931, p. 109.

223. MÜHLFELD, A. "Die Lenkung des Kraftrades" ("The Steering of a Motorcycle"), *ATZ*, No. 10, 1951.

224. EVANS, R. D. "Properties of Tires Affecting Riding, Steering, and Handling," *SAE Journal*, February, 1935.

225. BULL, A. W. "Tire Behavior in Steering," *SAE Journal*, August, 1939.

226. HUBER, L. "Die Fahrrichtungsstabilität des schnellaufenden Kraftwagens" ("The Stability of Travel Direction of a Fast Motor Vehicle"), *Deutsche Kraftf.-Forsch.*, No. 44, Berlin, 1940.

227. RIEKERT, P., LEUNIG, G., AND GAUSS, F. "Die Strassenlage des Kraftwagens" ("Roadability of a Motor Vehicle"), *ATZ*, No. 2, 1951.

228. MÖLLER, E. "Windkanalmessungen an Kraftfahrzeugen bei Seitenwind" ("Wind-Tunnel Measurements of Motor Vehicles Subjected to a Sidewind"), *ATZ*, No. 4, 1951.

229. SAWATZKI, E. "Die Luftkräfte und Luftkraftmomente am Kraftwagen..." ("Wind Loads and Moments Acting Upon a Vehicle"), *Deutsche Kraftf.-Forsch.*, No. 50, Berlin, 1940.

230. SARTORIS, G. L. "Weight Distribution. A Mathematical Consideration of the Various Factors Involved," *The Automobile Engineer*, January, 1930, p. 21.

231. ZIMIELEV, G. V. "Steering Phenomena. A Study of the Factors Affecting Stability on Curves," Trans. of a paper by the Acad. of Science of the USSR, *The Automobile Engineer*, August, 1946, p. 341.

232. BUEN DE LOZANO, V. "Prédétermination de la condition de stabilité

des véhicules automobiles" ("Determination of the Stability of Motor Vehicles"), *SIA*, 1950, p. 127.
233. Julien, A. "Convergences des théories françaises et des études et réalisations anglo-saxonnes concernant la stabilité de la route" ("Convergence between French Theories and Anglo-Saxon Practice Related to Road Stability"), *SIA*, 1948, p. 231.
234. Sezé, V. G. "Stabilité de route des voitures" ("Road Stability of Vehicles"), *SIA*, October, 1937.
235. Gratzmüller, G. "Théorie de la tenue de route" ("The Theory of Road Stability"), *SIA*, May, 1942.
236. Olley, M. "National Influence on American Passenger Cars," *Proc. Inst. of Automobile Engineers*, Vol. XXXII, 1938.
237. Beck, G. "Massnahmen zur Steigerung der Geländegängigkeit bei Sonderfahrzeugen" ("Ways of Increasing the Cross-Country Performance of Motor Vehicles"), *ATZ*, No. 15, 1939.
238. Koessler, P., and Glaubitz, H. "Zum Wendevorgang bei Gleiskettenfahrwerken" ("Steering of Tracked Vehicles"), *ATZ*, No. 8, 1951.
239. Merritt, H. E. "Some Considerations Influencing the Design of High Speed Track Vehicles," *Proc. Inst. of Automobile Engineers*, Vol. XXXIII, 1938–39.
240. Steeds, W. "Tracked Vehicles," *The Automobile Engineer*, April, 1950.
241. Serruys, M. *La suspension et la direction des véhicules routiers* ("Suspension and Steering of Motor Vehicles"), Paris, 1947.
242. Rowell, H. S. "Principles of Vehicle Suspension," *The Automotive Engineer*, Vol. 13, 1923, p. 118.
243. Guest, J. J. "The Main Free Vibrations of an Autocar," *Proc. Inst. of Automobile Engineers*, Vol. 20, 1926, p. 505.
244. Jacklin, H. M. "Human Reaction to Vibration," *SAE Trans.*, Vol. 31, 1936, p. 401.
245. Moss, F. A. "Three Low Priced Cars Vary in Ride Comfort," *SAE Journal*, 1933, p. 24.
246. Brown, R. W., and Dickinson, H. C. "Criteria are Set for Riding Comfort," *SAE Journal*, 1935.
247. Janeway, R. N. "Vehicle Vibration Limits to Fit the Passengers," *SAE Journal*, 1948.
248. "Ride and Vibration Data," SP–6, Special Publication Dept., SAE, 1951.
249. Lehr, E. "Schwingungsfragen der Fahrzeugfederung" ("Problems of Vibration of Vehicle Suspension"), *ZVDI*, 1930, p. 1117.
250. Lehr, E. "Die schwingungstechnischen Eigenschaften des Kraftwagens" ("Vibrational Properties of a Motor Vehicle"), *ZVDI*, 1937, p. 329.
251. Lehr, E. "Einfluss einer Flüssigkeitdämpfung der Fahrzeugfederung" ("Effect of Fluid Damping upon Vehicle Suspension"), *ZVDI*, 1934, p. 721.
252. Lehr, E., and Bertschinger, U. "Über den Zusammenhang zwischen Schwingungseigenschaften und Fahreigenschaften von Kraftfahr-

zeugen" ("Relationship between Vibrational Properties and Ride Qualities of Motor Vehicles"), *ATZ*, 1946, p. 34.

253. Bangold, R. A. *The Physics of Blown Sand in Deserts*, London, 1941.

254. Cornish, V. *Waves of Sand and Snow*, London, 1917.

255. Kühner, K. "Das Kraftfahrzeug im Gelände" ("Motor Vehicle and Terrain"), *VDI*, No. 34, 1935.

256. Chudakov, E. A. *Teoria Avtomobilya* (The Theory of an Automobile), Moscow, 1950.

257. Buckmiller, G., Kimmel, A., and Riekert, P. "Erfahrungen mit dem Propellerschlitten der Firma Wideroes, Oslo" ("Experience Gained with Propeller Sleds, Make Wideroes, Oslo"), *FKFS*, Stuttgart, 1942.

258. Augustin, J. A. "Luftschrauben Schlitten" ("Propeller Sleds"), *ATZ*, 1942, p. 447.

259. Nuttall, C. J., and Hecker, L. G. "Supplementary Final Report on Contract OEM-154, Emphasizing Amphibious Vehicle Design," OSRD, 1945.

260. Attwood, E. L., and Pengelly, H. S. *Theoretical Naval Architecture*, London, 1931.

261. Robb, A. M. *Theory of Naval Architecture*, London, 1952.

262. Van Lammeren, W. P. A. *Resistance, Propulsion, and Steering of Ships*, Stam-Haarlem, 1948.

263. Proctor, R. R. "Notes on Soil Compaction," *Eng. News Record*, Vol. 53, 1933.

264. Corps of Engineers, Mississippi River Comm. *Trafficability of Soils*, Vicksburg, 1945.

265. Corps of Engineers, Mississippi River Comm. *Trafficability of Soils, Pilot Tests, Self-Propelled Vehicles*," Tech. Memo No. 3–240, Vicksburg, 1947.

266. McKibben, E. G., and Hull, D. O. "Soil Penetration Tests as a Means of Predicting Rolling Resistance," *Agr. Eng.*, No. 6, 1940.

267. Haefeli, R., Bader, H., and Bucher, E. "Das Zeitprofil; eine graphische Darstellung der Entwicklung der Schneedecke" ("The Time-Profile; a Graphic Presentation of the Development of Snow Cover"), *Geologie der Schweiz, Schnee und seine Metamorphose*, No. 3, Zurich, 1938.

268. Gross, W. A., and Elliott, A. D. "Traction as Influenced by Soils and Their Condition," paper presented at the SAE Advisory Committee Meeting, Aberdeen Proving Ground, 1946.

269. Bekker, M. G. "Proposed Field Soil Testing Device," Tech. Memo. No. 1, Associate Committee on Soil and Snow Mechanics, National Research Council of Canada, Ottawa, 1945.

270. Weiss, S. J. "Soil Truss, Portable Test Apparatus, Used in Vehicle Mobility Studies," BUDOCKS Navy Docks P–23, Technical Digest No. 23, Washington, 1952.

271. Wilson, C. W., and Nuttall, C. J. "A Uniform Procedure for Measuring Soil Mechanical Properties and Reporting Soil Data in Conjunction with Quantitative Vehicle Mobility Data in Homogeneous

Soils," joint report by Aberdeen Proving Ground and Stevens Institute of Technology, TTI–698, ETT–370, Hoboken, 1950.

272. Touring Club Italiano. *Institute sperimentale del touring club italiano a del R automobil club d'Italia,* Vol. 2, 1932.

273. KNIPPING, GÖLZ, AND MITTMEYER. "Strassenbahn und Fahrzeug" ("The Road and the Motor Vehicle"), *VDI,* 1933.

274. PETERSEN, H. "Zusammenstellung von Geräten zur Messung der Ebenflächigkeit von Fahrbahnen" ("Symposium on Instruments for Measuring the Smoothness of the Road"), *Die Strasse,* No. 12, 1939.

275. BUCHANAN, J. A., AND CATUDAL, A. L. "Standardizable Equipment for Evaluating Road Surface Roughness," *Proc. XX Annual Meeting,* Highway Res. Board, 1940.

276. MOYER, R. A., AND SHUPE, J. W. *Roughness and Skid Resistance Measurements on Paved Roads in California,*" Highway Res. Board Bulletin No. 37, Washington, 1951.

277. BROULHIET, M. G. "L'emploi du podomètre dans l'appréciation du confort" ("The Use of the Podometer in the Evaluation of Ride Comfort"), *Journal SIA,* 1937, p. 15.

278. KÜHNER, K. "Fahrwiderstand und Leistungsbedarf geländegängiger Fahrzeuge" ("Motion Resistance and Power Requirement of Off-the Road Vehicles"), *ZVDI,* No. 8, 1938.

279. CASAGRANDE, A. "Classification and Identification of Soil," *Proc. ASCE,* June, 1947.

280. HOGENTOGLER, C. A. *Engineering Properties of Soil,* New York, 1937.

281. BAVER, L. D. *Soil Physics,* New York, 1940.

282. ROBINSON, G. W. *Soils, Their Origin, Construction, and Classification,* London, 1932.

283. SCHAEFER, V. J., DEQUERVAIN, M. R., AND KLEIN, G. J. *Tentative Snow Classification,* Committee on Snow and Ice, Int. Assn. of Hydrology, Published by NRC of Canada, Ottawa, 1950.

284. JARRET, Col. J. B. *Armoured Car Journal,* Vol. LIX, No. 2, 1950.

285. FLÖSSEL, W. *Bergsteigfähigkeit und Literleistung* (Climbing Ability of a Vehicle and Engine Capacity), Stuttgart, 1950.

286. KÜHNER, K. "Die Leistungsabmessungen beim Fahrzeugmotor; Anpassung des Triebwerks an den Geländebetrieb" ("Power Output of the Engine; Adjustment of the Transmission to Terrain Operation"), *VDI,* No. 39, 1938.

287. SPANNHAKE, W., KLUGE, H., AND UNRUH, H. "Kritische Untersuchung der Leistungsübertragung durch Zahnradwechselgetriebe und hydrodynamische Getriebe auf Strassenfahrzeugen mit Antrieb durch Brennkraftmaschinen" ("Critical Investigation of the Power Transfer by Gear and Hydrodynamic Transmissions for Motor Vehicles"), *Deutsche Kraftfahrforschung,* Zwischenbericht Nr. 41, Part III & IV.

288. KUGEL, F. "Strömungsgetriebe und Kupplungen in der Kraftfahrtechnik" ("Hydraulic Transmissions and Clutches in Automobile Engineering"), *ATZ,* No. 11, 1938.

289. PROKOFIEV, W. I. *Avtomobilnye Gidroperedatchi* (Hydraulic Transmissions for Automobiles), Mashgiz, Moscow, 1948.

290. Herendon, W. B., Deimel, A. H., Youngs, P. R., Schaefer, R. M., Kelley, O. K., McFarland, F. R., and Sicklesteel, D. T. "SAE Symposium on Automatic Transmissions and Torque Converters," papers presented at the SAE Summer Meeting, French Lick, Ind., 1951.
291. Wislicenus, G. F. *Fluid Mechanics of Turbo Machinery*, New York, 1947.
292. Reichenbächer, H. "Verzweigungsgetriebe und Leistungregelung bei motorgetriebenen Fahrzeugen" ("Powersplitting Transmissions and Power Control for Motor Vehicles"), *ZVDI*, Nos. 45/46, 1943.
293. Leunig, G. "Zwei Möglichkeiten kraftfahrtechnischer Entwicklung" ("Two Possibilities of the Automotive Development"), *ATZ*, No. 2, 1951.
294. Thüngen, H. "Grundlagen für selbsttätige Regelung von Kraftfahrzeuggetrieben" ("Fundamentals of Automatic Control of Motor Vehicle Transmission"), *ZVDI*, No. 10, 1934.
295. Arnold, R. "Die selbsttätige Regelung von Kraftfahrzeugen" ("Automatic Control of Motor Vehicles"), *ATZ*, Nos. 3 and 5, 1948.
296. Eckert, B. "Bauliche Gestaltung von Kraftfahrzeuggetrieben" ("Structural Forms of Motor Vehicle Transmissions"), *ATZ*, 1941, p. 342.
297. Maier, A. "Konstruction und Entwicklung der Kraftfahrzeug Stufengetriebe" ("Construction and Development of a Motor Vehicle Gearbox"), *ATZ*, 1941, p. 417.
298. Maier, A. "Grundlagen für die Berechnung und Grössenbestimmung der Fahrzeugwechselgetriebe" ("Fundamentals of Calculation and Determination of the Size of a Motor Vehicle Transmission"), *ATZ*, 1943, p. 494.
299. Eckert, B. "Das Kraftfahrzeug Wechselgetriebe" ("The Transmission of a Motor Vehicle"), *ATZ*, 1941, p. 225.
300. Stoll, H. "Der Kraftstoffverbrauch von Fahrzeug Ottomotoren" ("Fuel Consumption of Otto Engines"), *ATZ*, Nos. 5 and 6, 1944.
301. Zimelev, G. V. *Mashinostroyenye* (Machine Design), Vol. 11, Moscow, 1949.
302. Jante, A. "Darstellung und Wertung der Fahrfähigkeiten und die Verfahren zur Fahrplanung" ("A Presentation and Evaluation of Performance, and the Method for Planning the Travel"), *ATZ*, 1936, p. 151.
303. Jante, A. "Der Anfahrvorgang" ("The Start Performance"), *ATZ*, 1940, p. 226.
304. Lowag, G. "Dynamik des Antriebes von Kraftfahrzeugen mit Verbrennungsmotoren" ("Dynamics of the Propulsion of Motor Vehicles Equipped with Internal Combustion Engines"), *ATZ*, 1942, p. 156.
305. Jante, A. "Über die zweckmässige Darstellung von Kennlinien-Diagramen für Fahrzeugmotoren und Kraftfahrzeuge" ("A Functional Preparation of Characteristics-Diagrams of Engines and Motor Vehicles"), *ATZ*, 1936, p. 326.
306. Kamm, W. "Dimensionslose Betrachtung und die Grenzwerte der Fahrzeugbeschleunigung" ("Dimensionless Consideration and the

Limit Values of the Acceleration of a Motor Vehicle"), *ATZ*, No. 1, 1952.

307. EBERHORST, VON, AND EBERAN, R. "Der Brennstoffverbrauch des Kraftfahrzeuges bei Wechselfahrt. Einfluss der Fahrweise und der Getriebebauart" ("Fuel Consumption of a Motor Vehicle during a Nonuniform Ride. Effect of the Way of Driving and of the Type of Transmission"), *ATZ*, No. 9, 1951.
308. RIXMANN, W. "Der Einfluss der Fahrweise und der Fahrgeschwindigkeit auf die Gesamtwirtschaftlichkeit des Kraftfahrzeugs" ("Effect of the Way of Travel and Speed upon the Over-all Economy of the Motor Vehicle"), *ATZ*, No. 3, 1952.
309. WIMMER AND WIRBITZKY. *Fahrweise und Wirtschaftlichkeit von Nutzkraftwagen* (The Way of Travel and the Economy of Trucks), Munich, 1951.
310. ANDREAU, J. "L'énergie éolienne et l'automobile" ("Wind Energy and the Automobile"), *SIA Journal*, 1949, p. 413.
311. STEFANSSON, V. *Arctic Manual*, New York, 1950.
312. AUGUSTIN, J. A. "Der Einsatz von Kraftfahrzeugen in der Sovjetunion" ("The Introduction of Motor Vehicles in the Soviet Union"), *ATZ*, No. 3 and 4, 1943.
313. WARLAM, A. A., AND CAPOZZOLI, L. J. "Forecasting of Soil Trafficability," progress report No. 107–02, Res. Div. New York Univ., New York, 1948.
314. JACOBS, W. C. "Wartime Development in Applied Climatology," *Meteorological Monographs*, Am. Meteorological Soc., Vol. 1, No. 1, 1947.
315. BEKKER, M. G. "Two Stroke Multi-Bank Gasoline Engine," a report for Canadian Ministry of Supply, Ottawa, 1943.
316. KAMM, W. "Einfluss der Reichsautobahnen auf die Gestaltung der Kraftfahrzeuge" ("The Effect of the Super-Highway upon the Development of an Automobile"), *ATZ*, No. 13, 1934.
317. TURUNEN, W. A. "Gas Turbines in Automobiles," *SAE Quarterly Trans.*, Vol. 4, No. 1, 1950.
318. BARR, R. H. H. "Gas Turbines — Consideration of Prospects in Road Transport," *The Automobile Engineer*, 1948, p. 305.
319. SCHOPPE, F. "Gas Turbinen zum Antrieb von Strassenfahrzeugen" ("Gas Turbines for Motor Vehicles"), *ATZ*, No. 1, 1952.
320. LANCHESTER, F. W. "The Horse Power of the Petrol Motor in Its Relation to Bore, Stroke, and Weight," The Incorporated Institution of Automobile Engineers, *Proc. IAE*, London, 1907, p. 155.
321. TAYLOR, C. FAYETTE. "New Common Denominators Yield Significant Engine Comparisons," *SAE Journal*, Jan., 1951.
322. "Turbocompound Engine Wright R 3350," *Aviation Week*, Feb. 18, 1952.
323. ZWICKY, F. "Morphology of Aerial Propulsion," *Helvetica Phys. Act.*, Vol. XXI, Fasc. 5, 1948.
324. LANGHAAR, H. L. *Dimensional Analysis and Theory of Models*, New York, 1951.

325. MURPHY, G. *Similitude in Engineering*, New York, 1950.
326. BRIDGMAN, P. W. *Dimensional Analysis*, Yale Univ. Press, 1922.
327. CONRAD, R. O. *Structural Models, Part I: Theory*, N.D. Bureau of C. & R., Bull. No. 13, 1938.
328. LUTZ, O. "Grundlagen für Modellversuche an Fahrzeugen" ("Fundamentals of Scale-Model Tests of Motor Vehicles"), *ATZ*, p. 211, 1934.
329. BEKKER, M. G. "Scale Model Research of Motor Vehicles," paper submitted to the Canadian Ministry of Supply and NRC, Ottawa, 1944.
330. KAMM, W., AND SCHMIDT, C. *Das Versuchs- und Messwesen auf dem Gebiete des Kraftfahrzeugs* (Measurements and Testing of Motor Vehicles), Berlin, 1939.
331. DIXON, W. J., AND MASSEY, F. J., JR. *Introduction to Statistical Analysis*, New York, 1951.
332. BINGHAM, E. C. *Fluidity and Plasticity*, New York, 1922.
333. BRADLEY, J., AND WOOD, S. A. "Factors Affecting the Motion of a Four-Wheeled Vehicle with Locked Wheels," *The Automobile Engineer*, 1931, p. 34.
334. HUBER, L., AND DIETZ, O. "Pendelbewegungen von Lastkraftwagen-Anhängern und ihre Vermeidung" ("Oscillations of Truck Trailers and Their Avoidance"), *VDI*, No. 16, 1937.
335. DIETZ, O. "Pendelerscheinungen an Strassen-Anhängerzügen" ("Oscillation Phenomena of Highway Trailer-Trains"), *VDI-DKF*, Forschungsheft No. 16.
336. MARQUARD, E. "Untersuchung der Federungeigenschaften von Kraftfahrzeugen an Modellen" ("Investigation of Springing Characteristics of Motor Vehicles by Means of Models"), *ATZ*, No. 17, 1937.
337. BEKKER, M. G. "WAB Moving Stand for Research Purposes. Project of Stand Reproducing Vehicle Vibrations," paper submitted to the Canadian Ministry of Supply, Ottawa, 1942.
338. SHEWHART, W. A. *Economic Control of Quality of Manufactured Product*, New York, 1931.
339. MOOD, A. M. *Introduction to the Theory of Statistics*, New York, 1950.
340. CRAMER, H. *Mathematical Methods of Statistics*, Princeton, N. J., 1946.
341. HOEL, P. G. *Introduction to Mathematical Statistics*, New York, 1947.
342. EZEKIEL, M. *Methods of Correlation Analysis*, New York, 1930.
343. FISHER, R. A. *The Design of Experiments*, Edinburgh, 1942.
344. COCHRAN, W. G., AND COX, G. M. *Experimental Designs*, New York, 1950.
345. KEMPTHORNE, O. *The Design and Analysis of Experiments*, New York, 1952.
346. MORRISON, M., AND NUTTALL, C. J. "Wheels vs. Tracks for Military Vehicles, Part IV, Further Investigations of the Bernstein Equation for the Rolling Resistance of Wheels in Soil," ETT Report No. 456, Hoboken, N. J., July, 1952.

INDEX

Printed and bound by CPI Group (UK) Ltd, Croydon, CR0 4YY
16/09/2026
14963492-0005